International Review of Science

Physical Chemistry
Series Two

Consultant Editor
A. D. Buckingham, F.R.S.

Publisher's Note

The International Review of Science is an important venture in scientific publishing, presented by Butterworths. The basic concept of the Review is to provide regular authoritative reviews of entire disciplines. Chemistry was taken first as the problems of literature survey are probably more acute in this subject than in any other. Biochemistry and Physiology followed naturally. As a matter of policy, the authorship of the Review of Science is international and distinguished, the subject coverage is extensive, systematic and critical.

The Review has been conceived within a carefully organised editorial framework. The overall plan was drawn up and the volume editors appointed by seven consultant editors. In turn, each volume editor planned the coverage of his field and appointed authors to write on subjects which were within the area of their own research experience. No geographical restriction was imposed. Hence the 500 or so contributions to the Review of Science come from many countries of the world and provide an authoritative account of progress. The publication of Physical Chemistry Series One was completed in 1973 with thirteen text volumes and one index volume; in accordance with the stated policy of issuing regular reviews to keep the series up to date, volumes of Series Two will be published between the middle of 1975 and early 1976; Series Two of Organic Chemistry will be published at the same time, while Inorganic Chemistry Series Two was published during the first half of 1975. Volume titles are the same as in Series One but the articles themselves either cover recent advances in the same subject or deal with a different aspect of the main theme of the volume. In Series Two an index is incorporated in each volume and there is no separate index volume.

Butterworth & Co. (Publishers) Ltd.

PHYSICAL CHEMISTRY SERIES TWO

Consultant Editor
A. D. Buckingham, F.R.S.,
Department of Chemistry University of Cambridge

Volume titles and Editors

1 **THEORETICAL CHEMISTRY**
Professor A. D. Buckingham, F.R.S., *University of Cambridge* and Professor C. A. Coulson, F.R.S., *University of Oxford*

2 **MOLECULAR STRUCTURE AND PROPERTIES**
Professor A. D. Buckingham, F.R.S., *University of Cambridge*

3 **SPECTROSCOPY**
Dr. D. A. Ramsay, F.R.S.C., *National Research Council of Canada*

4 **MAGNETIC RESONANCE**
Professor C. A. McDowell, F.R.S.C., *University of British Columbia*

5 **MASS SPECTROMETRY**
Professor A. Maccoll, *University College, University of London*

6 **ELECTROCHEMISTRY**
Professor J. O'M Bockris, *The Flinders University of S. Australia*

7 **SURFACE CHEMISTRY AND COLLOIDS**
Professor M. Kerker, *Clarkson College of Technology, New York*

8 **MACROMOLECULAR SCIENCE**
Professor C. E. H. Bawn, C.B.E., F.R.S., *formerly of the University of Liverpool*

9 **CHEMICAL KINETICS**
Professor D. R. Herschbach *Harvard University*

10 **THERMOCHEMISTRY AND THERMO-DYNAMICS**
Dr. H. A. Skinner, *University of Manchester*

11 **CHEMICAL CRYSTALLOGRAPHY**
Professor J. Monteath Robertson, C.B.E., F.R.S., *formerly of the University of Glasgow*

12 **ANALYTICAL CHEMISTRY —PART 1**
Professor T. S. West, *Imperial College, University of London*

13 **ANALYTICAL CHEMISTRY —PART 2**
Professor T. S. West, *Imperial College, University of London*

INORGANIC CHEMISTRY SERIES TWO

Consultant Editor
H. J. Emeléus, C.B.E., F.R.S.
Department of Chemistry University of Cambridge

Volume titles and Editors

1 **MAIN GROUP ELEMENTS—HYDROGEN AND GROUPS I–III**
Professor M. F. Lappert, *University of Sussex*

2 **MAIN GROUP ELEMENTS—GROUPS IV AND V**
Dr. D. B. Sowerby, *University of Nottingham*

3 **MAIN GROUP ELEMENTS—GROUPS VI AND VII**
Professor V. Gutmann, *Technical University of Vienna*

4 **ORGANOMETALLIC DERIVATIVES OF THE MAIN GROUP ELEMENTS**
Professor B. J. Aylett, *Westfield College, University of London*

5 **TRANSITION METALS—PART 1**
Professor D. W. A. Sharp, *University of Glasgow*

6 **TRANSITION METALS—PART 2**
Dr. M. J. Mays, *University of Cambridge*

7 **LANTHANIDES AND ACTINIDES**
Professor K. W. Bagnall, *University of Manchester*

8 **RADIOCHEMISTRY**
Dr. A. G. Maddock, *University of Cambridge*

9 **REACTION MECHANISMS IN INORGANIC CHEMISTRY**
Professor M. L. Tobe, *University College, University of London*

10 **SOLID STATE CHEMISTRY**
Dr. L. E. J. Roberts, *Atomic Energy Research Establish-ment, Harwell*

ORGANIC CHEMISTRY SERIES TWO

Consultant Editor
D. H. Hey, F.R.S., *formerly of the Department of Chemistry, King's College, University of London*

Volume titles and Editors

1 **STRUCTURE DETERMINATION IN ORGANIC CHEMISTRY**
Professor L. M. Jackman, *Pennsylvania State University*

2 **ALIPHATIC COMPOUNDS**
Professor N. B. Chapman, *Hull University*

3 **AROMATIC COMPOUNDS**
Professor H. Zollinger, *Eidgenossische Technische Hochschule, Zurich*

4 **HETEROCYCLIC COMPOUNDS**
Dr. K. Schofield, *University of Exeter*

5 **ALICYCLIC COMPOUNDS**
Professor D. Ginsburg, *Technion-Israel Institute of Technology, Haifa*

6 **AMINO ACIDS, PEPTIDES AND RELATED COMPOUNDS**
Professor N. H. Rydon, *University of Exeter*

7 **CARBOHYDRATES**
Professor G. O. Aspinall, *York University, Ontario*

8 **STEROIDS**
Dr. W. F. Johns, *G. D. Searle & Co., Chicago*

9 **ALKALOIDS**
Professor K. Wiesner, F.R.S., *University of New Brunswick*

10 **FREE RADICAL REACTIONS**
Professor W. A. Waters, F.R.S., *formerly of the University of Oxford*

BIOCHEMISTRY SERIES ONE

Consultant Editors
H. L. Kornberg, F.R.S.
Department of Biochemistry University of Leicester and
D. C. Phillips, F.R.S., *Department of Zoology, University of Oxford*

Volume titles and Editors

1 **CHEMISTRY OF MACRO-MOLECULES**
Professor H. Gutfreund, *University of Bristol*

2 **BIOCHEMISTRY OF CELL WALLS AND MEMBRANES**
Dr. C. F. Fox, *University of California, Los Angeles*

3 **ENERGY TRANSDUCING MECHANISMS**
Professor E. Racker, *Cornell University, New York*

4 **BIOCHEMISTRY OF LIPIDS**
Professor T. W. Goodwin, F.R.S., *University of Liverpool*

5 **BIOCHEMISTRY OF CARBO-HYDRATES**
Professor W. J. Whelan, *University of Miami*

6 **BIOCHEMISTRY OF NUCLEIC ACIDS**
Professor K. Burton, F.R.S., *University of Newcastle upon Tyne*

7 **SYNTHESIS OF AMINO ACIDS AND PROTEINS**
Professor H. R. V. Arnstein, *King's College, University of London*

8 **BIOCHEMISTRY OF HORMONES**
Professor H. V. Rickenberg, *National Jewish Hospital & Research Center, Colorado*

9 **BIOCHEMISTRY OF CELL DIFFER-ENTIATION**
Professor J. Paul, *The Beatson Institute for Cancer Research, Glasgow*

10 **DEFENCE AND RECOGNITION**
Professor R. R. Porter, F.R.S., *University of Oxford*

11 **PLANT BIOCHEMISTRY**
Professor D. H. Northcote, F.R.S., *University of Cambridge*

12 **PHYSIOLOGICAL AND PHARMACO-LOGICAL BIOCHEMISTRY**
Dr. H. K. F. Blaschko, F.R.S., *University of Oxford*

PHYSIOLOGY SERIES ONE

Consultant Editors
A. C. Guyton,
Department of Physiology and Biophysics, University of Mississippi Medical Center and
D. F. Horrobin,
Department of Physiology, University of Newcastle upon Tyne

Volume titles and Editors

1 **CARDIOVASCULAR PHYSIOLOGY**
Professor A. C. Guyton and Dr. C. E. Jones, *University of Mississippi Medical Center*

2 **RESPIRATORY PHYSIOLOGY**
Professor J. G. Widdicombe, *St. George's Hospital, London*

3 **NEUROPHYSIOLOGY**
Professor C. C. Hunt, *Washington University School of Medicine, St. Louis*

4 **GASTROINTESTINAL PHYSIOLOGY**
Professor E. D. Jacobson and Dr. L. L. Shanbour, *University of Texas Medical School*

5 **ENDOCRINE PHYSIOLOGY**
Professor S. M. McCann, *University of Texas*

6 **KIDNEY AND URINARY TRACT PHYSIOLOGY**
Professor K. Thurau, *University of Munich*

7 **ENVIRONMENTAL PHYSIOLOGY**
Professor D. Robertshaw, *University of Nairobi*

8 **REPRODUCTIVE PHYSIOLOGY**
Professor R. O. Greep, *Harvard Medical School*

MTP International Review of Science

Physical Chemistry
Series Two

Volume 2
Molecular Structure and Properties

Edited by **A. D. Buckingham, F.R.S.**
University of Cambridge

Butterworths · London and Boston

THE BUTTERWORTH GROUP

ENGLAND
Butterworth & Co (Publishers) Ltd
London: 88 Kingsway, WC2B 6AB

AUSTRALIA
Butterworths Pty Ltd
Sydney: 586 Pacific Highway 2067
Melbourne: 343 Little Collins Street, 3000
Brisbane: 240 Queen Street, 4000

CANADA
Butterworth & Co (Canada) Ltd
Toronto: 2265 Midland Avenue, Scarborough, Ontario M1P 4S1

NEW ZEALAND
Butterworths of New Zealand Ltd
Wellington: 26–28 Waring Taylor Street, 1

SOUTH AFRICA
Butterworth & Co (South Africa) (Pty) Ltd
Durban: 152–154 Gale Street

U.S.A.
Butterworth (Publishers) Inc
Reading: 161 Ash Street, Mass., 01867

Library of Congress Cataloging in Publication Data

Buckingham, Amyand David.
Molecular structure and properties.
(Physical chemistry, series two; v. 2) (International review of science)
Includes index.
1. Molecules. 2. Chemistry, Physical and theoretical. I. Title. II. Series. III. Series: International review of science.
QD450.2.P59 vol. 2 [QD461] 541′.3′08s [539′.6]
ISBN 0 408 70601 5 75–4593

First Published 1975 and © 1975

Typeset, printed and bound in Great Britain by
REDWOOD BURN LIMITED
Trowbridge & Esher

Consultant Editor's Note

The International Review of Science was conceived as a comprehensive, critical and continuing survey of progress in research. The difficult problem of keeping up with advances on a reasonably broad front makes the idea of the Review especially appealing, and I was grateful to be given the opportunity of helping to plan it.

Physical Chemistry Series One was published in 1972/1973. Its success was assured by the very great distinction of its editors and authors. Our need for critical reviews at a high level has not diminished. In the rather difficult times being experienced in most parts of the world, research workers should seek to broaden the range of their expertise; it is hoped that this Series will be of use in this connection.

Like its forerunner, Series Two consists of thirteen volumes covering Physical and Theoretical Chemistry, Chemical Crystallography and Analytical Chemistry. Each volume has been edited by a distinguished chemist; in several cases the person responsible for Series One has acted again as editor. The editors have assembled a strong team of authors, each of whom has assessed and interpreted recent progress in a specialised field in terms of his own experience. I believe that their efforts have again produced useful and timely articles which will help us in our efforts to keep abreast of progress in research.

It is my pleasure to thank all those who have collaborated in this venture —the volume editors, the authors and the publishers.

Cambridge A. D. Buckingham

Preface

Every chemist is interested in molecular structure and properties. To a few—and I readily confess that I am one such—it is a subject of great fascination. Some use modern techniques to measure molecular constants to an accuracy of at least one part in ten thousand; others favour qualitative generalisations based on results for a variety of molecules of different types. It is intended that this series of reviews of Molecular Structure and Properties will assist all chemists by providing them with critical and up-to-date articles on all aspects of this central area in our subject.

The first volume was edited by Professor G. Allen of Manchester and published in 1972. He is a contributor to this second volume. It contained seven articles on such topics as infrared refraction, neutron and electron diffraction, crystalline polymers, diamagnetic susceptibilities, acoustic studies of conformation and dielectric polarisation.

This second volume contains nine reviews covering a wide range from high-resolution molecular-beam studies to the relationship of the properties of liquids and polymers to molecular structure. The authors are distinguished scientists; they have written with authority and clarity about their own specialities. Their expert touch may be discerned. I believe that their contributions will help many research workers in their efforts to overcome the disadvantage of specialisation. As editor, I wish to express to them my admiration and gratitude.

Cambridge A. D. Buckingham

Contents

Molecular structure determination by high-resolution spectroscopy D. R. Lide, Jr., *National Bureau of Standards, Washington*	1
The properties of molecules from molecular beam spectroscopy J. S. Muenter, *University of Rochester, New York,* and T. R. Dyke, *University of Oregon*	27
The structure and properties of van der Waals molecules B. J. Howard, *University of Southampton*	93
Structure determination by n.m.r. spectroscopy K. A. McLauchlan, *University of Oxford*	119
Electric dipole polarisibilities of atoms and molecules M. P. Bogaard and B. J. Orr, *University of New South Wales*	149
Bonding features in magnetochemical models M. Gerloch, *University of Cambridge*	195
Hyperfine interactions and molecular structure M. G. Clark, *Royal Radar Establishment, Worcestershire*	239
Equilibrium properties of molecular fluids P. A. Egelstaff, and C. G. Gray, *University of Guelph,* and K. E. Gubbins, *University of Florida*	299
Structure/property relationship in high polymers G. Allen and D. C. Watts, *University of Manchester*	349
Index	395

1
Molecular Structure Determination by High-resolution Spectroscopy

D. R. LIDE, JR.
National Bureau of Standards, Washington

1.1	INTRODUCTION	2
	1.1.1 *Definition and scope*	2
	1.1.2 *Historical background*	2
1.2	DERIVATION OF STRUCTURAL INFORMATION	4
1.3	THEORY OF THE VIBRATING ROTOR	5
1.4	TYPES OF STRUCTURE DETERMINATIONS	7
	1.4.1 *Equilibrium structure* (r_e)	7
	1.4.2 *Effective structure* (r_0)	9
	1.4.3 *Substitution structure* (r_s)	10
	1.4.4 *Variants on the substitution structure*	13
	1.4.5 *Average structure* (r_z or $\langle r \rangle$)	15
	1.4.6 *Comparison with electron diffraction*	17
1.5	SPECIAL PROBLEMS OF STRUCTURE DETERMINATION	19
	1.5.1 *Small coordinates*	19
	1.5.2 *Large-amplitude motions*	21
1.6	STRUCTURE COMPARISONS	22

1.1 INTRODUCTION

1.1.1 Definition and scope

The term 'molecular structure' has been used to describe various features of individual molecules and crystals that are susceptible to experimental measurement. In this review the term is used in a restricted sense to include all information pertinent to the geometric arrangement of atoms in an isolated, gas-phase molecule. On a quantitative level the structure is described in terms of a set of interatomic distances and bond angles that specify the location of each of the atoms in the molecule. In addition, gross structural features such as the presence or absence of symmetry elements will be considered. The description of the forces between atoms will not be addressed as such but will be discussed from the viewpoint of the influence of internal vibrations on the measurement of the geometric parameters.

The experimental techniques pertinent to this review are the various types of molecular spectroscopy carried out on gas-phase molecules. These include measurements in the ultraviolet, visible, infrared and microwave regions of the spectrum. In order to gain useful structural information, the spectroscopic resolution must be high enough to distinguish individual rotational features. The other important technique for structural investigation of gas-phase molecules, electron diffraction, will be discussed for comparison purposes, but no comprehensive treatment will be given. For that purpose the reader is referred to a review by Kuchitsu in this series[1] and to a recent review by Hedberg[2].

1.1.2 Historical background

The observation of fine structure in the visible and ultraviolet band spectra of molecules under high spectral resolution goes well back into the nineteenth century. As early as 1886, Deslandres noted the regularity of these patterns, which were later to be identified with rotational structure. The association of i.r. absorption spectra with vibrations of the atoms within a molecule developed early in this century. In 1912, Bjerrum suggested that the spectrum observed in the near- and medium-i.r. region was not determined solely by vibrations, but was also influenced by molecular rotation. At about the same time, Nernst proposed that the quantum theory must be involved in the rotation of molecules. Thus when Bohr introduced his theory of the hydrogen atom the groundwork had been laid for the quantitative application of quantum concepts to the rotation of molecules.

A review by Von Bahr[3], prepared in 1914, provides an interesting account of the understanding of molecular rotation at that time. The author calculated moments of inertia of several molecules from the contour of i.r. vibration–rotation bands. Interatomic distances were derived in this way for several diatomic molecules, although the values were in error by $\sqrt{2}$ because of inadequacies in the theory. When this correction is applied, interatomic distances* of 1.3 and 1.6 Å are obtained for HCl and HBr, respectively. By 1919, refinements in the theory and experimental advances, which now permitted individual rotational lines in the vibration–rotation band to be resolved, led to much improved values. Imes[4] reported distances of 1.28 Å in HCl and 1.42 Å in HBr, which are quite close to the currently accepted values of 1.274 and 1.414 Å.

The derivation of structural information from measurements in other spectral regions proceeded somewhat more slowly. The pure rotational spectrum of HCl in the far-i.r. region was resolved by Czerny[5] in 1925; his results agreed well with the previous measurements from the vibration–rotation spectrum. Although vibration–rotational fine structure on visible and u.v. emission bands could be clearly resolved and accurately measured, the difficulty in identifying the molecule responsible for the spectrum and in selecting its ground electronic state delayed the general application to measurement of structure.

With the major advances in theory that occurred after the introduction of the new quantum mechanics, as well as the gradual refinement of spectroscopic instrumentation, molecular spectroscopy became a widely used tool for structure determination. At the time of the publication of Herzberg's classic volume on i.r. and Raman spectra[6] in 1945, about 30 polyatomic molecules had been studied at a high enough resolution to yield some quantitative structural information. Interatomic distances had also been determined in a large number of diatomic molecules from the visible–u.v. and, to a lesser extent, i.r. spectra. At this time a major experimental advance took place, namely, the opening of the microwave region of the spectrum. Microwave spectroscopy enabled the pure rotational spectrum to be studied at a resolution which exceeded that in the i.r. by a factor of 10 000. The fact that isotope shifts could be well resolved, even with heavy elements, made it feasible to determine the structure of very complex molecules where the number of structural parameters greatly exceeded the information obtainable from a single isotopic species. This technique quickly became the most valuable spectroscopic tool for structure determination and still remains so.

Post-war advances in other spectroscopic regions should also be mentioned since they provide a useful complement to microwave spectroscopy. Resolution in the i.r. has been greatly improved by the use of more refined grating spectrometers and interferometric (Fourier transform) instruments. The structures of many important non-polar molecules, which are not susceptible to study by microwave spectroscopy, have been determined by these techniques. Much valuable structural information has been obtained from rotational Raman spectra, a field pioneered by Stoicheff in the 1950s. Finally, the introduction of flash-photolysis techniques made it possible to obtain high-resolution absorption spectra of short-lived polyatomic molecules in the visible and u.v. regions. Many molecular species that are not accessible to study by the other techniques mentioned here have been investigated in this way. The recent development of tunable i.r. lasers opens new possibilities for structural studies, although few results are yet available.

* Throughout this review interatomic distances will be stated in ångstrom units (Å). For conversion to SI units, 1 Å = 10^{-10} m.

1.2 DERIVATION OF STRUCTURAL INFORMATION

Before considering the detailed treatment, it may be useful to summarise the methods by which structural information can be extracted from high-resolution spectroscopic data. This discussion applies to data obtained from any region of the spectrum—microwave i.r., visible, or u.v.

Quantitative information on molecular geometry is derived primarily from measured moments of inertia. The rotational contribution to the spectrum is a function of the principal moments of inertia, which depend upon the masses of the atoms and their locations within the molecule. Assuming the masses are accurately known, each moment of inertia provides a relationship among the atom coordinates; if a sufficient number of these relations can be established, a unique solution for the interatomic distances and angles is possible. However, a single isotopic species provides, at most, three independent principal moments; in the case of a symmetric rotor only two moments exist and a linear rotor or spherical rotor is described by one moment of inertia. Thus, it is only for the most simple molecular types that a single isotopic species provides enough information for a unique determination of the complete structure. These include diatomic molecules, linear or bent XY_2 molecules, trigonal XY_3 molecules and tetrahedral XY_4 molecules.

In the general case it is necessary to have data on more than one isotopic species in order to fix the structural parameters uniquely. Kraitchman[7] has discussed the minimum number of isotopic species required for a structure determination on various molecular types in the ideal case of a perfectly rigid rotor. However, as will be explained in the next section, non-rigidity effects have a major influence on the accuracy of structural parameters derived from measured moments of inertia. Thus it frequently happens that data on a particular set of isotopic species, whilst sufficient in principle for a unique structure determination, do not actually permit a meaningful calculation.

In situations where moments of inertia are not available for a sufficient number of isotopic species, it is sometimes possible to reduce the degrees of freedom by assuming values for some of the structural parameters. For example, a distance involving a hydrogen atom can often be assumed without introducing a significant uncertainty into a derived distance between two heavier atoms.

Certain conclusions of a more general nature may sometimes be reached from spectroscopic data without carrying out a quantitative structure determination. A classical example is the missing centre line in vibration–rotation bands of linear molecules[6]. If such a 'zero gap' is observed in a spectrum, one can conclude unambiguously that the molecule has a linear configuration, even without quantitative measurements. Likewise, the presence of an axis of symmetry in a molecule leads to intensity alternations in the rotational (and vibration–rotational) spectrum as a result of nuclear spin statistical weights. The type of symmetry axis and the identification of the nuclei that lie off the axis can often be inferred from relative intensity measurements alone. Another method of detecting either a plane or axis of symmetry is through measurement of the Stark effect. If sufficiently sensitive transitions can be found, one can demonstrate that a component of the permanent electric dipole moment is vanishingly small, which provides strong evidence for the existence of a symmetry axis or plane.

The inertial defect provides another basis for conclusions on the gross structure of a molecule. This is defined by:

$$\Delta = I_c - I_a - I_b \tag{1.1}$$

where I_a, I_b and I_c are the principal moments of inertia in order of increasing magnitude. A well-known theorem states that $\Delta = 0$ for a rigid collection of co-planar mass points. For real molecules with a planar equilibrium configuration, Δ is found to have a small positive value which, for an assumed model, can be estimated from vibration–rotation interaction theory[8,10] or from experience with related molecules. Thus a measurement of the principal moments of a single isotopic species may be sufficient to establish the planarity of a quite complex molecule. An extention to molecules with a plane of symmetry may be made through the relation:

$$\Delta' = I_c - I_a - I_b + 2\sum_i m_i z_i^2 \tag{1.2}$$

where z_i is the out-of-plane distance of each atom of mass m_i that lies off the symmetry plane and Δ' is a pseudo-inertial defect[11] of magnitude comparable with the Δ discussed above. In this way, out-of-plane atoms may be located without carrying out a complete structure determination. Alternatively, for a model in which the location of the out-of-plane atoms can be assumed, the planarity of the remainder of the molecule may be demonstrated.

1.3 THEORY OF THE VIBRATING ROTOR

The central problem in the extraction of accurate interatomic distances and angles from rotational spectra is the proper handling of vibrational effects. The moments of inertia obtained directly from a spectral analysis necessarily represent averages of some type over the vibrational motion. Unfortunately the nature of this average is quite complicated, and considerable information on the vibrational potential function is required to correct the observed effective moments to the moments appropriate to a hypothetical non-vibrating molecule or to some other model which is physically meaningful. In this section the basic theory of vibration–rotation interactions is outlined. This theory has been developed over a period dating from the basic paper of Wilson and Howard[12]. The contributions of Nielsen and co-workers[13] have been particularly important.

The coordinate system conventionally chosen for describing a vibrating, rotating set of mass points is that defined by the Eckart conditions, which specify that (a) the origin of the coordinate system lies at the instantaneous centre of mass and (b) the vibrational angular momentum vanishes as all atoms pass through their equilibrium positions. On the assumption that the vibrational amplitudes are small (which excludes special effects such as internal rotation and inversion), the vibrational potential energy may be expanded as:

$$V = \tfrac{1}{2}hc\sum_{s}\omega_s q_s^2 + hc\sum_{stu} k_{stu}q_s q_t q_u + hc\sum_{stuv} k_{stuv}q_s q_t q_u q_v + \ldots \quad (1.3)$$

Here the ω_s quantities are the harmonic 'frequencies', expressed in wave-number (cm^{-1}) units, of the fundamental vibrational modes; the k_{stu} and k_{stuv} quantities are anharmonic potential constants (cubic and quartic, respectively); and q_s is a dimensionless normal coordinate related to the actual normal coordinate Q_s by

$$q_s = (\hbar/2\pi c\omega_s)^{\frac{1}{2}}Q_s \quad (1.4)$$

The instantaneous moments and products of inertia may likewise be expanded as power series in the normal coordinates:

$$\begin{aligned} I_{\alpha\alpha} &= I_{\alpha\alpha}^{e} + \sum_{s} a_s^{\alpha\alpha}Q_s + \sum_{st} A_{st}^{\alpha\alpha}Q_sQ_t + \ldots \\ I_{\alpha\beta} &= -\sum_{s} a_s^{\alpha\beta}Q_s - \sum A_{st}^{\alpha\beta}Q_sQ_t + \ldots \end{aligned} \quad (1.5)$$

where $I_{\alpha\alpha}^{e}$ is the equilibrium moment and

$$\begin{aligned} a_s^{\alpha\alpha} &= (\partial I_{\alpha\alpha}/\partial Q_s)_e \\ A_{st}^{\alpha\alpha} &= \tfrac{1}{2}(\partial^2 I_{\alpha\alpha}/\partial Q_s \partial Q_t)_e \end{aligned} \quad (1.6)$$

The indices α, β take values x, y, z, which are the axes defined by the Eckart conditions. The derivatives $a_s^{\alpha\alpha}$ and $A_{st}^{\alpha\alpha}$ depend upon the molecular model (generally in a complicated way) and the masses. They are also functions of the atomic positions at the equilibrium configuration, but since we are dealing with small corrections it is usually adequate to evaluate these derivatives with an assumed set of equilibrium interatomic distances and angles.

The formulation of the problem requires setting up the Hamiltonian in terms of the quantities defined above and making suitable approximations which allow the vibration–rotation energy levels to be expressed in these terms. The procedure is described in many places[13,14] and will not be repeated here. The basic result is that the rotational part of the energy is given to a first approximation by the energy of an effective rigid rotor described by rotational constants A_v, B_v and C_v. These constants depend on the vibrational state in a manner that is described by:

$$B_v = B_e - \sum_{s} \alpha_s^{(B)}(v_s + \tfrac{1}{2}) + \sum_{st} \gamma_{st}^{(B)}(v_s + \tfrac{1}{2})(v_t + \tfrac{1}{2}) + \ldots \quad (1.7)$$

where the v_s are vibrational quantum numbers and

$$B_e = h^2/8\pi^2 I_B^e \quad (1.8)$$

Thus we see that the observable quantities B_v must be corrected (even when all $v_s = 0$) for the vibrational effects described by $\alpha_s^{(B)}$, $\gamma_{st}^{(B)}$, ... in order to obtain the equilibrium moments I_e. Similar expressions apply to A_v and C_v.

There are two ways to carry out this correction. If the rotational spectrum can be measured in each of the $3N - 6$ vibrational states (N is the number of atoms in the molecule) in which a normal mode is singly excited ($v_s = 1$), then all the $\alpha_s^{(B)}$ terms in equation (1.7) can be experimentally determined (assuming that the $\gamma_{st}^{(B)}$ terms are negligibly small). Provided these measurements are accurate enough, we then have a good value of B_e. However, this is a difficult experimental problem and has been achieved only on a limited number of relatively simple molecules. Microwave spectra in excited states may be difficult to measure because of the low population of these states, and accurate high-resolution infrared measurements are limited to relatively light molecules.

The alternative approach is to calculate the $\alpha_s^{(B)}$ from other molecular information. In terms of the formulation outlined above, the $\alpha_s^{(B)}$ terms are given by

$$\alpha_s^{(B)} = -\frac{2B_e^2}{c\omega_s}\left[3A_{ss}^{xx} + 4\sum_t (\zeta_{st}^x)^2 \frac{\omega_t^2}{\omega_s^2 - \omega_t^2}\right] - 2B_e^2\left[3ck_{sss}(c\omega_s)^{-\frac{3}{2}}\frac{2\pi a_s^{xx}}{h^{\frac{1}{2}}} + \sum_{t\neq s} ck_{sst}(c\omega_t)^{-\frac{3}{2}}\frac{2\pi a_t^{xx}}{h^{\frac{1}{2}}}\right] \tag{1.9}$$

Here we have identified the rotational constant B with the x axis. The quantities ζ_{st}^x are Coriolis coupling constants which depend only on the harmonic part of the potential function. Thus we see that a large number of potential constants, both quadratic and cubic, are required to calculate the $\alpha_s^{(B)}$ term. Although some information on these constants can be obtained from purely vibrational spectral data, it is usually impossible to determine the anharmonic constants completely without using experimental values for the α terms. Thus the problem of determining potential constants is intimately coupled with that of determining equilibrium structures, and data on excited vibrational states are essential.

The determination of the $\gamma_{st}^{(B)}$ and higher terms in equation (1.7) is a formidable job which has been approached in only a few cases. Fortunately, the series appears to converge rapidly enough that these terms have a very minor effect on structure determinations. A more serious problem is the existence of accidental vibrational resonances that perturb the effective rotational constants. A Fermi type of resonance[14] can mix two or more excited states so that the observed rotational constants can no longer be interpreted by the simple relation of equation (1.7). If all of the interacting states can be studied, the perturbations can be handled with no basic difficulty. However, it has been found that even very weak resonances, involving levels that are well separated in energy, can often have a non-trivial effect on the rotational constants. Thus it is hard to exclude a contribution from weak resonances with unobserved levels. The effect of Fermi resonance on structure determinations has been thoroughly studied by Morino *et al.*[15,16,59].

1.4 TYPES OF STRUCTURE DETERMINATIONS

1.4.1 Equilibrium structure (r_e)

As pointed out in Section 1.3, the calculation of interatomic distances and angles of the equilibrium configuration of a molecule requires the undoing of the vibrational averaging, which can only be done through measurements of vibrationally excited states. Determination of r_e structures is thus restricted to relatively small molecules, where the number of normal vibrational modes is not too large. Equilibrium structures are known for about 300 diatomic molecules[17], a fair number of triatomics and a few three-, four- and five-atom molecules.

The equilibrium structure of a molecule should be isotopically invariant to a very high approximation. In diatomic molecules this is found to be true down to the level (<0.0001 Å), where a breakdown of the Born–Oppenheimer approximation may occur[18-20]. In polyatomic molecules the calculation of equilibrium moments of inertia is complicated by higher-order terms in the expansion of the rotational constants and by resonances among excited levels (see Section 1.3). In the few cases where sufficient data are available for an accurate comparison, the invariance of the r_e structure is confirmed reasonably well. A recent study of several isotopic species of HCN by Winnewisser *et al.*[21] gave r_e(CH) and r_e(CN) which agreed within 0.0005 Å or better with values previously obtained by Rank *et al.*[22] from different species and a different set of levels. Somewhat larger variations have been found by Maki and Johnson[23] for OCS. Sets of distances calculated from various combinations of isotopic species are shown in Table 1.1. The spread of values

Table 1.1 Equilibrium bond distances (in Å) calculated from B_e values for different pairs of isotopic species of OCS (data from Ref. 23)

	r_e(C—O)	r_e(C—S)	r_e(O—S)
$^{16}O^{12}C^{32}S$—$^{16}O^{13}C^{32}S$	1.153 94(42)	1.563 21(34)	2.717 15(12)
$^{16}O^{12}C^{32}S$—$^{16}O^{12}C^{34}S$	1.157 37(69)	1.560 43(56)	2.717 80(13)
$^{16}O^{12}C^{32}S$—$^{18}O^{12}C^{32}S$	1.155 17(17)	1.562 21(14)	2.717 39(3)
$^{16}O^{13}C^{32}S$—$^{16}O^{12}C^{34}S$	1.148 75(104)	1.567 35(83)	2.716 10(21)
$^{16}O^{13}C^{32}S$—$^{18}O^{12}C^{32}S$	1.154 83(12)	1.562 50(10)	2.717 33(2)
$^{18}O^{12}C^{32}S$—$^{16}O^{12}C^{34}S$	1.155 59(14)	1.561 86(11)	2.717 45(4)

is larger than the uncertainty in the primary experimental measurements, indicating that higher-order terms or weak resonances are having a perceptible effect. Even so, it can be concluded that the equilibrium structure of OCS has been defined to an accuracy of from ±0.001 to 0.002 Å, the best values being $r_e(CO) = 1.154_3$ Å and $r_e(CS) = 1.562_8$ Å. Values very close to these had previously been obtained by Morino and Nakagawa[59].

It is apparent from this discussion that the determination of accurate equilibrium structures is limited to relatively small molecules. Even there, an appreciable effort is required. Methods have been developed, therefore, to obtain useful structural information from the available spectroscopic data, even though a complete equilibrium structure cannot be derived. The various types of structure determinations in common use will be described in the remainder of this section.

1.4.2 Effective structure (r_0)

The simplest type of structure calculation makes use of the effective moments of inertia in the ground vibrational state, which are defined in terms of the observed ground-state rotational constants by a relation analogous to equation (1.8), i.e.

$$B_0 = h^2/8\pi^2 I_B^\circ \tag{1.10}$$

The effective moments I_A°, I_B°, I_C° (which we shall denote collectively by I_0) are parameters determined directly from the spectrum (either rotation, vibration–rotation or fine structure in electronic bands), without the insertion of any auxilliary data. In general the I_0 values may be measured with very high precision, especially by microwave techniques. For a few simple molecular types, the I_0 values for a single isotopic species are sufficient to determine the structure uniquely. In most cases, however, it is necessary to have results for several isotopic species.

The first drawback of an r_0 structure is the lack of unique operational definition. Problems exist even in a simple case like the bent XY_2 molecule, where there are two structural parameters (the X—Y distance and the YXY angle) and three observed I_0 values. Because of the existence of a finite inertial defect, described by equation (1.1), the three possible ways of combining the moments I_A°, I_B°, I_C° as pairs lead to three distinct sets of structural parameters. This is illustrated by data on the normal isotopic species of ozone[24] in Table 1.2. The differences between the sets are not trivial and there is no basis for choosing one over the other.

Table 1.2 Variation of r_0 structure of ozone with choice of moments

Moments used	r_0(OO)/Å	∠OOO/degrees
I_A°, I_B°	1.2759	116.98
I_A°, I_C°	1.2771	117.03
I_B°, I_C°	1.2794	116.47

The lack of uniqueness becomes even more apparent when isotopic substitution is considered. In the most general case of a molecule of N atoms with no element of symmetry, the minimum requirement for a complete structure determination is the three principal moments for $N - 2$ isotopic species. If an isotopic substitution can be made at every atomic position, thereby providing data on $N + 1$ different species, the data can be combined in $N(N^2 - 1)/6$ ways. For a molecule of four atoms this means that 10 distinct calculations are possible, and each will in general yield a different set of structural parameters. In practice it is often difficult or impossible to substitute at every position because of the lack of a stable isotope or difficulties with synthesis of enriched compounds. On the other hand, multiply-substituted species are frequently available. Thus the data set available for an r_0-structure determination tends to be rather arbitrary and no objective criteria exist for choosing the results from a particular subset. A general least-squares fit to all the available ground-state moments is equally arbitrary. Experience has shown that addition of data on a new isotopic species can change the solution by a significant (and unpredictable) amount.

The range of r_0 structures obtained from different combinations of isotopic species is illustrated in Table 1.3 for some simple molecules. An ambiguity of the order of 0.01 Å is common. However, there are exceptions such as HCN in Table 1.3 where the r_0 value of the CN distance is not very sensitive

Table 1.3 Variation of r_0 structures with choice of isotopic species

Molecule	*Parameter*	r_0/Å*	r_e/Å
OCS	r(CO)	1.155–1.165	1.154
	r(CS)	1.558–1.565	1.563
ClCN	r(CCl)	1.627–1.634	1.629
	r(CN)	1.157–1.166	1.160
HCN	r(CH)	1.058–1.069	1.066
	r(CN)	1.155–1.158	1.153
N_2O	r(NN)	1.12 –1.14†	1.126
	r(NO)	1.18 –1.20†	1.186

* The range of r_0 values represents the results obtained from various pairs of isotopic species
† Certain combinations of isotopic species lead to imaginary distances

to the choice of input data. This will tend to occur when one pair of atoms makes the dominant contribution to the moment of inertia. Thus r_0 structures can provide useful information, but caution must be exercised in drawing conclusions from them.

1.4.3 Substitution structure (r_s)

In an effort to remove the ambiguities of the r_0 structure, Costain proposed in 1958 the use of a 'substitution' or r_s structure. This calculation is based on ideas originally proposed by Kraitchman[7]. The r_s coordinates of a given atom in a molecule are determined by the isotope shifts in the moments of inertia when that atom (and no others) is substituted. A complete r_s structure thus requires single isotopic substitutions on every atom. In contrast to an r_0-structure determination, straightforward rules can be established for carrying out the calculation.

The general expression relating the Cartesian coordinates of an atom to the isotope shifts in the moments of inertia when that atom is substituted may be written

$$x^2 = \mu^{-1}\Delta P_x\{1 + [\Delta P_y/(I_x - I_y)]\}\{1 + [\Delta P_z/(I_x - I_z)]\} \quad (1.11)$$

where

$$\Delta P_x = \tfrac{1}{2}(-\Delta I_x + \Delta I_y + \Delta I_z)$$

Here ΔI_x, ΔI_y and ΔI_z are isotope shifts in the respective moments; I_x, I_y and I_z are moments of the original molecule; and $\mu = M\Delta m/(M + \Delta m)$, where M is the mass of the original molecule and Δm the mass change on substitution. The expressions for y^2 and z^2 and for ΔP_y and ΔP_z are obtained by cyclic permutation of x, y and z. In the special case of a linear molecule (applicable also to an atom lying on the axis of a symmetric rotor), the above expression reduces to

$$z^2 = \mu^{-1}\Delta I_x \tag{1.12}$$

where z is the symmetry axis. The origin of the coordinate system is at the centre of mass of the original molecule.

From a computational viewpoint, determination of an r_s structure is simple and relatively unambiguous. The Cartesian coordinates may be used to calculate interatomic distances and angles, provided their signs, which are not specified by equation (1.11), can be established on other grounds. This can usually be done by requiring that the distances and angles have reasonable values. In cases of real ambiguity in signs, multiple isotopic substitutions may be necessary.

The r_s parameters will not reproduce the ground-state moments of inertia but generally give calculated moments which are smaller. From this and other arguments, Costain[25] reasoned that the r_s parameters usually provide a closer approximation to the r_e values than do the r_0 values. The relation can be given explicitly for a diatomic molecule, where

$$\begin{aligned} r_s &= r_e(1 + a/4B_e) \\ &= \tfrac{1}{2}(r_e + r_0) \end{aligned} \tag{1.13}$$

No such simple relation exists for polyatomic molecules, although Watson[30] has recently shown that the relation

$$I_s = \tfrac{1}{2}(I_e + I_0) \tag{1.14}$$

holds to a first approximation (here I_s is the moment calculated from the r_s parameters). The lack of a well-defined relation between r_s and r_e for a given bond is one of the drawbacks of substitution structures. However, there is a reasonable body of experience showing that r_s distances usually tend to lie between the r_e value and the median of the range of r_0 values that would be obtained from the same set of input data.

In molecules containing elements of symmetry there may be redundancy in the data available for an r_s calculation, i.e. different combinations of ΔI values may be used in calculating a given coordinate. If z is a symmetry axis, the coordinates of an atom on this axis are

$$\begin{aligned} x^2 &= y^2 = 0 \\ z^2 &= \mu^{-1}\Delta I_x = \mu^{-1}\Delta I_y \end{aligned} \tag{1.15}$$

Thus the z coordinate can be calculated from either ΔI_x or ΔI_y, which are independent experimental quantities. In most cases that have been studied the agreement of the results from different computational routes is fairly good, the range of solutions being much less than commonly encountered with r_0 structures. An example is shown in Table 1.4. It is seen that the C—C distance in propane is internally consistent to 0.001 or 0.002 Å. The range of C—H distances is somewhat greater, especially in the CH_3 group, where the range is 0.018 Å. This is probably associated with the large amplitude of the torsional oscillation of the CH_3 group.

Table 1.4 Substitution parameters (r_s) of propane calculated from various combinations of moments of inertia

	I_x, I_y, I_z	I_x, I_y	I_y, I_z	I_x, I_z
r(CC)/Å	1.5277	1.5263	1.5252	1.5256
∠CCC/degrees	112.24	112.36	112.50	112.48
CH_2 *group*				
r(CH)/Å	1.0943	1.0952	1.0971	1.0963
∠HCH/degrees	106.32	106.16	105.90	106.04
CH_3 *group*				
r(CH)/Å	1.0863	1.0970	1.0799	1.0930
∠CCH/degrees	111.90	111.36	112.81	111.28

Good internal consistency at a level of a few thousandths of an ångstrom unit is also found in cases where enough isotopic species have been studied to permit the calculation to be done with various subsets of the data. Some examples have been given by Costain[25, 26]. In CH_3Cl and CH_3Br, Schwendeman and Kelly[27] showed that the carbon–halogen distance differs by about 0.001 Å when the calculation is based on the CD_3X species rather than the CH_3X species. However, it should be emphasised that this difference is well outside the precision of the experimental data, so that isotopic invariance of r_s structures cannot be claimed at a level below 0.001 Å.

Unfortunately, some exceptions to the internal consistency of r_s structures have been found in more recent work. For example, in ethynyldifluoroborane[28], BF_2CCH, the C—C and C—H distances are almost independent of the choice of moments (and of isotopic species) used in the structure calculation, but the B—C distance varies over a range of 0.008 Å. The ambiguity in locating the B atom may be associated with its proximity to the centre of mass (see Section 1.5.1), although it is sufficiently removed (0.38 Å) that one would not normally expect any difficulty.

The calculation of a complete r_s structure requires substitution at each position in the molecule. This raises problems with certain important atoms such as fluorine and phosphorus, where there is only one stable isotope. However, if only one atom has not been substituted, one can determine its coordinates by invoking the first-moment (i.e. centre of mass) conditions:

$$\sum m_i x_i = 0, \text{ etc.} \qquad (1.16)$$

Checks made on molecules where all the r_s coordinates are known have shown that this condition is usually satisfied to a high degree of approximation; thus if one coordinate is missing, it can be calculated from equation

(1.16) with the reasonable assurance that the value is not too different from what would have been obtained if that atom could have been isotopically substituted. It has become common practice to use the term 'r_s structure' to include those cases where the first-moment condition was used.

It is also possible to use the second-moment condition, i.e. the fact that the atomic coordinates should in principle reproduce the moment of inertia:

$$\sum m_i(y_i^2 + z_i^2) = I_{xx}, \text{ etc.} \tag{1.17}$$

However, the r_s coordinates generally do not satisfy this condition as well as they do the first-moment condition. Typically, the calculated moments obtained by equation (1.17) are 0.1–0.5% less than the observed ground-state moments[25]. Thus the use of equation (1.17) to determine a missing coordinate can produce significant errors. An alternative is to scale the ground-state moments by some factor based on experience with related molecules before applying equation (1.17). However, the arbitrary nature of this correction is not very satisfactory.

Substitution structures have been used extensively for comparing interatomic distances and angles in related molecules and drawing conclusions from the observed trends. The confidence placed in such comparisons is based partly on the internal consistency of most r_s structures, although that consistency does not insure the absence of systematic effects which would invalidate the comparison with other molecules. Another basis for confidence is the relatively smooth trends that have been found when r_s parameters are used in correlations of bond length with other molecular properties. An example is the variation in C—C single bond lengths with hybridisation of the carbon orbitals[29]. Finally, r_s values for bonds that are expected to be relatively invariant to molecular environment have been found to be remarkably constant. Thus C≡N and C≡C bonds, which, because of their high strength should be insensitive to changes in the remainder of a molecule, show very little variation from one molecule to another when r_s values are compared. In the series FCN, ClCN, BrCN and ICN, the C≡N r_s bond length is constant to the nearest 0.001 Å at 1.159 Å, even though the dynamics of the internal motions, the location of the centre of mass and the vibrational potential constants vary over a wide range. The length of the C≡N bond in most other cyanides falls within 0.002 Å of this value[26], the major exception being HCN, where it is 0.004 Å shorter. Similar constancy is found for C≡C triple bonds. This type of experience lends credence to the use of r_s parameters for structural correlations, even though the exact relation of these parameters to the true equilibrium structure is uncertain.

1.4.4 Variants on the substitution structure

Certain refinements and extensions of the substitution structure have been proposed. Watson[30] has shown how data on several isotopic species can be extrapolated to infinite mass, thereby giving an estimate of the r_e structure. This 'mass-dependent' or r_m method has the advantage that it requires only ground-state moments and is therefore not affected by resonances among excited vibrational states. However, accurate data on a number of isotopic species must be available.

The basis of the method is the relation given in equation (1.14), which may be rewritten as

$$I_m = 2I_s - I_0 \tag{1.18}$$

to emphasise that I_m is only a first-order approximation to I_e. An r_s structure must be calculated for each isotopic species through the use of equation (1.11) Then equation (1.18) gives a set of I_m values which, if the number of isotopic species is sufficient, can be used for a least-squares determination of a new set of structural parameters. This set is designated as r_m parameters, and with certain restrictions they should provide a good approximation to the r_e parameters.

Watson has applied this method to several diatomic molecules and to N_2O, OCS, SO_2 and HCN. In most cases the r_m values are in excellent agreement with the best r_e values determined by other methods. A discrepancy exists for OCS, although it is not clear which calculation is at fault. The result for HCN is poor, as expected, because the approximation on which the method is based breaks down for very light atoms.

It is possible that this method will find further uses. However, a large number of isotopic species must be studied and the demands on the precision of the input data are very high. The problem of handling hydrogen–deuterium substitution is also a limitation.

Schwendeman[31] has explored the calculation of structural parameters from differences in moments of inertia through a different procedure from that employed in the usual substitution structure. Instead of using equation (1.11) to calculate Cartesian coordinates from the differences in moments of singly-substituted species, a least-squares fit to all the ΔI values is carried out. The result has been described as a pseudo-Kraitchman or p-Kr structure. While the input data and the basic model are the same as used for an r_s structure, the effective weighting of the data is different. A comparison is shown in Table 1.5 for a moderately complex molecule, 2-chloropropane.

Table 1.5 Comparison of r_s and p-Kr structures in 2-chloropropane (data from Ref. 31)

*Parameter**	*r_s values*		*p-Kr(I) values†*		*p-Kr(P) values‡*	
	Bond length /Å	*Bond angle* /degrees	*Bond length* /Å	*Bond angle* /degrees	*Bond length* /Å	*Bond angle* /degrees
CCl	1.797		1.800		1.803	
CC	1.520		1.517		1.521	
CH_s	1.091		1.099		1.091	
CH_2	1.099		1.094		1.091	
$\angle H_sCCl$		105.2		104.9		105.0
$\angle CCC$		112.8		113.0		112.7
$\angle CCH_1$		111.0		110.9		110.7
$\angle H_1CH_3$		109.2		109.0		108.9

* The H_1, H_2 and H_3 atoms are in the CH_3 group; H_s is attached to the central C atom
† Calculated using ΔI values (see text)
‡ Calculated using ΔP values

The two sets of p-Kr parameters given in Table 1.5 result from using the moments of inertia directly and using the P_x, P_y, P_z quantities defined in equation (1.11). Again, the use of different linear combinations of the input data in the least-squares procedure is seen to influence the results significantly. Variations of several thousandths of an ångstrom unit in heavy-atom bond distances, and even more for bonds involving hydrogen, must be regarded as typical.

The procedure of Schwendeman would appear to have computational advantages in some situations. For example, data on multiply-substituted species cannot be utilised directly for (r_s-structure determinations since equation (1.11) applies to a species in which only one atom differs isotopically from the parent species. However, one feels intuitively that equivalent information is contained in these multiply-substituted species, and a least-squares fit of the differences in moments (with respect to the parent species) is probably the best way to extract the information. A similar procedure has been used previously, for example, in vinyl chloride[32], where data on several multiply-substituted deuterium species are available. In cases where some atoms cannot be substituted and the first-moment condition or some other constraint must be used, the calculation of a p-Kr structure probably permits a more informative assessment of the uncertainties in the result.

1.4.5 Average structure (r_z or $\langle r \rangle$)

The r_z or $\langle r \rangle$ structure[33-36] may be regarded as a compromise between the practical difficulty of determining r_e structures and the lack of a well-defined physical meaning for r_s structures. This structure refers to the nuclear positions averaged over the vibrational motion in a particular state. The determination of the average structure requires some information on the vibrational potential function but not as much as is needed to obtain the equilibrium structure.

It was shown by Morino *et al.*[35] and by Herschbach and Laurie[33] that removal of the harmonic part of the vibrational dependence of the effective moments I_0 yields the moments that would be obtained for a rigid molecule in which each atom is frozen at its average position. Therefore, if the harmonic part of the vibrational potential function (i.e. the usual vibrational force constants) is accurately determined, the effective moments I_0 can be corrected to give a set of moments I^* from which the average structure can be calculated. Laurie and Herschbach[34] have given explicit formulae for this correction in simple molecular types such as linear XY_2, bent XY_2, linear XYZ, linear X_2Y_2 and tetrahedral XY_4 molecules. The computation is more difficult in more complex molecules, but it can always be carried out if the harmonic potential function is accurately known. Obviously, the same number of isotopic species is required for an average structure determination as for any other type of structure.

Average structures have been determined for a number of triatomic molecules[34,37] and for some larger molecules such as CH_4[34] and NH_3[38]. In bonds involving hydrogen, r_z values generally fall from 0.010 to 0.015 Å

Table 1.6 Distances determined from various isotopic species of ClCN*

Isotopic pair	Cl—C			C—N			Cl···N		
	r_e	r_z	r_0	r_e	r_z	r_0	r_e	r_z	r_0
$^{35}Cl^{12}C^{14}N$ $^{37}Cl^{12}C^{14}N$	$1.629_3 \pm 0.002$	$1.627_6 \pm 0.002$	$1.627_5 \pm 0.001$	$1.160_2 \pm 0.007$	$1.166_5 \pm 0.002$	$1.166_0 \pm 0.001$	$2.789_5 \pm 0.002$	$2.794_1 \pm 0.0007$	$2.793_4 \pm 0.0003$
$^{35}Cl^{12}C^{15}N$ $^{37}Cl^{12}C^{15}N$	—	$1.627_2 \pm 0.003$	$1.627_0 \pm 0.002$	—	$1.166_8 \pm 0.002$	$1.166_3 \pm 0.002$	—	$2.794_0 \pm 0.0008$	$2.793_3 \pm 0.0006$
$^{35}Cl^{12}C^{15}N$ $^{35}Cl^{12}C^{14}N$	—	$1.634_9 \pm 0.0009$	$1.634_0 \pm 0.0005$	—	$1.157_0 \pm 0.0005$	$1.157_4 \pm 0.0003$	—	$2.791_8 \pm 0.0007$	$2.791_3 \pm 0.0004$
$^{37}Cl^{12}C^{15}N$ $^{37}Cl^{12}C^{14}N$	—	$1.634_9 \pm 0.0009$	$1.634_1 \pm 0.0005$	—	$1.156_8 \pm 0.0005$	$1.157_1 \pm 0.0003$	—	$2.791_8 \pm 0.0007$	$2.791_3 \pm 0.0004$
	r_s(Cl—C) = 1.631			r_s(C—N) = 1.159			r_s(Cl—N) = 2.790		

* Distances in Å

higher than r_e values. The difference is smaller (0.001–0.005 Å) with bonds involving two heavy atoms. Some comparisons are given in Section 1.6.

In practice, there are problems in determining r_z structures with high accuracy. The value of r_z is not isotopically invariant; this can be seen for a diatomic molecule by noting the asymmetric shape of the potential function, which implies that a heavier isotopic species will have its energy levels lower in the well where the average interatomic separation is smaller. The extreme case occurs for distances involving hydrogen; for example, r_z for DCl is 0.005 Å shorter than for HCl. Thus, in an over-determined situation, the choice of isotopic species to be used in the calculation has a bearing on the result, just as in the case of r_0 structures. In a study of ClCN, Lafferty *et al.*[37] found a range of 0.007 Å in the C—Cl distance and 0.010 Å in the C—N distance, depending on the pair of isotopic species used (see Table 1.5). Furthermore, application of harmonic corrections to the effective moments leads to some degradation in accuracy because the uncertainty in the force constants must be taken into account. It is seen from Table 1.6 that the r_z structure of ClCN differs very little from the r_0 structure determined from the same pair of isotopic species. Thus it appears that the average structures mirror at least some of the problems encountered with effective structures.

1.4.6 Comparison with electron diffraction

It is frequently of interest to compare structures determined from spectroscopic data with those obtained by the electron-diffraction technique. The diffraction of electrons by a molecule is determined by the complete set of interatomic distances, both bonded and non-bonded, within the molecule. The vibrations of the molecule influence the diffraction pattern, leading, in effect, to a time-averaged distribution of the instantaneous atomic separations. The nature of this averaging differs from the vibrational effects in the spectroscopic technique, although it is ultimately a function of the same set of vibrational potential constants.

As in spectroscopy, several distinct types of distances derived from diffraction experiments have been reported in the literature[1,39]. The most widely used is $r_g(0)$, frequently written simply r_g, which is the centre of gravity of the radial distribution function $P(r)$[40]:

$$r_g = r_g(0) = \int_0^\infty rP(r)\mathrm{d}r \Big/ \int_0^\infty P(r)\mathrm{d}r \tag{1.19}$$

It may be shown that r_g represents the average of instantaneous internuclear distance over all the vibrations of the molecule. Since the averaging is different for different vibrational states, r_g is a function of the distribution of molecules over vibrational states and is therefore temperature dependent. This temperature dependence can be removed by extrapolating the results of the analysis to $T = 0$.

The relation of r_g to the spectroscopic distance parameters may be obtained in the following way. To a good approximation r_g can be written[39]

$$r_g = r_e + \langle \Delta z \rangle_T + (1/2r_e)(\langle \Delta x^2 \rangle_T + \langle \Delta y^2 \rangle_T) \tag{1.20}$$

Here we have set up a local coordinate system *xyz* such that the *z* axis lies on the line joining the equilibrium positions of the two atoms in question. The averages are taken over the Boltzmann distribution of vibrational states. If the harmonic and anharmonic vibrational potential constants are known, the averages in equation (1.20) may be calculated and a value of r_e derived from r_g. This can readily be done for diatomic molecules and, with more difficulty, for a few simple polyatomic molecules. However, when larger molecules are involved, one encounters the same limitations from incomplete knowledge of the potential constants which are present in the spectroscopic technique.

One of the most careful analyses of this type has been done on methane[1,41,42]. The results are shown in Table 1.7. The agreement of r_e distances between

Table 1.7 Comparison of distances* in methane (data from Refs. 1, 34, 41 and 42)

	CH_4	CD_4
Electron diffraction:		
r_g	1.107	1.103
r_a	1.098	1.096
r_e (from r_g)	1.085	1.086
Spectroscopy:		
r_0	1.094	1.092
r_z	1.099	1.096
r_e (from r_0)	1.085	1.085

* Values in Å

the two methods and the two isotopic species is quite satisfactory. This table also illustrates the magnitude of the hydrogen–deuterium isotope effect on both r_g and r_0 values.

An important feature of r_g distances is the absence of the usual geometric constraints on a set of related bonded and non-bonded distances. This comes about because of the nature of the averaging process. Thus, in a linear BAB molecule, the r_g value of the B—B distance is somewhat less that twice the A—B distance. The physical reason for this is easily seen by noting that the bending of the molecule must necessarily reduce the instantaneous value of the B—B distance from its equilibrium value. This 'shrinkage effect' has been extensively studied[1,43,44]. It must be taken into account in order to avoid incorrect deductions regarding linearity or planarity of molecules.

If the vibrational potential constants are not known well enough to convert r_g values to r_e values through equation (1.20), it may still be possible to compare with the average distance (r_z) from a spectroscopic determination. In terms of projections on the equilibrium internuclear axis (rather than the actual instantaneous separation), the average distance derived from a diffraction experiment is, to a good approximation,

$$r_\alpha = r_e + \langle \Delta z \rangle_T \quad (1.21)$$

Hence, we have from equations (1.20) and (1.21),

$$r_{\alpha} = r_{g} - (1/2r_{e})(\langle \Delta x^2 \rangle_T + \langle \Delta y^2 \rangle_T) \qquad (1.22)$$

This correction term involves only the harmonic potential constants. When r_g is extrapolated to $T = 0$, the r_α value in equation (1.22) approaches r_z, the spectroscopic parameter discussed in Section 1.4.5. However, the extrapolation is somewhat dependent on anharmonic constants, which introduces a degree of uncertainty in the comparison of average distances obtained by the two techniques. The good agreement of the average distances for methane is shown in Table 1.7.

In some cases it is difficult to make an accurate structure determination from either spectroscopic or diffraction data alone. A considerable improvement can often be made by combining the two types of data, after making suitable corrections for the inherent differences in definitions of structural parameters. Kuchitsu[1] has discussed this procedure and given many examples of its successful application.

1.5 SPECIAL PROBLEMS OF STRUCTURE DETERMINATION

1.5.1 Small coordinates

One of the most serious limitations on the accuracy of structure determinations from spectroscopic data is the difficulty in locating atoms that lie close to the centre of mass or to a principal axis. In such a case, one or more coordinates of the atom in question is small and the contribution of that atom to the moments of inertia is correspondingly small. The fact that the moments depend upon the square of the Cartesian coordinates accentuates the problem. When the fractional contribution of a given atom to the total moment is so small that vibrational effects are of comparable magnitude, there is an inherent difficulty in determining the precise location of that atom.

An early example of this problem was encountered[45] in N_2O. The ground-state moment of inertia of the $^{15}N^{15}N^{16}O$ species is smaller than that of the $^{15}N^{14}N^{16}O$ species, in spite of the larger mass. Thus, a calculation of an r_0 structure from the data on these two species leads to imaginary bond distances. What has happened is that the vibrational contribution to the moment changes sufficiently upon isotopic substitution to mask the small change in the equilibrium moment that is produced by the larger mass.

The use of substitution structures permits a quantitative estimate of the errors that can result from these effects. In the case of N_2O, Costain[25] showed that use of the first-moment condition, equation (1.16), to locate the central nitrogen atom gave a set of distances that were consistent to better than 0.001 Å, regardless of the combination of isotopic species used. In the five species studied the distance of this N atom from the centre of mass varies from 0.04 to 0.12 Å. If, however, the coordinate of this atom is calculated directly from the isotope shift by means of equation (1.12), errors ranging from 0.01 to 0.05 Å are found, the errors increasing inversely as the distance from the centre of mass. Costain has suggested[26] that the uncertainty in the coordinate of an atom near the centre of mass as determined from equation (1.12) might be estimated roughly as

$$|\delta z| = |0.0012/z| \tag{1.23}$$

with a maximum value of about 0.05 Å if z is very small (the constant is sometimes given as 0.0015). This relation indicates that serious problems will occur for coordinates less than 0.15–0.20 Å. In such cases it is preferable to use the first-moment condition if that is possible.

It should be pointed out that these problems do not arise when the coordinate of an atom is small by virtue of the fact that the centre of mass is fixed primarily by the large mass of that atom. To take a simple example, the Br atom in HBr is only 0.02 Å from the centre of mass yet the r_s coordinate is perfectly well determined. Thus a single very heavy atom in a molecule can often be located quite accurately even though its coordinates are necessarily small.

A method of circumventing this problem has been suggested by Pierce[46]. This 'double-substitution' method requires data on four isotopic species rather than two in order to fix the coordinate of the atom in question. Starting with a parent molecule, the atom with the small coordinate is first substituted and the change in moment of inertia determined. Another atom well removed from the centre of mass is then substituted, followed by a further substitution of the atom with the small coordinate. Although there may be large vibrational contributions to each of these ΔI values, Pierce showed that they will tend to cancel when the second difference is taken. When applied to N_2O, the double-substitution method gives N—N and N—O distances of 1.1297 and 1.1846 Å, respectively; this compares with 1.1286 and 1.1875 Å determined by Costain[25] by the use of the first-moment equation.

The double-substitution method has been used to advantage in several cases. However, the requirements on precision of the experimental data are very high.

The underlying cause of this problem has been analysed by Laurie and Herschbach[34]. As discussed in Section 1.4.5, the average bond length is not isotopically invariant. Various evidence indicates that the change in an interatomic distance r_z between two heavy atoms is typically of the order of 10^{-5}–10^{-4} Å. If this decrease in bond length is designated as δ, then equation (1.12), which applies to a linear molecule, must be replaced by

$$\Delta I_x = \mu z^2 \pm 2mz\delta \tag{1.24}$$

Quadratic terms in δ have been neglected in equation (1.24), and m is the mass of the substituted atom. We thus see that the fractional error in $|z|$ is approximately $|m\delta/\mu z|$. Consideration of the signs of the coordinates, with reference to any given molecular model, shows that a coordinate calculated from equation (1.12) will always be smaller in absolute value because of this error. However, a bond length is the difference of two coordinates which may have opposite signs, so that the error in the bond length may be either positive or negative.

As long as the coordinate z is not too small in absolute value, the error $m\delta/\mu z$ will not be serious. For example, for $m = 20$ a.m.u., $\mu = 1$ a.m.u., $z = 1$ Å and $\delta = 5 \times 10^{-5}$ Å, the error in z is only 0.001 Å, which is at the limit of reliability of any r_s structure. However, as z becomes small this error is greatly magnified, and it is easy to see how errors of 0.01 Å or larger can be introduced when z is as small as 0.10 Å. In other words, the empirical rule of equation (1.12) is consistent with a reasonable set of parameters such as $m = 20$ a.m.u. and $\delta = 6 \times 10^{-5}$ Å.

One might hope that a correction for this source of error could be made by establishing the value of δ independently. The observed moments could then be suitably modified before the r_s calculation is performed. Schwendeman[31] has described such a calculation for 2-chloropropane with assumed values for δ for each bond. Unfortunately, there appears to be no simple way to determine the δ values independently and no rules for estimating them have been proposed. Thus the approach of Laurie and Herschbach is principally useful for assessing the magnitude of the possible errors in an r_s structure.

1.5.2 Large-amplitude motions

The theory of vibration–rotation interactions outlined in Section 1.3 is based on the assumption of infinitesimal vibrational displacements. There are a number of situations involving large-amplitude motions where this assumption is not valid. Several cases of this type will be discussed here.

Internal rotation about a single bond has been much studied, especially through its effects on the microwave spectrum. Tunnelling of the rotating group through a hindering potential barrier leads to a splitting of energy levels and spectral lines. When the barrier is high, theory[47] shows that the spectrum can be described in terms of effective rigid rotors for each of the sub-levels resulting from this splitting, plus certain additional terms. An analysis of the spectrum thus yields rotational constants that have been modified by the coupling of internal and overall rotation. The correction terms in the rotational constants are functions of the barrier height and the geometry of the molecule. Once the effects of internal rotation are analysed, the corrections can be easily calculated.

In the high-barrier case these corrections are small and generally are not very sensitive to isotopic substitution of a heavy atom. They do change considerably, however, on replacement of a CH_3 group by a CD_3 group. Therefore, it is wise to make the corrections before using the rotational constants in a structure calculation. The corrections are larger in the intermediate barrier region and difficulties can arise in calculating them accurately. In molecules with very low barriers, the effective rotational constants may depend on the moments of inertia of only one part of the molecule[48].

Another type of large-amplitude motion is inversion of a group such as NH_2 or NH. The familiar inversion in NH_3 does not complicate the structure determination because of the fairly high barrier and the high symmetry of the molecule. However, in molecules such as formamide[49-51], NH_2CHO, and cyanamide[52,53], NH_2CN, the barrier to inversion is much lower and the large amplitude of the NH_2 wagging mode leads to special problems. There is a difficulty in determining whether the equilibrium configuration is planar or whether there are two shallow minima for the hydrogen atoms symmetrically placed on each side of the plane of the rest of the molecule. The case of formamide is illustrative. The first study of its microwave spectrum[49] led to the conclusion that the molecule is planar. A re-investigation[50] seemed to establish that the equilibrium configuration is non-planar, with the hydrogen atoms of the NH_2 group lying 0.15 Å out of the plane of the heavy atoms. However, a third study[51] has reversed this conclusion and yielded a planar structure. It is still not clear whether this question has been finally settled.

A similar situation exists in four- and five-membered rings where a large-amplitude puckering motion is possible. Lafferty[54] has recently reviewed this class of molecules. Again, it is sometimes difficult to distinguish a planar ring configuration with a large-amplitude, highly anharmonic motion from a puckered ring with a low potential barrier. The effective rotational constants are also influenced in a rather complex way by the vibrational interaction.

The final example to be discussed is a large-amplitude bending vibration in a linear (or 'pseudo-linear') molecule. The first case of this type to be studied was caesium hydroxide[55]. When an r_0 structure is calculated from the ground-state moments of CsOH and CsOD, an OH distance of 0.92 Å is obtained. This is 0.04–0.05 Å shorter than the values typically found for OH bonds, and it would be difficult to devise a satisfying explanation for such a short bond in CsOH. Similar results were found[56] in RbOH.

A study of excited vibrational states of these molecules has shown that the bending mode has an unusually low frequency, which implies a large vibrational amplitude. In fact, the mean displacement from linearity in the ground vibrational state is calculated to be about 20°. Some reflection shows that the r_0 structure is similar to the r_z structure in the sense that the OH distance resulting from this calculation is really an average of the projection of the OH bond on to the equilibrium internuclear axis. Since the instantaneous value of this projection is necessarily less than its equilibrium value, the average will also be less. With a mean amplitude of 20°, the average will be significantly shorter than the true bond length.

By modifying the customary theory of vibration–rotation interactions to account for the large amplitudes, Lide and Matsumura[57] were able to explain the complex way in which the effective rotational constants vary as the bending mode is excited. This permitted an extrapolation to equilibrium moments, which, when used in a structure calculation, yield an OH bond length of 0.96 Å. This is a much more reasonable value.

In dealing with a molecule of this type, it may not be easy to answer the question of linearity of the equilibrium configuration. A linear molecule with a large-amplitude bending motion should show a similar behaviour, in many respects, to a non-linear molecule with a small potential hump at the linear configuration. The good fit of the CsOH and RbOH microwave spectra to the theory described in Ref. 57 provides fairly strong evidence for linearity, although it is difficult to exclude completely the possibility of a small potential hump at the linear position. A more detailed treatment of this problem has been given by Hougen *et al.*[58].

1.6 STRUCTURE COMPARISONS

The question may be raised of whether it is really necessary to deal with so many types of distances derived from spectroscopic data (and a comparable number from electron-diffraction data). If we had a better understanding of vibrational effects and better data on vibrational potential functions, it would be possible to calculate one type from another, and the particular choice of a distance would be a matter of convenience. We are far from that state, and the type of distance available for a given molecule often depends on rather arbitrary factors.

Fortunately this confusion is not as serious as it might seem, because most of the uses to which structural data are put involve a comparison of bond lengths and angles, either within the same molecule or in different molecules. As long as one uses bond lengths that are defined in the same way, such a comparison should be valid to a higher level of precision than the absolute accuracy of any given distance. However, if different types of distances are mixed in making a comparison of this type, the inherent systematic differences can mask the real trend that one hopes to find.

Data are given in Table 1.8 for CH, CO and CN bonds in a few selected molecules. Where available, the r_0, r_z, r_s and r_e values are listed for each distance considered. For a given molecule, the spread of these four values varies from 0.002 to 0.007 Å (median 0.004 Å) for CO and CN bonds. As

Table 1.8 Selected structure comparisons* (data from Refs. 1, 17, 21, 23, 34, 37, 60 and 61)

	r_0	r_z	r_s	r_e
r_{CH}(CH)	1.1303	1.1388	1.1250	1.1198
r_{CH}(CH_4)	1.0940	1.099	1.0938	1.0850
r_{CH}(HCN)	1.064	1.0739	1.0632	1.0655
Δr(CH−CH_4)	0.036_3	0.040	0.031_2	0.034_8
Δr(CH−HCN)	0.066_3	0.064_9	0.061_8	0.054_3
Δr(CH_4−HCN)	0.030	0.025	0.030_6	0.019_5
r_{CO}(CO)	1.1309	1.1323	1.1295	1.1283
r_{CO}(CO_2)	1.1621	1.1625	—	1.1600
r_{CO}(OCS)	1.160	—	1.1602	1.1543
r_{CO}(CH_2O)	1.2051	1.2073	1.2042	1.203
Δr(CO_2−CO)	0.031_2	0.030_2	—	0.031_7
Δr(OCS−CO)	0.029_1	—	0.030_7	0.026_0
Δr(CH_2O−CO)	0.074_2	0.075_5	0.074_7	0.075
Δr(CO_2−OCS)	0.002	—	—	0.005_7
Δr(CH_2O−CO_2)	0.043_0	0.045_3	—	0.043
Δr(CH_2O−OCS)	0.045_1	—	0.044_0	0.049
r_{CN}(HCN)	1.1564	1.1574	1.1551	1.1532
r_{CN}(ClCN)	1.1660	1.1665	1.159	1.1602
r_{CN}(CN)	1.1747	1.1763	1.1733	1.1720
Δr(ClCN−HCN)	0.009_6	0.009_1	0.004	0.007_0
Δr(CN−HCN)	0.018_3	0.018_9	0.018_2	0.018_8
Δr(CN−ClCN)	0.008_7	0.009_8	0.014	0.011_8

* All distances in Å

expected, the spread for CH bonds is larger, ranging from 0.011 to 0.019 Å. Table 1.8 also gives differences in bond lengths calculated with each of the four types of distance. The spread is significantly narrowed; for changes in CO and CN bonds the spread is 0.001 to 0.005 Å (median 0.002 Å), and for CH bonds, 0.009 to 0.012 Å. In fact, with a few conspicuous exceptions, the stability of the Δr values in Table 1.8 is very encouraging. One can say with reasonable confidence that the changes in the CO and CN bonds are defined to an accuracy of $\pm$0.001–0.002 Å, and the changes in CH bonds to perhaps $\pm$0.005–0.010 Å.

A few comments can be made about the exceptions. The spread in Δr(CH) is due in large part to the fact that r_0 and r_s are less than r_e in HCN, while the reverse is true in CH and CH_4 ($r_z - r_e$ is also smaller in HCN). The physical reason for this is clear; the bending vibration in HCN, which is a relatively large amplitude motion, tends to reduce the values of r_0, r_z and r_s through the effect described in Section 1.5.2 for CsOH. The rather low value of r_s(CN) in ClCN may be due to the same effect.

Our conclusion is that distance comparisons can be made to a confidence level of a few thousandths of an ångstrom unit for bonds involving heavy atoms, as long as the same type of distance is used in the comparison. The uncertainty is somewhat larger, but still no more than 0.01 Å, for bonds involving hydrogen. However care must be exercised in certain cases, particularly in those molecules with large-amplitude bending vibrations.

References

1. Kuchitsu, K. (1972). *MTP International Review of Science, Physical Chemistry Series One*, Vol. 2, p. 203, (G. Allen, editor) (London: Butterworths)
2. Hedberg, K. (1974). *Critical Evaluation of Chemical and Physical Structural Information*, (D. R. Lide and M. A. Paul, editors) (Washington: National Academy of Sciences)
3. Von Bahr, E. (1914). *Phil. Mag.*, **28,** 71
4. Imes, E. S. (1919). *Astrophys. J.*, **50,** 251
5. Czerny, M. (1925). *Z. Phys.*, **34,** 227
6. Herzberg, G. (1945). *Infrared and Raman Spectra of Polyatomic Molecules*, (New York: Van Nostrand)
7. Kraitchman, J. (1953). *Amer. J. Phys.*, **21,** 17
8. Oka, T. and Morino, Y. (1961). *J. Mol. Spectrosc.*, **6,** 472
9. Oka, T. and Morino, Y. (1962). *J. Mol. Spectrosc.*, **8,** 9
10. Herschbach, D. R. and Laurie, V. W. (1964). *J. Chem. Phys.*, **40,** 3142
11. Laurie, V. W. (1958). *J. Chem. Phys.*, **28,** 704
12. Wilson, E. B. and Howard, J. B. (1936). *J. Chem. Phys.*, **4,** 260
13. Nielsen, H. H. (1959). *Handbuch der Physik*, **37/1,** 173
14. Lide, D. R. (1974). *Methods of Experimental Physics*, Vol. 3, 11, (D. Williams, editor) (New York: Academic Press)
15. Morino, Y., Nakamura, J. and Yamamoto, S. (1967). *J. Mol. Spectrosc.*, **22,** 34
16. Morino, Y. and Matsumura, C. (1967). *Bull. Chem. Soc. Jap.*, **40,** 1095
17. Rosen, B (1970). *Spectroscopic Data Relative to Diatomic Molecules*, (Oxford: Pergamon)
18. Bunker, P. R. (1968). *J. Mol. Spectrosc.*, **28,** 422
19. Bunker, P. R. (1970). *J. Mol. Spectrosc.*, **35,** 306
20. Watson, J. G. K. (1973). *J. Mol. Spectrosc.*, **45,** 99
21. Winnewisser, G., Maki, A. G. and Johnson, D. R. (1971). *J. Mol. Spectrosc.*, **39,** 149
22. Rank, D. H., Skorinko, G., Eastman, D. P. and Wiggins, T. A. (1960). *J. Opt. Soc. Amer.*, **50,** 421

23. Maki, A. G. and Johnson, D. R. (1973). *J. Mol. Spectrosc.*, **47,** 226
24. Hughes, R. H. (1956). *J. Chem. Phys.*, **24,** 131
25. Costain, C. C. (1958). *J. Chem. Phys.*, **29,** 864
26. Costain, C. C. (1966). *Trans. Amer. Cryst. Assoc.*, **2,** 157
27. Schwendeman, R. H. and Kelly, J. D. (1965). *J. Chem. Phys.*, **42,** 1132
28. Lafferty, W. J. and Ritter, J. J. (1971). *J. Mol. Spectrosc.*, **38,** 181
29. Lide, D. R. (1962). *Tetrahedron*, **17,** 125
30. Watson, J. K. G. (1973). *J. Mol. Spectrosc.*, **48,** 479
31. Schwendeman, R. H. (1974). *Critical Evaluation of Chemical and Physical Structural Information*, (D. R. Lide and M. A. Paul, editors) (Washington: National Academy of Sciences)
32. Kivelson, D., Wilson, E. B. and Lide, D. R. (1960). *J. Chem. Phys.*, **32,** 205
33. Herschbach, D. R. and Laurie, V. W. (1962). *J. Chem. Phys.*, **37,** 1668
34. Laurie, V. W. and Herschbach, D. R. (1962). *J. Chem. Phys.*, **37,** 1687
35. Morino, Y., Kuchitsu, K. and Oka, T. (1962). *J. Chem. Phys.*, **36,** 1108
36. Toyama, M., Oka, T. and Morino, Y. (1964). *J. Mol. Spectrosc.*, **13,** 193
37. Lafferty, W. J., Lide, D. R. and Toth, R. A. (1965). *J. Chem. Phys.*, **43,** 2063
38. Morino, Y., Kuchitsu, K. and Yamamato, S. (1967). *Spectrochim. Acta*, **24A,** 335
39. Kuchitsu, K. and Cyvin, S. J. (1972). *Molecular Structures and Vibrations*, Chap. 12, (S. J. Cyvin, editor) (Amsterdam: Elsevier)
40. Bartell, L. S. (1955). *J. Chem. Phys.*, **23,** 1219
41. Bartell, L. S., Kuchitsu, K. and de Neui, R. J. (1961). *J. Chem. Phys.*, **35,** 1211
42. Kuchitsu, K. and Bartell, L. S. (1962). *J. Chem. Phys.*, **36,** 2470
43. Bastiansen, O. and Traetteberg, M. (1960). *Acta Crystallogr.*, **13,** 1108
44. Morino, Y., Nakamura, J. and Moore, P. W. (1962). *J. Chem. Phys.*, **36,** 1050
45. Coles, D. K. and Hughes, R. H. (1949). *Phys. Rev.*, **76,** 178
46. Pierce, L. (1959). *J. Mol. Spectrosc.*, **3,** 575
47. Herschbach, D. R. (1959). *J. Chem. Phys.*, **31,** 91
48. Wilson, E. B., Lin, C. C. and Lide, D. R. (1955). *J. Chem. Phys.*, **23,** 136
49. Kurland, R. J. and Wilson, E. B. (1957). *J. Chem. Phys.*, **27,** 585
50. Costain, C. C. and Dowling, J. M. (1960). *J. Chem. Phys.*, **32,** 158
51. Hirota, E., Sugisaki, R., Nielsen, C. J. and Sorensen, G. O. (1974). *J. Mol. Spectrosc.*, **49,** 251
52. Lide, D. R. (1962). *J. Mol. Spectrosc.*, **8,** 142
53. Millen, D. J., Topping, G. and Lide, D. R. (1962). *J. Mol. Spectrosc.*, **8,** 153
54. Lafferty, W. J. (1974). *Critical Evaluation of Chemical and Physical Structural Information*, (D. R. Lide and M. A. Paul, editors) (Washington: National Academy of Sciences)
55. Lide, D. R. and Kuczkowski, R. L. (1967). *J. Chem. Phys.*, **46,** 4768
56. Matsumura, C. and Lide, D. R. (1969). *J. Chem. Phys.*, **50,** 71
57. Lide, D. R. and Matsumura, C. (1969). *J. Chem. Phys.*, **50,** 3080
58. Hougen, J. T., Bunker, P. R. and Johns, J. W. C. (1970). *J. Mol. Spectrosc.*, **39,** 136
59. Morino, Y. and Nakagawa, T. (1968). *J. Mol. Spectrosc.*, **26,** 496
60. Olson, W. B. (1972). *J. Mol. Spectrosc.*, **43,** 190
61. Oka, T. (1960). *J. Phys. Soc. Jap.*, **15,** 2274
62. Takagi, K. and Oka, T. (1963). *J. Phys. Soc. Jap.*, **18,** 1174

2
The Properties of Molecules from Molecular Beam Spectroscopy

T. R. DYKE
University of Oregon
and
J. S. MUENTER
University of Rochester, New York

2.1	INTRODUCTION	28
2.2	EXPERIMENTAL	29
	2.2.1 *Molecular beam electric resonance spectroscopy*	29
	2.2.2 *Molecular beam magnetic resonance spectroscopy*	31
	2.2.3 *Molecular beam maser spectroscopy*	32
	2.2.4 *Molecular beam absorption spectroscopy*	33
	2.2.5 *Molecular beam deflection measurements*	33
2.3	DIATOMIC MOLECULES	34
	2.3.1 *Rotational spectroscopy*	34
	2.3.2 *Electric dipole moment and electric polarisability interaction*	36
	2.3.3 *Nuclear hyperfine interactions*	39
	2.3.4 *Quadrupole coupling interaction*	40
	2.3.5 *Spin–rotation interaction*	42
	2.3.6 *Spin–spin interaction*	43
	2.3.7 *Zeeman studies of diatomic molecules*	44
	2.3.8 $X^2\Pi$ *Diatomics*	48
	2.3.9 *Metastable molecules*	50
2.4	POLYATOMIC MOLECULES	53
	2.4.1 *Introduction*	53
	2.4.2 *Linear polyatomic molecules*	56
	2.4.3 *Tetrahedral molecules*	59
	2.4.4 C_{3v} *Pyramidal molecules, ammonia and phosphine*	61
	2.4.5 C_{3v} *Substituted methane molecules*	63

2.4.6 *Asymmetric top molecules* 64
2.4.7 *Discussion* 68

2.5 STUDIES OF WEAKLY BOUND COMPLEXES 70
2.5.1 *Electric deflection experiments on dimers* 71
2.5.2 *Alkali metal dimers* 72
2.5.3 *Hydrogen-bonded molecules* 75
2.5.4 *van der Waals molecules* 79

2.6 CONCLUSIONS 82

2.7 APPENDIX 83

2.1 INTRODUCTION

Many classic experiments in chemistry and physics have involved atomic and molecular beams. The Stern–Gerlach experiment, discovery of nuclear quadrupole moments, invention of the maser, and atomic 'clocks' are among the many results of this research. In this review, we are concerned with the molecular properties which have been obtained from molecular beam experiments, whose prototype was carried out by Rabi and co-workers[1] in the late 1930s. In recent years, techniques that were originally restricted to a small number of molecules of very specific types, e.g. alkali halides, have been extended to the full range of chemical species. Well over 100 molecules, in a wide variety of isotopic species and vibrational states, have been studied by molecular beam spectroscopic methods. The purpose of this review article is to focus on the molecular properties which have been obtained from these experiments.

Of the more than 100 molecules that have been studied in beam spectroscopic experiments, roughly one third have been stable diatomic species. The majority of these molecules are alkali halides and were the subject of many early experiments. In general, these have been thoroughly reviewed and we refer readers to these articles[2-5] for details we have not covered. The section on diatomic molecules in this article, therefore, only briefly discusses the older experiments. New results are presented, and the diatomic molecule introduces the necessary theory and models used to obtain molecular constants from spectroscopic data. For this latter reason the diatomic molecule section is organised in terms of molecular interactions rather than by the type or class of molecule. However, recent studies of open-shell diatomic molecules are discussed.

Much of the recent activity in molecular beam spectroscopy has been devoted to the study of polyatomic molecules and this area had not been extensively reviewed. The polyatomic molecule section of this article is organised by molecular type. Each molecule is very briefly discussed and the molecular properties obtained are presented in tabular form. One of the most exciting developments in molecular beam spectroscopy has been the study of weakly bound chemical species. These molecules include alkali-halide dimers, alkali-metal diatomic molecules, hydrogen-bonded molecules, and van der Waals molecules. These molecules are also presented by chemical type.

A very brief discussion of experimental techniques precedes the sections dealing with individual molecules. It has been impossible to present a complete set of literature citations. The excellent bibliographies of earlier reviews are cited here, particularly with respect to a diatomic species. We hope the references to more recent work are reasonably complete through the early part of 1974.

2.2 EXPERIMENTAL

The data presented in this article come from five different types of experiments: molecular beam electric resonance spectroscopy (MBER), molecular beam magnetic resonance spectroscopy (MBMR), molecular beam maser spectroscopy, molecular beam absorption spectroscopy, and molecular beam deflection experiments. Each of these techniques is very briefly discussed. The distinguishing feature of these experiments is, of course, that the molecules under investigation are in the form of a beam of isolated molecules travelling essentially collision-free through a vacuum. Until recently, the vast majority of molecular beams were formed by simple effusion sources[6], in which all molecular degrees of freedom have equilibrium energy distributions characterised by the temperature of the effusive source. A relatively new molecular beam source is the supersonic nozzle[7,8]. Hydrodynamic flow in the nozzle produces non-equilibrium conditions which generate molecular beams ideally suited to many spectroscopic studies. In addition, weakly bound, unstable species can be generated in these sources[9].

2.2.1 Molecular beam electric resonance spectroscopy

In an MBER experiment, after the beam is formed a specific quantum state is selected by passing the beam through a non-uniform field. The use of electrostatic state selection fields limit MBER experiments to molecules having a permanent electric dipole moment, although in special cases this moment can be extremely small. The state selection process normally utilises very large electric fields, and the Stark energies are larger than any hyperfine splittings in the states of interest. Thus, the molecules will be selected according to their J and M_J rotational quantum numbers[10]. The chosen $|J,M_J\rangle$ state serves as the initial level in an electric-dipole-induced spectroscopic transition. Since it is the rotational quantum numbers which are being selected, the transitions are either pure rotational transitions or transitions within a single rotational state[11]. Following the transition region, the beam passes through a second non-uniform electrostatic field which serves to analyse the molecular state after the molecule has been exposed to the spectroscopic radiation. The intensity of the molecular beam that passes through these three regions of the spectrometer is then monitored by a molecular beam detector.

There are different types and configurations of MBER spectrometers, differing in the type of non-uniform fields used for state selection. Electrostatic dipoles, quadrupoles and hexapoles have been used. The most common example uses quadrupole fields for state selection and analysis. In this type of spectrometer the quadrupole fields serve as electrostatic lenses[12]. Using this lens analogy, the first field focuses the diverging molecules in the selected state to parallel rays, and the second field focuses the same state into the entrance aperture of the detector. Ideally, the only molecules reaching the detector are in the desired state. If a transition changes the state between the two fields, the second field will divert the molecules in the new state away from the detector. The occurrence of a transition is, therefore, observed by a change in the number of molecules detected. The need to detect the molecules in the beam originally limited this type of experiment to alkali halide molecules which could be surface ionised. However, sensitive and reliable universal detectors using electron bombardment ionisation are now readily available[13-17]. In addition, it is possible to detect selectively molecules in excited metastable electronic states if the electronic energy is sufficient to produce an Auger electron from a clean metal surface[18].

Electric dipole transitions have been observed by this technique in a frequency range extending from less than 100 kHz to 370 GHz. The lower limit is imposed by the need for the molecules to make adiabatic transitions from one field region to another while the upper limit is determined only by the existence of suitable radiation sources[19]. The ability to observe the very low frequency electric dipole transitions is unique to the method and is advantageous in many situations. Since the spectroscopic radiation is not being detected, the sensitivity of the experiment is independent of the transition frequency. Similarly, the sensitivity is independent of the transition moment involved since an arbitrarily large radiation amplitude can be used. This means that normally forbidden transitions can often readily be observed. Perhaps the most common MBER transition, going from $|J,M_J\rangle$ to $|J,M_J + 1\rangle$ in a linear molecule, utilises both these advantages.

The most common reason for using molecular beam techniques is to obtain very high resolution. In MBER experiments the resolution is limited by the time, through the uncertainty principle, available for the measurement. This time is fixed by the time of flight through the radiation fields. Resonance regions are usually of the order of 10 cm in length and FWHM linewidths of 1–10 kHz normally obtain. The resonance region includes a uniform d.c. electric field for Stark measurements and can also have a magnetic field for Zeeman measurements[20-22]. In general, the Stark effect is not sufficiently dependent on the molecular vibrational state to allow the state selection of different vibrational states. However, the transition frequencies for different vibrational states can always be resolved and the method is well suited for observing vibrational dependence of molecular parameters.

The sensitivity of an MBER spectrometer is determined by a variety of parameters, but not on the transition moment or frequency as already mentioned. In any spectroscopic experiment the observed intensity is related to the population difference between the two levels involved in the transition. Experiments using state selection obtain a significant advantage by maximising the population difference. This is particularly important for experiments where the transition energy is much less than kT. Because of the importance of state selection to MBER spectroscopy, any energy levels which cannot be selected with high purity are poor choices for this technique. A notable example is the symmetric top, where the Stark effect varies slowly from state to state.

The advantages of MBER spectroscopy include high resolution, the ability to observe low frequency and forbidden transitions, and the ability to search for transitions of unknown frequency. This last point is important in extending high resolution to systems that have not already been studied by more conventional techniques such as microwave absorption spectroscopy. Rapid searching for MBER transitions is made possible by the absence of resonant radiation circuits and the ability to lower the instrument resolution by broadening the spectral distribution of the radiation[23]. This can be done with no sacrifice in signal strength since molecules and not radiation are being detected. The total input power of the radiation must be increased, however. It is therefore possible to search for transitions quickly using low resolution and then measure them accurately with very high resolution.

For more details of the experimental aspects of MBER spectroscopy one should read the suitable portions of Ramsey[24], and the review articles by Kusch and Hughes[3] and English and Zorn[4].

2.2.2 Molecular beam magnetic resonance spectroscopy

MBMR techniques are completely analogous to those of MBER with the electric interactions, electric fields and the electric dipole transitions replaced by their magnetic counterparts. Since all molecules have magnetic moments, MBMR is a universal method and is not limited to polar molecules. However, magnetic properties of molecules are different from electric properties, and the actual experiments using MBMR are quite different from MBER experiments. Whilst a molecule has only a single electric dipole moment, it can have more than one magnetic moment. The state selection process normally utilises the largest magnetic moment present: in a paramagnetic molecule, the electron moment: in a molecule with a non-zero nuclear spin, the nuclear moment; and in a spinless molecule, the rotational magnetic moment[25]. The greatest difference between MBMR and MBER is the fact that the spacing between adjacent magnetic sublevels is independent of J, the rotational quantum number[25a]. The magnetic properties obtained are, therefore, a rotational average of a large number of J states. The specific average in question depends on the magnetic state selecting fields and relative magnitudes of molecular interactions. This requires lineshape analysis and adds considerable difficulty to the data reduction. The problem is further complicated if any excited vibrational levels are populated at the beam temperature. These problems do not arise in very light molecules where a very small number of rotational states are populated. The classic example is Ramsey's extensive study of the hydrogen molecule[2].

A few MBMR experiments have been carried out in the presence of external electric fields as well as magnetic fields[26]. These have been on non-polar molecules to observe their polarisability anisotropy. MBMR experiments have been done with quite long resonance regions for very high resolution. This could present an extreme difficulty of requiring uniform magnetic fields over a very large volume. This problem is circumvented by using the method of Ramsey resonance, which employs separated oscillating fields at the beginning and end of the resonance region[27]. Any inhomogeneities in the magnetic field in the drift region between the radiation regions are effectively averaged by the molecule.

Greater detail of magnetic resonance techniques are contained in the comprehensive text by Ramsey[28] and in review articles by English and Zorn[4].

2.2.3 Molecular beam maser spectroscopy

The beam maser spectrometer[29] begins very similarly to an MBER spectrometer with a beam source followed by an electrostatic state selecting field. Beam maser experiments are limited to molecules having permanent electric dipole moments, as are MBER experiments. The molecules then enter a resonant cavity which is tuned to the frequency of the electric dipole transition to be observed. The transition is stimulated by coupling into the cavity a small amplitude of radiation of the correct frequency. The occurrence of the transition is observed by monitoring the power level in the cavity with a sensitive receiver. The vacuum chambers and pumping requirements for a beam maser are much simpler than for either MBER or MBMR experiments, and not having to detect the molecules in the beam is a great simplification. However, the electronics requirements are greater. A single cavity cannot be tuned over a very wide frequency region and a number of cavities are required for good spectral coverage. Searching for transitions of unknown frequency is difficult. Low frequencies become hard to deal with both because of the size of the cavity required and for sensitivity reasons[30]. Thus, most maser studies are carried out in the microwave region of the spectrum and pure rotational transitions are usually observed. The sensitivity of microwave masers can be very high, particularly by employing large nozzle beam sources. The beam in a maser can be much less well defined than in either MBER or MBMR experiments. At the present time the maser appears to be the best form of beam spectroscopy to study large molecules because of this sensitivity advantage.

Resolution of a maser is normally determined by the cavity length. While cavities of arbitrary size can be used by employing different modes of excitation, it becomes difficult to maintain uniform radiation fields in large, high-frequency cavities. Kukolich[31] has extensively used two cavity masers, with each cavity serving as a Ramsey separated oscillating field to obtain extremely high resolution. The drift space between the cavities can also be used for scattering in energy-transfer experiments[32]. There is no difficulty in applying external magnetic fields to the resonating molecules as long as a magnet is available into which the entire cavity can be inserted[33]. It is very much more difficult to apply an electric field, however, since the cavity must be an excellent conductor. A detailed review of maser techniques can be found in the review article by Lainé[34].

2.2.4 Molecular beam absorption spectroscopy

The simplest application of molecular beams to reduce linewidth is to carry out standard molecular absorption transverse to the beam[35, 36]. This eliminates Doppler width as well as collisional broadening. The main difficulty with this technique is poor sensitivity. Compared with conventional microwave absorption experiments, the sample density is extremely low. Compared with MBER or MBMR spectroscopy, beam absorption suffers from the nearly equal population of the two levels involved in the transition. MBER and MBMR experiments avoid this problem through state selection. For molecules with very strong microwave absorptions, however, it is an effective technique. Multiple beams are normally employed requiring large amounts of sample and high pumping speed. The latter requirement usually limits this method to molecules that can be cryo-pumped on a liquid-nitrogen-cooled surface. Resolution is again determined by the time the molecules spend in the radiation field. A suitable configuration to achieve the microwave radiation field necessary is an interferometer[37]. External fields have been used with this configuration of microwave cell.

2.2.5 Molecular beam deflection measurements

The first three spectroscopic beam techniques discussed employ non-uniform fields to deflect molecules in the state selection process. The deflection force arises from the interaction of the field gradient with a molecular moment. A simple non-spectroscopic experiment utilising this deflection can readily observe the presence or absence of the molecular moment. Whilst all molecules have some form of magnetic moment, the deflection technique can readily distinguish between open shell and closed shell molecules[38]. The method has been applied most frequently to observe whether a molecule has an electric dipole moment[39, 40]. In molecules with completely unknown geometry, the presence or absence of an electric dipole moment provides important qualitative molecular structure information. An obvious example is to distinguish between a linear or bent ABA molecular structure. In addition to the lack of quantitative information, a problem with this type of measurement is to distinguish between a very small moment and a moment identically equal to zero. The problem arises because induced Stark effects from molecular polarisabilities can mask small permanent Stark effects in the deflection experiment. In addition, large energy denominators in second-order Stark coefficients can make small permanent moments unobservable. For example, CO is observed not to deflect in normal circumstances because it has a small moment and a relatively large rotational energy which appears in the denominator of the Stark coefficient.

The overwhelming advantage of these experiments lies in the ease with which they are performed. It is literally possible to obtain definitive structural information on a molecule of unknown geometry in a few minutes.

2.3 DIATOMIC MOLECULES

A large body of precise information on diatomic molecules has been gathered by MBMR and MBER spectroscopy. Rotational constants, electric dipole moments and electric polarisabilities, nuclear and rotational magnetic moments, magnetic susceptibilities and nuclear hyperfine interactions are some of the properties which have been discussed.

In Table 2.1, several articles and reviews are listed which contain useful

Table 2.1

	Reference
Rotational spectroscopy	2, 5, 41, 42
Electric dipole moments	41, 43
Quadrupole coupling interaction	2, 4, 44
Nuclear and rotational magnetic moments, hyperfine interactions	2, 45
MBMR and MBER reviews	2–4

collections of data found primarily, or in part, by molecular beam spectroscopy. In the following section we do not attempt to give a complete collection of data for diatomic molecules, but instead refer the reader to the references in Table 2.1. In particular, we do not discuss much of the excellent high-temperature microwave absorption experiments, which are often performed under nearly collision-free conditions. Rotational spectra have been studied for alkali halides, Group IIIA/VII diatomics, Group IVA/VI diatomics and some transition-metal halides and oxides by this method, and the results reviewed by Lovas and Lide[5].

2.3.1 Rotational spectroscopy

The Hamiltonian for a molecule is most conveniently divided into the following terms:

$$H = H_e + H_{VR} + H_{E,H} + H_{HFS} \tag{2.1}$$

These terms represent the electronic terms, vibration–rotation terms, the interaction with external electric and magnetic fields, and the nuclear hyperfine interactions.

For the first two terms in equation (2.1) for $^1\Sigma$ diatomic molecules, Dunham's treatment[46] is generally used. For an effective potential

$$V = a_0\xi^2(1 + a_1\xi + a_2\xi^2 + \cdots) + B_eJ(J+1)(1 - 2\xi + 3\xi^2 - \cdots) \tag{2.2}$$

the energy levels can be found with a WKB approximation as

$$W_{V,J} = \sum_{l,k} Y_{l,k}(v + \tfrac{1}{2})^l J^k (J+1)^k \tag{2.3}$$

$\xi = (r - r_e)/r_e$ and B_e is the equilibrium rotational constant. The first few $Y_{l,k}$ values are related to the spectroscopic band constants[42]:

$$\begin{array}{ll} Y_{10} \approx \omega_e & Y_{11} \approx -\alpha_e \\ Y_{01} \approx B_e & Y_{20} \approx -\omega_e x_e \\ Y_{02} \approx -D_e & \text{etc.} \end{array} \tag{2.4}$$

$\Delta J = 1$ transitions for the first few rotational and vibrational states can be measured by microwave absorption or microwave MBER spectroscopy. These measurements typically can be made to 1:10^6 accuracy and the appropriate coefficients in equation (2.3) determined.

From the spectroscopic constants it is possible to determine r_e and several of the potential constants, a_i in equation (2.2). For alkali halides it is interesting to compare the results with an ionic model proposed by Rittner[47], and extended by others[48-50]. The potential energy of an alkali halide is represented as the interaction of two polarisable ions:

$$U = Ae^{-r/\rho} - C/r^6 - e^2/r - e^2(\alpha_1 + \alpha_2)/r^4 - 2e^2\alpha_1\alpha_2/r^7 \tag{2.5}$$

The first term is an exponential repulsion, and the second gives the van der Waals attraction. The remaining terms give the electrostatic energy of two ions with polarisability α_1 and α_2. The constant C can be calculated from the

Table 2.2 **Potential coefficients a_0, a_1, a_2 determined from the experimental data; A_1, A_2, and A_3 are obtained from Rittner's theoretical expressions[41]**

	a_0	a_1	A_1	a_2	A_2	a_3	A_3
LiF							
LiCl							
LiBr	1.02	−2.45	−1.71	5.5	+1.18	−14	+1.54
LiI	1.14	−2.56	−1.75	6.0	1.68	−15	+0.28
NaF							
NaCl	1.66	−3.14	−3.01	9.2	5.71	−31	−8.0
NaBr	1.64	−3.16	−3.09	7.4	6.0	−16	−8.2
NaI	1.74	−3.23	−3.32	8.8	7.0	−26	−11.0
KF	1.41						
KCl	1.81	−3.43	−3.46	10.0	7.7	−32	−12.8
KBr	1.63	−3.35	−3.35	9.1	7.4	−26	−11.5
KI	1.64	−3.41	−3.50	9.1	7.9	−25	−13.4
RbF	1.78						
RbCl	2.08	−3.66	−3.92	11.8	10.0	−42	−19.1
RbBr	1.73	−3.49	−3.53	10.3	8.1	−33	−13.7
RbI	1.64	−3.49	−3.55	10.2	8.0	−32	−13.6
CsF	2.01	−3.21	−3.35	8.4	7.0	−26	−10.8
CsCl	2.00	−3.59	−3.83	10.6	9.5	−35	−17.6
CsBr	2.03	−3.72	−4.03	11.7	10.5	−41	−20.8
CsI	1.53	−3.45	−3.53	9.2	7.9	−26	−13.2

alkali and halide ion ionisation potentials[51,52]. The repulsion constants A and ρ can be determined from the condition

$$\left(\frac{dU}{dr}\right)_{r_e} = 0$$

$$\left(\frac{d^2U}{dr^2}\right)_{r_e} = \text{force constant}$$

and the ion polarisabilities. This potential gives good values for alkali halide binding energies[47]. By expanding equation (2.5) in powers of $\xi = (r - r_e)/r_e$, the a_i in equation (2.2) can be calculated. Comparison of these results with experiment is shown in Table 2.2. The agreement of these results indicates that the ionic model give a good approximate potential for alkali halides.

The equilibrium internuclear separation can be found with high accuracy from experimental B_e values. *Ab initio* calculations by Matcha[53], McLean[54,55] and Yoshimine[55,56] agree very well, within 2% of the experimental values[5], for a number of alkali halides.

2.3.2 Electric dipole moment and electric polarisability interaction

The interaction of a molecule with an external electric field gives the following terms in (2.1):

$$H_E = -\boldsymbol{\mu}\cdot\boldsymbol{E} - \tfrac{1}{2}\boldsymbol{E}\cdot\boldsymbol{\alpha}\cdot\boldsymbol{E} \tag{2.6}$$

$\boldsymbol{\mu}$ and $\boldsymbol{\alpha}$ are the molecular electric dipole moment and electric polarisability. For a diatomic molecule, the leading terms[57,58] in a perturbation treatment of (2.6) are

$$\begin{aligned} H_E = \frac{\mu^2 E^2}{B_v} & \left[\frac{J(J+1) - 3M^2}{2J(J+1)(2J-1)(2J+3)} \right. \\ & \left. - \frac{D_v}{B_v}\left(1 + \frac{r_e}{\mu}\frac{d\mu}{dr}\right)\frac{J^2 + J + M^2 - 1}{(2J+3)(2J-1)}\right] \\ & - \frac{a_a E^2(2J^2 + 2J - 1 - 2M^2)}{2(2J+3)(2J-1)} - \tfrac{1}{2}(a_s - \tfrac{1}{3}a_a)E^2 \\ & + E^4 \text{ terms} \ldots \end{aligned} \tag{2.7}$$

The term involving Dv represents corrections to the dipole moment and rotational energy levels from centrifugal distortion effects; a_a is the polarisability anisotropy, $a_{\parallel} - a_{\perp}$, and a_s is the spherical polarisability, $\frac{1}{3}(a_{\parallel} + 2a_{\perp})$.

In an electric resonance experiment, $\Delta J = 0, \Delta M_J = 1$ transitions for different vibration–rotation states can be measured, and the parameters in equation (2.7) determined. Dipole moments typically are accurate to 1:10^4, and precisions of 1:10^5 can be achieved. The centrifugal distortion and polarisability terms are generally very small, and often they can be neglected, or values from other sources used to calculate them. However, there have been several accurate determinations of a_a by this method, and they are given in Table 2.3. The spherical polarisability term cannot be measured in a resonance experiment since it is independent of the (J,M) state, but has been measured in beam deflection experiments.

Table 2.3 Polarisability anisotropy, $\alpha_a = \alpha_{\parallel} - \alpha_{\perp}$

Molecule	α_a/Å^3	*Ref.*
H_2	0.3016(5)	59a
D_2	0.2917(4)	59a
HCl	0.31(10)	60
HF	0.220(6)	61
FCl	1.0(5)	62
LiNa	24(2)	63

In Tables 2.4 and 2.5 a collection of μ_e values from MBER experiments are given. For comparison, *ab initio* values are also listed. In general, the calculations give good results, varying by 0.1–0.2 D from experimental results. However, calculations on LiNa were off by a factor of 2 (see Table 2.19).

The ionic model proposed by Rittner also gives values for alkali halide dipole moments[47]:

$$\mu_e = er_e - [r_e^4 e(\alpha_1 + \alpha_2) + 4er_e\alpha_1\alpha_2]/(r_e^6 - 4\alpha_1\alpha_2) \tag{2.8}$$

Using experimental equilibrium internuclear distances[5] and ion polarisabilities[47], dipole moments can be calculated by equation (2.8) and are given in Table 2.4. Fair agreement with experiment is found, although the calculated moments are all too low, particularly for the lithium halides. The discrepancy has been lowered to a few per cent by de Wijn[49], who suggests that the short-range repulsive forces in the molecule tend to quench the polarisation of the halide ions along the internuclear axis. Thus the value of α should be lowered by 2/3 for the halide ion, and improved agreement with experiment results. The change in dipole moment with vibrational state has also been predicted with some success by this model[50].

The high precision of dipole moment measurements is useful for determining the variation of μ with vibrational state[90]:

$$\mu_v = \mu_0 + \mu_{\text{I}}(v + \tfrac{1}{2}) + \mu_{\text{II}}(v + \tfrac{1}{2})^2 + \ldots \tag{2.9}$$

where

$$\mu_0 = \mu_e + O(B_e^2/\omega_e^2) \tag{2.10}$$

The terms of order $B_e{}^2/\omega_e{}^2$ in (2.10) are generally negligible, and μ_e can be found from the experimentally determined coefficients in equation (2.9). The dipole moment can also be expanded in a Taylor series about r_e:

$$\mu(r) = \mu_e + \sum_{i=1}^{\infty} \left(\frac{d^i\mu}{dr^i}\right)_{r_e} (r - r_e)^i (i!)^{-1} \tag{2.11}$$

The relations between the coefficients of equations (2.9) and (2.11) are known[91] and may be used to test values of the dipole derivatives from calculations[71] or other experiments. The terms in (2.11) cannot be found directly from μ_v

Table 2.4 Electric dipole moments from MBER experiments (in D)‡ and calculated values for alkali halides

Molecule	μ_e*(exp.)**	μ_e*(Rittner)*†	μ_e*(ab initio)*	*Ref.*
$^{6}Li^{19}F$	6.284 09(25)	5.31	6.30	43, 54
$^{6}Li^{35}Cl$	7.085 3(13)	5.24	7.22	43, 53
$^{6}Li^{79}Br$	7.226 2(16)	5.41	6.98	53, 64
$^{6}Li^{127}I$	7.428 5(10)	5.38		65
$^{23}Na^{19}F$	8.123 5(15)	7.49	8.34	53, 66
$^{23}Na^{35}Cl$	8.972 1(6)	7.77	9.15	43, 53
$^{23}Na^{79}Br$	9.091 8(13)	7.96	8.72	43, 53
$^{23}Na^{127}I$	9.210 3(30)	7.99		43
$^{39}K^{19}F$	8.558 3(8)	8.08	8.69	53, 67
$^{39}K^{35}Cl$	10.239 1(10)	9.18	10.46	43, 53
$^{85}Rb^{19}F$	8.513 1(7)	8.00	8.76	43, 53
$^{85}Rb^{35}Cl$	10.483 (6)	9.45		43
$^{133}Cs^{19}F$	7.848 6(13)	7.28		43
$^{133}Cs^{35}Cl$	10.358 (5)	9.36		43

* Ref. 43 is a useful compilation of dipole moments as a function of vibrational state, and contains the pertinent experimental references

† Calculated as described in the text from the ionic model of Rittner

‡ 1 D = 3.33564×10^{-30} C m

Table 2.5 Electric dipole moments (in D) for diatomic molecules from MBER experiments

Molecule	μ_e*(exp.)*	μ_e*(ab initio)*	*Ref.*
DF	1.818 805(5)*	1.816	68, 69
$H^{35}Cl$	1.093 3(5)	1.043	60, 70
^{35}ClF	0.888 1(2)*	1.099	62, 309
^{28}SiO	3.088 2(8)	3.68	71, 72
^{74}GeO	3.272 0(8)		71
PN	2.751 1(6)	3.23	72, 73
^{88}SrO	8.913 (3)	10.2	72, 74
^{138}BaO	7.933 (3)		75
^{138}BaS	10.86 (2)*		76
^{205}TlF	4.228 2(8)*		77, 77a
^{7}LiH	5.828 (3)	5.886	78, 79
^{6}LiO†	6.84 (3)*	6.874	80, 81
OH†	1.667 58(10)	1.780	82, 83
SH†	0.758 0(1)	0.861	83, 84
NO†	0.158 72(2)		85
CO	0.122 2(1)	0.17‡	86, 87
CO§	1.373 62(7)	1.43	88, 89

* $v = 0$ state, not μ_e

† $X^2\Pi$, $v = 0$ state

‡ C^-O^+, which agrees with experiment

§ $a^3\Pi$ electronic state

values, but Kaiser[60] was able to determine the first four dipole derivatives for HCl and DCl in (2.11) by combining MBER and i.r. intensity measurements (Table 2.6). The small difference in the dipole moment functions for HCl and DCl is caused by a breakdown in the Born–Oppenheimer approximation.

Table 2.6 HCl/DCl **dipole derivatives***

Derivative	HCl	DCl	*ab initio calc.*
μ_e	+1.0929	1.0919	+1.18
$(d\mu/dr)_e$	+0.925(20)	+0.935(25)	+1.47
$(d^2\mu/dr^2)_e$	+0.16(11)	+0.14(13)	+0.93
$(d^3\mu/dr^3)_e$	−3.83(90)	−3.81(11)	
$(d^4\mu/dr^4)_e$	−9.3(45)	−7.6(48)	

* Ref. 60; μ in Debye and r in Å

2.3.3 Nuclear hyperfine interactions

The largest terms in the nuclear hyperfine energy for a singlet electronic state are

$$\sum_{\substack{i \\ j \neq i}} \nabla E_i \cdot \boldsymbol{Q}_i + \boldsymbol{I}_i \cdot \boldsymbol{M}_i \cdot \boldsymbol{J} + \boldsymbol{I}_i \cdot \boldsymbol{D}_{ij} \cdot \boldsymbol{I}_j + \delta_{ij} \boldsymbol{I}_i \cdot \boldsymbol{I}_j \qquad (2.12)$$

where the subscripts i and j refer to the nuclei of the molecule. $\boldsymbol{I}_i$ refers to the nuclear spin angular momentum and $\boldsymbol{J}$ to the rotational angular momentum. The first term in (2.12) gives the electrostatic interaction of the nuclear quadrupole moment with the field gradient near the nucleus caused by the charge distribution in the rest of the molecule. The next term is the spin–rotation term, which gives the interaction of the nuclear magnetic moment with the magnetic field caused by the rotation of the molecule. The last two terms are the tensor and scalar spin–spin interactions. The tensor spin–spin interaction is dominated by a classical dipole–dipole interaction, but also contains a contribution from the electron-coupled (exchange-coupled) spin–spin interaction. The scalar spin–spin interaction is entirely due to the electron-coupled mechanism and its magnitude is observed in non-molecular beam n.m.r. experiments. The tensor term is observed in beam experiments, but vanishes in isotropic n.m.r. experiments owing to averaging over all orientations.

With the aid of the Wigner–Eckart theorem, equation (2.12) can be reduced to an effective Hamiltonian for calculating matrix elements diagonal in J. Also, for diatomics, the tensors in (2.12) have only one independent element. Thus (2.12) simplifies to

$$eq_1Q_1F(I_1) + eq_2Q_2F(I_2) + C_1\boldsymbol{I}_1\cdot\boldsymbol{J}$$

$$+ C_2\boldsymbol{I}_2\cdot\boldsymbol{J} + \frac{C_3[3(\boldsymbol{I}_1\cdot\boldsymbol{J})(\boldsymbol{I}_2.\boldsymbol{J}) + 3(\boldsymbol{I}_2\cdot\boldsymbol{J})(\boldsymbol{I}_1\cdot\boldsymbol{J} - 2(\boldsymbol{I}_1\cdot\boldsymbol{I}_2)J(J+1)]}{(2J-1)(2J+3)} + C_4\boldsymbol{I}_1\cdot\boldsymbol{I}_2$$

and $$F(I_i) = \frac{3(\boldsymbol{I}_i\cdot\boldsymbol{J})^2 + \frac{3}{2}(\boldsymbol{I}_i\cdot\boldsymbol{J}) - I_i(I_i+1)(J)(J+1)}{2I_i(2I_i-1)(2J-1)(2J+3)} \qquad (2.13)$$

The various interaction constants are discussed in more detail in sections that follow. Equation (2.13) is valid only for matrix elements diagonal in J. This is often sufficient since hyperfine energies are much smaller than rotational energies for all but a few cases involving the quadrupole coupling interaction (eqQ terms). In such cases equation (2.12) must be reconsidered[101].

Methods for calculating the eigenvalues of (2.12) have been thoroughly discussed[2,3,42]. MBER and MBMR radiofrequency spectroscopy can be used to determine the hyperfine splittings, and the constants in (2.13) are usually calculated by least-squares fitting to the experimental spectra. It must be noted that if a large number of rotation–vibration states of a molecule are excited, as for alkali halides, the MBMR spectrum is a statistical average of these states. MBER spectra are usually well resolved for states of different (v,J) and are therefore somewhat more reliable.

2.3.4 Quadrupole coupling interaction

The quadrupole coupling term in the hyperfine energy, equation (2.13), is the electrostatic interaction of the nuclear quadrupole moment, Q, with the field gradient, $-q$, near that nucleus, caused by the charge distribution of the rest of the molecule. If accurate electronic wavefunctions are available, the field gradient can be calculated from the experimental eqQ, and the nuclear quadrupole moment can be obtained. Some quadrupole moments found in this manner are given in Table 2.7. Ratios of eqQ for different isotopes of the

Table 2.7 Nuclear quadrupole moments from molecular beam *eqQ* measurements and *ab initio* field gradients

Molecule	*Nucleus*	$Q/10^{-24}$ cm^2	*Reference*
D_2	D	0.002798(3)	92
^{7}LiH	^{7}Li	−0.0375	93
^{6}LiF	^{6}Li	−0.00080(8)	94
LiBr	^{7}Li	−0.0429(2)	95
	^{79}Br	0.2677	95
	^{81}Br	0.2236	95
P^{14}N	^{14}N	0.0147	73

same element provide a particularly simple way of determining the quadrupole moment, if Q is already known for one of the isotopes. In Table 2.7 the ^{6}Li quadrupole moment was determined in this manner from the ratio of eqQ for ^{6}Li and ^{7}Li in LiF.

More typically, the nuclear quadrupole moments calculated from atomic beam experiments can be combined with molecular eqQ values to obtain the field gradient. The field gradient and its dependence on vibrational state can be used to test *ab initio* calculations and semi-empirical models. In Table 2.8 some data comparing experimental and calculated field gradients for alkali halides are given.

Table 2.8 Alkali halide field gradients (units are 10^{15} e.s.u.). Values from experiment, *ab initio* (a.i.) calculations and from anti-shielding (a.s.) factors§

Nucleus	*Molecule*	$q_e(exp.)$*	$q_e(a.i.)$†	$q_e(a.s.)$‡	*Ref.*
^{7}Li	LiCl	−0.0589	−0.0677	−0.0873	96
	LiBr	−0.0647	−0.0690	−0.0695	95
^{23}Na	NaF	−1.17	−1.25	−0.686	66
	NaCl	−0.787	−0.805	−0.372	97
	NaBr	−0.668	−0.586	−0.313	5
^{39}K	KF	−2.08	−1.96	−1.31	99
	KCl	−1.48	−1.41	−0.709	99
^{85}Rb	RbF	−3.61	−2.81	−5.91	98
^{35}Cl	LiCl	+0.493	+0.386	+6.64	5
	NaCl	+0.965	+0.680	+4.16	97
	KCl	−0.0244	−0.0771	+2.89	99
^{79}Br	LiBr	+1.59	+1.89	−10.62	64
	NaBr	+2.50	+3.40	+6.93	5

* Calculated from the experimental references with the nuclear quadrupole moments[100]: $Q(^7Li) = -0.0454$ $Q(^{23}Na) = +0.10(1)$, $Q(^{39}K) = +0.053(8)$, $Q(^{85}Rb) = +0.27(2)$, $Q(^{35}Cl) = -0.0795(5)$ and $Q(^{79}Br) = +0.32(2) \times 10^{-24}$ cm^2

† From Ref. 53

‡ $|q_e(a.s.)| = 2e[1 - \gamma(\infty)]/R_e^3$. $1 - \gamma(\infty)$ values[44] are the following: $Li^+ = 0.74$, $Na^+ = 5.1$, $K^+ = 14$, $Rb^+ = 72$, $Cl^- = 57$ and $Br^- = 113$

§ See also Ref. 129

The Townes–Dailey theory[102] has been used to explain quadrupole coupling interactions. Basically, this theory assumes that the field gradient at the nucleus is dominated by the contribution of the lowest unclosed electron shells, and that this contribution can be related to the atomic fine structure splitting. The magnitude of this contribution depends on the hybridisation and degree of ionic character of the bond. Since alkali halides are nearly 100% ionic, both the alkali and halogen nucleus are surrounded by closed shells[102a]. The field gradient should vanish at the nucleus for these spherical shells, and only a small $+2p/r^3$ gradient due to the other ion (of net charge p) gives a non-zero contribution. In fact, measured field gradients at the alkali nucleus are 10–100 times larger than this. This has been explained by Sternheimer[103] as a polarisation of the closed electron shells by the other ion. This polarisation leads to an enhancement of the field gradient by a factor $\gamma(r)$ (r is the separation between the two ions). For the positive ion, the field gradient at the nucleus is given by

$$q = -e[1 - \gamma(r)]/r^3 \qquad (2.14)$$

where the anti-shielding factor, γ, is -10 to -100 for alkali nuclei. With γ calculated for the free ion $[\gamma(\infty)]$, agreement with experiment is reasonable at the alkali nucleus (Table 2.8), but is very poor at the halogen nucleus, and in some cases has the incorrect sign. de Wijn[49] was able to remove some of the discrepancy by considering the effects of the short-range repulsive forces on the polarisability of the ions. Although these theories give considerable insight into the nature of the quadrupole coupling interaction, the *ab initio* calculations would seem to give the most accurate results for a wide range of molecules although even here some discrepancy exists.

Listed in Table 2.9 are experimental and *ab initio* quadrupole coupling constants for some first- and second-row hydrides. The agreement is quite good, although the question of vibrational averaging must be considered for DF and LiD. Table 2.9 is by no means a complete listing of MBER and MBMR quadrupole coupling determinations. It would be useful to have further field gradient calculations to compare with the precise beam data.

Table 2.9 Quadrupole coupling constants (in MHz) **from molecular beam experiments and** *ab initio* **calculations‡**

Nucleus	*Molecule*	*eqQ(exp)*	*eqQ(a.i.)*	*Ref.*
D	DF	0.35424(8)*	0.363	68
	DCl	0.189(3)	0.213	60, 104
	LiD	0.033(1)	0.0344	78, 79
Cl†	HCl	66.785(3)	65.6	60

* $v = 0$ state, not the equilibrium value
† The experimental and calculated first and second derivatives of the field gradient also agree to better than 10%
‡ See also Ref. 129

It has been suggested that the nuclear quadrupole moment may depend to a slight extent on its chemical environment because of a nuclear polarisation effect. In addition, higher-order interactions[42] (pseudo-quadrupole effect) can have the same operator dependence in equation (2.13) as the quadrupole interaction. These effects can be tested for by obtaining ratios of eqQ for isotopes of the same element in different molecules. However, recent determinations of these ratios have not shown any statistically significant effects. In one of the most accurate experiments, Hillborn *et al.*[95] measured the ratio of $eqQ(^{79}Br)/eqQ(^{81}Br)$ in ^7LiBr as 1.197 061 5(40). The value for HBr was determined by van Dijk and Dymanus[105] as 1.197 052(5). The value for atomic bromine[106] is 1.197 056 8(15). Thus the false quadrupole and polarisation effects are limited to $4{:}10^6$ or less in these cases. Other experiments[107] have also yielded null differences for eqQ ratios.

2.3.5 Spin–rotation interaction

The spin–rotation term in the hyperfine Hamiltonian is the interaction of the nuclear magnetic moment with the magnetic field caused by the rotation of the molecule, and therefore is proportional to $\boldsymbol{I}\cdot\boldsymbol{J}$. The interaction constant C_1 (or C_2) in equation (2.13) is given by (following Hindermann and Cornwell[108])

$$C_1 = C_1(\text{nucl.}) + C_1(\text{el.}) + C_1(\text{accel.})$$

where

$$C_1(\text{nucl.}) = e\mu_0\mu_N g_1 BZ_2/hR \tag{2.15}$$

and

$$C_1(\text{el.}) = -\frac{\mu_0\mu_B\mu_N g_1 B}{\pi}\sum_n{}' \langle 0|\sum_i (L_{1i})_x r_{1i}^{-3}|n\rangle\langle n|\sum_i (L_{1i})_x|0\rangle/(E_n^0 - E_0^0)$$

C_1(accel.) is a small relativistic correction[109]. MKSA units are employed, so μ_0 is the permitivity of free space. μ_B and μ_N are the Bohr and nuclear magnetons. B is the rotational constant, and R the internuclear separation of the molecule. Z_i is the charge number of the ith nucleus. $\boldsymbol{L}_{1i}$ and $\boldsymbol{r}_{1i}$ are the angular momentum and position vectors of the ith electron with nucleus 1 as origin, and the sum n is over electronic states. In recent papers, the sign convention, $H = +C\,\boldsymbol{I}\cdot\boldsymbol{J}$, has been used, and has been employed in (2.15).

The nuclear contribution can be relatively easily calculated, but the electronic portion must be found by *ab initio* calculations. For example, Lipscomb and Stevens[110] calculate $C_F = 312.4$ kHZ and $C_H = -70.65$ kHZ for HF compared with $C_F = 284.6(7)$ and $C_H = -73.1(9)$ kHZ from an MBER experiment[68]. Accurate spin–rotation constants have been determined in a wide range of molecules and can be used to test *ab initio* wavefunctions. Values from early work are compiled in Refs. 42 and 2. More recent values may be found in the papers cited in this review.

The electronic contribution to the spin–rotation term is related to the 'paramagnetic' part of the nuclear magnetic shielding tensor [see equation (2.19)]. The isotropic shielding constant, which is observed in n.m.r. chemical shift work, is the following[108, 111]:

$$\sigma_1 = \sigma_1^{\mathrm{d}} + \sigma_1^{\mathrm{p}}$$

$$\sigma_1^{\mathrm{d}} = \left(\frac{e^2\mu_0}{8\pi m}\right)\langle 0|\sum_i (r_{1i})^{-1}|0\rangle \qquad (2.16)$$

$$\sigma_1^{\mathrm{p}} = -\left(\frac{\mu_0\mu_B^2}{2\pi}\right)\sum_n{}' \langle 0|\sum_i (L_{1i})_x r_{1i}^{-3}|n\rangle\langle n|\sum_i (L_{1i})_x|0\rangle/(E_n^0 - E_0^0)$$

The subscript 1 refers to nucleus 1 in equation (2.15). The diamagnetic term, σ_1^{d}, can be theoretically calculated, and the paramagnetic term σ_1^{p}, can be obtained from a measurement of the spin–rotation constant, C_1, and (2.15). In this way, as outlined by Ramsey[111], the shielding constant can be determined on an absolute scale for a nucleus in a given molecule, and then n.m.r. chemical shift data for other molecules put on an absolute scale. This has been done for H_2 and fluorine in HF, for example[108, 112, 113].

2.3.6 Spin–spin interaction

The tensor spin–spin interaction in the hyperfine Hamiltonian is dominated by a classical magnetic dipole–dipole interaction[2]:

$$H = (\mu_0/4\pi)\,[\boldsymbol{\mu}_1\cdot\boldsymbol{\mu}_2 - 3(\boldsymbol{\mu}_1\cdot\boldsymbol{r})(\boldsymbol{\mu}_2\cdot\boldsymbol{r})/r^2]/r^3 \qquad (2.17)$$

With the Wigner–Eckart theorem, $\boldsymbol{\mu}$ and $\boldsymbol{r}$ can be shown to be proportional to $\boldsymbol{I}$ and $\boldsymbol{J}$, and equation (2.13) results with

$$C_3(\mathrm{Dir.}) = (\mu_0/4\pi)(\mu_1/I_1)(\mu_2/I_2)\langle 1/r^3\rangle_{v,J} \qquad (2.18)$$

$\langle 1/r^3\rangle_{v,J}$ is the expectation value of $1/r^3$ in the vibration–rotation state $|v,J\rangle$, and r is the separation of the two spins.

C_3 also can have a contribution from the electron-coupled (indirect) spin–spin interaction[114]:

$$\boldsymbol{I}_1 \cdot \boldsymbol{\delta} \cdot \boldsymbol{I}_2$$

where $\boldsymbol{\delta}$ is a traceless, second rank tensor. This term arises from the magnetic interaction of the nuclei with the electrons surrounding them. The interaction is transmitted through the exchange coupling of the electrons, and general expressions have been derived[114]. Since $\boldsymbol{\delta}$ has the same orientation dependence as the direct spin–spin interaction, the experimental value of C_3 is the sum of the two interactions.

In Table 2.10, some MBMR and MBER experimental values of C_3 and C_4 are given, where C_4 is the scalar, electron-coupled spin–spin constant [equation (2.13)]. Also shown is the theoretical value of C_3(Dir), which can be calculated if the potential function for the molecule is known. In the fourth column of Table 2.10, the value for the electron-coupled contribution to C_3 is given,

$$C_3(\text{exp.}) = C_3(\text{Dir.}) + C_3(\text{e.c.})$$

Table 2.10 shows that C_3(e.c.) can often be ignored in analysing the spin–spin hyperfine interactions, but the case of thallium fluoride, where both terms contributing to C_3 are large, illustrates that considerable caution must be used in obtaining $\langle 1/r^3 \rangle$ by this method.

Table 2.10 Spin–spin interaction (in kHz) for some fluorides*. H_2 is included for comparison

Molecule	$C_4(=\delta_{12})$	C_3(exp)	C_3(Dir)	C_3(e.c.)
H_2†		288.36(6)	288.29(18)	0.07(14)
^{7}LiF	0.21(4)	11.390(15)	11.33	0.06(15)
^{23}NaF	0.0(4, −1)	3.85(25)	4.16	−0.31(25)
^{39}KF	0.3 or −0.6	0.29(12)	0.51	−0.22(12)
^{87}RbF	0.66(10)	3.16(18)	3.15	0.01(18)
^{85}RbF	0.23(6)	0.93(5)	0.93	0.00(5)
^{133}CsF	0.61(9)	0.92(8)	1.15	−0.23(8)
^{205}TlF‡	−13.30(72)	3.50(15)	7.19	−3.69(16)

* Ref. 115 † Ref. 116 ‡ Ref. 77

2.3.7 Zeeman studies of diatomic molecules

The magnetic field dependent, or Zeeman, term in the general molecular Hamiltonian [equation (2.1)] can be written, for a molecule with no unpaired electrons, as follows[117]:

$$H_{\mathrm{H}} = -\mu_{\mathrm{N}}\boldsymbol{J}\cdot\boldsymbol{G}\cdot\boldsymbol{B} - \tfrac{1}{2}\boldsymbol{B}\cdot\boldsymbol{\chi}\cdot\boldsymbol{B} - \mu_{\mathrm{N}}\sum_K g_K \boldsymbol{I}_K\cdot[\mathbf{1} - \boldsymbol{\sigma}(K)]\cdot\boldsymbol{B} \qquad (2.18)$$

$\boldsymbol{G}$ is the rotational magnetic moment tensor, $\boldsymbol{B}$ is the external magnetic field, $\boldsymbol{\chi}$ is the magnetic susceptibility tensor, and $\boldsymbol{\sigma}(K)$ is the nuclear shielding tensor for the Kth nucleus.

These three separate contributions to the Zeeman Hamiltonian arise from the interaction of the applied field with the three different magnetic moments present: the rotational magnetic moment, $\mu_N \boldsymbol{J}\cdot\boldsymbol{G}$; the induced magnetic moment, $\boldsymbol{B}\cdot\boldsymbol{\chi}$; and the nuclear magnetic moment, $\mu_N g_K \boldsymbol{I}_K$. The nuclear moment must be corrected by the shielding of the nucleus by the other charges in the molecule. The shielding is represented by $\boldsymbol{\sigma}$.

Molecular energies can be calculated by obtaining matrix elements of H_H with suitable wavefunctions. For a linear molecule a convenient effective Hamiltonian, considering only matrix elements diagonal in J, may be written as follows[21]:

$$H_H = \left[(\chi_{\parallel} - \chi_{\perp})B^2 - 2\mu_N B(\sigma_{\parallel} - \sigma_{\perp})\sum_K g_K M_K\right]\frac{3M_J^3 - J(J+1)}{3(2J-1)(2J+3)} - g_J\mu_N M_J G - \mu_N B[1 - \tfrac{1}{3}(2\sigma_{\perp} + \sigma_{\parallel})]\sum_K g_K M_K \qquad (2.19)$$

The parallel and perpendicular subscripts refer respectively to the tensor component along the molecular axis and the component perpendicular to the axis. A term proportional to the spherical susceptibility, which does not depend on molecular orientation and cannot be observed spectroscopically, has been omitted from equation (2.19).

It is frequently necessary to consider another interaction arising from the presence of a large external magnetic field in a beam experiment. This term[117], $-\boldsymbol{\mu}\cdot(\boldsymbol{v} \times \boldsymbol{B})$, is a relativistic effect and $\boldsymbol{v} \times \boldsymbol{B}$ is an electric field due to the molecular charge distribution crossing the magnetic field at the velocity, $\boldsymbol{v}$, of the beam molecules. In an experiment utilising both large external magnetic and electric fields, this additional field causes only slight modification to the electric field and does not significantly alter the observed spectrum. In the usual experimental configuration, with the two external fields parallel to one another, $\boldsymbol{v} \times \boldsymbol{B}$ is perpendicular to the applied fields and to the beam axis. The resultant electric field will therefore be slightly rotated from the direction of the applied field. In experiments using only magnetic fields, the $\boldsymbol{v} \times \boldsymbol{B}$ field introduces the Stark effect, generating splittings which can be as large as several kilohertz and must be considered in analysing the spectra. The $\boldsymbol{v} \times \boldsymbol{B}$ field can introduce problems in studies of molecules with very large Stark coefficients because a Maxwellian velocity distribution in the beam produces a non-uniform field which can cause line broadening. A monoenergetic supersonic nozzle beam can be used to minimise this problem. Ramsey has used this field to advantage in Stark–Zeeman MBER experiments where the only electric field present was this motional field[118].

In a linear molecule the Zeeman Hamiltonian depends on four molecular properties, the rotational magnetic moment, g_J, the magnetic susceptibility anisotropy, $x_{\parallel} - x_{\perp}$, the nuclear shielding anisotropy, $\sigma_{\parallel} - \sigma_{\perp}$, and the average or spherical nuclear shielding, $\frac{1}{3}(2\sigma_{\perp} + \sigma_{\parallel})$. It is these interactions that make magnetic state selection, and, therefore, MBMR spectroscopy, possible[2]. MBMR experiments have been used to measure some, or all, of these four molecular properties for a wide variety of molecules. MBER experiments can also obtain this type of information by employing an external magnetic field in the resonance region of the spectrometer. As is the case for so many molecular interactions, the first observation of many of these Zeeman properties was in the extensive studies of the hydrogen molecule by Ramsey and co-workers[2]. Since there is no problem with overlapping rotational states in these MBMR experiments, all of the properties in the Hamiltonian are well known for H_2, D_2 and HD. The most recent MBMR measurements on hydrogen include magnetic properties (Code and Ramsey[92]) and polarisability anisotropy measurements (English and MacAdam[59]).

MBMR studies of a large number of heavier diatomic molecules have also been carried out; however, more precise data are available for polar molecules from MBER Stark–Zeeman experiments. The largest body of MBER magnetic work is from Gräff and co-workers[77] on the alkali halides and other ionic compounds that can be surface-ionised. These data are sumarised in table 2 in the review article by Flygare and Benson[45] and in more recent work from Gräff's laboratory[119, 120]. A high precision spectrometer with a very long *C*-field magnet has recently been built in Ramsey's laboratory and used to study alkali halides[118]. MBER spectrometers with universal beam detectors and external magnetic fields are in use at the Catholic University of Nijmegen[21] and the University of Rochester[22]. Recent diatomic molecule studies from these two laboratories include HF [21], HCl [21], SiO [22] and GeO[22].

MBMR studies of several molecules provide definitive Zeeman properties. These include LiH[121] , CO [122, 123], N_2 [124], F_2 [125, 126] and alkali dimers[127, 128] as well as H_2. Magnetic properties of LiH and CO have not been observed in MBER[129] experiments and extensive MBMR work has been performed. The other molecules are, of course, non-polar. The alkali dimer studies are presented in more detail in Section 2.5. The nitrogen molecule studied was $^{15}N^{15}N$

Table 2.11 Magnetic properties of diatomic molecules AB

	σ(A)/ p.p.m.	σ(B)/ p.p.m.	g_J	$\chi_{\parallel} - \chi_{\perp}$/ 10^7 J kG^{-2} mol^{-1}	θ/ 10^{-26} e.s.u.	*Ref.*
^{13}CO(exp)	−5.0		−0.2569(2)	−7.76		23
CO(theor)	−11.5	64.2	−0.2424	−6.97		130
LiH(exp)			−0.654(7)			121
LiH(theor)			−0.668			131
HF(exp)	108.9(8)		0.74104(15)	1.342(25)	2.21	21
HF(theor)	108.4	481.6	0.738			110
HF(theor)*	108.5	482.2	−0.026	−2.4	2.35	133
HF(theor)*	108.5	482.2	−0.033	−2.4	2.35	134
HF(theor)	108.5	482.1			2.20	135
HCl(exp)		942(200)	0.45935(9)	3.49(10)	3.53(14)	21
HCl(theor)*	141.8	1150.3	0.005	−3.3	3.80	133
HCl(theor)*	141.8	1150.3	0.005	−3.3	3.80	134
F_2(exp)			−0.121			125
F_2(theor)			−0.101			132
N_2(exp)			0.2593(5)			124
SiO(exp)			0.1536(1)	14.0(4)	−1.1(5)	22
GeO(exp)			0.1409(1)	16.8(4)	−4.4(6)	22
TlF			0.053582(75)	5.4(24)	13.1	77

* High frequency term not calculated, bulk susceptibility used

which, like fluorine, has nuclear spins of $\frac{1}{2}$, and therefore the Hamiltonian for these molecules is similar to that for H_2. It should be mentioned that the magnetic properties obtained from MBMR experiments with overlapping J states introduce serious difficulties only when the Zeeman interactions cannot be made large relative to other molecular interactions. This 'strong field' condition can normally be obtained as long as their are no large nuclear quadrupole interactions, such as are present in the alkali halides.

A representative sample of Zeeman properties of diatomic molecules is listed in Table 2.11, along with theoretical values where available. Other, and more complete, compilations of this type can be found in Refs. 21, 45 and 136.

The usefulness of these magnetic properties can be seen from their detailed definition[117]. Each of the tensors in equation (2.18) or the constants in (2.19) can be divided into a sum of two terms, the first depending on only the ground electronic state and the second depending on a sum over all electronic states. This separation is the same as that already discussed for the spin–rotation interaction (2.15) and nuclear shielding (2.16). The ground electronic state terms of this type are referred to as: nuclear, Lamb, low frequency or diamagnetic. The excited state terms are variously known as: electronic, high frequency or paramagnetic. Definitions for these terms for the general tensors of (2.18) are given below (using the notation of Dymanus[117]):

$$\begin{aligned}
\boldsymbol{G} &= \boldsymbol{G}^{\mathrm{n}} + \boldsymbol{G}^{\mathrm{e}};\ \boldsymbol{\chi} = \boldsymbol{\chi}^{\mathrm{d}} + \boldsymbol{\chi}^{\mathrm{p}};\ \boldsymbol{\sigma} = \boldsymbol{\sigma}^{\mathrm{d}} + \boldsymbol{\sigma}^{\mathrm{p}} \\
G^{\mathrm{n}}_{gg'} &= 2(\mu_{\mathrm{B}}/\mu_{\mathrm{N}})A_g m \sum_K Z_K[r_K^2\delta(gg') - (\boldsymbol{r}_K)_g(\boldsymbol{r}_K)_{g'}] \\
G^{\mathrm{e}}_{gg'} &= 4(\mu_{\mathrm{B}}/\mu_{\mathrm{N}})A_g {\sum_n}' (E_0 - E_n)^{-1}[\langle 0|L_g|n\rangle\langle n|L_{g'}|0\rangle] \qquad (2.20) \\
\sigma(K)^{\mathrm{d}}_{gg'} &= (\mu_0/4\pi)e\mu_{\mathrm{B}}\langle 0|\sum_i r_{iK}^{-3}[r_{iK}^2\delta(g,g') - (\boldsymbol{r}_{iK})_g(\boldsymbol{r}_{iK})_{g'}]|0\rangle \\
\sigma(K)^{\mathrm{p}}_{gg'} &= (\mu_0\mu_{\mathrm{B}}e)/(2\pi m){\sum_n}' (E_0 - E_n)^{-1}\{\langle 0|\sum_i [(\boldsymbol{L}_i^K)_g/r_{iK}^3]|n\rangle\langle n|\sum_i (\boldsymbol{L}_i^K)_{g'}|0\rangle\} \\
\chi^{\mathrm{d}}_{gg'} &= -\mu_{\mathrm{B}}^2 m\langle 0|\sum_i r_i^2[\delta(g,g') - (\boldsymbol{r}_i)_g(\boldsymbol{r}_i)_{g'}]|0\rangle \\
\chi^{\mathrm{p}}_{gg'} &= -2\mu_{\mathrm{B}}^2 {\sum}' (E_0 - E_n)^{-1}[\langle 0|L_g|n\rangle\langle n|L_{g'}|0\rangle]
\end{aligned}$$

A_g is the principal rotational constant about the g axis, e and m are the charge and mass of the electron, $\boldsymbol{r}_i$ and $\boldsymbol{r}_K$ are the position vectors of the ith electron and Kth nucleus, and $\boldsymbol{L}_g$ is the total electronic orbital angular momentum referenced to the g axis. The other quantities have been defined [see equation (2.15)].

The electronic, or paramagnetic, terms are closely related to one another and provide information about the charge distribution which is not readily available from other sources. In general, the nuclear, or diamagnetic, term can be separated from the electronic terms. This is done by combining data from separate interactions and, in some cases, through calculating the nuclear term. An example of this has already been discussed in relating spin–rotation interactions to magnetic shielding [see equations (2.15) and (2.16)], from which absolute shielding can be determined. Ramsey and co-workers[113] have also used this relationship to demonstrate 'antishielding' in F_2 and N_2. In these molecules the magnetic field at the nucleus is, in fact, enhanced by shielding because the magnitude of σ^p is greater than the magnitude of σ^d. A particularly extreme case of shielding occurs[137] at the fluorine nucleus in ClF. The observed[62] sign of C_F requires σ^p to be greater than zero (and is therefore really 'diamagnetic') in this molecule. A discussion of this case is given by Cornwell[138].

One of the more useful electronic properties that can be obtained from Zeeman data and (2.20) is the quadrupole moment of the molecule. It was first noted by Ramsey[139] that the quadrupole moment of a molecule can be written in terms of the magnetic susceptibility anisotropy and the paramagnetic susceptibility. This can be seen from the definition of the quadrupole moment, Θ and equation (2.20):

$$\Theta_{gg'} = \tfrac{1}{2}\sum_i \langle 0|e_i(3\boldsymbol{r}_{ig}\boldsymbol{r}_{ig'} - r_i^2\delta_{gg'})|0\rangle$$

At present, this provides one of the few methods for the experimental determination of molecular quadrupole moments[139a]. It also follows that the rotational magnetic moment allows the susceptibility to be separated into its diamagnetic and paramagnetic components.

Perhaps surprisingly, knowledge of the quadrupole moment can also provide information about the electric dipole moment. This obtains since only the lowest multipole moment of a charge distribution is independent of the chosen origin. The change in the quadrupole moment on isotopic substitution, $\Delta\Theta$, equals $2\mu\Delta R_{CM}$ where ΔR_{CM} is the change in the centre of mass. As first discussed by Townes *et al.*[140], this allows the experimental determination of the sign of the electric dipole moment. In practice, it is not necessary to obtain quadrupole moments, and only the rotational magnetic moments of two isotopic species are necessary. The sign of g_J is necessary, however. This experiment was first carried out on CO using g_J values from microwave absorption spectroscopy[141]. The result $-$CO$+$ is contrary to chemical intuition and *ab initio* calculations which were considered to be accurate at the time. The sign of g_J had been assumed, however, and a controversy developed. The validity of the assumption and the moment sign was proved by[122, 123] the MBMR determination of the sign of g_J. Current calculations, using configuration interaction, predict the correct sign of this moment[87, 142] (see Table 2.5). The signs of many electric dipole moments have been determined in this way[45].

The experimental data on magnetic properties are also sensitive tests of *ab initio* computational methods. This is particularly true for the properties depending on the sum over electronic states. Various theoretical methods are available to avoid the necessity to obtain all the requisite wave functions to carry out the summation[130-136a]. Lipscomb and co-workers have been particularly active in this area.

2.3.8 $X^2\Pi$ Diatomics

Recently, a number of molecules with $^2\Pi$ ground electronic states have been studied by MBER spectroscopy: ^{14}NO[85, 143], ^{15}NO[143], ^{7}LiO[80], OH[82], OD[144] and SH[84]. With the exception of NO, beams of these molecules are difficult to produce and high-temperature ovens or chemical reactions are necessary. Much fruitless effort has been spent attempting to produce beams of other radicals such as CH and CN.

The Hamiltonian appropriate for a $^2\Pi$ molecule is the following[145-147]:

$$\begin{aligned} H &= H_0 + H_1 + H_2 \\ H_0 &= B[(J_x - L_x - S_x)^2 + (J_y - L_y - S_y)^2] + A\boldsymbol{L}\cdot\boldsymbol{S} \\ H_1 &= a\boldsymbol{I}\cdot\boldsymbol{L} + (b + c)I_xS_x + \tfrac{1}{2}b(I^+S^- + I^-S^+) \\ &\quad + \tfrac{1}{2}d(\mathrm{e}^{2\mathrm{i}\phi}I^-S^- + \mathrm{e}^{-2\mathrm{i}\phi}I^+S^+) \\ &\quad + e[\mathrm{e}^{\mathrm{i}\phi}(S^-I_z + I^-S_z) + \mathrm{e}^{-\mathrm{i}\phi}(S^+I_z + I^+S_z)] \\ H_2 &= eQ\{\tfrac{1}{4}q_1(3I_z^2 - I^2) + \tfrac{3}{8}q_2[\mathrm{e}^{2\mathrm{i}\phi}I^{-2} + \mathrm{e}^{-2\mathrm{i}\phi}I^2] \\ &\quad + \tfrac{3}{4}q_0[(I_zI^- + I^-I_z)\mathrm{e}^{\mathrm{i}\phi} + (I_zI^+ + I^+I_z)\mathrm{e}^{-\mathrm{i}\phi}]\} \end{aligned} \qquad (2.21)$$

H_0 gives the rotational and spin–orbit terms, where $\boldsymbol{J}$, $\boldsymbol{L}$ and $\boldsymbol{S}$ are the total angular momentum and the electronic orbital and spin angular momentum. The second term, H_1, is the Frosch and Foley hyperfine Hamiltonian[145] (with interaction constants a, b, c, d and e) amended by Dousmanis[146]. H_2 gives the nuclear quadrupole coupling interaction. In these terms, $\boldsymbol{I}$ refers to the nuclear spin, and the interaction constants are

$$\begin{aligned} q_1 &= e[(3\cos^2\chi - 1)/r_1^3]\mathrm{av} \\ q_2 &= e[\sin^2\chi/r_1^3]\mathrm{av} \\ q_0 &= e[\sin\chi\cos\chi/r_1^3]\mathrm{av} \\ a &= 2\mu_0\mu_1/I[1/r^3]\mathrm{av} \\ b &= 2\mu_0\mu_1/I\left[8\pi\psi^2(0)/3 - \left(\frac{3\cos^2\chi - 1}{2}\right)\right]\mathrm{av} \\ c &= 3\mu_0\mu_1/I[3\cos^2\chi - 1)/r_1^3]\mathrm{av} \\ d &= \mu_0\mu_1/I[3\sin^2\chi/r_1^3]\mathrm{av} \\ e &= 3\mu_0\mu_1/I[\sin\chi\cos\chi/r_1^3]\mathrm{av} \end{aligned} \qquad (2.22)$$

$\boldsymbol{r}_1$ refers to the vector from the magnetic nucleus to the interacting electron, and χ is the angle between $\boldsymbol{r}_1$ and the internuclear axis. The symbol [ ]av indicates an integral over appropriate electronic wavefunctions[145-147]. The first term in the expression for b is the Fermi contact term, and ψ^2 (0) is the unpaired electron density at the nucleus. The interaction constants in (2.21) are determined in the MBER experiment from the lambda doubling transitions, which typically fall in the radiofrequency or microwave spectrum.

Neumann[85] and Meerts and Dymanus[143] have studied $^2\Pi$ NO. The hyperfine constants and dipole moment are in Table 2.12. Meerts and Dymanus pointed out that it was necessary to use at least third-order perturbation theory in hyperfine interactions to fit the results. The degenerate perturbation theory of Freed was used[148].

^{7}LiO was investigated by Freund, Herbst, Mariella and Klemperer[80]. To produce LiO beams, Li_2O was heated to 1800°C in an iridium tube oven, with a small amount of oxygen admitted to oxidise any lithium metal formed. The hyperfine constants (Table 2.12) were in reasonable agreement with those

Table 2.12 **Hyperfine constants and dipole moments for $X^2\Pi$ molecules from MBER experiments. The constants are the conventional Frosch and Foley constants (but see Ref. 143 and 146). These values are for the vibrational ground state. All units are MHz, except μ which is in D**

	^{14}NO	^{15}NO	^{7}LiO	OD	SH
a	84.195(2)	−118.189(3)	6.12(2)	13.297(2)	‡
b	41.79(5)	−59.78(5)	−19.67(14)	−17.962(6)	−63.54(4)
c	−58.66(5)	83.51(5)	−15.40(15)	20.234(6)	
d	112.600(1)	−157.929(1)	1.94(10)	8.768(1)	27.386(60)
eqQ_1	−1.852(2)		0.444(8)*	0.143(2)	
eqQ_2	24.23(8)		0.109(18)	−0.122(6)	
μ	0.1587(2)		6.84(3)	1.65312(14)†	0.7580(1)
Ref.	85, 113	143	80	82, 144	84

* Denoted eqQ_o in Ref. 80 † μ_{OH} = 1.6676(9) D from Ref. 82 ‡ $a + \frac{1}{2}(b + c)$ = 17.075(6) MHz

calculated with a model assuming a completely ionic bond with the unpaired electron on the oxygen nucleus. However, the negative Fermi contact constant [−14.54(15) MHz] indicates at least a small amount of covalent character to the bond. An *ab initio* calculation[149] by Wahl gave a dipole moment of 6.87 D, in good agreement with experiment, as is 0.397 MHz for eqQ, calculated by Cade[149].

Meerts and Dymanus have reported hyperfine constants and dipole moments for OD [144], OH [82] and SH [84] (Table 2.11). Molecular beams of these radicals were produced by the reaction

$$H + NO_2 \rightarrow OH + NO$$

The hydrogen atoms were generated by a microwave discharge of H_2O. SH was produced by the reaction of hydrogen atoms on H_2S. Dipole moments calculated by Cade and Huo[83] gave 1.78 D for OH and 0.86 D for SH, which agree reasonably well with experiment.

2.3.9 Metastable molecules

Several metastable molecules in excited triplet states have been studied by beam techniques: $c^3\Pi_u$ H_2 by Lichten[150-152], $a^3\Pi$ CO by Freund and Klemperer[88, 153-156] and $a^3\Sigma_u^+$ N_2 by Freund, Miller, Santia and Lurio[157, 158]. The H_2 and N_2 metastables were studied by MBMR techniques and CO by MBER spectroscopy. In all cases the metastables were produced by electron impact and detected by secondary electron emission (Auger detection). It must be noted that electron impact excitation is an excellent means for producing electronic states forbidden in optical excitation, but only a small number of molecules are produced in any given rovibronic state. The resulting spectra have very low signal-to-noise ratios, and the success of these experiments is a considerable achievement.

The apparatus used by Lichten[150] (Figure 2.1) was kept to about 20 cm long because of the short lifetime of metastable H_2. $c^3\Pi_u$ H_2 comforms to

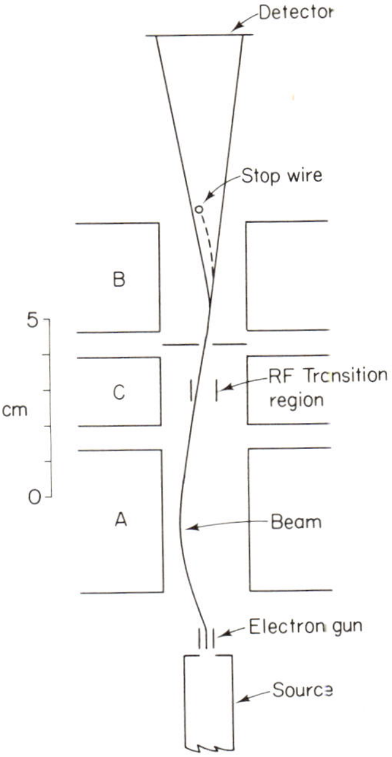

Figure 2.1 The MBMR apparatus for $c^3\Pi_u$ H_2. The 'A' and 'B' fields are magnetic deflection fields. The C region is the r.f. transition region and also has a homogeneous magnetic field. The length of the apparatus was kept very small to prevent radiative decay of the metastable molecules. (From Lichten[150], by courtesy of the American Institute of Physics)

Hund's case *b*; i.e. the electron orbital angular momentum along the H_2 axis (Λ) is coupled to the end-over-end rotation ($\boldsymbol{O}$) to form a resultant, $\boldsymbol{N}$. In turn, $\boldsymbol{N}$ is coupled with the electron spin $\boldsymbol{S}$, to form $\boldsymbol{J}$, which is the total angular momentum for ortho-H_2 ($I = 0$). For para-H_2 ($I = 1$), the nuclear spin $\boldsymbol{I}$ couples with $\boldsymbol{J}$ to form the total angular momentum $\boldsymbol{F}$. The energy for the fine structure[159] is

$$\begin{aligned} E_{fs} &= \{\Lambda A + [N(N+1) - \Lambda]C\}\boldsymbol{N}\cdot\boldsymbol{S}/N(N+1) \\ &+ [3N\cdot S(2N\cdot S + 1) - 4N(N+1)]/2N(N+1)(2N-1)(2N+3) \\ &\times \{[N(N+1) - 3\Lambda^2]B_0 \pm (-1)^{N+1}(\tfrac{3}{2})^{1/2}N(N+1)B_2\} \end{aligned} \tag{2.23}$$

The spin–orbit constant A and spin–rotation constant C are -3741 and 24 MHz, respectively[152, 159], and the spin–spin constants B_0 and B_2 -1420.80 and -4483.91 MHz, respectively[152]. Hyperfine constants for ortho-H_2, $c^3\Pi_u$, are given in Ref. 151. *Ab initio* calculations[160, 161] agree with the experimental results to within about 10%. Recently, Kagann and English[162] have measured the Stark effect for this state of hydrogen, and found reasonable agreement (10%) with theory.

It is also interesting to note that odd N states of para-H_2 and even N states of ortho-H_2 can undergo an allowed predissociation[163] to the repulsive $^3\Sigma_u^+$, and these states are effectively non-metastable.

The MBER r.f. spectrum of $a^3\Pi$ CO was first observed by Freund and Klemperer[153], and thoroughly investigated in following work[88,154-156]. $a^3\Pi$ CO conforms to Hund's case *a* for low rotational states. Both the electronic orbital and spin angular momentum are coupled to the internuclear axis,

$$\Omega = \Lambda + \Sigma$$

where Λ and Σ are the projections of the orbital and spin angular momentum along the internuclear axis. However, for high J states the molecule tends towards Hund's case b as the electron spin becomes de-coupled from the internuclear axis. For molecular beam precision, 2 kHz in these experiments, the molecule must be considered as intermediate between case a and b.

For $^{12}C^{16}O$, $v = 0$–5 for $\Omega = 0, 1, 2$ and low J states were studied. Spectra for $v = 0$–3 for $^{13}C^{16}O$ were also obtained. Electronic–rotation interaction, spin–orbit and spin–spin interactions cause the lambda doubling of $^3\Pi$ levels. The two levels have opposite parity, and electric dipole transitions between these doublets were directly observed. From the Stark effect of these transitions the electric dipole moment was obtained[88], $\mu_e = 1.373\,62(7)$ D. The variation of lambda doubling and dipole moment with Ω, v and J were obtained. Of particular interest was the perturbation of $v = 4$ and 5 by $v = 0$ and 1 of the nearly coincident a$'\,^3\Sigma^+$ state. The perturbation analysis of the Stark effect allows the dipole moment of the a$'\,^3\Sigma^+$ state to be determined as $-1.06(20)$ D ($-$CO$+$, if a $^3\Pi$ is $+$CO$-$), and the $^3\Pi$–$^3\Sigma^+$ transition moment as 0.50(1) D.

The dipole moment of CO in its ground state (Table 2.5) is small and has the sign $-$CO$+$. The electron configuration for $X^1\Sigma^+$ CO is

$$(1\sigma)^2(2\sigma)^2(3\sigma)^2(4\sigma)^2(5\sigma)^2(1\pi)^4$$

The 5σ orbital is a non-bonding orbital located on the carbon, and apparently counteracts the expected carbon–oxygen polarity. In the a$^3\Pi$ state an electron is excited from the 5σ lone pair orbital to a 2π antibonding orbital. Thus, upon excitation to the a$^3\Pi$, some charge density is shifted to the oxygen, causing a large change in the dipole moment, in agreement with experiment. Calculations[89] indicate that the dipole moment for a$^3\Pi$ is $+$CO$-$, which supports this simple argument. The a$'\,^3\Sigma^+$ state is formed by excitation of an electron from a 1π bonding orbital centred on the oxygen to the 2π orbital which is centred more on the carbon. The a$'\,^3\Sigma^+$ dipole moment might be expected to be $-$CO$+$ and experimentally it is found to have the opposite sign from the a$^3\Pi$ moment.

The dipole moment and lambda doubling fine structure for a$^3\Pi$ $^{13}C^{16}O$ agree very well with $^{12}C^{16}O$, although a systematic variation about 2×10^{-4} D for the dipole moment may be due to Born–Oppenheimer breakdown or coupling between fine and hyperfine interactions in the Hamiltonian[155]. The nuclear hyperfine parameters were measured for $^{13}C^{16}O$. The Fermi contact interaction was found to be quite large, 1935.1(5) MHz, and in surprisingly good agreement with the unrestricted Hartree–Fock calculation of Huo[164], 2023 MHz. For the direct nuclear–electronic spin–spin interaction, the experimental result is 7.9(1) MHz [$\frac{1}{2}c$ in equation (2.22)], but Huo's calculation gave 16.4 MHz.

The lowest excited state of N_2, a$^3\Sigma_u^+$, has been studied with an MBMR spectrometer by Freund *et al.*[157, 158]. Nitrogen in this state obeys Hund's case b (as does c$^3\Pi$ hydrogen) since $\Lambda = 0$ for a Σ state, and the electron spin is not coupled to the internuclear axis. Instead, the end-over-end rotational angular momentum, $\boldsymbol{N}$, and $\boldsymbol{S}$ are coupled together to form $\boldsymbol{J}$, the total angular momentum exclusive of nuclear spin. $\boldsymbol{F}$ is the total angular momentum including nuclear spin. In this study fine and hyperfine parameters were determined for several states of low (N, J, F) and for $v = 0$–12. $v = 12$ is roughly half way up the potential well for this electronic state of N_2, and thus considerable information about the vibrational variation of the fine and hyperfine parameters was obtained. An extensive theoretical treatment is given in Ref. 158. The constants $\alpha = 13.184(13)$ MHz and $\beta = -12.660(9)$ MHz were determined, where $\beta = \frac{1}{3}c$ and $\alpha - \beta = b$ [b and c defined in (2.22)]. The quadrupole coupling is $-2.571(4)$ MHz. This value is similar to that for ground state N_2, if allowance for the different internuclear distance is made. This is reasonable, since the dominant electron configuration for these two states differ only by the excitation of a $2p\pi_u$ electron to a $2p\pi_g$ orbital. This is also illustrated by the similarity of the nuclear spin–rotation interaction for metastable nitrogen, 12(5) kHz, for ground state nitrogen, 15.7(7) kHz, and 10.13(3) kHz for the isoelectronic HCN. It should be emphasised that the fine and hyperfine parameters and their vibrational dependence have been measured, and will make excellent tests for *ab initio* wavefunctions for $a^3\Sigma_u^+$ N_2.

The $X^3\Sigma_g^-$ (ground) state of molecular oxygen has also been investigated by MBMR spectroscopy. Hendrie and Kusch[165] determined the g value of the rotational magnetic moment to be $+1.22(15) \times 10^{-4}$ Bohr magneton. The g-value for the electron spin is almost that for a free electron, as was the case for $a^3\Sigma_u^+$ N_2. The ratio of the electron spin g-value to that of a free electron spin was found to be 0.999 810(13) for the oxygen ground state.

It is difficult in the space allotted to discuss fully these interesting open-shell molecules. The theoretical treatments of the spectra, in particular, are quite complex. In fact, to fit the beam data to its full accuracy (~10 kHz or less), it is necessary to introduce more constants from higher orders of perturbation theory than can be determined directly from the spectra. However, these constants can often be estimated or obtained from other experiments. The original papers should be consulted for further discussion (see also Ref. 165b).

2.4 POLYATOMIC MOLECULES

2.4.1 Introduction

The introduction of universal beam detectors to MBMR and MBER spectroscopy[12-17], and the use of nozzle beam sources[7-9] in all forms of molecular beam spectroscopy, have been significant factors in extending these techniques to larger molecules. The basic problems of extremely high resolution spectroscopy of polyatomic molecules are associated with very dense spectra and very few molecules in any given quantum state. In particular, the rotational partition function increases very rapidly as the molecule becomes larger, and it is the low rotational temperature of a supersonic nozzle beam that makes this source so attractive to large-molecule beam spectroscopy. The increasing complexity of the spectra must be attacked theoretically with very strong emphasis on the use of group theory[162a] to set up the problem and high speed computers to realise the solution. The use of machine calculations are also vital in obtaining theoretical spectra which can be compared with the experimental results. The differences are then minimised in a least-squares sense by adjusting the molecular parameters used in the theoretical calculation. In some situations it is also necessary to carry out lineshape and line-blending calculations to deal with unresolved spectra[163a]. The problems of large-molecule spectroscopy are very apparent in complicated unresolved spectra, even with instrumental linewidths much less than one kilohertz.

A large number of polyatomic molecules have been studied by a variety of molecular beam methods. Individual molecules are briefly described, with emphasis placed on aspects of the experiments and/or results that are special to that work. For molecular properties that have been well characterised, data are given in the text or in tables. In a few cases, where either the data are scattered throughout the literature or only available in theses, more complete data presentations are contained in the appendix. In general, raw data or constants associated with a specific transition, rather than a molecular property, are not given; original literature references are provided for this type of information.

The molecules are discussed in groups, with the groupings determined primarily by molecular symmetry. The vast majority of the polyatomic molecules studied are included in the following groups: linear, tetrahedral, C_{3v} symmetric tops, and C_{2v} asymmetric tops. The molecules considered in this section are stable species whose geometry is already well known from other, lower resolution, studies. The emphasis of the studies here are on details of the electronic structure of the molecules made available from hyperfine, Zeeman and Stark data. Polyatomic molecules having weak bonds, such as hydrogen bonds or van der Waals bonds, are presented in a separate section. Here the main emphasis is in determining molecular geometry, and beam techniques are used not for high resolution but because these molecules are best produced in beam sources and studied in the collisionless environment of a molecular beam.

A suitable hyperfine Hamiltonian for a general polyatomic molecule can be written[117]:

$$H_{\text{hyp}} = \sum_K \boldsymbol{Q}_K \cdot \nabla \boldsymbol{E} + \sum_K \boldsymbol{I}_K \cdot \boldsymbol{M} \cdot \boldsymbol{J} + \sum_{K,L} \boldsymbol{I}_K \cdot \boldsymbol{D} \cdot \boldsymbol{I}_L \tag{2.24}$$

The summations are over the nuclei in the molecule. The first term is the nuclear quadrupole coupling interaction between the nuclear quadrupole moment tensor, $\boldsymbol{Q}$, and the field gradient tensor $\nabla \boldsymbol{E}$. The spin–rotation interaction is described by the tensor $\boldsymbol{M}$, and $\boldsymbol{D}$ specifies the direct spin–spin interaction. The indirect, or electron-coupled, spin–spin interaction has been assumed to be negligible. Equation (2.24) can be cast into a form completely analogous to the diatomic Hamiltonian in (2.13) with appropriate substitutions for q, C_1, and C_3. For the polyatomic case[164a]:

$$q = \frac{2}{J(J+1)} \sum_g \langle J_g^2 \rangle q_{gg}$$

$$C_1 = \frac{1}{J(J+1)} \sum_g \langle J_g^2 \rangle M_{gg}$$

$$C_3 = \left(\frac{\mu_0}{4\pi}\right) \frac{(\mu_K/I_K)(\mu_L/I_L)}{J(J+1)} \sum_g \langle J_g^2 \rangle R_{gg}$$

The sum is over the components in the inertial principal axis system. $\langle J_g^2 \rangle$ is the average value of the square of the molecule's angular momentum. q_{gg} is the field gradient at the nucleus in question. M_{gg} is the spin–rotation tensor component and $\boldsymbol{R} = (r^2 1 - 3\boldsymbol{rr})/r^5$, where $\boldsymbol{r}$ is the vector between the two nuclei responsible for the spin–spin interaction in question. The average squares of the angular momenta can be calculated from asymmetric rotor functions. However, in many cases, $\langle J_g^2 \rangle$ can be obtained with sufficient accuracy from $E(\kappa)$ tables and the following expressions[164a, 165a]:

$$\langle J_a^2 \rangle = \tfrac{1}{2}\left[J(J+1) + E(\kappa) - (\kappa+1)\frac{\partial E(\kappa)}{\partial \kappa}\right] \tag{2.26}$$

$$\langle J_b^2 \rangle = \frac{\partial E(\kappa)}{\partial \kappa}, \text{ and } \langle J_c^2 \rangle = \tfrac{1}{2}\left[J(J+1) - E(\kappa) + (\kappa-1)\frac{\partial E(\kappa)}{\partial \kappa}\right]$$

From any single transition, labelled J_τ, it is only possible to obtain constants $q_{J\tau}$, $C_{J\tau}$ and $R_{J\tau}$, describing the quadrupole, spin–rotation and spin–spin interactions for that transition. These constants depend on a linear combination of the appropriate diagonal tensor elements. The individual tensor elements can be separated only if a sufficient number of transitions have been observed to obtain independent combinations. For many of the polyatomic molecules studied this much information is not available and only transition-dependent constants are available. In this article, only the molecular properties, i.e. tensor elements, are reported.

Equations (2.6) and (2.18), the Hamiltonian for external electric and magnetic fields, respectively, are valid for polyatomic molecules. The Stark effect in polyatomic molecules is well discussed[166,166a]. The high precision of MBER measurements normally requires careful attention to higher-order terms in the perturbation expansion of (2.6), however. The Stark effect has not yet been used to determine polarisabilities of non-linear molecules, but polarisability effects were considered in the dipole moment measurement of water[167]. To determine the tensor elements in the Zeeman Hamiltonian, equation (2.18), several transitions must be observed, as in the case for hyperfine interactions. Where sufficient information is available, the diagonal tensor elements of the magnetic interactions, and the molecular quadrupole moment components calculated from them, are given.

One additional difficulty arises because different authors use different conventions for different molecules. The greatest confusion concerns spin–rotation tensors which ideally would be specific as C_{aa}, C_{bb} and C_{cc} or C_x, C_y and C_z. The subscripts refer to molecule-fixed principal axes. In tetrahedral or symmetric top molecules, at least two of the three components are equal and the diagonal elements of the spin–rotation tensor can be written C_x, C_x and C_z or $C_\perp$, $C_\perp$ and $C_\parallel$. However, since data are often best fit with combinations of the parallel and perpendicular elements, different combinations appear in the literature. In tetrahedral molecules, C_a and C_d are frequently reported, where $C_a \equiv \tfrac{1}{3}(C_\parallel + 2C_\perp)$ and $C_d \equiv C_\perp - C_\parallel$ [168]. Symmetric-top molecule studies often report $C_\alpha \equiv \tfrac{1}{2}(C_\parallel + C_\perp)$ and $C_\beta \equiv C_\parallel$ [169]. A useful discussion on interconverting the tetrahedral and symmetric-top notation is given in Ref. 169.

2.4.2 Linear polyatomic molecules

The simplest polyatomic molecules are linear, but only a small number have been studied by beam spectroscopy. In addition, much of this work is available only from unpublished theses. Carbonyl sulphide (OCS) has been extensively studied with MBER spectroscopy by both Dymanus[170-172] and Muenter[173-175]. Hydrogen cyanide (HCN) has been investigated in Gordy's laboratory using millimeter-wave beam maser methods[37, 176], and the thesis of Tomasevich[177] contains large amounts of MBER data on HCN. Detailed MBER studies of lithium hydroxide (LiOH) are contained in the thesis of Freund[178]. MBMR experiments on carbon dioxide, OCS, carbon disulphide and acetylene have been performed in Ramsey's laboratory[179, 180].

OCS is by far the most thoroughly studied linear polyatomic molecule, with many isotopic species and several vibrational states having been investigated. The molecular beam value of the OCS dipole moment is the accepted internal standard for Stark effect measurements[170, 173]. Accurate dipole moments are available for singly-substituted ^{13}C, ^{34}S, ^{33}S and ^{18}O species, and for excited vibrational bending modes, $v_2 = 1$ and 2[171, 172, 175]. The transitions observed in the excited bending vibrational states were between l-type doublets. For $v_2 = 1$, Dymanus and co-workers[171] measured the l-type doubling for the first twelve J states. These transition frequencies were fit with the expression[181]

$$v_1 = \tfrac{1}{2}q(v_2 + 1)J(J + 1),$$
$$\text{where } q = 6361.413(5) - 4.27(5) \times 10^{-3}J(J + 1)\ \text{kHz}$$

The J^4 coefficient is known to be less than 10^{-6} and the vibrational dependence of q has been measured[175]. A similar investigation of the much smaller splittings of the $v_2 = 2$, l-type doublets has been made[175] on $J = 4$–20. These measurements are accurately predicted[181] by $v_2 = -\frac{1}{2}\delta + \frac{1}{2}[\delta^2 + 4q^2 \times (J + 2)(J + 1)(J)(J - 1)]^{\frac{1}{2}}$; δ is equal to $E^0_{\Sigma^-} E^0_{\Delta}$, the separation of the $l = 0$ and $l = 2$ states in $v_2 = 2$. Numerically, $\delta = 1.734\,64(5) \times 10^8 - 2.506(3) \times 10^3 J(J + 1)$ kHz and q has the value given above, minus the vibrational correction of 86.52(9) kHz.

Hyperfine measurements have been made on the isotopic species containing ^{13}C and ^{33}S in the ground vibrational and in the $v_2 = 1$ excited state[172]. Zeeman data have been taken on a variety of isotopic species in the ground vibrational state[170] and on the normal species[171] for $v_2 = 1$. There has been a substantial amount of controversey over properties which can be obtained from magnetic measurements on OCS. Molecular quadrupole moment data from microwave absorption linewidths are in very poor agreement with the magnetic data from microwave and MBER Zeeman measurements. The accepted value, $\Theta = -0.786(14) \times 10^{26}$ e.s.u., is the MBER value[170]. The most accurate rotational g value, $g_J = -0.028\,839(6)$, is also the MBER result[170]. In this case, microwave absorption data conflicted with MBMR results[180]. The difference was attributed to centrifugal distortion; however, MBER experiments[170, 174] showed that the MBMR results were, in fact, correct. Systematic errors in magnetic field measurement were evidently present in the microwave experiment[170, 174], pointing to the need for an internal calibrant for magnetic field measurements. A detailed compilation of OCS data, with references, is included in the appendix.

HCN and DCN have been extensively studied in the ground vibrational state as well as in excited bending modes[176, 177]. Published data are from millimetre beam maser studies[176] on the ground vibrational state of several isotopes of HCN. Magnetic shielding calculations from spin–rotation constants are presented in Ref. 176. Unpublished data from the Tomasevich thesis are given in the appendix. In general, the MBER hyperfine data is about an order of magnitude more precise than the maser information, and includes excited bending vibrational states. The MBER uncertainty quotations are for 95% certainty with careful consideration given to possible correlations between parameters. The two studies are complimentary in that rotational constant data are provided from the maser work which are not available from the r.f. MBER investigation, while the MBER work contains dipole moment information not available from the zero field maser experiments (see also Ref. 177a).

Alkali hydroxides have been studied by microwave absorption spectroscopy[182-184], and they present interesting problems in molecular structure. Very unusual vibration–rotation interactions for the bending modes are observed, and it is extremely difficult to say whether or not the molecule is truly linear at equilibrium. While the geometries are clearly not that of rigid, bent molecules, the possibility of potential barriers at linear configurations cannot be ruled out. LiOH is the lightest alkali hydroxide and is, therefore, the most amenable to calculation. However, it is not well suited for microwave absorption studies. Not only are quite high temperatures necessary for vaporisation of solid LiOH, but the equilibrium of the disproportionation reaction

$$2\,\mathrm{LiOH} \rightarrow \mathrm{Li_2O} + \mathrm{H_2O}$$

lies far to the right at low pressures. Molecular beams of LiOH can readily be generated by heating Li_2O in an oven with a large overpressure of water, reversing the above reaction.

MBER spectroscopy on beams produced in this manner has been carried out on ^{7}LiOH and ^{7}LiOD in a number of vibrational states[178]. Microwave, as well as r.f., experiments were performed and pure rotation, *l*-type doubling and pure Stark transitions were observed. The rotational constant data, which are listed in the appendix, were fitted by the following expression:

$$\begin{aligned} B_e = B_{v_1 v_2^l v_3} &+ (v_1 + \tfrac{1}{2})a_1 + (v_2 + 1)a_2 + (v_3 + \tfrac{1}{2})a_3 \\ &- (v_1 + \tfrac{1}{2})(v_2 + 1)\gamma_{12} - (v_1 + \tfrac{1}{2})(v_3 + \tfrac{1}{2})\gamma_{13} \end{aligned}$$

The results are $B_e = 35\,742.8$, $a_1 = 497.8$, $a_2 = 109.9$, $a_3 = 126.3$, $\gamma_{12} = 18.5$ and $\gamma_{13} = 49.4$ MHz. However, caution should be taken since these results are obtained from fitting six parameters to six data. Using B_0 for LiOH and LiOD gives r_0(Li—O) = 1.594 Å and r_0(O—H) = 0.921 Å. Many *l*-type doubling transitions were observed in LiOH and LiOD. The doubling constant q was found to be 295.8, 305.3 and 298.7 MHz for the 01^10 and 03^10 states of LiOH and the 01^10 state of LiOD, respectively. Doublings observed in the 02^20 state were used to obtain[181] the separation $E^0_\Delta - E^0_\Sigma = 27.5$ cm^{-1}. Similarly, 03^30 doublings provided $E^0_\phi - E^0_\pi = 50.2$ cm^{-1}. The dipole moment of the ground state of ^{7}LiOH is 4.755(2) D. This moment increases with excitation of both v_1 and v_2 vibrational states, and decreases on deuteration. Hyperfine information obtained include eqQ and C for ^7Li and this data, along with the dipole moments, are also in the appendix.

MBMR spectroscopy has been used to observe the rotational magnetic moments of several linear molecules, including molecules with no nuclear magnetic moments[179,180]. These measurements are successful because sufficient deflection of high J states can be obtained by using the relatively large rotational magnetic moment of states with large M_J values. This can be done by inducing multiple quantum transitions in the resonance region so the initial state cascades from the top of the M_J manifold of Zeeman states to the bottom. The results[180] obtained are: CO_2, $g_J = 0.055\,08$; CS_2, $g_J = 0.022\,74(2)$; OCS, $g_J = 0.028\,89(2)$; and HCCH, $g_J = 0.049\,03(4)$. The sign of the OCS moment is known to be negative and this work showed the sign of the acetylene rotational g value to be the same as OCS. In addition, the spin–rotation interaction in acetylene was measured to be 3.58(1) kHz.

The linear molecules that have been studied by beam spectroscopy are all rather different from one another, both physically and chemically, and intercomparisons are not particularly useful. LiOH can best be compared with LiF, by considering the compound to be ionic and observing that F^- is not drastically different from OH^-. The LiF and LiO bond lengths are quite similar, with the LiF length being 0.03 Å shorter. The dipole moment and lithium eqQ of LiOH are both much smaller than the corresponding LiF properties. This is in agreement with the longer bond length and also with LiOH being less ionic than LiF, but before this type of comparison is pushed too far it should be observed that these two properties of LiF are much closer to LiOH in the 02^00 vibrational state. Clearly, intramolecular motions and vibrational averaging are very important and cannot be ignored when making comparisons between molecules, particularly between diatomic and triatomic species.

The linear molecules are the only class of polyatomic molecules for which there exist much data on excited vibrational states. A variety of excited states have been observed, and it is very tempting to try to extract equilibrium configuration data and perhaps derivatives of properties as a function of normal coordinates. At this time, there clearly is not enough information available for this purpose and, considering the difficulty encountered in diatomic cases, it is unlikely that such results will be available. However, the success in obtaining vibrational potential information from the excited bending modes of these molecules points strongly to the desirability of studies of additional excited vibrational states of linear molecules[185].

The magnetic hyperfine properties give a variety of different descriptions of the electronic distribution in each molecule and can usefully be correlated with n.m.r. data. An example of this is the combining of spin–rotation constants with chemical shift data to obtain the high frequency or paramagnetic portion of the magnetic shielding[176]. Properties such as molecular quadrupole moments and polarisabilities are important in determining intermolecular potential functions. These molecules are not too large for high quality *ab initio* calculations, and the hyperfine properties provide rigorous tests for theoretical wavefunctions. This is particularly true for those magnetic properties that depend on sums over all electronic states.

2.4.3 Tetrahedral molecules

The most studied and best understood diatomic molecule is certainly hydrogen. One of the most studied polyatomic molecules is methane, CH_4. Methane has been the subject of a large number of molecular beam spectroscopic experiments, most of which have been by MBMR[168, 186-193]. Following the original work of Anderson and Ramsey[168], several of Ramsey's co-workers, most notably Ozier, worked on both theoretical and experimental aspects of the problem. Early experiments observed reorientation of the sum of the proton magnetic moments relative to a strong external magnetic field. These data were interpreted in terms of the spin–rotation tensor, the proton spin–spin interaction and the rotational magnetic moment in the form of the rotational g value, g_J. Later experiments[188] in weak external magnetic fields improved the knowledge of these molecular constants. The spin–rotation constants obtained for CH_4 are $C_a = \frac{1}{3}(C_{\parallel} + 2C_{\perp}) = 10.4(1)$ kHz and $C_d = C_{\perp} - C_{\parallel} = 18.5(5)$ kHz. The spin–spin interaction constant of 20.9(3) kHz can be explained entirely in terms of the classical dipole–dipole interaction.

Applying the Pauli principle to the proton spins in methane require the total nuclear spin angular momentum to be 0, 1 or 2 to give ortho, para and meta spin states. Contrary to the diatomic hydrogen case, the symmetry properties of tetrahedral wavefunctions permit different spin states to be sublevels of the same rotational level. The much smaller energy separations of these sublevels permit hyperfine interactions in isolated methane molecules to mix these spin states, and allow transitions which interconvert them. The transition probability in this case is normally extremely small; however, Ozier *et al.*[188a] observed these transitions by taking advantage of level crossing. At the magnetic field where the $M_I = 1$ Zeeman component of the $I = 1$ state is just equal to the energy of the $I = 0$ state, avoided crossings occur and strong interactions between the normally isolated spin states result. There has been some confusion on the interpretation of these transitions, and all of the observations are not fully explained[189]. However, inversion in the methane molecule is not being observed, as suggested in the original paper[189]. Perhaps the most fascinating result from MBMR studies of methane is the direct observation of the permanent electric dipole moment of the ground vibrational state of CH_4. It has long been known that certain degenerate excited vibrational states of tetrahedral symmetry support a permanent electric dipole moment. Centrifugal distortion transfers a small part of this excited state moment to the ground vibrational state. Ozier[190] directly observed the Stark effect of ground state MBMR transitions and determined this moment to be $5.38(10) \times 10^{-6}$ D in the $J = 1$ state.

Deuterium substitution in methane[191] generates a slightly polar, symmetric-top molecule, CH_3D. Even though the dipole moment, which arises from different hydrogen and deuterium motions, is only 0.005 D, the first-order Stark effect allows rotational state selection and the application of MBER spectroscopy. The electric resonance spectra for $J = 1$ and $J = 2$ rotational states were fitted with spin–rotation, spin–spin, deuterium quadrupole coupling and dipole moment parameters. The spin–rotation parameters (in the same notation as for CH_4, for comparison with the MBMR work) are $C_a = 10.5$ and $C_d = 18.1$ kHz. eqQ_D was found to be 191.48(77) kHz. Because of the small absolute value of the electric dipole moment, centrifugal distortion was easily observed, with the $J = 1, K = 1$ and $J = 2, K = 2$ moments being $5.6409(5) \times 10^{-3}$ and $5.6794(5) \times 10^{-3}$ D, respectively.

MBMR spectroscopy has also been used[187,192] to study CF_4, SiH_4, SiF_4, GeH_4 and GeF_4. These experiments were similar to, though less extensive than, the methane work. The results in kHz are: CF_4, $C_a = -6.81$, $-13 < C_d < 17$; SiH_4, $C_a = +3.71$, $C_d = 9.0(35)$; SiF_4, $C_a = -2.46(8)$, $C_d < 3$; GeH_4, $C_a = +3.79$, $C_d = 5.5(50)$; and GeF_4, $C_a = -1.84(8)$, $C_d < 3$. Where signs are specifically given for the above data, the sign was determined.

All the tetrahedral molecule studies mentioned so far were performed on the ground vibrational state. In fact, one of the reasons that the detailed spectroscopic measurements could be carried out on these species was that they all could be cooled to quite low temperatures, leading to relatively small partition functions and a small number of populated states. Because of the possibility that excited vibrational modes of F_2 symmetry can have permanent dipole moments, excited vibrational states of tetrahedral molecules are also of considerable interest. The first observation of such a moment was by Uehara *et al.*[193] who took advantage of a He–Ne laser coincidence with a suitable methane rotation–vibration transition. Molecular beam electric deflection has been used to observe the Stark effect in excited vibrational levels of several tetrahedral molecules[194,195]. Electric deflection has been observed in CF_4, CCl_4, SiF_4, $SiCl_4$, $GeCl_4$, TiF_4, $TiCl_4$, VF_4 and VCl_4 as a function of temperature[196]. The most detailed studies were carried out on CCl_4, CF_4 and $SiCl_4$. These molecules exhibited no deflection at room temperature, and the change in deflection as a function of temperature, shown in Figure 2.2, strongly suggests that the vibrational mode responsible for the effect is ν_3. The other mode of F_2 symmetry, ν_4, appeared not to have sufficiently large Stark coefficients to cause observable deflection. The origin of this Stark effect was unequivocally shown to be from excited vibrational states by observing the various isotopic species of the tetrachlorides. Criteria for the presence of excited state moments, whether a first- or second-order Stark effect obtains, and for the effect of isotopic substitution, have been deduced from group theoretical considerations[194,197]. This work indicates that the isotopic species with T_d and C_{3v} symmetry have first-order Stark effects and show deflection, while the doubly substituted molecules of C_{2v} symmetry (i.e. X $^{35}Cl_2$, $^{37}Cl_2$) have a second-order Stark effect which will not lead to observed deflection. The experimental observations are in agreement with these findings.

The beam spectroscopic studies of methane have been extremely productive both in determining molecular properties and also in developing theoretical models for interpreting hyperfine structure in large-molecule spectra. The spin–rotation interaction parameters are important in the magnetic relaxation of methane and relate directly to n.m.r. experiments[198]. The general study of hyperfine interactions and the specific observation of ortho–para transitions provide important information concerning nuclear spin isomerisation in methane[199]. The molecular beam studies of methane have been important in stimulating other experiments and theoretical considerations for tetrahedral

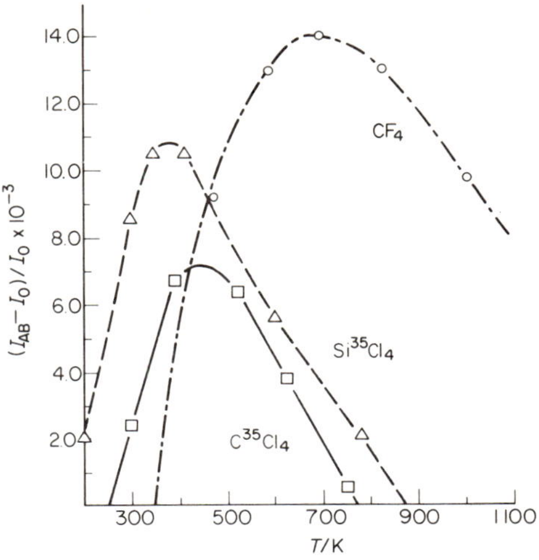

Figure 2.2 The temperature dependence of the refocused beam, I_{AB}, relative to the unobstructed beam, I_0, for tetrahedral molecules

molecules. These include Ozier's observation of pure rotational transitions in CH_3D in the i.r.[200], microwave absorption from $\Delta J = 0$ transitions in the ground vibrational state of CH_4 [201] and a variety of studies of other tetrahedral molecules[202-207]. Theoretical work, particularly on excited vibrational states and their electric dipole moments, has been done by Mills, Watson and Fox[208-212]. Better understanding of the spectroscopic properties of methane has direct application to the study of the earth's, and other planetary, atmospheres.

2.4.4 C_{3v} Pyramidal molecules, ammonia and phosphine

The ammonia molecule, NH_3, is another example of an extremely well studied molecule. Many aspects of ammonia, most importantly its inversion doubling of rotational levels, make this molecule ideal for a variety of high-resolution studies. For example, microwave absorption was first observed in ammonia[213], and the first maser utilised ammonia. Since the original maser experiment[214] by Gordon *et al.* in 1954, ammonia has been studied in many beam maser experiments[31, 33, 215-220]. The highest resolution of these has been the extensive studies by Kukolich using a two-cavity maser with a 350 Hz linewidth[31, 216, 217]. The complexity of the hyperfine spectra of ammonia observed at this resolution is immense, and there have been difficulties in interpreting all the information obtained. The theoretical work of Hougen[218] appears to have answered all of the presently outstanding questions. The Zeeman effect in ammonia has also been studied in beam masers by having the entire maser cavity in the magnetic field[33]. These experiments have been performed in a conventional magnet and also in an extremely high-field solenoid at the National Magnet

Laboratory. The most complete picture of the Zeeman work is obtained by combining results obtained from microwave absorption experiments[222] with the maser results.

The transitions observed in ammonia experiments are between inversion levels of a given J,K rotational state. The data are, therefore, different from those obtained from high-resolution transitions involving pure rotational energy levels. In particular, spin–rotation and spin–spin interaction constants have not been reported for ammonia in a form compatible with data from other polyatomic molecules. The most extensive, correct listing of hyperfine data for NH_3 is given by Hougen in Ref. 218. In general, properties for each inversion level are obtained, and centrifugal distortion and isotopic dependence are easily observed. Hougen presents a very interesting discussion of higher order interactions which dominate the $J,K = 3,2$ rotational state. This state is unique since many interactions, which depend on $J(J + 1) - 3K^2$ to first order, vanish for this transition.

While a single constant can specify the nitrogen nuclear quadrupole coupling, the deuterium eqQ values are more detailed[219-221]. The three principal components for the deuterium eqQ tensor are: $eq_{xx}Q = -147$, $eq_{yy}Q = -144$ and $eq_{zz}Q = 291$ kHz. Zeeman data[33] provide $g_J(\|) = 0.500(2)$, $g_J(\perp) = 0.568(2)$ and $\chi_{\|} - \chi_{\perp} = 0.32(7)$ kHz kG^{-2}. These data can be combined[33] to obtain the component of the molecular quadrupole moment parallel to the C_{3v} axis: $\Theta_{\|} = -3.3(4) \times 10^{-26}$ e.s.u. cm^2.

The other C_{3v} pyramidal molecule that has been studied[223] in detail with beam spectroscopy is phosphine, PH_3. Since phosphorus is immediately below nitrogen in the Periodic Table, it is expected to have many similarities to ammonia. From a spectroscopic point of view, however, a very large difference is the height of the barrier to inversion, which is approximately 2070 cm^{-1} in ammonia but is calculated to be greater than 9000 cm^{-1} in phosphine[224]. This means that the inversion frequency for the ground vibrational state of phosphine should be less than 1 Hz compared to the 22 GHz splitting that dominates all the high resolution spectroscopy of ammonia.

MBER spectroscopy is well suited to the study of PH_3, as shown by Davies *et al.*[223]. The most distinctive feature of the MBER spectrum is a series of intense r.f. transitions at zero external field. These result from transitions between the states $J,K = +3$ and $J,K = -3$. The two $K = 3$ rotational wavefunctions belong to the same non-degenerate representation in the C_{3v} point group and can only be accidentally degenerate. Fourth-order Coriolis interactions lift this degeneracy. As in the case of l-type doubling in excited vibrational states, the J dependence of this splitting is the same as that for asymmetry doublets in slightly asymmetric top molecules and is given by a constant times $(J + K)!/(J - K)!$, where K must here be a multiple of three. The empirically found coefficient for the $K = 3$ doubling in phosphine is 85.5 Hz. This is a factor of approximately 4 larger than predicted by Nielson and Dennison[225], indicating that interactions not considered by these authors are present.

The observed hyperfine structure on the phosphine transitions was fitted with calculated spin–spin interactions and two spin–rotation parameters for both the protons and the phosphorus nucleus. The values for phosphorus are $C_{\perp} = -114.90(13)$ and $C_{\|} = -116.38(32)$; for hydrogen, $C_{\perp} = 8.01(8)$ and

$C_{\parallel} = 7.69(19)$ kHz. The electric dipole moment was measured, yielding $\mu = 0.5740(2)$ D.

Phosphine was only the second phosphorus-containing compound for which accurate spin–rotation constants have been obtained. The first was PN, which is a high-temperature species and not suitable for gas-phase n.m.r. spectroscopy. Therefore, phosphine was the first compound for which gas-phase n.m.r. average shielding data, and also spin–rotation data, were available. This permitted the separation of the diamagnetic and paramagnetic shielding and the establishing of an absolute magnetic shielding scale for phosphorus[226].

2.4.5 C_{3v} Substituted methane molecules

Experimental considerations for the different types of beam spectroscopy on symmetric top molecules change drastically when the molecule contains more than one heavy atom. In the case of MBMR spectroscopy the transitions re-orienting nuclear spins involve large numbers of thermally populated rotational levels with different hyperfine parameters, and broad unresolved spectra result. It is difficult to obtain precise molecular properties from these data because accurate lineshapes and intensities are required to calculate blended lineshapes. There has been one unpublished study[227] of spin–rotation interactions in the heavier symmetric top molecules using MBMR spectroscopy. Fortunately, these problems do not plague MBMR measurements of rotational magnetic moments where the Zeeman splitting is independent of J. MBMR utilising multiple quantum transitions has been used to study[25] nickel tetracarbonyl and iron pentacarbonyl to obtain $g_J = 0.0179(5)$ and 0.0210(5), respectively.

There is no difficulty in interpreting data obtained from MBER spectroscopic studies of heavier symmetric tops, since rotational or pure Stark transitions can readily be resolved for the different J,K states. However, the state selection process in a symmetric top, with linear Stark effect, is very indiscriminate. If a large number of J,K states are thermally populated, the Stark coefficients almost form a continuous classical distribution. In this situation, any given set of voltages applied to the state selection fields in the MBER spectrometer will focus a rather wide distribution of J,K states into the detector. Thus the majority of molecules detected will introduce noise, and only a small fraction will be removed by the resonance process on the single J,K state being studied. Thus, MBER spectroscopy suffers from a serious sensitivity problem for the heavier symmetric top molecules. Methyl fluoride[169] is the only symmetric top containing two heavy atoms which has been extensively studied with MBER, and it is indeed the most favourable case. Yet it was necessary to employ long time averaging to obtain good spectra. The spin–rotation constants obtained for the hydrogen atoms are $C_{\perp} = 0.8(15)$ and $C_{\parallel} = 14.7(1)$ kHz and for the fluorine atom, $C_{\perp} = 4.0(19)$ and $C_{\parallel} = -51.1(13)$ kHz. The dipole moment is 1.8585(5) D. Wofsy *et al.*[169] give a detailed discussion of the relationship between the spin–rotation data and magnetic shielding in CH_3F. This discussion includes comparisons of theoretical calculations, beam data and data from oriented n.m.r. samples.

Fortunately, the difficulties of MBMR and MBER spectroscopy in observing symmetric top spectra do not affect beam maser spectroscopy. Good signal-to-noise ratios can be obtained for resolved spectra of a variety of symmetric top molecules, particularly those with large A rotational constants and large dipole moments. Molecules of this type have been extensively studied in the laboratories of Kukolich and Dymanus. Some of these molecules are: CH_3Cl[228], CH_3CN[229], CH_3NC[230], CH_3OH[231, 232] and CF_3H[233, 234]. Many deuterated species of these molecules have also been studied with particular emphasis on obtaining deuterium eqQ values. Methyl alcohol[231, 232] has been included in this section even though it is not truly a symmetric top, both because it is a singly substituted methane, closely related to the other molecules, and because the spectra were treated as if the molecule were a symmetric top. With the case of methanol in mind, Dymanus has considered in detail how internal rotation effects hyperfine spectra and their interpretation[235].

The hyperfine data for these molecules are usually obtained from a single transition. This means that spin–rotation tensor components are not available. In addition, the spin–spin interactions can usually be calculated from the known structure, i.e. any contribution of the tensor portion of the electron coupled spin–spin interaction is negligible. The most interesting data from these maser studies are the nuclear quadrupole coupling constants. The eqQ values for deuterium are tabulated and discussed at the end of this section. Kukolich[228] has made very accurate measurements for chlorine eqQ values in several isotopic species of CH_3Cl. The observed changes in eqQ(Cl) are related to changes in the acidity of the hydrogen atoms which have been replaced by deuterium. An interesting nitrogen quadrupole coupling interaction occurs in CH_3NC, where $eqQ(N) = 489.4(4)$ kHz. This unusually small value indicates the nearly spherical electronic structure at nitrogen atoms having four bonds[236].

2.4.6 Asymmetric top molecules

The majority of asymmetric top molecules that have been studied by beam spectroscopy belong to the C_{2v} point group and most of these molecules have very similar nuclear spin distributions. The molecules H_2O, H_2S, H_2Se, H_2CO, H_2CCO (ketene) and F_2CO all have two nuclei with spin 1/2 off the symmetry axis, and the remaining nuclei are on the axis and have a spin of zero. Two additional C_{2v} molecules which do not have this nuclear spin structure are furan and CH_2F_2. Asymmetric top molecules that have been studied in very high resolution that are not in the C_{2v} point group include H_2NCHO, HFCO, HNCO and formic acid.

These molecules have been investigated by either beam maser or MBER spectroscopy, and in some cases, both. The majority of work on this class of molecule, however, has been done with beam maser spectroscopy. The most studied of these molecules is formaldehyde (H_2CO), which was extensively investigated in the early days of maser spectroscopy[165, 237-240] and more recently with higher resolution and greater sensitivity[241-244]. Almost all conceivable isotopically substituted species have been studied[241, 244], either by enrichment or in natural abundance. A sufficient number of transitions have been observed in zero field and in external magnetic fields to obtain the diagonal tensor elements for spin–rotation[243], electric field gradient at the hydrogen nucleus[244], rotational g value[242], magnetic susceptibility[242] and molecular quadrupole moment[242]. These values are listed in Table 2.13. The field gradients at the hydrogen nuclei can be transformed from the inertial axes to bond axes to obtain $q_{\xi\xi} = 8.04(6)$, $q_{\chi\chi} = -3.86(6)$ and $q_{\zeta\zeta} = -4.18(5)$; ξ is along the C—D bond, χ is perpendicular to ξ and in the molecular plane and ζ is perpendicular to the molecular plane. To within experimental error, the Z principal axis of the field gradient tensor is coincident with the

Table 2.13 Tensor properties of formaldehyde

	aa	*bb*	*cc*	*Ref.*
M_{gg}/kHz	−4.0(7)	2.4(7)	−2.1(7)	243
$q_{gg}/10^{14}$ statvolt cm^{-2}	−0.618(5)	4.797(5)	−4.179(5)	244
G_{gg}	−2.9024(6)	−0.2245(1)	−0.0994(1)	242
χ_{gg}/kHz kG^{-2}	−0.06(16)	0.33(12)	−0.27(20)	242
$\Theta_{gg}/10^{-26}$ e.s.u.	8.4	1.4	−9.8	242

C—D bond direction, but the charge distribution is not cylindrically symmetric. Theoretical values[245] for the field gradient are in fair agreement with experiment, giving $q_{\xi\xi} = 8.50$, $q_{\chi\chi} = -4.30$ and $q_{\zeta\zeta} = -4.20$. An MBER experiment[246], studying the electric dipole moment of formaldehyde, has been performed. Preliminary results are: ground vibrational state, $\mu = 2.3315$; $v_6 = 1$ (out-of-plane bend), $\mu = 2.309$; $v_5 = 1$ (in-plane bend), $\mu = 2.329$ D. In related molecules, eqQ(D) has been measured[247] to be 205(4) kHz in DFCO, and in F_2CO the spin–rotation tensor elements[248] are $M_{aa} = -19.77(21)$, $M_{bb} = -13.46(14)$ and $M_{cc} = -7.80(26)$ kHz.

The other obvious class of molecules in this group are the water-related molecules. The extreme importance of water to all aspects of man's existence makes it one of the most studied molecules[117, 165, 167, 249-253]. Maser studies of a variety of isotopic species, including ^{17}O, in zero field and in magnetic fields[117] have yielded many parameters describing the electronic distribution of the water molecule. Table 2.14 summarises the various tensor components that are available for water from the work of Dymanus and co-workers[117, 249-252]. The eqQ(D) tensor components reported in deuterated water are for the principle axis system of the field gradient and not the inertial axis system. Although equation (2.24) shows that only diagonal elements of the field gradient tensor in the inertial axis system can be measured from the quadrupole coupling interaction, by combining data from various isotopic species both the principal components and the axis of this tensor can be determined[252]. The z axis of the gradient coordinate system does not coincide with the O—D bond direction but is offset 1°16′ from the bond axis system. In addition, the gradient is far from being cylindrically symmetric about the O—D direction. This result, along with similar evidence from formaldehyde,

Table 2.14 Tensor properties of water[117, 252]

	aa	*bb*	*cc*
$M_{gg}(D_2O)$/kHz	−1.78	−1.41	−1.57
$G_{gg}(H_2O)$	0.657(1)	0.718(7)	0.645(6)
$G_{gg}(HDO)$	0.618(6)	0.413(4)	0.437(4)
$G_{gg}(D_2O)$	0.32530(10)	0.36009(22)	0.32513(15)
$eq_{gg}Q(D)$/kHz*	307.95(14)	−174.78(20)	−133.13(14)
$eq_{gg}Q(^{17}O)$/kHz	−7877.8(80)	−2297(72)	10175(67)
$\chi_{gg} - \chi_{av}(D_2O)/10^{-2}$ kHz kG^{-2}	3.88(20)	−1.66(40)	−2.23(3)
$\Theta_{gg}(D_2O)/10^{-26}$ e.s.u.	2.724(14)	−0.321(27)	−2.402(21)

* These axes are the field gradient tensor principal axes; *aa* is 1°16′ off the OD direction in the molecular plane, *bb* is perpendicular to the molecular plane and *cc* is in the plane

places in doubt quadrupole coupling constants obtained by assuming cylindrical symmetry. Dymanus[117] has compared his experimental results with the results of several *ab initio* calculations. The agreement between experimental and theoretical one-electron properties for water is very good. The electric dipole moment of H_2O and D_2O have been measured in MBER experiments[167] to obtain moments for the rotationless molecules. This was accomplished by using the dipole moment function of Clough *et al.*[254]. This is necessary due to the large centrifugal distortion present, and the fact that only high J transitions are accessible to microwave Stark measurements. The rotationless moments obtained are $\mu_0(H_2O) = 1.8546(4)$ and $\mu_0(D_2O) = 1.8545(4)$ D.

Hydrogen sulphide, H_2S, has been studied by beam maser[165] and MBER[256] techniques. Spin–rotation constants and eqQ(D) have been obtained, but sufficient numbers of transitions have not been observed to determine tensor elements. Assuming cylindrical symmetry[165], $eqQ(D) = 154.7(16)$ kHz. The MBER study of H_2S is most notable because of the frequency of the transitions observed. This investigation extended MBER techniques into the submillimetre region of the spectrum and clearly demonstrated that the only frequency limits on this type of spectroscopy are availability of suitable radiation sources and the inherent limitation of the radiative lifetime of the states involved. This work also answered many questions on the feasibility of observing MBER transitions at zero field and on the requirements for observing very high frequency transitions using separated oscillating fields (Ramsey resonance). The beam maser used to study HCN has also recently been extended to submillimetre wavelengths[255].

Hydrogen selenide is sufficiently heavy that more transitions are available than in H_2O and H_2S. Dymanus and co-workers[249, 257] have determined the spin–rotation and eqQ tensor elements listed in Table 2.15.

Methylene fluoride (CH_2F_2) has been studied by Kukolich and co-workers using maser spectroscopy both at zero field[258, 259] and in 100 kG magnetic fields[260]. The first study provided the components of the spin–rotation tensor. The Zeeman study provided rotational magnetic moments and susceptibility anisotropies from which accurate molecular quadrupole moments were obtained. In addition, magnetic shielding, or chemical shift, anisotropies

Table 2.15 Tensor properties[249,257] of H_2Se

	aa	*bb*	*cc*
M_{gg}(H)/kHz	−38.77(10)	2.37(49)	−8.95(95)
M_{gg}(D)/kHz	0.51	−2.46(1)	−2.13(2)
$eq_{gg}Q$(D)/kHz	123.6(6)	−56.5(60)	−66.8(71)

were observed. The results are not in agreement with values obtained from n.m.r. studies of molecules oriented in liquid crystals and raise questions about the nature of this property in the condensed phase compared with isolated molecules. Table 2.16 summarises the results of the CH_2F_2 beam experiments.

Table 2.16 Tensor properties[259] of methylene fluoride*

	aa	*bb*	*cc*	*Ref.*
M_{gg}(H)/kHz	6.5(30)	1.0(14)	0.3(13)	258
M_{gg}(D)/kHz	−20.4(10)	−4.9(5)	−13.1(5)	258
$3(\chi_{gg} - \chi_{av})$/erg kG^{-2} mol^{-1}	−4.16(2)	0.58(1)	3.59(2)	260
G_{gg}	−0.725	−0.0411	−0.0398	260
$\Theta_{gg}/10^{-26}$ e.s.u.	−3.95(14)	2.05(16)	1.90(18)	260

* $eq_{zz}Q$(D) = 186(10) kHz where z is the C—D axis, $\eta = (q_{xx} - q_{yy})/q_{zz} = -0.15(5)$.

The last of the C_{2v} molecules considered is furan (CH═CH—O—CH═CH, the two terminal CH groups joined) which has been studied by Thadeus *et al.* using maser spectroscopy[261]. This work is significant to beam spectroscopy since it is now obvious that these techniques need not be limited to small molecules with intense transitions. The transitions observed have absorption intensities of the order of 10^{-8} cm^{-1}, line strengths which are generally considered to be weak for microwave absorption spectroscopy. The combination of nozzle beam sources, improved electronics (particularly in the area of mixer design) and the use of computer averaging make maser studies feasible for large complicated molecules. The success of the furan work also depended on the ability to analyse the spectra which could not have been done without the use of an extremely large digital computer. This research group is continuing to work on large molecules related to furan, and primary results are available for pyrrole[262]. There are only two different types of hydrogens in furan and the spin–rotation tensor elements have been determined for each. The spectral data were first fitted with spin–rotation constants, using the calculated spin–spin interaction, but the results were not as good as the experimental measuring accuracy. Using spin–rotation constants plus the three independent proton–proton separations which determine the spin–spin interaction resulted in a significantly better statistical fit of the data. The resulting proton–proton distances are close to, but not in complete agreement with, the distances obtained from the rotational constant data. It is normally assumed that the tensor portion of the electron-coupled spin–spin interaction is negligibly small in molecules containing only light atoms[77]; this may not be correct in furan. In addition, spin–spin interactions depend on the vibrational average of $1/R^3$ while rotational constants depend on the vibrational average of $1/R^2$. It is not clear whether the differences observed are significant. The availability of many spin–spin constants does point out the ability to obtain molecular geometry from this source which could be very useful in special cases. The spin–rotation tensor elements obtained for furan are as follows: protons 1 (α to oxygen), $M_{aa} = -537(3)$, $M_{bb} = 611(5)$, $M_{cc} = -548(13)$; and protons 3 (β to oxygen), $M_{aa} = 253(4)$, $M_{bb} = -189(5)$, $M_{cc} = -512(11)$ Hz.

The remaining asymmetric top molecules, which do not fit into the molecular point group classifications used, are: fluoroformaldehyde (HFCO)[248], formamide (NH_2CHO)[263], isocyanic acid (HNCO)[264] and formic acid (HCO_2-H)[221, 265]. Fluoroformaldehyde has already been discussed. Formamide is also related to formaldehyde as well as to ammonia since it is an addition product of these two molecules. Formamide's nitrogen quadrupole coupling diagonal tensor elements have been determined to be: $eqQ_{aa} = 1960(2)$, $eqQ_{bb} = 1888(3)$, and $eqQ_{cc} = 3848(4)$ kHz. Isocyanic acid involves nitrogen in an unusual bonding configuration and hyperfine data are available for one transition. Assuming cylindrical symmetry, $eqQ(\mathrm{D}) = 345(2)$ kHz. Formic acid has provided difficulties in spectral interpretation and several errors exist in the literature[221, 232, 265]. The most recent work[221], however, provides extensive eqQ_{D} information, including the diagonal elements of the tensor for the acidic deuterium; $eq_{aa}Q = -119.3(20)$, $eq_{bb}Q = 267.5(30)$, and $eq_{cc}Q = -148.2(20)$ kHz. Transforming to the bond axis system shows small deviations from cylindrical symmetry with $eq_{xx}Q = -148$, $eq_{yy}Q = -124$, and $eq_{zz}Q = 272$ kHz. The eqQ value for the carbon-bonded deuterium is 166(2) kHz, assuming cylindrical symmetry.

2.4.7 Discussion

The polyatomic molecules that have been studied by beam spectroscopy cover a wide range of chemical types. The molecular properties obtained from these studies can be viewed in a variety of ways. Different molecular constants provide different pictures of the electronic distribution of the molecule. For example, nuclear quadrupole coupling constants depend on $1/r^3$ and describe details of the electronic distribution close to the nucleus in question; electric dipole moments, of course, provide the first moment of the charge distribution; magnetic susceptibilities can be related to the second moment of the charge distribution, etc. The relationship between different properties such as spin–rotation, magnetic shielding, rotational magnetic moments and magnetic susceptibilities has already been discussed for diatomic molecules. Obviously, similar relationships hold for polyatomic molecules. An example of gaining additional information in this manner is the use of the spin–rotation constants of methyl fluoride to obtain magnetic shielding information[169]. The application of phosphine data to determining an absolute chemical shift scale for ^{31}P is an excellent example of this[223, 226]. One of the most useful applications of measured properties is the determination of molecular quadrupole moments. A detailed review of these aspects of magnetic and hyperfine data is provided by Flygare and Benson[45].

One molecular property which is available primarily from molecular beam studies is deuterium quadrupole coupling constants. eqQ_D attracts attention because a large percentage of molecules contain hydrogen and because chemically useful information has been obtained from alkali, halogen and nitrogen atom quadrupole coupling constants via the Townes and Dailey theory[102]. Deuterium quadrupole couplings are quite small both because of the small nuclear quadrupole moment of deuterium and the fact that the electronic distribution at hydrogen atoms is dominated by s orbitals and is, therefore, rather spherical. Table 2.17 lists sixteen examples of eqQ_D. The

Table 2.17 Deuterium quadrupole coupling constants (in kHz)

DCO_2H	166
CD_3CH_3	167
CD_3CN	166
CF_3D	171
CD_3CCH	176
HDCO	163
CH_3D	191
CD_2F_2	186
DCN	200
DCCD	202
CH_3OD	303
HCO_2D	272
HDO	308
NH_2D	291
DF	354
DCl	187

first six of these have been listed together in the literature[221] to point out that this quantity does not appreciably vary in atoms bonded to carbon. However, the seventh entry, deuteromethane, upsets this picture. It is particularly difficult to reconcile the difference between methane (191 kHz) and ethane (167 kHz) with any simple model. Yet, there certainly are similarities; DCN *vs.* DCCD or the OD-containing compounds, for example. *Ab initio* calculations have been carried out on the smaller molecules in Table 2.17, including CH_3D [266], HDO [267], HDCO [245], HF [133, 134] and HCl [133, 134]. The agreement obtained for eqQ_D is generally good, within five or ten percent. But these calculations are not readily extended to larger molecules. With the general availability of high quality semi-empirical calculations it appears that it is appropriate to consider a systematic study of deuterium quadrupole coupling constants by this means. Coordination between such a theoretical study and the choice of new molecules to study would be very useful at this time.

Salem has provided a relationship between eqQ and the vibrational force constant in diatomic molecules[268]. Flygare[269] and Kukolich[265] have investigated the application of this expression to polyatomic molecules and have found only marginal success. One difficulty encountered is the variation among force constants obtained from different assumed force fields.

For nuclei that do not lie on a symmetric top axis, *eqQ* can provide information about the symmetry of the electronic distribution of the chemical bond. This information is available only if the three diagonal elements of the coupling tensor have been determined. These data are available for ammonia[219-221], formaldehyde[244], water[117], difluoromethane[258-260] and the acidic deuteron in formic acid[221]. The electron distribution at the deuteron in NH_2D is essentially cylindrically symmetric, as expected from simple considerations. In formaldehyde, the deviations from cylindrical symmetry are significant, and the OD bond in water and formic acid are both quite nonsymmetric. A greater number of examples are clearly necessary to obtain better understanding of chemical bonding from this source. In addition, the assumption of cylindrical symmetry which is often used to obtain *eqQ* for bond axes from minimal data must be made with care.

One final application of high-resolution spectroscopic data obtained from beam measurements should be mentioned. The significant number of polyatomic molecules observed by radio astronomy in extraterrestrial space creates a need for very specific information. Hyperfine patterns can be important for unequivocal identification. Observed population inversions require detailed information on intermolecular potential functions for proper explanation. Finally, the collisionless atmosphere of molecular beams provide an ideal laboratory technique to attempt to study the presently unknown species, such as Xogen, detected by radio astronomy[271].

2.5 STUDIES OF WEAKLY BOUND COMPLEXES

Dimers and polymers of stable molecules occur very frequently in nature. Understanding the collective properties of molecules often requires knowing the degree of polymerisation and the nature of the polymers. For example, statistical theories of liquid water[272] are based on the intermolecular potential of two water molecules. Thus, it is useful to know the structure of the water dimer in constructing this potential. Theories concerning the nature of condensed phases, intermolecular potentials and energy transfer processes in lasers often involve dimeric species or the intermolecular forces involved in these complexes.

Our discussion below is complementary to a more extensive review[273] of van der Waals molecules and hydrogen-bonded molecules in this volume.

The problem in studying these molecules is that their stabilisation energies are generally low, 24 kcal mol^{-1} for Li_2 down to less than 1 kcal mol^{-1} for van der Waals molecules. It is often difficult to generate these species, and the conditions which produce dimers frequently produce a broad distribution of polymers, particularly for hydrogen-bonded species. Conventional techniques of vibrational spectroscopy and electron diffraction are made difficult by the complicated mixtures of polymers which often occur. Conventional microwave absorption spectroscopy would generally be the method of choice for obtaining the structure of polar dimers. However, producing a measurable concentration of dimers generally involves a high concentration of monomers Then collision (e.g. dipolar) broadening[274] washes out the spectrum.

Molecular beam spectroscopy, particularly MBER experiments utilising a universal detector, is an ideal technique for solving these problems. The mass spectrometer permits analysis of complicated mixtures, and the collision-free environment of a molecular beam eliminates collision-broadening of spectra. By the use of supersonic nozzle beams, even loosely bound van der Waals molecules can be produced in reasonable concentrations[275]. From microwave and r.f. spectra, the moments of inertia, electric dipole moments and nuclear hyperfine interactions can be obtained. Since the charge distribution of the monomer 'sub-molecule' is often only slightly altered in the dimer, the dipole moments and hyperfine interactions can be calculated from the orientation of the monomer units relative to the dimer inertial axes. Thus many of the properties measured by beam spectroscopy can be used to determine structures. It should be noted that this is the reverse situation from most MBMR and MBER experiments, in which the molecular structure is very well known from other techniques, and the high spectral resolution of beam experiments used to obtain very precise values for some quantity, such as the electric dipole moment or nuclear hyperfine interactions.

2.5.1 Electric deflection experiments on dimers

Alkali halide dimers were originally observed in beam experiments as complications in MBMR lineshapes[276]. The lithium halide[277] and TlF[278] dimers have been found to be non-polar by molecular beam electric deflection experiments, and presumably have planar, rhombic structures. Recent electric deflection experiments on a wide range of loosely bound complexes are presented in Table 2.18. The possibilities for future spectroscopic and structural studies can be seen from this table.

Table 2.18 The polarity of van der Waals molecules determined by molecular beam electric deflection experiments[275, 279, 280]

Polar molecules	*Non-polar molecules*
ArNO	$(NO)_{3,4}$
ArHCl	NeHCl
NeDCl	COH_2
XeHCl	$(BF_3)_{2,3}$
ArHF	
$ArBF_3$	
$KrBF_3$	$(HF)_{3-6}$
$(NO)_2$	$(H_2O)_{5,6}$
BF_3NO	ArXe
BF_3CO	$(ethylene)_2$
$(CO)_2$	
$(HF)_2$	
$(H_2O)_2$	
$(NH_3)_2$	
$(HCl)_2$	
Buta-1,3-diene–ethylene	
$(Cl_2)_2$	

A detailed discussion of the results in Table 2.18 can be found in the references, but a few generalisations can be made. The largest electric dipole moments obviously occur in complexes whose 'sub-molecules' have large dipole moments and are oriented in such a way as to prevent cancellation of the moment (note the alkali halide dimers where cancellation occurs). Molecules such as ArXe and $(CO_2)_2$, in which charge transfer or molecular quadrupole-induced dipoles are the mechanisms which produce the dipole moment of the complex, have small dipole moments. Finally, if one of the sub-molecules is a hydride, large-amplitude motion of the hydrogen must be considered. This is demonstrated[275] by the behaviour of NeHCl, which is apparently non-polar, implying that the vibration of the HCl sub-molecule is nearly isotropic with respect to the intertial axes of the complex. The Ne–HCl potential is not completely isotropic, however, since NeDCl does show a small refocusing. The deuterium has less vibrational amplitude than the proton, and the angular dependence of the potential is not completely averaged out (see Ref. 275 for a more elegant interpretation).

2.5.2 Alkali metal dimers

The alkali-metal dimers have dissociation energies from 24 kcal mol^{-1} for Li_2 to 10 kcal mol^{-1} for Cs_2 and can be produced by effusion as well as supersonic nozzle sources. Logan *et al.*[281] have determined the magnitude and sign of the quadrupole coupling interaction from the nuclear resonances (MBMR) of the homonuclear alkali dimers (Table 2.19). Brooks *et al.*[282, 283] used MBMR experiments involving multiple quantum transitions to measure g_J for several alkali-metal dimers (Table 2.19). Since the alkali dimers have

Table 2.19 Alkali dimer data from MBMR and MBER experiments

AB	g_J/nuclear magnetons	$g_J(H_2)\times M_p / M_{alk}$	eqQ_A/ MHz*	eqQ_B/ MHz*	μ/D§	*Ref.*
$^{7}Li_2$	0.10797(11)	0.12677	+0.060			281, 282
$^{23}Na_2$	0.03892(10)	0.03868	−0.423			281, 282
$^{39}K_2$	0.02163(6)	0.02283	−0.158			281, 282
$^{85}Rb_2$	0.00953(4)	0.01047	−1.10			281, 282
$^{85}Rb^{87}Rb$	0.00940(6)					282
$^{87}Rb_2$			−0.580			281
$^{133}Cs_2$	0.00547(3)	0.00669	+0.23			281, 282
$^{23}Na^{7}Li$				0.028(4)†	0.463(2)	284
$^{39}K^{7}Li$				$\leqslant$0.009	3.45(10)	284
$^{85}Rb^{7}Li$			−9.116(14)	0.030(13)	4.00(10)	284
$^{39}K^{23}Na$	0.0253(2)		−0.718(2)	0.171(3)‡	2.76(10)	283, 284
$^{23}RbNa$					3.1(3)	284
$^{133}Cs^{23}Na$					4.75(20)	284

* If no sign is given, it has not been determined

† Ref. 283 gives 0.054(5) MHz from MBMR

‡ Ref. 283 gives 0.134(8) MHz from MBMR

§ Except for NaLi, the dipole moments were calculated from μ^2/B with r_0 estimated from Badger's rule (Ref. 284)

g_J of $0.1\mu_N$ or less, it was necessary to use multiple quantum transitions ($\Delta\mu_J > 1$) to ensure a large enough change in the magnetic moment to cause an appreciable deflection in the beam. They found that g_J values for the homonuclear alkalis were approximately given by $g_J(\text{alkali}_2) = g_J(H_2) \times M_{\text{proton}}/M_{\text{alkali}}$. This can be interpreted with a hydrogenic model in which the only effect of the closed electron shells is to reduce the effective nuclear charge to $+1$, and in which the valence electrons are distributed as in hydrogen. The contribution of the $+1$ ions is easily shown to be μ_N/M_{alkali} where μ_N is the nuclear magneton. Apart from a $1/M_{\text{alkali}}$ factor, the contribution of the valence electrons to g_J will be similar to hydrogen, and the above equation results. Column 2 in Table 2.19 shows the agreement with experiment.

Dagdigian and Wharton[284] have produced supersonic nozzle beams of heteronuclear alkali dimers and have obtained their r.f. spectra with MBER experiments. The spectra were analysed for quadrupole coupling constants and for electric dipole moments (Table 2.19). Badger's rule[285] was used to estimate r_e, which was in turn used to obtain μ_0 from the experimental values of μ_0^2/B_0. With the exception of NaLi, whose B_0 is known from optical data, the error limits for the dipole moments are due to the uncertainty in B_0. Values of B_0 from optical data or MBER microwave experiments would reduce the uncertainty.

The trend in the electric dipole moments (Table 2.19) can be accounted for with an ionic model. For most of the alkali dimers, AB, the binding energy of the corresponding ion pair of lowest energy, A^+B^-, is around 10 kcal mol^{-1} (if the ion separation is taken as the dimer internuclear separation). This is an appreciable fraction of the actual binding energy of the dimers, and thus the electronic structure should have a significant ion-pair contribution. The binding energy of the reversed ion pair, A^-B^+, ranges from 0 to 40 kcal mol^{-1} less than for A^+B^-. Under the assumption that both ion states can contribute to the electronic structure of the molecule, it is reasonable to assume that the degree of ionic character, μ_0/eR, correlates with the difference in binding energy of the two ion pairs. This is shown in Figure 2.3. For example, Na^+Li^- is only 7.6 kcal mol^{-1} more stable than Na^-Li^+ and the resulting ionic character (3.4%) and dipole moment are low. Cs^+Na^- is 32 kcal mol^{-1} more stable than Cs^-Na^+ and the ionic character (27%) and dipole moment are much larger.

The quadrupole coupling constants derived from MBMR and MBER can be used to obtain the electric field gradient at the nucleus (Table 2.20), with known values of the nuclear quadrupole moment. As previously discussed (Section 2.3.3), the alkali nucleus in an alkali halide sees a field gradient much larger than $-2e/R^3$ owing to the polarisation of the electron shells of the alkali ion by the halide ion. Such an effect does not occur in the homonuclear alkali dimers, and their charge distribution is relatively spherical at the nucleus, giving a small eqQ. The heteronuclear alkali dimers do have some ionic character, and the field gradient near the positively charged nucleus is intermediate between the homonuclear q and that of a corresponding alkali halide.

Laser and white-light induced fluorescence from supersonic nozzle beams of Na_2 and K_2 have been studied by Sinha *et al.*[286]. They determined the vibrational temperature, 153(5) K, and rotational temperature, 55(10) K, for

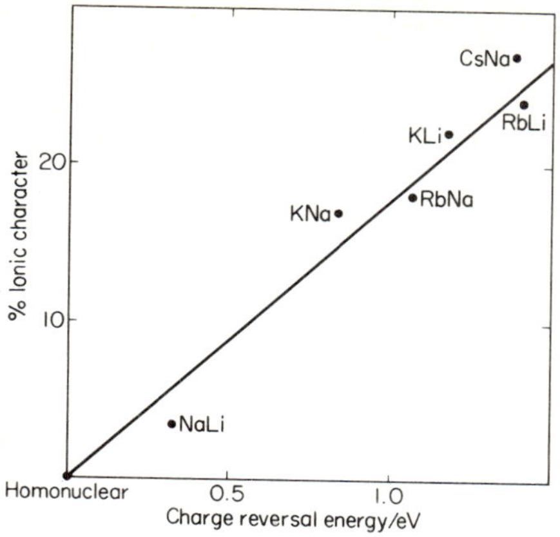

Figure 2.3 Correlation of the ionic character (defined as μ_0/er_0) with the charge reversal energy. (From Dagdigian and Wharton[284], by courtesy of the American Institute of Physics.)

Table 2.20 Electric field gradients at the nuclei of the alkali dimers*

Nucleus	*Molecule*	$q/10^{16}$ e.s.u.†
^{7}Li	$^{7}Li_2$	$-(2.20 \pm 0.16) \times 10^{-3}$
	$^{23}Na^{7}Li$	$(\pm)(1.04 \pm 0.13) \times 10^{-3}$
	$^{39}K^{7}Li$	$\leqslant(\pm)3.2 \times 10^{-4}$
	$^{85}Rb^{7}Li$	$(\pm)(1.10 \pm 0.45) \times 10^{-3}$
	$^{7}Li^{35}Cl$	$-(9.0 \pm 0.3) \times 10^{-3}$
^{23}Na	$^{23}Na^{7}Li$	$-(1.04 \pm 0.10) \times 10^{-2}$
	$^{23}Na_2$	$-(5.8 \pm 0.3) \times 10^{-3}$
	$^{39}K^{23}Na$	$(\pm)(2.33 \pm 0.19) \times 10^{-3}$
	$^{23}Na^{35}Cl$	$-(7.4 \pm 0.3) \times 10^{-2}$
^{39}K	$^{39}K^{7}Li$	$-(2.88 \pm 0.29) \times 10^{-2}$
	$^{39}K^{23}Na$	$-(2.01 \pm 0.19) \times 10^{-2}$
	$^{39}K_2$	$-(4.5 \pm 0.3) \times 10^{-3}$
	$^{39}K^{35}Cl$	$-(1.59 \pm 0.16) \times 10^{-1}$
^{85}Rb	$^{85}Rb^{7}Li$	$-(4.80 \pm 0.06) \times 10^{-2}$
	$^{85}Rb_2$	$-(5.77 \pm 0.06) \times 10^{-3}$
	$^{85}Rb^{35}Cl$	$-(2.76 \pm 0.03) \times 10^{-1}$
^{133}Cs	$^{133}Cs_2$	$-(1.07 \pm 0.10) \times 10^{-1}$
	$^{133}Cs^{19}F$	$-(5.8 \pm 0.6) \times 10^{-1}$

* The alkali halides are included for comparison

† From Ref. 285, after converting from atomic units to e.s.u. (1 a.u. = 0.324 123 × 10^{-16} e.s.u.). Ref. 285 used $Q(^{7}Li) = -0.0375 \times 10^{-24}$ cm^2, $Q(^{23}Na) = +(0.101 \pm 0.08) \times 10^{-24}$ cm^2, $Q(^{39}K) = +(0.049 \pm 0.04) \times 10^{-24}$ cm^2, $Q(^{85}Rb) = +(0.263 \pm 0.002) \times 10^{-24}$ cm^2 and $Q(^{133}Cs) = -(0.0030 \pm 0.0011) \times 10^{-24}$ cm^2

an Na_2 beam from a source nozzle kept at 920 K. Highly vibrationally excited dimers were not observed, although velocity distribution measurements by Gordon *et al.*[287] indicated that for Rb_2 and Cs_2, roughly half the heat of dimerisation goes into vibrational excitation.

Further fluorescence–molecular beam experiments by Sinha *et al.*[288] have shown Na_2 molecules to be aligned in supersonic nozzle beams. Measurements of the degree of polarisation gave the ratio of 2:3 for molecules with $\boldsymbol{J}$ parallel and $\boldsymbol{J}$ perpendicular to the beam axis. Johnson *et al.*[289] have reported on the alignment of I_2 molecules in a similar experiment, although the measured alignment was only a fraction of a percent. The physical model for the alignment results is that molecular orientations with the smallest inelastic or reactive collision cross-section along the beam axis are more prevalent. The effect is enhanced if a mixture of gases is expanded through a nozzle, since at least at the beginning of the expansion there is a velocity differential between molecules with different masses. This may explain the greater degree of alignment in Na_2 than in I_2, since the sodium nozzle beam contains a large fraction of monomer.

These experiments show considerable promise for measuring rotational and vibrational distributions in molecular beams by laser-induced fluorescence. Their application to crossed-beam and solid surface-beam experiments is an exciting prospect[290].

2.5.3 Hydrogen-bonded molecules

The structure of $(HF)_2$ has been determined by Dyke *et al.*[280] and the structure of $(H_2O)_2$ by Dyke and Muenter[291], both studies being carried out by microwave and r.f. MBER spectroscopy. These molecules are of great importance in hydrogen-bonding theories as model compounds for larger molecules and for comparison with the many extant *ab initio* calculations[292, 293]. Water dimer is important in constructing theories of liquid water[272] and has been observed in the atmosphere[294].

The HF sub-molecules in $(HF)_2$ can be considered as rigid ($D_e \approx 140$ kcal mol^{-1}). If the two monomers are labelled 1 and 2, two isoenergetic conformations exist, with either monomer 1 or monomer 2 as the proton donor molecule. These conformations can interchange by breaking the hydrogen bond and rotating the monomers (Figure 2.4a). This interchange is primarily a double-minimum proton motion and is feasible, since the barrier will not be larger than 4–5 kcal mol^{-1}, the hydrogen bond energy. This double-minimum motion causes a splitting of vibration–rotation energy levels, which is normally interpreted as the rate of barrier penetration. In $(HF)_2$, this tunnelling doubling is 19.776(12) GHz and drastically affects[280] the microwave spectrum. The permutation-inversion group theory, discussed by Longuet-Higgins[295], proved to be valuable in understanding the electric dipole selection rules and nuclear spin statistics for the tunnelling sub-levels.

Table 2.21 gives the spectroscopic constants from the microwave and r.f. MBER spectrum of $(HF)_2$, $(DF)_2$ and HF· · ·DF. These results show that the tunnelling doubling is much smaller for $(DF)_2$ than for $(HF)_2$. Since the motion is primarily that of the hydrogen, deuteration increases the reduced mass by a factor of two, causing the large decrease in tunnelling frequency[296]. In addition, the partially deuterated molecule does not show a tunnelling doubling since HF· · ·DF and DF· · ·HF are distinguishable and have different zero-point vibrational energies. Only the deuterium-bonded species was observed, implying that it is more stable than DF· · ·HF. The signal-to-noise ratio for HF· · ·DF was such that a factor of five fewer DF· · ·HF molecules would have been observed. This cannot be used to establish a bound on the energy difference, however, since the supersonic nozzle beam temperature is not easily characterised.

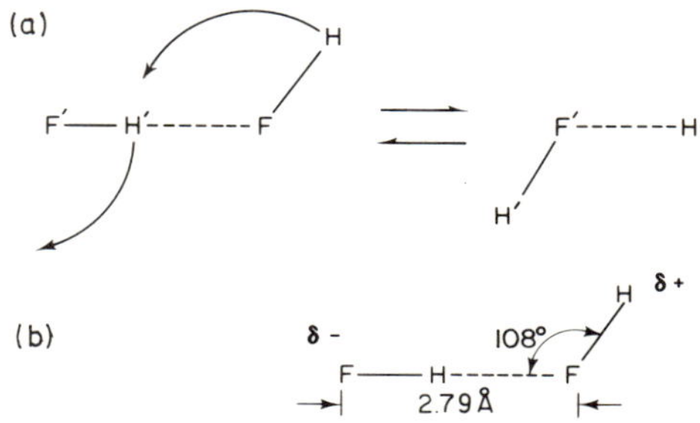

Figure 2.4 (a) Proton tunnelling in $(HF)_2$ has the effect of interchanging the proton donor and proton acceptor HF molecules. (b) The $(HF)_2$ structure. The charge polarisation shown would tend to enhance further hydrogen bond formation

Table 2.21 Hydrogen fluoride dimer spectroscopic constants* §

		$(HF)_2$	$(DF)_2$	HFDF
	$(B+C)/2$	6504(2)	6252.333(12)	6500.383(10)
	D_J		0.0699(15)	0.0603(9)
$K=0$	ν_0		1580.522(20)	
	a_0†		−0.645(8)	
	eqQ_D		0.110(8)	0.270(15)
	μ_a	2.9886(8)	2.9919(6)	3.029(3)
	$(B+C)/2$	6541(2)	6268.486(5)	6532.717(10)
	D_J		~0.06	0.0816(6)
$K=1$	$B-C$	97(1)	135.545(6)	91.23(2)
	ν_1	~32100	2062.796(12)	
	a_1†		−0.622(3)	
	b_1‡		−0.570(8)	
	μ_a	2.9886(8)	2.9454(2)	2.8896(2)

* All units MHz except μ_a which is in D
† 'a' gives the difference in $(B+C)/2$ for the two tunnelling states
‡ 'b' gives the difference in $B-C$ for the two tunnelling states
§ Data not found in Ref. 280 are to be published by those authors

The structure of HF dimer was determined with observed rotational constants and the following model. It was assumed that the monomer internuclear distance was unchanged on hydrogen bonding. From the values of $(B + C)/2$ for the various isotopic species, $R_{FF} = 2.79(5)$ Å and the proton acceptor makes an angle of 108(10)° with the F—F axis (Figure 2.4b). It should be noted that the error limits reflect uncertainties in the model and not in the experimental data.

The deuterium quadrupole coupling constants were found to be in agreement with the above results if the field gradient near the deuterium nucleus was assumed to be unchanged on hydrogen bonding. Then the component of the quadrupole coupling tensor along the *a*-inertial axis (virtually coincident with the F—F axis), due to one of the DF submolecules, is just $(eqQ_D)_{DF} \times [(3\cos^2\theta - 1)/2]$, where $(eqQ_D)_{DF}$ has the monomer value, and θ is the angle which DF makes with the *a*-inertial axis. For HF· · ·DF, θ is 23°, which is within the estimated vibrational amplitude of the deuteron, if its equilibrium position is on the F—F axis (i.e. a linear hydrogen bond). Because of the tunnelling motion, in the DF dimer $(eqQ_D)_a$ is the average of the quadrupole coupling interaction for the two different environments. Combining this value with HF· · ·DF $(eqQ_D)_a$, the deuteron acceptor submolecule is found to be at an angle of 119(5)°, in reasonable agreement with the angle obtained from the rotational constants.

The dipole moment along the F—F axis is measured to be 3.00 D. The sum of the dipole moment components along this axis for the two HF units is 2.38 D for a linear hydrogen bond and ∠FFH = 108°. This is enhanced by electrostatic effects such as dipole–induced-dipole terms and charge-transfer effects. An *ab initio* calculation[297] gives this enhancement as 0.40 D, and the resulting 2.78 D moment is in reasonable agreement with the experimental value. Increasing the FFH angle to 115° makes the calculated result 3.00 D.

Ab initio calculations are in good agreement with the above results, particularly for the F—F internuclear distance, given as 2.78 Å by Kollman and Allen[292]. The F—F internuclear distance in the dimer is 2.79 Å, compared with 2.49 Å for crystalline HF [298]. The decrease in F—F distance can be interpreted as evidence of a cooperative effect. The formation of one hydrogen bond causes a polarisation of charge (Figure 2.4b) which enhances the formation of another hydrogen bond, and thus causes a stronger and shorter hydrogen bond as the HF polymer adds more HF units. The ∠FFH = 108° in the dimer is very nearly tetrahedral and can be compared with the ∠FFF = 120° in crystalline HF [298].

The authors of this review are interested in the structure of the water dimer, and have used microwave and r.f. MBER spectra in their studies. The possible number of proton tunnelling motions for $(H_2O)_2$ is much larger than for $(HF)_2$. In addition to the proton donor–acceptor interchange as in $(HF)_2$, the protons of an H_2O submolecule can interchange by an internal rotation around the H_2O symmetry axis. Permutation-inversion group theory was invaluable in understanding the electric dipole selection rules for the tunnelling sub-levels (the *PI* group for water dimer is isomorphic with the D_{4h} point group).

In general, the tunnelling–rotational spectrum for the water dimer is quite complex. However, the group-theoretical results suggest that the doubly degenerate (*E*) tunnelling sub-levels have rigid-rotor type selection rules, at least for *a*-type transitions. Thus embedded in a complex tunnelling-rotational spectrum are series of transitions which follow a rigid-rotor pattern, i.e.

$$\nu = [(B + C)/2](2J + 2) \qquad (2.27)$$

for $\Delta J = +1$ and $\Delta K = 0$.

In the region of the spectrum from 8 to 50 GHz, over 50 transitions, irregularly spaced, have been observed for the water dimer. Two series of transitions obeying (2.27) have been assigned, corresponding to the doubly degenerate, E tunnelling states. In the microwave spectrum, $J = 2 \leftrightarrow J = 3$, $J = 3 \leftrightarrow J = 4$ for $K = 2$, and $J = 3 \leftrightarrow J = 4$ for $K = 3$ have been observed. In the r.f. spectrum, $K = 2$ asymmetry doubling transitions have been observed and assigned by their characteristic J^4 dependence. $\Delta M_J = \pm 1$ pure Stark transitions have been measured for the lowest few J states of $K = 2$, 3 and 4. Radiofrequency–microwave double resonance experiments were performed to double check the quantum number assignments. The spectral constants are given in Table 2.22.

Table 2.22 Water dimer spectroscopic constants* †

	$(H_2O)_2$		$(D_2O)_2$	
	E_1	E_2	E_1	E_2
$K = 2$				
$(B + C)/2$	6144.77(1)	6155.70(1)	5421(2)	5424(2)
D_J	0.0474(1)	0.0533(1)		
C_2‡	0.0091(2)	0.0355(2)	0.0073(2)	0.0274(2)
μ_a	2.5870(15)	2.6327(9)	2.6010(10)	2.6132(10)
$K = 3$				
$(B + C)/2$	6200(2)			
μ_a	2.5658(11)	2.6238(8)	2.5901(9)	2.6037(6)

* Data not found in Ref. 291 to be published by those authors
† All units in MHz, except μ_a, which is in D
‡ For a rigid rotor, $C_2 = (C - B)^2/(2A - B - C)$

The basic structure of $(H_2O)_2$ can be determined from the data in Table 2.22 and is shown in Figure 2.5. It is a good assumption that the H_2O submolecules have the same geometry as the free monomer. The values of $(B + C)/2$ for $(H_2O)_2$ and $(D_2O)_2$ allow two coordinates to be determined, and the component of the electric dipole moment along the O—O axis (*a*-inertial axis) allows a third coordinate to be determined since

$$(\mu_a)_D = \mu_0(\cos\theta_1 + \cos\theta_2) + \Delta\mu \qquad (2.28)$$

where $\cos\theta_1$ and $\cos\theta_2$ project the monomer dipole moments along the *a*-inertial axis, and $\Delta\mu$ is an enhancement due to electrostatic interactions (e.g. dipole–induced-dipole) and charge transfer. An *ab initio* calculation[299] gives $\Delta\mu = 0.53$ and is assumed to be accurate to ± 0.1 D. The result (Figure 2.5) is $R_{OO} = 2.99(1)$ Å, $\angle OH_3O = 0(5)°$ (linear hydrogen bond) and $\theta = 61(5)°$. It was also assumed that the molecular conformation possessed a plane of symmetry. The $K = 2$ asymmetry doubling constants support this. The conformation in Figure 2.5 is very nearly an accidental symmetric top and is consistent with the small asymmetry doubling constants for $(D_2O)_2$. The fact that the $(H_2O)_2$ asymmetry doubling is slightly larger, rather than much smaller, than that in $(D_2O)_2$ is caused by the greater vibrational amplitude of protons. Partial deuteration and ^{18}O studies in progress should help to refine the structure.

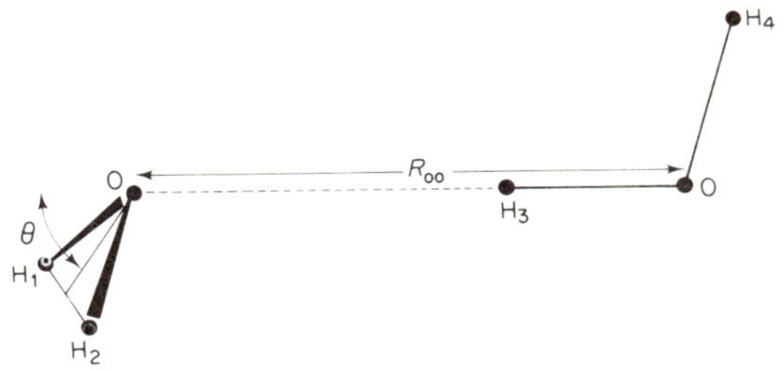

Figure 2.5 The water dimer structure. The conformation shown has a plane of symmetry. The proton acceptor angle is 61°, near the angle of 54.7° for tetrahedral directivity. $R_{OO} = 2.99$ Å and $\angle OH_3O \approx 0$. (From Dyke and Muenter[291], by courtesy of the American Institute of Physics.)

The agreement between *ab initio* calculations[293] and experiment is quite good. Most of the calculations give $R_{OO} \approx 3.0$ Å, a linear hydrogen bond and the conformation with a plane of symmetry. The angle θ (Figure 2.5) is generally calculated as considerably less than 60°, however. The experimental structures for $(H_2O)_2$ and $(HF)_2$ show several important trends. The proton acceptor molecule shows tetrahedral directivity in these molecules, and the heavy-atom internuclear distance decreases by 0.2–0.3 Å in the crystalline phase[298, 300]. Electron diffraction results[301] suggest that the lower polymers of HF also have shortened FF distances, 2.55 Å. This confirms the idea that a cooperative effect occurs on hydrogen bonding. The formation of one hydrogen bond causes a polarisation of charge which enhances the formation of further hydrogen bonds (Figure 2.4b). The result is stronger and shorter hydrogen bonds.

2.5.4 van der Waals molecules

Molecular complexes of an inert gas atom and a polar diatomic submolecule are bound together by electrostatic interactions and London dispersion forces and have stabilisation energies of < 1 kcal mol^{-1}. Using a simple nozzle with 0.03 mm orifice but no shaping, it is possible to generate molecular beams of these complexes[302-304]. Stagnation pressures of 1–3 atm are used, and the gas mixtures have a large excess of inert gas. The nozzle conditions are such that the effective internal temperatures are very low, roughly 20 K, since neither excited vibrational states nor rotational states with (prolate) $K > 0$ are observed. The structure of these complexes is determined from r.f.

and microwave MBER spectra. Because there is very little charge transfer in these complexes, the electric dipole moment and nuclear hyperfine interactions yield considerable structural information.

Novick *et al.*[302] have determined the ArHCl spectrum, and the spectral constants are given in Table 2.23. These results show that the HCl submolecule is undergoing large-amplitude vibrational motion but is not a free

Table 2.23 van der Waals molecules spectroscopic constants* †

Species	$(B + C)/2$	D_J	μ_a	eqQ_a
$ArH^{35}Cl$	1678.511(5)	0.0203(2)	0.8144(10)	−23.027(10)
$ArH^{37}Cl$	1631.566(5)			
$ArD^{35}Cl$	1657.596(10)		1.0036(7)	$-36.25(2)_{Cl}$ $+0.102(5)_{D}$
$ArD^{37}Cl$	1611.876(10)			
ArHF‡	3065.719(2)	0.0721(2)	1.332(2)	
$Ar^{35}ClF$	1327.113(5)	0.0047(2)	1.053(3)	−140.869(15)
$Ar^{37}ClF$	1319.650(5)	0.0047(2)		−111.053(15)

* Data from Refs. 302–304 † All units MHz except μ_a ‡ Spin–spin interaction, S_{HF} = 29.2(15) kHz

rotor with respect to the ArHCl inertial axis. The rotational constant $(B + C)/2$ primarily gives information about $1/R^2$, where R is the Ar—Cl distance, and about $\langle\cos^2\theta\rangle$, where $\theta = \angle$ArClH. Furthermore, the component of the electric dipole moment and the chlorine quadrupole coupling constant along the a-inertial axis (Ar—Cl axis) are given by

$$(eqQ)_a = \frac{3\cos^2\theta - 1}{2}\, eqQ$$

$$\mu_a = \mu\langle\cos\theta\rangle + \alpha_{Ar}\left[\frac{2\mu\langle\cos\theta\rangle}{R^3} + \frac{3}{2}\Theta\frac{(3\langle\cos^2\theta\rangle - 1)}{R^4}\right] \qquad (2.29)$$

eqQ, μ and Θ are the chlorine or deuterium quadrupole coupling constant, dipole moment and molecular quadrupole moment of the free HCl (DCl) molecule. The result of this analysis is shown in Figure 2.6.

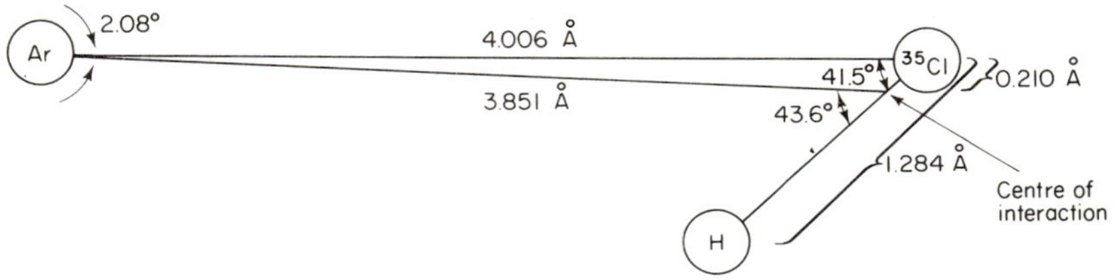

Figure 2.6 Average structures of ArHCl. The distance between Ar and the centre of interaction is relatively invariant to isotopic substitution. The angle is the average of $\cos^2\theta$, chosen as the best rigid approximation to a distribution in angle. The position of the 'centre of interaction' is obtained by calculating the position of that point (assumed to be on the H—Cl axis) whose distance to all three atoms is equal in ArHCl and ArDCl. (From Novick *et al.*[302], by courtesy of the American Institute of Physics.)

The ArClH and ARClD angles are found to be acute. From $\langle\cos\theta\rangle$, $\angle$ArClD is 33.5° and $\angle$ArClH is 47.5°. This difference is reasonable since these are vibrationally averaged quantities, and a vibrational potential which has $\angle$ArClH = 0 at equilibrium gives a reasonable fit to $\langle\cos\theta\rangle$ and $\langle\cos^2\theta\rangle$. The Ar—Cl distance is 4.01 Å and differs by about 0.02 Å for ArHCl and ArDCl. If R is computed from a point 0.21 Å from the chlorine nucleus along the HCl axis, a value of 3.851 Å is found, and the difference between ArHCl and ArDCl is only 0.0015 Å. This 'centre of interaction' (C.I.) is in agreement with other calculations and experiments[305] and is represented in Figure 2.6. Further, the Ar—C.I. distance of 3.851 Å is very close to the sum of the van der Waals radii for Ar and Cl.

The centrifugal distortion constant, D_J, can be used to estimate the Ar—HCl vibrational stretching frequency, since the observed change in $\angle$ArClH with J is very small. Then from the diatomic molecule formula,

$$\omega_s = \sqrt{\frac{4B^3}{D_J}} = 32.2\ \mathrm{cm^{-1}} \tag{2.30}$$

This value agrees very well with a calculated value of 32 $\mathrm{cm^{-1}}$ by Bratoz and Martin[306] and 32.4 $\mathrm{cm^{-1}}$ from inelastic scattering studies by Farrar and Lee[307].

Harris *et al.*[303] have found the structure of ArHF to be very similar to that of ArHCl (see Table 2.23). The Ar—HF stretching frequency calculated from D_J is 42.1 $\mathrm{cm^{-1}}$ and again agrees with the calculation of Bratoz and Martin[306], which gives 45 $\mathrm{cm^{-1}}$. The Ar—F distance in the vibrational ground state is 3.540 Å. Using a potential which fits ω_s and $\langle\cos\theta\rangle$, the equilibrium Ar—F distance is found to be 3.458 Å and the potential well-depth 116 $\mathrm{cm^{-1}}$.

By analyzing the HF nuclear spin–spin interaction, the value of $\langle 1/R_{HF}{}^3\rangle$ was calculated and found[68] to be the same as for free HF. The upper limit to the change in HF internuclear distance upon formation of ArHF is 0.015 Å.

The non-hydride ArClF molecule does not have the large-amplitude vibrational motion associated with molecules such as ArClH and ArHF, and the structure averaged over the vibrational ground state is very close to the equilibrium geometry. Harris *et al.*[304] have found ArClF to have a nearly linear average structure, with Cl in the middle and with $\angle$ArClF = 168.87(1)° (determined from the chlorine quadrupole coupling constant). The structural parameters are in Table 2.24. From the centrifugal distortion parameter, the Ar—ClF stretching frequency is calculated to be 47 $\mathrm{cm^{-1}}$, assuming the diatomic molecule relation between D_J, B and ω_s [equation (2.30)]. Table 2.24 shows that the ArClF angle is slightly different for ^{35}Cl and ^{37}Cl. The ratio of these angles can be calculated under the assumption of a linear equilibrium geometry and harmonic-oscillator vibrational wavefunctions. Since this

Table 2.24 ArClF **structural constants**[304]

	Ar^{35}ClF	Ar^{37}ClF
ω_s	47.2(10) $\mathrm{cm^{-1}}$	46.5(10) $\mathrm{cm^{-1}}$
R_0	3.3301(1) Å	3.3290 Å
R_e*	3.286 Å	
π-$\angle$ArClF†	11.134(12)°	11.064(12)°

* Calculated from the 6–12 potential which reproduces ω_s and R_0
† Actually $\cos^{-1}(\langle\cos^2\psi\rangle)^{\frac{1}{2}}$ where $\psi = \pi - \angle$ArClF

harmonic oscillator model gives the proper mass dependence of the supplement of the ArClF angle, ψ, it can be shown that

$$\langle \psi^2 \rangle \approx \hbar / m_b \omega_b \tag{2.31}$$

where m_b is the reduced mass appropriate for the bending motion, and the bending frequency, ω_b, is found to be 40 cm^{-1}.

The structure for ArClF predicted by adding Ar—Cl and Ar—F Lennard-Jones or Morse potentials gives an ArClF angle far from linearity, so the assumption of pairwise additivity of atom–atom potentials to represent van der Waals interactions in a molecule such as ArClF breaks down. In understanding the conformation of large molecules, the stereospecificity of van der Waals interactions may play an important role, and further experimental studies are needed, since simple theories involving pairwise additivity of atom–atom potentials would lead to incorrect results.

The alkali–inert gas diatomic KAr has been studied by Mattison *et al.*[308] with MBMR spectroscopy. This type of molecule is present in optical pumping experiments of alkali metals in an inert-gas buffer. In the beam experiment a 0.03 mm nozzle with 100 Torr potassium and 2000 Torr argon stagnation pressure was used to generate KAr. It was possible to remove the alkali atoms from the beam by magnetic deflection. Although K and KAr both have the same magnetic moment and virtually the same velocity from a supersonic nozzle, the lighter atom is more easily deflected. A Zeeman transition was observed at essentially a free atomic resonance frequency for weak magnetic fields ($ha\boldsymbol{I}\cdot\boldsymbol{S} > g_s m_0 \boldsymbol{S}\cdot\boldsymbol{H}$). Fine structure was observed due to the spin–rotation term

$$h\gamma \boldsymbol{S}\cdot\boldsymbol{N} \tag{2.32}$$

where $\boldsymbol{S}$ is the electron spin, $\boldsymbol{N}$ the molecular rotational angular momentum, and γ the coupling constant; $\bar{\gamma}$ was determined as 0.24(1) MHz. $\bar{\gamma}$ is an average value for all the vibration–rotation states present in the KAr beam, and further analysis may lead to the dependence of γ on internuclear separation. In recent work[310] changes have been observed in the hyperfine coupling constant in KAr.

2.6 CONCLUSIONS

As we have shown in this review, a large body of very precise experimental information concerning molecular properties has been obtained by molecular beam spectroscopy. From the initial experiments of the late 1930s, this field has developed a variety of techniques and capabilities. Molecular beam spectroscopy is no longer a narrow, specialised method but is now a general experimental technique.

Applications of beam spectroscopy cover an extremely wide range of interests, including basic atomic physics experiments looking for the polarisation of nuclei, high-resolution spectroscopy seeking to define better the electronic structure of small molecules, quantum state selection in molecular beam scattering experiments, and the study of weak chemical interactions such as hydrogen and van der Waals bonds. Universal beam detectors and supersonic nozzle beam sources, introduced relatively recently, have certainly not been fully exploited. The rapid growth and development of new experiments during the past five years should continue.

2.7 APPENDIX

Properties of $^{16}O^{12}C^{32}S$, $^{16}O^{13}C^{32}S$ and $^{16}O^{12}C^{34}S$

	$^{16}O^{12}C^{32}S$	*Ref.*	$^{16}O^{13}C^{32}S$	*Ref.*	$^{16}O^{12}C^{34}S$	*Ref.*
μ, $v_2 = 0$/D	0.71519(3)	170, 173	0.71531(4)	170	0.71541(4)	170
μ, $v_2 = 1$/D	0.70433(3)	171	0.70480(3)	172	0.70456(3)	172
μ, $v_2 = 2$/D	0.6936(3)	175				
C/KHz			3.1(2)	170		
g, $v_2 = 0$	0.028839(6)	170	0.028710(15)	170	0.028242(10)	170
$g_\perp$, $v_2 = 1$	0.02930(4)	171				
$g_\parallel - g_\perp$, $v_2 = 1$	0.0905(5)	171				
$g_{xx} - g_{yy}$, $v_2 = 1$	$2.3(3) \times 10^{-4}$	171				
$\chi_\parallel - \chi_\perp$, $v_2 = 0$/kHz kG^{-2}	−2.348(3)	170	−2.360(9)	170	−2.342(5)	170
$\chi_\parallel - \chi_\perp$, $v_2 = 1$/kHz kG^{-2}	−2.38(10)	171				
$\chi_{xx} - \chi_{yy}$, $v_2 = 1$/kHz kG^{-2}	−0.03(3)	171				
θ, $v_2 = 0/10^{-26}$ e.s.u.	−0.786(14)	170	−0.716(40)	170	−0.858(23)	170

Properties[172] of $^{18}O^{12}C^{32}S$ and $^{16}O^{12}C^{33}S$

	μ/D	eqQ/kHz	eqQ'/kHz	$C_\perp$/kHz	$C_\parallel - C_\perp$/kHz	$C_{xx} - C_{yy}$/kHz
$^{18}O^{12}C^{32}S$, $v_2 = 0$	0.71450(3)					
$^{18}O^{12}C^{32}S$, $v_2 = 1$	0.70367(3)					
$^{16}O^{12}C^{33}S$, $v_2 = 0$	0.71536(21)	29118.4(12)		0.87(5)		
$^{16}O^{12}C^{33}S$, $v_2 = 1$		28682.5(18)	1180.8	1.16(8)	11.5(5)	−0.12(11)

MBER properties[177] of HCN

	μ/D*	$a_\parallel - a_\perp$/Å^3	eqQ(N)/kHz	eqQ(D)/kHz	C(H)/kHz	C(N)/kHz
HCN, $v_2 = 0$	2.98459(1)	1.0(2)	−4707.9(3)		4.3(1)	10.13(3)
HCN, $v_2 = 1$	2.94164(2)		−4808(2)†		−27.4(6)‡	12.1(3)§
HCN, $v_2 = 2$, $l = 0$	2.89865(11)		−4899(2)		−3.8(15)	10.7(8)
DCN, $v_2 = 0$	2.99020(2)		−4704.0(5)	202.2(5)	−0.6(3)	8.2(2)

* Uncertainty quoted is precision; absolute accuracy is ±0.05%

† eq_zQ; $eq_xQ - eq_yQ = 295(3)$

‡ If $C_\perp$(H) for $v_2 = 1$ equals C(H) for $v_2 = 0$, then $C_\parallel$(H) = 50(1) for $v_2 = 1$

§ If $C_\perp$(N) for $v_2 = 1$ equals C(N) for $v_2 = 0$, then $C_\parallel$(N) = 14 for $v_2 = 1$

LiOH rotational frequencies[178], $J = 1 \rightarrow J = 0$

Vib. state	^{7}LiOH/MHz	^{7}LiOD/MHz
000	70684.88(10)	62954.78(10)
100	69775.52(10)	
$02^{0}0$	70331.76(10)	
$12^{0}0$	69446.88(20)	
001	70481.6(10)	
101	69672(2)	

Properties[178] of LiOH

Vib. state	μ(LiOH)/D	μ(LiOD)/D	$eqQ(^{7}\mathrm{Li})$/kHz	$C(^{7}\mathrm{Li})$/kHz
000	4.755(2)	4.711(1)	295.8(15)	1.70(29)
$02^{0}0$	5.031(2)	4.89(1)	346.9(21)	1.18(40)
100	4.851(2)	4.80(1)	299.7(31)	1.17(58)
$01^{1}0$	4.899(5)	4.815(5)		
$03^{1}0$	5.133(5)	5.002(5)		

References

1. E.g., Rabi, I., Millman, S., Kusch, P. and Zacharias, J. (1939). *Phys. Rev.*, **55,** 529
2. Ramsey, N. F. (1955). *Molecular Beams* (London: Oxford U. P.)
3. Kusch, P. and Hughes, V. (1959). *Handbuch der Physik* (S. Flugge, editor) Vol. 37, part 1 (Berlin: Springer)
4. Zorn, J. C. and English, T. C. (1973). *Advances in Atomic and Molecular Physics* (D. Bates and I. Estermann, editors) Vol. 9 (New York: Academic); English, T. C. and Zorn, J. C. (1973). *Methods of Experimental Physics* (D. Williams, editor) Vol. 3 (New York: Academic)
5. Lovas, F. and Lide, D. (1971). *Advances in High Temperature Chemistry* (L. Eyring, editor) Vol. 3 (New York: Academic)
6. Ref. 2, Chap. 6.
7. Anderson, J., Andres, R. and Fenn, J. (1966). *Advan. Chem. Phys.*, **10,** 275
8. Dyke, T. R., Tomasevich, G. R., Klemperer, W. and Falconer, W. (1972). *J. Chem. Phys.*, **57,** 2277
9. See Section 2.5 of this review.
10. In the case of molecules with unpaired electrons the state selection can also be sensitive to electronic contributions to the total angular momentum. See Sections 2.3.8 and 2.3.9
11. Several different types of transition within a single rotational state can occur: pure Stark, Stark-hyperfine, inversion (as in ammonia), or *l*-type doubling transitions
12. Bennewitz, H. G., Paul, W. and Schlier, C. (1955). *Z. Phys.*, **141,** 6
13. Weiss, R. (1961). *Rev. Sci. Instr.*, **32,** 397
14. Kaufman, M. (1964). *Thesis*, Harvard University
15. Brink, G. O. (1966). *Rev. Sci. Instr.*, **37,** 857
16. Lee, Y. T., McDonald, J. D., LeBreton, R. R. and Herschbach, D. R. (1969). *Rev. Sci. Instr.*, **40,** 1402
17. A commercial beam detector suitable for spectroscopy is manufactured by Extra Nuclear Co., Pittsburgh, Pa.
18. Freund, R. S. and Klemperer, W. A. (1965). *J. Chem. Phys.*, **43,** 2422
19. An ultimate limit on high-frequency transitions is the radiative lifetime of the states involved
20. Dreschler, W. and Gräff, G. (1961). *Z. Phys.*, **163,** 165
21. deLeeuw, F. and Dymanus, A. (1973). *J. Mol. Spectrosc.*, **48,** 427
22. Davis, R. E. and Muenter, J. S. (1974). *J. Chem. Phys.*, **61,** 2940
23. Kaufman, M., Wharton, W. and Klemperer, W. (1965). *J. Chem. Phys.*, **43,** 943
24. Ref. 2, Chapters 13 and 14
25. Cederberg, J. W. and Ramsey, N. F. (1964). *Phys. Rev.*, **136,** 960

25a. This statement applies only to diamagnetic molecules

26. English, T. C. and MacAdam, K. B. (1970). *Phys. Rev. Lett.*, **24,** 555
27. Ref. 2, Section 5, 4
28. Ref. 2, Chapter 5
29. Thaddeus, P. and Krisher, L. (1961). *Rev. Sci. Instr.*, **32,** 1083
30. Although the energy, *hν*, involved in each transition decreases linearly with frequency, the sensitivity of a maser spectrometer falls off more slowly with decreasing frequency owing to the increasing sensitivity of electronic components.
31. Kukolich, S. G. (1965). *Phys. Rev.*, **138A,** 1322
32. Kukolich, S. G. (1973). *Chem. Phys. Lett.*, **20,** 519
33. Kukolich, S. G. (1970). *Chem. Phys. Lett.*, **5,** 401; Kukolich, S. G. and Casleton, K. H. (1973). *ibid.*, **18,** 408; see also Kukolich, S. G. (1971). *ibid.*, **12,** 216
34. Lainé, D. C. (1970). *Reports Progr. Phys.*, **33,** 1001
35. Townes, C. H. and Schawlow, A. L. (1955). *Microwave Spectroscopy*, Section 15-10 (New York: McGraw-Hill)
36. Huiszoon, C. (1971). *Rev. Sci. Instr.*, **42,** 477
37. DeLucia, F. and Gordy, W. (1969). *Phys. Rev.*, **187,** 58
38. Code, R. F., Falconer, W. E., Klemperer, W. and Ozier, I. (1967). *J. Chem. Phys.*, **47,** 4955
39. Büchler, A., Stauffer, J. and Klemperer, W. (1964). *J. Amer. Chem. Soc.*, **86,** 4544

40. Novick, S., Davies, P., Dyke, T. and Klemperer, W. (1973). *J. Amer. Chem. Soc.*, **95,** 8547
41. Honig, A., Mandel, M., Stitch, M. L. and Townes, C. H. (1954). *Phys Rev.*, **96,** 629
42. Ref. 35, Chapters 1, 6 and 8
43. Hebert, A. J., Lovas, F. J., Melendres, C. A., Holowell, C. D., Story, T. L. and Street, K. (1968). *J. Chem. Phys.*, **48,** 2824
44. Lucken, E. A. C. (1969). *Nuclear Quadrupole Coupling Constants*, (New York: Academic Press)
45. Flygare, W. H. and Benson, R. C. (1971). *Mol. Phys.*, **20,** 225
46. Dunham, J. L. (1932). *Phys. Rev.*, **41,** 721
47. Rittner, E. S. (1951). *J. Chem. Phys.*, **19,** 1030
48. Rice, S. A. and Klemperer, W. (1957). *J. Chem. Phys.*, **27,** 573; and (1956). *ibid.*, **26,** 618
49. de Wijn, H. W. (1966). *J. Chem. Phys.*, **44,** 810
50. Maltz, C. (1969). *Chem. Phys. Lett.*, **3,** 707
51. London, F. (1930). *Z. Phys.*, **63,** 245
52. Buckingham, A. D. (1965). *Discuss. Faraday Soc.*, **40,** 232
53. Matcha, R. L. (1970). *J. Chem. Phys.*, **53,** 4490; (1970). *ibid.*, **53,** 485; (1968). *ibid.*, **49,** 1264; (1968). *ibid.*, **48,** 335; (1967). *ibid.*, **47,** 5295; (1967). *ibid.*, **47,** 4595
54. McLean, A. D. (1963). *J. Chem. Phys.*, **39,** 2653
55. McLean, A. D. and Yoshimine, M. (1968). *IBM J. Res. Develop.*, **12,** 206
56. Yoshimine, M. (1968). *J. Phys. Soc. Jap.*, **25,** 1100
57. Wharton, L. and Klemperer, W. (1963). *J. Chem. Phys.*, **39,** 1881
58. Scharpen, L., Muenter, J. and Laurie, V. (1967). *J. Chem. Phys.*, **46,** 2431
59. English, T. C. and MacAdam, K. B. (1970). *Phys. Rev. Lett.*, **24,** 555
59a. MacAdam, K. B. and Ramsey, N. F. (1972). *Phys. Rev. A*, **6,** 898
60. Kaiser, E. W. (1970). *J. Chem. Phys.*, **53,** 1686
61. Muenter, J. S. (1972). *J. Chem. Phys.*, **56,** 5409
62. Davis, R. E. and Muenter, J. S. (1972). *J. Chem. Phys.*, **57,** 2836
63. Graff, J., Dagdigian, P. J. and Wharton, L. (1972). *J. Chem. Phys.*, **57,** 710
64. Hebert, A. J., Breivogel, F. W. and Street, K. (1964). *J. Chem. Phys.*, **41,** 2368
65. Breivogel, F. W., Hebert, A. J. and Street, K. (1965). *J. Chem. Phys.*, **42,** 1555
66. Holowell, C. D., Hebert, A. J. and Street, K. (1964). *J. Chem. Phys.*, **41,** 3540
67. Van Wachem, R. and Dymanus, A. (1967). *J. Chem. Phys.*, **46,** 3749
68. Muenter, J. S. and Klemperer, W. (1970). *J. Chem. Phys.*, **52,** 6033
69. Bender, C. F. and Davidson, E. R. (1969). *Phys. Rev.*, **183,** 23
70. Grimaldi, F., LeCourt, A. and Moser, C. (1968). *Symp. Faraday Soc.*, **2,** 59
71. Raymonda, J. W., Muenter, J. S. and Klemperer, W. (1970). *J. Chem. Phys.*, **52,** 3458
72. McLean, A. D. and Yoshimine, M. (1967). *Int. J. Quantum Chem.*, **1S,** 313
73. Raymonda, J. and Klemperer, W. (1971). *J. Chem. Phys.*, **55,** 232
74. Kaufman, M., Wharton, L. and Klemperer, W. (1965). *J. Chem. Phys.*, **43,** 943
75. Wharton, L. and Klemperer, W. (1963). *J. Chem. Phys.*, **38,** 2705
76. Melendres, C. A., Hebert, A. J. and Street, K. (1969). *J. Chem. Phys.*, **51,** 855
77. von Boekh, R., Gräff, G. and Ley, R. (1964). *Z. Phys.*, **179,** 285
77a. Buckingham, A. D. and Love, I. (1970). *J. Magn. Resonance*, **2,** 338
78. Wharton, L., Gold, L. P. and Klemperer, W. (1962). *J. Chem. Phys.*, **37,** 2149
79. Docken, K. K. and Hinze, J. (1972). *J. Chem. Phys.*, **57,** 4936
80. Freund, S. M., Herbst, E., Mariella, R. P. Jr. and Klemperer, W. (1972). *J. Chem. Phys.*, **56,** 1467
81. Wahl, A. C., quoted in Ref. 80
82. Meerts, W. and Dymanus, A. (1973). *Chem. Phys. Lett.*, **23,** 45
83. Cade, P. E. and Huo, W. (1966). *J. Chem. Phys.*, **45,** 1063
84. Meerts, W. and Dymanus, A. (1974). *Astrophys. J.*, **187,** L45
85. Neumann, R. M. (1970). *Astrophys. J.*, **161,** 779
86. Muenter, J. S. (1975). *J. Mol. Spectrosc.*, in press
87. Grimaldi, F., LeCourt, A. and Moser, C. (1967). *Int. J. Quantum Chem*., **1S,** 153
88. Wicke, B. G., Field, R. W. and Klemperer, W. (1972). *J. Chem. Phys.*, **56,** 5758
89. Green, S. (1972). *J. Chem. Phys.*, **57,** 2830
90. Schlier, C. (1961). *Fortschr. Phys.*, **9,** 455
91. Rice, S. A. and Klemperer, W. (1957). *J. Chem. Phys.* **27,** 573

92. Code, R. F. and Ramsey, N. F. (1971). *Phys. Rev.*, **A4**, 1945
93. Docken, K. K. and Hinze, J. (1972). *J. Chem. Phys.*, **57**, 4936
94. Wharton, L., Gold, L. P. and Klemperer, W. (1964). *Phys. Rev.*, **133**, B270
95. Hillborn, R. C., Gallagher, T. F. and Ramsey, N. F. (1972). *J. Chem. Phys.*, **56**, 955
96. Kusch, P. (1949). *Phys. Rev.*, **75**, 887
97. DeLeeuw, F. H., van Wachem, R. and Dymanus, A. (1970). *J. Chem. Phys.*, **53**, 981
98. Zorn, J. C., English, T. C., Dickinson, J. T. and Stephenson, D. A. (1966). *J. Chem. Phys.*, **45**, 3731
99. van Wachem, R. and Dymanus, A. (1967). *J. Chem. Phys.*, **46**, 3749
100. Fuller, G. H. and Cohen, V. W. (1969). *Nuclear Data Tables*, **A5**, 433
101. Ref. 35, p. 155
102. Townes, C. H. and Dailey, B. P. (1949). *J. Chem. Phys.*, **17**, 782
102a. Buckingham, A. D. (1962). *Trans. Faraday Soc.*, **58**, 1277
103. Sternheimer, R. M. (1966). *Phys. Rev.*, **146**, 140 and references contained therein
104. Dixon, M. and Davidson, E. R. (1969). *Phys. Rev.*, **183**, 23
105. van Dijk, F. A. and Dymanus, A. (1969). *Chem. Phys. Lett.*, **4**, 170
106. Brown, H. H. and King, J. G. (1966). *Phys. Rev.*, **142**, 53
107. Hammerle, R. H., Dickinson, J. T., Van Ausdal, R. G., Stephenson, D. A. and Zorn, J. C. (1969). *J. Chem. Phys.*, **50**, 2086
108. Hindermann, D. K. and Cornwell, C. D. (1968). *J. Chem. Phys.*, **48**, 4148
109. Ramsey, N. F. (1953). *Phys. Rev.*, **90**, 232
110. Stevens, R. M. and Lipscomb, W. N. (1964). *J. Chem. Phys.*, **41**, 184
111. Ramsey, N. F. (1950). *Phys. Rev.*, **78**, 699
112. Ref. 2, p. 162
113. Baker, M. R., Anderson, C. H. and Ramsey, N. F. (1964). *Phys. Rev.*, **133**, A1533
114. Ramsey, N. F. (1953). *Phys. Rev.*, **91**, 303
115. English, T. C. and Zorn, J. C. (1967). *J. Chem. Phys.*, **47**, 3896
116. Ramsey, N. F. (1952). *Phys. Rev.*, **87**, 1075
117. E.g., Verhoeven, J. and Dymanus, A. (1970). *J. Chem. Phys.*, **52**, 3222
118. Cecchi, J. and Ramsey, N. (1974). *J. Chem. Phys.*, **60**, 53
119. Heitbaum, J. and Schonwasser, R. (1972). *Z. Naturforsch.*, **27a**, 92
120. Ley, R. and Schauer, W. (1972). *Z. Naturforsch.*, **27a**, 77
121. Lawrence, T., Anderson, C. and Ramsey, N. (1963). *Phys. Rev.*, **130**, 1865
122. Ozier, I., Crapo, L. and Ramsey, N. (1968). *Phys. Rev.*, **172**, 63
123. Ozier, I., Crapo, L. and Ramsey, N. (1968). *J. Chem. Phys.*, **49**, 2314
124. Chan, S., Baker, M. and Ramsey, N. (1964). *Phys. Rev.*, **136**, A1224
125. Ozier, I., Crapo, L., Cederberg, J. and Ramsey, N. (1964). *Phys. Rev. Lett.*, **13**, 482
126. Ozier, I. (1965). *Thesis*, Harvard University
127. Brooks, R., Anderson, C. and Ramsey, N. (1964). *Phys. Rev.*, **136**, A62
128. Brooks, R., Anderson, C. and Ramsey, N. (1972). *J. Chem. Phys.*, **56**, 5193
129. Docken, K. K. and Freeman, R. R. (1974). *J. Chem. phys.*, **61**, 4217
130. Stevens, R. and Karplus, M. (1968). *J. Chem. Phys.*, **49**, 1094
131. Stevens, R. and Lipscomb, W. (1964). *J. Chem. Phys.*, **40**, 2238
132. Stevens, R. and Lipscomb, W. (1964). *J. Chem. Phys.*, **41**, 3710
133. Cade, P. and Huo, W. (1967). *J. Chem. Phys.*, **47**, 614 and 648
134. McLean, A. and Yoshimine, M. (1967). *J. Chem. Phys.*, **47**, 3256
135. Bender, C. and Davidson, E. (1969). *Phys. Rev.*, **183**, 23
136. Lipscomb, W. (1966). *Advances in Magnetic Resonance*, Vol. 2 (New York: Academic)
136a. Cohen, H. D. and Roothan, C. C. J. (1965). *J. Chem. Phys.*, **43**, S34
136b. Pople, J. A., McIver, J. W., Jr. and Ostland, N. S. (1968). *J. Chem. Phys.*, **49**, 2960
137. Alexakos, L. and Cornwell, C. (1964). *J. Chem. Phys.*, **41**, 2098
138. Cornwell, C. (1966). *J. Chem. Phys.*, **44**, 874
139. Ramsey, N. (1950). *Phys. Rev.*, **78**, 221 and Ref. 2
139a. Buckingham, A. D. (1959). *J. Chem. Phys.*, **30**, 1580; Buckingham, A. D., Disch, R. L. and Dunmee, D. A. (1968). *J. Amer. Chem. Soc.*, **90**, 3104; Angel, J. R. P., Sanders, P. G. H. and Woodgate, G. K. (1967). *J. Chem. Phys.*, **74**, 1552
140. Townes, C., Dousmanis, G., White, R. and Schwarz, R. (1955). *Discuss. Faraday Soc.*, **19**, 56
141. Rosenblum, B., Nethercot, A. and Townes, C. (1958). *Phys. Rev.*, **109**, 400

142. For a discussion of the CO moment, see Billingsley, F. and Krauss, M. (1974). *J. Chem. Phys.*, **60,** 4130
143. Meerts, W. L. and Dymanus, A. (1972). *J. Mol. Spectrosc.*, **44,** 320
144. Meerts, W. L. and Dymanus, A. (1973). *Astrophys. J.*, **180,** L93
145. Frosch, R. A. and Foley, H. M. (1952). *Phys. Rev.*, **88,** 1337
146. Dousmanis, G. C. (1955). *Phys. Rev.*, **97,** 967
147. Gallagher, J. J. and Johnson, C. M. (1956). *Phys. Rev.*, **103,** 1727
148. Freed, K. F. (1966). *J. Chem. Phys.*, **45,** 4214
149. As reported in Ref. 80.
150. Lichten, W. (1960). *Phys. Rev.*, **120,** 848
151. Lichten, W. (1962). *Phys. Rev.*, **126,** 1020
152. Brooks, P. R., Lichten, W. and Reno, R. (1971). *Phys. Rev.*, **A4,** 2217
153. Freund, R. S. and Klemperer, W. (1965). *J. Chem. Phys.*, **43,** 2422
154. Stern, R. C., Gammon, R. H., Lesk, M. E., Freund, R. S. and Klemperer, W. (1970). *J. Chem. Phys.*, **52,** 3467
155. Gammon, R. H., Stern, R. C., Lesk, M. E., Wicke, B. G. and Klemperer, W. (1971). *J. Chem. Phys.*, **54,** 2136
156. Gammon, R. H., Stern, R. C. and Klemperer, W. (1971). *J. Chem. Phys.*, **54,** 2151
157. Freund, R. S., Miller, T. A., DeSantis, D. and Lurio, A. (1970). *J. Chem. Phys.* **53,** 2290
158. DeSantis, D., Lurio, A., Miller, T. A. and Freund, R. S. (1973). *J. Chem. Phys.*, **58,** 4625
159. Jette, A. N. (1974). *Chem. Phys. Lett.*, **25,** 590
160. Chiu, L. Y. (1966). *Phys. Rev.*, **145,** 1; (1967). *ibid.*, **159,** 190
161. Lombardi, M. (1973). *J. Chem. Phys.*, **58,** 797
162. Kagann, R. H. and English, T. C. (1974). *Phys. Rev. Lett.*, **33,** 995
162a. See, for example, Refs. 186, 189 and 280
163. Herzberg, G. (1950). *Molecular Spectra and Molecular Structure I, Spectra of Diatomic Molecules*, 416 (New York: Van Nostrand)
163a. A particularly impressive example is Ref. 261
164. Huo, W. (1966). *J. Chem. Phys.*, **45,** 1554
164a. Ref. 35, p. 161
165. Hendrie, J. M. and Kusch, P. (1957). *Phys. Rev.*, **107,** 716
165a. Thaddeus, P., Krisher, L. and Loubser, J. (1964). *J. Chem. Phys.*, **40,** 257
165b. (1966). *La Structure Hyperfine Magnétique des Atoms et des Molécules*, *Colloq. Int. CNRS*, No. 164, 78 (Paris: CNRS)
166. Ref. 35, Chapter 10
166a. (1972). *MTP International Review of Science, Physical Chemistry Series One*, Vol. 3, Chap. 3, *Spectroscopy* (D. A. Ramsey, editor) (London: Butterworths)
167. Dyke, T. and Muenter, J. (1973). *J. Chem. Phys.*, **59,** 3125
168. Anderson, C. and Ramsey, N. (1966). *Phys. Rev.*, **149,** 14
169. Wofsy, S., Muenter, J. and Klemperer, W. (1971). *J. Chem. Phys.*, **55,** 2014
170. de Leeuw, F. and Dymanus, A. (1970). *Chem. Phys. Lett.*, **7,** 288
171. Reinartz, J., Meerts, W. and Dymanus, A. (1972). *Chem. Phys. Lett.*, **16,** 576
172. Reinartz, J. and Dymanus, A. (1974). *Chem. Phys. Lett.*, **24,** 346
173. Muenter, J. S. (1968). *J. Chem. Phys.*, **48,** 4544
174. Davis, R. E. and Muenter, J. S. (1974). *Chem. Phys. Lett.*. **24,** 343
175. Fabricant, B. and Muenter, J. S. (1974). *J. Mol. Spectrosc.*, **53,** 57
176. Garvey, R. and DeLucia, F. (1974). *J. Mol. Spectrosc.*, **50,** 38
177. Tomasevich, G. R. (1970). *Thesis*, Harvard University
177a. Radford, H. E. and Kuntz, C. V. (1970). *J. Res. Nat. Bur. Stand.*, **74A,** 791
178. Freund, S. (1970). *Thesis*, Harvard University
179. Ramsey, N. (1961). *Am. Sci.*, **49,** 509
180. Cederberg, J., Anderson, C. and Ramsey, N. (1964). *Phys. Rev.*, **136,** A960
181. Maki, A. and Lide, D. (1967). *J. Chem. Phys.*, **47,** 3206
182. Kuczkowski, R. L., Lide, D. R. and Krisher, L. C. (1966). *J. Chem. Phys.*, **44,** 3131
183. Lide, D. R. and Kuczkowski, R. L. (1967). *J. Chem. Phys.*, **46,** 4768
184. Matsumura, C. and Lide, D. R. (1969). *J. Chem. Phys.*, **50,** 71
185. Lide, D. R. Jr. (1970). *J. Mol. Spectrosc.*, **33,** 448
186. Yi, P. Y., Ozier, I. and Ramsey, N. F. (1968). *Phys. Rev.*, **165,** 92

187. Ozier, I., Crapo, L. and Lee, J. S. (1968). *Phys. Rev.*, **172,** 63
188. Yi, P. N., Ozier, I. and Ramsey, N. F. (1971). *J. Chem. Phys.*, **55,** 5215
188a. Ozier, I., Yi, P. Y., Khosla, A. and Ramsey, N. F. (1970). *Phys. Rev. Lett.*, **24,** 642
189. Hougen, J. T. (1971). *J. Chem. Phys.*, **55,** 1122
190. Ozier, I. (1971). *Phys. Rev. Lett.*, **27,** 1329
191. Wofsy, S. C., Muenter, J. S. and Klemperer, W. A. (1970). *J. Chem. Phys.*, **53,** 4005
192. Lee, S., Ozier, I. and Ramsey, N. F. (1973). *J. Korean Nucl. Soc.*, **5,** 38
193. Uehara, K., Sakurai, K. and Shimoda, K. (1969). *J. Phys. Soc. Jap.*, **26,** 1018
194. Muenter, A. A. (1972). *Thesis*, Harvard University
195. Muenter, A., Dyke, T., Falconer, W. and Klemperer, W., to be published
196. From the magnitude of the observed electric deflection, it is not certain that TiF_4, VF_4 and VCl_4 have tetrahedral symmetry
197. Muenter, A. and Dyke, T., to be published
198. Oosting, P. and Trappeniers, N. (1971). *Physica*, **51,** 395
199. Curl, R., Kasper, J. and Pitzer, K. (1967). *J. Chem. Phys.*, **46,** 3220
200. Ozier, I., Ho, W. and Birnbaum, G. (1969). *J. Chem. Phys.*, **51,** 4873
201. Holt, C., Gerry, M. and Ozier, I. (1973). *Phys. Rev. Lett.*, **31,** 1033
202. Rosenberg, A., Ozier, I. and Kudian, A. (1972). *J. Chem. Phys.*, **57,** 568
203. Rosenberg, A. and Ozier, I. (1973). *Chem. Phys. Lett.*, **19,** 400
204. Rosenberg, A. and Ozier, I. (1973). *J. Chem. Phys.*, **58,** 5168
205. Ozier, I. and Rosenberg, A. (1973). *Can. J. Phys.*, **51,** 1882
206. Curl, R. and Oka, T. (1973). *J. Mol. Spectrosc.*, **46,** 518
207. Curl, R. (1973). *J. Mol. Spectrosc.*, **48,** 165
208. Mills, I. M., Watson, J. K. G. and Smith, W. L. (1969). *Mol. Phys.*, **16,** 329
209. Dorney, A. J. and Watson, J. K. G. (1972). *J. Mol. Spectrosc.*, **42,** 135
210. Watson, J. K. G. (1974). *J. Mol. Spectrosc.*, **50,** 281
211. Fox, A. K. (1972). *Phys. Rev.*, **A6,** 907
212. Fox, A. K. (1974). *J. Chem. Phys.*, **60,** 337
213. Cleaton, C. E. and Williams, N. H. (1934). *Phys. Rev.*, **45,** 234
214. Gordon, J. P., Zeiger, H. J. and Townes, C. H. (1954). *Phys. Rev.*, **95,** 2822
215. Thaddeus, P., Krisher, L. and Cahill, P. (1964). *J. Chem. Phys.*, **41,** 1542
216. Kukolich, S. G. (1967). *Phys. Rev.*, **156,** 83; see also Ref. 36
217. Kukolich, S. G. (1968). *Phys. Rev.*, **172,** 59; see also Ref. 36
218. Hougen, J. (1972). *J. Chem. Phys.*, **57,** 4207
219. Kukolich, S. G. (1968). *J. Chem. Phys.*, **49,** 5523
220. Kukolich, S. G. (1969). *J. Chem. Phys.*, **50,** 4601
221. Ruben, D. and Kukolich, S. (1974). *J. Chem. Phys.*, **60,** 100
222. Kukolich, S. G. and Flygare, W. H. (1969). *Mol. Phys.*, **17,** 127
223. Davies, P. B., Neumann, R. M., Wofsy, S. C. and Klemperer, W. A. (1971). *J. Chem. Phys.*, **55,** 3564
224. Lehn, J. M. and Munsch, B. (1969). *Chem. Commun.*, 1327
225. Nielsen, H. H. and Dennison, D. M. (1947). *Phys. Rev.*, **72,** 1101
226. Gillen, K. T. (1972). *J. Chem. Phys.*, **56,** 1573
227. Follett, T. F. (1970). *Thesis*, Harvard University
228. Kukolich, S. G. and Nelson, A. C. (1972). *J. Chem. Phys.*, **57,** 4052; Kukolich, S. G. and Nelson, A. C. (1973). *J. Amer. Chem. Soc.*, **95,** 680
229. Kukolich, S. G., Ruben, D. J. and Williams, J. R. (1973). *J. Chem. Phys.*, **58,** 3155
230. Kukolich, S. G. (1972). *J. Chem. Phys.*, **57,** 869
231. Heuvel, J. E. M. and Dymanus, A. (1973). *J. Mol. Spectrosc.*, **45,** 282
232. Casleton, K. H. and Kukolich, S. G. (1973). *Chem. Phys. Lett.*, **22,** 331
233. Reynders, J., Ellenbroek, A. and Dymanus, A. (1972). *Chem. Phys. Lett.*, **17,** 351
234. Kukolich, S. G. and Ruben, D. J. (1972). *J. Mol. Spectrosc.*, **44,** 607; see also Kukolich, S. G., Nelson, A. C. and Ruben, D. J. (1971). *J. Mol. Spectrosc.*, **40,** 33
235. Heuvel, J. and Dymanus, A. (1973). *J. Mol. Spectrosc.*, **47,** 363
236. Ref. 35, p. 239
237. Thaddeus, P., Loubser, J. and Krisher, L. (1959). *J. Chem. Phys.*, **31,** 1667
238. Takuma, H., Shimizu, T. and Shimoda, K. (1959). *J. Phys. Soc. Jap.*, **14,** 1595
239. Shimoda, K., Takuma, H. and Shimizu, T. (1960). *J. Phys. Soc. Jap.*, **15,** 2036
240. Takuma, H. (1961). *J. Phys. Soc. Jap.*, **16,** 309
241. Tucker, K., Tomasevich, G. and Thaddeus, P. (1971). *Astrophys. J.*, **169,** 429

242. Kukolich, S. G. (1971). *J. Chem. Phys.*, **54,** 8
243. Kukolich, S. G. and Ruben, D. J. (1971). *J. Mol. Spectrosc.*, **38,** 130
244. Tomasevich, G. and Tucker, K. (1973). *J. Mol. Spectrosc.*, **48,** 475
245. Neumann, D. and Moskowitz, J. (1969). *J. Chem. Phys.*, **50,** 2219
246. Krieger, D. and Muenter, J. S., to be published
247. Wang, J. H. S. and Kukolich, S. G. (1973). *J. Amer. Chem. Soc.*, **95,** 4138
248. Kukolich, S. G. (1971). *J. Chem. Phys.*, **55,** 610
249. Bluyssen, H., Dymanus, A. and Verhoeven, J. (1967). *Phys. Lett.*, **24A,** 482
250. Bluyssen, H., Verhoeven, J. and Dymanus, A. (1967). *Phys. Lett.*, **25A,** 214
251. Verhoeven, J., Bluyssen, H. and Dymanus, A. (1968). *Phys. Lett.*, **26A,** 424
252. Verhoeven, J., Dymanus, A. and Bluyssen, H. (1969). *J. Chem. Phys.*, **50,** 3330
253. Kukolich, S. G. (1969). *J. Chem. Phys.*, **50,** 3751
254. Clough, S. A., Beers, Y., Klein, G. P. and Rothman, L. S. (1973). *J. Chem. Phys.*, **59,** 2254
255. Garvey, R. M. and DeLucia, F. (1974). *29th Symposium on Molecular Structure and Spectroscopy*, Paper RB2
256. Cupp, R. E., Kempf, R. A. and Gallagher, J. J. (1968). *Phys. Rev.*, **171,** 60
257. Chandra, S. and Dymanus, A. (1972). *Chem. Phys. Lett.*, **13,** 105
258. Kukolich, S. G., Wang, J. H. S. and Ruben, D. J. (1973). *J. Chem. Phys.*, **58,** 5474
259. Nelson, A., Kukolich, S. and Ruben, D. (1974). *J. Mol. Spectrosc.*, **51,** 107
260. Kukolich, S. G. and Nelson, A. C. (1972). *J. Chem. Phys.*, **55,** 4446
261. Tomasevich, G., Tucker, K. and Thaddeus, P. (1973). *J. Chem. Phys.*, **59,** 131
262. Tomasevich, G. (1974). Personal communication
263. Kukolich, S. G. and Nelson, A. C. (1973). *Chem. Phys. Lett.*, **11,** 383
264. Kukolich, S. G., Nelson, A. C. and Yamanashi, B. S. (1971). *J. Amer. Chem. Soc.*, **93,** 6769
265. Kukolich, S. G. (1969). *J. Chem. Phys.*, **51,** 358
266. Caves, T. and Karplus, M. (1966). *J. Chem. Phys.*, **45,** 1670
267. Neumann, D. and Moskowitz, J. (1968). *J. Chem. Phys.*, **49,** 2056
268. Salem, L. (1963). *J. Chem. Phys.*, **38,** 1227
269. Flygare, W. (1967). *Rec. Chem. Progr.*, **28,** 63
270. Turner, B. E. (1973). *Sci. Amer.*, **228,** 50
271. (1974). *Quarterly Summary of Radio Telescope Observation, Unidentified Lines and Negative Searches* (Washington: U.S. Dept. of Commerce, Nat. Bur. of Stand.) Jan.
272. Rahman, A. and Stillinger, F. (1971). *J. Chem. Phys.*, **55,** 3336
273. Howard, B. J. (1975). *The Structure and Properties of van der Waals Molecules*, this volume, Chapter 3
274. The exception is CF_2HCO_2H–HCO_2H, Costain, C. C. and Srivastava, G. P. (1964). *J. Chem. Phys.*, **41,** 1620
275. Novick, S. E., Davies, P. B., Dyke, T. R. and Klemperer, W. (1973). *J. Amer. Chem. Soc.*, **95,** 8547
276. Ochs, S., Cote, E. and Kusch, P. (1953). *J. Chem. Phys.*, **21,** 459
277. Büchler, A., Stauffer, J. L. and Klemperer, W. (1964). *J. Amer. Chem. Soc.*, **86,** 4544
278. Muenter, J. S. (1974). *Chem. Phys. Lett.*, **26,** 97
279. Dyke, T. R. and Muenter, J. S. (1972). *J. Chem. Phys.*, **57,** 5011
280. Dyke, T. R., Howard, B. J. and Klemperer, W. (1972). *J. Chem. Phys.*, **56,** 2442
281. Logan, R., Cote, R. and Kusch, P. (1952). *Phys. Rev.*, **86,** 280
282. Brooks, R. A., Anderson, C. H. and Ramsey, N. F. (1964). *Phys. Rev.*, **136,** A62
283. Brooks, R. A., Anderson, C. H. and Ramsey, N. F. (1972). *J. Chem. Phys.*, **56,** 5193
284. Dagdigian, P. J. and Wharton, L. (1972). *J. Chem. Phys.*, **57,** 1487
285. Bader, R. M. (1934). *J. Chem. Phys.*, **2,** 128
286. Sinha, M. P., Schultz, A. and Zare, R. N. (1973). *J. Chem. Phys.*, **58,** 549
287. Gordon, R. J., Lee, Y. T. and Herschbach, D. R. (1971). *J. Chem. Phys.*, **54,** 2393
288. Sinha, M. P., Caldwell, C. D. and Zare, R. N. (1974). *J. Chem. Phys.*, **61,** 491
289. Johnson, S. E., Steinfeld, J. I. and Beenakker, J. J. M. (1973). *Northeast Regional Meeting of the American Chemical Society, Abstracts*, No. 21
290. Schultz, A., Cruse, H. W. and Zare, R. N. (1972). *J. Chem. Phys.*, **57,** 1354
291. Dyke, T. R. and Muenter, J. S. (1974). *J. Chem. Phys.*, **60,** 2029
292. Kollman, P. A. and Allen, L. C. (1970). *J. Chem. Phys.*, **52,** 5085; Del Bene, J. and Pople, J. A. (1971). *ibid.*, **55,** 2296

293. Morokuma, K. and Pederson, L. (1968). *J. Chem. Phys.*, **48,** 3275; Morokuma, K. and Winick, J. (1970). *ibid.*, **52,** 1301; Kollman, P. and Allen, L. C. (1969). *ibid.*, **51,** 3286; Hankins, D., Moskowitz, J. W. and Stillinger, F. H. (1970). *ibid.*, **53,** 4544; Del Bene, J. and Pople, J. A. (1970). *ibid.*, **52,** 4858; Diercksen, G. H. F. (1971). *Theoret. Chim. Acta*, **21,** 335; Popkie, H., Kistenmacher, H. and Clementi, E. (1973). *J. Chem. Phys.*, **59,** 1325
294. Gebbie, H. A., Burroughs, W. J., Chamberlain, J., Harris, J. E. and Jones, R. G. (1969). *Nature*, **221,** 143
295. Longuet-Higgins, H. C. (1963). *Mol. Phys.*, **6,** 445
296. Ref. 35, p. 302
297. Morokuma, K. (1974). Personal communication
298. Atoji, M. and Lipscomb, W. N. (1954). *Acta Crystallogr.*, **7,** 173
299. Morokuma, K. (1974). Personal communication; see also Ref. 20
300. Pimentel, G. C. and McClellan, A. L. (1960). *The Hydrogen Bond*, 294 (San Francisco: W. H. Freeman)
301. Janzen, J. and Bartell, L. (1969). *J. Chem. Phys.*, **50,** 3611. The mass spectral results of Ref. 280 imply that a distribution of lower polymers was observed here
302. Novick, S. E., Davies, P., Harris, S. J. and Klemperer, W. (1973). *J. Chem. Phys.*, **59,** 2273
303. Harris, S. J., Novick, S. E. and Klemperer, W. (1974). *J. Chem. Phys.*, **60,** 3208
304. Harris, S. J., Novick, S. E., Klemperer, W. and Falconer, W. E. (1974). *J. Chem. Phys.*, **61,** 193
305. Gordon, R. G. (1966). *J. Chem. Phys.*, **44,** 576 and references given in Ref. 302
306. Bratoz, S. and Martin, M. L. (1965). *J. Chem. Phys.*, **42,** 1051
307. Farrar, J. M. and Lee, Y. T., to be published
308. Mattison, E. M., Pritchard, D. E. and Kleppner, D. (1974). *Phys. Rev. Lett.*, **32,** 507
309. Green, S. (1973). *J. Chem. Phys.*, **58,** 3117
310. Freeman, R. E., Mattison, E. M., Pritchard, D. E. and Heppner, D. (1974). *Phys. Rev. Lett.*, **33,** 397

3
The Structure and Properties of van der Waals Molecules

B. J. HOWARD
University of Southampton

3.1	INTRODUCTION	93
3.2	METHODS USED IN THE FORMATION OF VAN DER WAALS MOLECULES	95
	3.2.1 *Equilibrium or thermodynamic sources*	95
	3.2.2 *Supersonic nozzles*	98
3.3	ENERGY LEVELS OF VAN DER WAALS MOLECULES	99
3.4	EXPERIMENTAL TECHNIQUES FOR THE STUDY OF VAN DER WAALS MOLECULES	102
	3.4.1 *Infrared spectroscopy*	103
	3.4.2 *Raman spectroscopy*	106
	3.4.3 *Ultraviolet spectroscopy*	107
	3.4.4 *Mass spectrometry*	109
	3.4.5 *Molecular beam spectroscopy*	110
	3.4.6 *Electron diffraction*	113
	3.4.7 *Other techniques*	113
3.5	CONCLUSION	115
	NOTES ADDED IN PROOF	115

3.1 INTRODUCTION

It has been realised for some time that an attractive potential exists between molecules in the gas phase. Some of the earliest work concerns the deviations from the ideal gas equation which could to a large extent be explained by the van der Waals equation of state,

$$(p + a/V^2)(V - b) = nRT$$

The term aV^{-2} arises from the long-range attractive force between molecules which reduces the observed pressure at the walls from that of the bulk gas. At the same time the b term represents an excluded volume due to the strong repulsive forces when molecules collide.

The ideas of a long-range attractive and short-range repulsive potential have been used to explain condensation to the liquid phase and at lower temperatures the formation of the solid phase. This type of potential between two molecules also permits the formation of bound 'double-molecules' (or dimers*) in the gas phase. The existence of such species was appreciated by Jeans[1] at the beginning of this century, but there has been little decisive work on their properties until fairly recently. A major hindrance to advancement has been the extreme weakness of the interaction potential between molecules; this is often much less than kT at room temperature so that the equilibrium concentration of dimers is quite small.

Historically the weak interactions between molecules have been known as van der Waals forces and the resulting dimers, trimers, etc. are consequently called van der Waals molecules. The detailed nature of van der Waals forces is well discussed in the literature on intermolecular forces[2-7] and they will thus only be mentioned briefly here. The forces occurring at large distances are typically either the long-range classical electrostatic interactions, such as those between two dipoles, or the purely quantum mechanical London dispersion forces. At shorter distances repulsive forces come into play; these arise mainly from the exchange forces due to the slight overlap of the electron distributions of the separate molecules. One important thing to note is that within this range of intermolecular distances there is very little redistribution of electron charge; consequently the individual components of the dimer retain much of their separate character even in the region of van der Waals bond formation. In addition, the bond strengths are rarely greater than 1 kcal mol^{-1} (equivalent to 350 cm^{-1} or kT at 503 K for a single molecule).

These properties may be compared with those of a normal chemical bond. In the latter case the bond lengths are much shorter than the sums of the van der Waals radii and there is considerable interaction between the electron distributions of the two separate species. The consequent large electron rearrangement results in the individual sections of the new molecule retaining few, if any, of their separate properties. At the same time the energy of this bond is typically of the order of 50–100 kcal mol^{-1}. This is considerably greater than that for a van der Waals bond, as might have been expected from the overlap criterion of bond strength.

There is a fairly recent review by Ewing[8] on the subject of van der Waals molecules. However, since then there have been a number of significant advances. Also we shall attempt to look at the subject from a different standpoint. First of all, because of the difficulty in producing complexes, Section 3.2 is devoted to the different methods of formation of van der Waals molecules and to how their concentrations may be optimised. In Section 3.3 we describe the internal motions of van der Waals molecules and obtain the energy level schemes of many types of complexes. Section 3.4 is concerned

* In this review article we shall use the terms dimer, trimer, etc. to describe the weakly bound entities formed from two, three, etc. molecules or atoms whether the separate species be like or unlike.

with the various experimental techniques that have been used to detect and study van der Waals molecules. In this section we also discuss the kind of information that has been obtained regarding the structure and properties of these species.

Although not strictly van der Waals molecules, we include in this review a brief discussion of the results for intermolecular hydrogen bonded species. In these examples the bond strengths may be as high as 5 kcal mol^{-1}, but are still weak compared with regular valence bonds. They share with van der Waals complexes the same difficulties of production and exhibit non-rigid behaviour.

3.2 METHODS USED IN THE FORMATION OF VAN DER WAALS MOLECULES

3.2.1 Equilibrium or thermodynamic sources

An essential requirement for the formation of a van der Waals complex is that the combined molecule be capable of possessing a potential energy less than that of the separate species. Such a potential energy curve as a function of intermolecular distance is shown in Figure 3.1a. This is correct only for the approach of two atoms but may be used for separate diatomic (or polyatomic) molecules provided their relative orientation is specified.

When two molecules approach each other with an energy greater than zero on our energy scale, they can approach to the distance of the inner repulsive potential. However, they will then separate again to an essentially infinite distance and are consequently unbound; their motions are thus just those of a pair of free molecules slightly perturbed by the intermolecular forces. If,

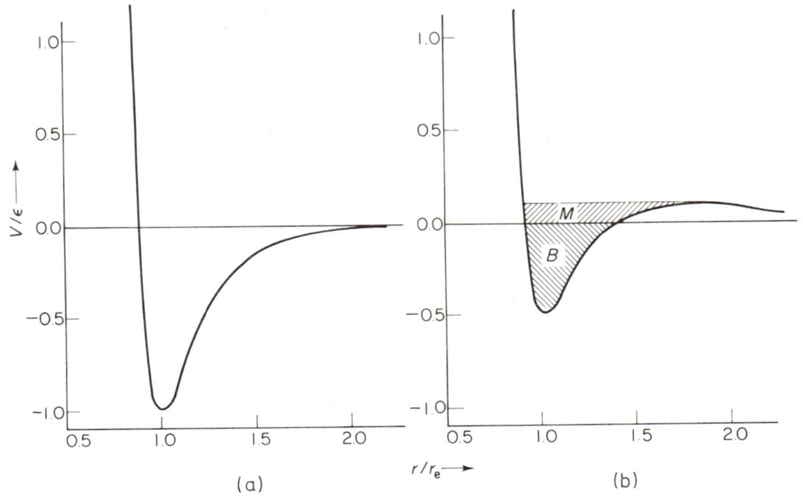

Figure 3.1 (a) Typical Lennard-Jones potential. (b) Lennard-Jones potential with centrifugal barrier showing the regions of bound (B) and metastable (M) dimers

on the other hand, two molecules (considered separate from the rest of the gas) have an energy that is less than zero, they must of necessity remain bound together and thus form a stable van der Waals molecule. It should be noted that in order to reach this state the two free molecules must, during close approach, collide with a third species which takes up the excess energy. Similarly, in order to break a van der Waals bond, energy must be absorbed through a collision with a third body (possibly the walls of the vessel). Hence, assuming a constant concentration of suitable 'third bodies', c_t, the law of mass action predicts that the rate of formation of the dimer is proportional to $c^2_{\text{monomer}}c_t$ and the rate of destruction of bound dimers is proportional to $c_{\text{dimer}}c_t$. Then at equilibrium the concentration or partial pressure of the dimer is, as expected, proportional to the square of monomer pressure. Thus the dimer concentration is enhanced at higher pressures. Naturally the formation of higher polymers is also favoured under these conditions.

Relative to the free molecules, the concentration of bound dimers is proportional to the Boltzmann factor exp $\Delta E/kT$ where ΔE is the energy separation and is positive in this example. This factor and hence the concentration of bound dimers is increased by reducing the temperature. Many of the studies of van der Waals complexes to be discussed in Section 3.4 have been performed using the conditions of high pressure and low temperature.

Our simple model for the dimer formation omits one important feature of the dynamics of approach of two particles. The potential energy curve in Figure 3.1a is correct if the two molecules approach along the line joining their centres of mass. However if they approach with collision parameter b (the distance of closest approach in the absence of intermolecular forces) and velocity v, the dimer will possess rotational angular momentum L. This has the value μbv where $\mu = m_1m_2/(m_1 + m_2)$ is the reduced mass of the complex. This extra rotational kinetic energy which is proportional to the square of the intermolecular distance r must be added to the Hamiltonian representing the stretching of the van der Waals bond. For a given value of L this behaves like an extra repulsion between the molecules and when added to the true potential energy yields the effective potential for relative motion of the molecules:

$$V_{\text{eff}}(r) = V(r) + L^2/(2\mu r^2)$$

Assuming the long-range attractive part of the potential $V(r)$ varies faster than r^{-2}, we may obtain an energy curve with a potential hump in addition to the well. This is shown in Figure 3.1b. As before, that portion of space with energy less than the energy of the separate molecules corresponds to bound dimers. However, a pair of molecules to the left of the potential hump with an energy greater than that of the separate molecules but less than the top of the hump are trapped together. Classically they are bound together, but quantum mechanically they can tunnel through the barrier and consequently this bound state has a finite lifetime. They are thus appropriately called metastably bound van der Waals molecules.

When calculating the equilibrium concentration of van der Waals dimers by statistical mechanics one should use the effective potential energy curve. Hill[9,10] has obtained the general partition function for such bound double-molecules. Stogryn and Hirschfelder[11] have extended this work to a number of molecular species. They assumed that the intermolecular interaction is given by a spherically symmetric Lennard-Jones (6–12) potential

$$V(r) = \varepsilon[(r_e/r)^{12} - 2(r_e/r)^6]$$

where ε is the well depth and r_e is the position of the potential minimum. By using the most probable values of ε and r_e they were able to calculate the mole fraction of bound and metastable dimers as a function of temperature and pressure. Some of their results are given in Table 3.1. These may be compared with the results of Buluggiu and Foglia[12], who used a Morse potential

$$V(r) + \varepsilon[\{1 - \exp[-\beta(r - r_e)]\}^2 - 1]$$

where ε and r_e have the same meaning as above and β is a force constant parameter. Their results using the molecular parameters of Konowalow and Hirschfelder[13] are shown in Table 3.2.

Table 3.1 **Lennard-Jones parameters and calculated mole fractions of dimers as a function[2,11] of $T^* = kT/\varepsilon$ at the standard density of** $(22\,414)^{-1}$ mol cm^{-3}

	$\frac{\varepsilon}{k}$K	r_e/Å	Mole fraction of dimer $T^* = 1$	$T^* = 2$	$T^* = 5$
Ne	34.9	3.086	0.001 93	0.000 588	0.000 137
Ar	119.8	3.822	0.003 54	0.001 08	0.000 252
Kr	171	4.041	0.004 19	0.001 28	0.000 298
Xe	221	4.602	0.006 18	0.001 89	0.000 298
O_2	117.5	4.02	0.004 11	0.001 25	0.000 292
CO_2	189	5.035	0.008 10	0.002 47	0.000 576

Table 3.2 **Morse curve parameters and calculated mole fractions of dimers as a function[12,13] of $T^* = kT/\varepsilon$ at the standard density of** $(22\,414)^{-1}$ mol cm^{-3}

	$\frac{\varepsilon}{k}$K	r_e/Å	β/Å^{-1}	Mole fraction of dimer $T^* = 1$	$T^* = 2$	$T^* = 5$
Ne	43.99	3.152	1.84	0.001 53	0.000 470	0.000 109
Ar	144.8	3.855	1.48	0.003 04	0.000 912	0.000 211
Kr	182.7	4.038	1.28	0.003 89	0.001 17	0.000 269
Xe	274.7	4.420	1.27	0.004 70	0.001 41	0.000 325

It will be noticed that, except at the very lowest temperatures, the calculated concentrations of dimers at near atmospheric pressure are in the region of 0.1% of the monomer. This is usually sufficient for study by a number of techniques, especially the many spectroscopic methods. One disadvantage of course is that we have to differentiate the properties of the dimer from those of the monomer. Also since the bond strengths are typically of the same order of magnitude as thermal energies, the high collision rate implies a short lifetime for the van der Waals complex. This results in broad spectral lines which can often hide some of the more interesting information about these molecules.

Stogryn and Hirschfelder[11] have also considered the lifetimes of the metastable dimers. They have shown that, except for energies near the potential hump, the mean lifetime is considerably longer than the period between collisions at atmospheric pressure. Thus just like the bound molecules their lifetime is dominated by the collision rate.

3.2.2 Supersonic nozzles

Supersonic nozzles were pioneered theoretically by Kantrowitz and Grey[14] and experimentally by Kistiakowsky and Slichter[15] as a source of high-intensity beams of molecules. Crudely a nozzle is just a small aperture through which gas expands from a high to a much lower pressure region. If the pressure of the gas is sufficiently low that the mean free path is greater than the dimensions of the aperture, we obtain molecular effusions. The gas molecules undergo a negligible number of collisions while expanding through the nozzle and possess a velocity distribution[16] of the form $u^3 \exp(-mu^2/2kT)$. This is basically the Maxwell–Boltzmann distribution of the bulk gas times a velocity factor to allow for the fact that faster molecules are more likely to effuse through the hole. In reality, even under effusion conditions the number of molecules with the lowest velocities is reduced, since they spend a much longer time in the region of the aperture and are likely to collide with faster moving species. This reduction in the number of low-velocity molecules is more evident when the pressure is increased to a value where the mean free path of the gas is considerably less than the dimensions of the nozzle. The molecules undergo many collisions in the regions of the aperture and tend to emerge with approximately the same velocity, U. The molecular velocity distribution is given[14, 17] by $\exp[-m(u - U)^2/2kT^1]$, where the effective Boltzmann temperature T^1 is typically only a few degrees Kelvin; mU^2 is very much greater than $2kT^1$ so that the molecular motion is dominated by mass flow.

This very low temperature in the frame moving with velocity U is real since there is considerable rotational and vibrational cooling. These modes of motion require several collisions for energy transfer so that their cooling lags behind that for translational motion. At these very low effective temperatures the small attractive forces between molecules become important and condensation may occur. Depending upon the conditions in the nozzle, aggregates like dimers or trimers may form; sometimes the particles approach the size of primitive liquid droplets and contain many thousands of molecules[18].

The first observations of these condensation effects appear to have been by Becker and co-workers[19-21]. Beyond a critical source pressure they found a rapid increase in the apparent beam intensity attributable to condensation. The later use of mass spectrometric detection of the molecular beams has confirmed the onset of polymerisation. For example, Bentley[22] and Henkes[23], working independently, observed the products of expanding carbon dioxide from a high-pressure reservoir through a small nozzle. Bentley showed that the observed polymer ions $(CO_2)_n^+$ with n up to 23 were unlikely to be formed in the mass spectrometer itself by ion–molecule reactions. He also demonstrated that the polymer concentrations showed the wrong pressure dependence to have been present in the gas before the expansion. Henkes thus concluded that the polymers were formed by condensation in the isentropic ($\Delta S = 0$) expansion process. Since that time there have been many applications of mass spectrometry to the detection of polymeric species formed in this manner. This forms the subject matter of Section 3.4.4.

It is useful to appreciate the experimental conditions required for the onset of nucleation in a supersonic nozzle. Certainly increased pressures and reduced temperatures aid polymerisation as in a thermal source. Also condensation is improved if a larger aperture nozzle is used. Bier and Hagena[24] suggest that at a given temperature the beginnings of significant condensation require a critical value of the product (p_0d) of the source pressure and nozzle diameter. Using mass spectrometric detection of argon dimer, Golomb *et al.*[25] observed that the maximum dimer beam intensity corresponded to a constant p_0d. However, the use of a plateau in the curve of total beam flux *versus* source pressure as a criterion of significant condensation corresponds to a constant p_0d^2. Milne *et al.*[26] prefer to use p_0^2d since their experiments show that the mole fraction of dimer formed in the isentropic expansion varies approximately linearly with this function. The results thus depend upon the criterion used.

Since the products of a supersonic nozzle source are not in equilibrium with each other, it is possible, by choosing the correct conditions, to form preferentially species which only occur in low concentrations in a thermodynamic (equilibrium) source. For example, hydrogen fluoride forms many polymers. It appears that at the high pressures and low temperatures required to form significant concentrations of the dimer, the higher polymers predominate[27]. Although dimers as well as hexamers were used to explain the *PVT* data[28], it is possible that agreement might be obtained with another polymer pattern. However, with a supersonic nozzle and a source pressure of a few hundred torr there are only sufficient intermolecular collisions to form the dimer; the higher polymers have low concentrations[29].

The supersonic nozzle is now a frequent source of van der Waals molecules in many fields of study. Several of these applications are discussed in Section 3.4.

3.3 ENERGY LEVELS OF VAN DER WAALS MOLECULES

Many of the techniques to be described in Section 3.4 are spectroscopic and involve transitions between different energy levels of the van der Waals complexes. It is thus desirable at this stage to have a knowledge of the expected energy levels for various types of complexes in order to describe their spectra.

By far the simplest examples of van der Waals molecules are those formed by the approach of two atoms. Neglecting electronic motions, the energy of the molecule in the centre of mass axis system is just the sum of the relative kinetic energy (along the interatomic axis), the rotational motion of the complex and the potential energy of interaction which only depends upon the interatomic distance r. The Hamiltonian describing the system is thus

$$\mathscr{H} = p_r^2/2\mu + (\boldsymbol{J}\hbar)^2/2\mu r^2 + V(r) \tag{3.1}$$

where μ is the reduced mass of the two atoms, $p_r = -i\hbar\partial/\partial r$ represents the linear momentum along the van der Waals bond and $\boldsymbol{J}\hbar$ is the rotational angular momentum of the complex. In a fairly rigid diatomic molecule it is usual to solve for the vibrational motion in the absence of the rotation[30, 31]. However, in the weakly bound van der Waals complexes, where the rotational and vibrational frequencies are often of the same order of magnitude, the strong rotation–vibration coupling makes it desirable to include the rotational term as an effective centrifugal potential, as shown in Figure 3.1b. The vibrational Hamiltonian is then solved for each value of $\boldsymbol{J}^2 = J(J+1)$. A typical energy level diagram of bound states is shown in Figure 3.2; the results[32] are for H_2–Ar with the hydrogen treated as an atom of mass 2.

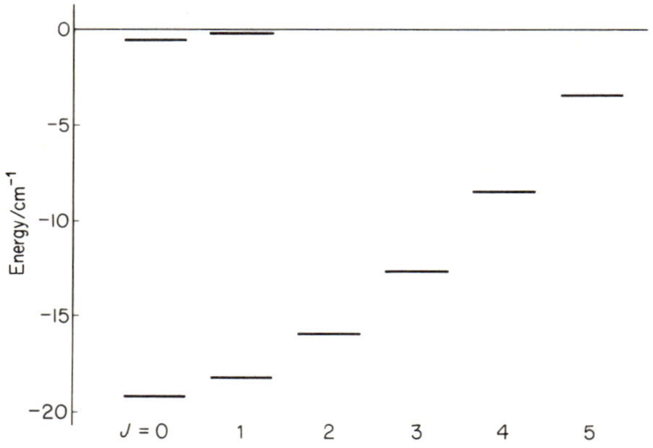

Figure 3.2 Energy level diagram of H_2–Ar assuming an isotropic potential. Energies are those calculated by Cashion[32]

When one of these atoms is replaced by a diatomic molecule, the molecular motion becomes rather more complex. The whole problem has been studied theoretically by Bratoz and Martin[33] and is discussed here. Let us first consider the case when the intermolecular potential is just a function of the separation of the atom A and the centre of mass of the molecule XY; it does not depend upon the orientation of the diatomic molecule. The Hamiltonian for such a system is

$$\mathscr{H} = p_r^2/2\mu + (\boldsymbol{R}\hbar)^2/2\mu r^2 + V(r) + \boldsymbol{B}\boldsymbol{G}_s^2 \tag{3.2}$$

where the first three terms correspond to the stretching of the van der Waals bond and have the same meaning as in equation (3.1), provided the mass of

XY is considered to be concentrated at its centre of mass. We use $\boldsymbol{R}\hbar$ for the rotational angular momentum of the pseudo-diatomic molecule and reserve the symbol $\boldsymbol{J}$ for the total angular momentum of the system. B is the rotational constant of XY and $\boldsymbol{G}_s\hbar$ is the angular momentum of XY measured in the space-fixed coordinate system. $\boldsymbol{G}_s$ should be considered as a vibrational angular momentum in the complex. We have omitted the stretching vibrational motion of the diatomic molecule from equation (3.2) because it is normally of a much higher frequency than the motions we are considering; it may be factored off in the same way as the electronic motion in the Born–Oppenheimer separation[34]. If we are interested in excitations of this vibration, we can accurately solve the problem by using a slightly different $V(r)$ for each vibrational level. The van der Waals stretching vibrational levels are similar to those in Figure 3.2 with J replaced by R. To each of these must be added a series of energy $BG(G + 1)$ due to the free rotation of XY. This scheme is shown in the first column of Figure 3.3. Cashion[32] has used this model to interpret the vibrational spectra of H_2–Ar.

In reality there is always some angle dependence of the intermolecular potential. This couples $\boldsymbol{G}_s$ and $\boldsymbol{R}$ together so that only $\boldsymbol{J} = \boldsymbol{G}_s + \boldsymbol{R}$ is a true constant of motion. The splitting of each level with a given value of G and R into the different possible values of J is shown in the second column of Figure 3.3. This scheme is analogous to the much discussed coupling of the electron

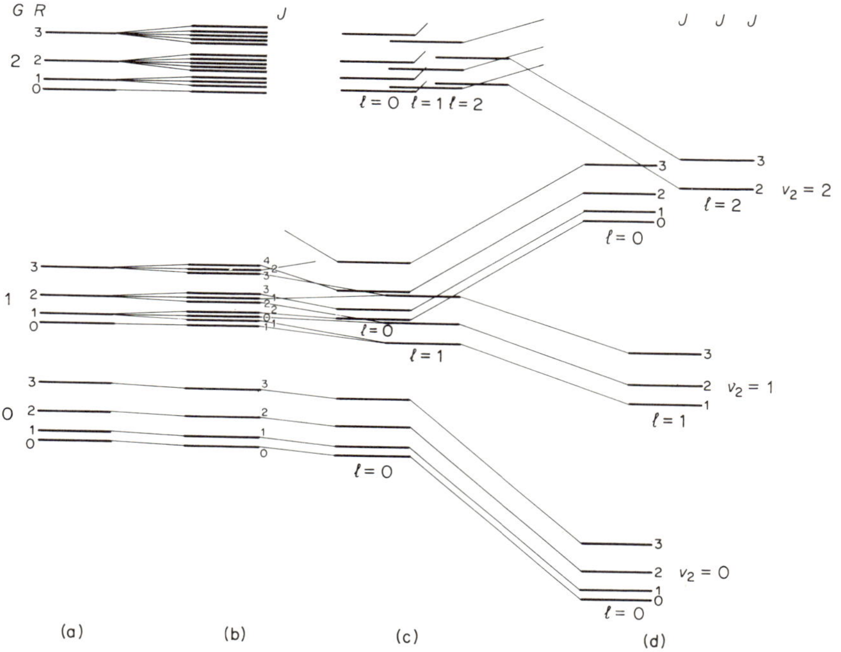

Figure 3.3 Correlation diagram for an A—XY type van der Waals molecule, showing (a) an isotropic intermolecular potential, (b) a slightly anisotropic potential, corresponding to Hund's coupling case *b*, (c) a stronger anisotropic potential, corresponding to Hund's case *a*, and (d) a linear triatomic molecule

spin to the molecular rotation in Hund's coupling case *b* [35]. It has been called case 1 by Bratoz and Martin[33], who obtain expressions for the energies.

As the anisotropy of the intermolecular potential increases so that the barrier to the rotation of XY is greater than $(\hbar^2/\mu r^2)JG$, we must consider $\boldsymbol{G}$ to be quantised in the molecular axis system; the Coriolis forces are now insufficient to overcome the intermolecular forces. The component l of $\boldsymbol{G}$ along the intermolecular axis is a good quantum number. So is G provided the barrier height is less than $BG(G + 1)$. At the same time, $\boldsymbol{R}$ is no longer a good description of the end-over-end rotational motion of the complex and should be replaced by $\boldsymbol{J} - \boldsymbol{G}_m$, giving $(1/2\mu r^2)(\boldsymbol{J} - \boldsymbol{G}_m)^2$ as the rotational term in equation (3.2). $\boldsymbol{G}_m$ is the rotation-like vibrational angular momentum of XY measured with respect to the rotating molecular axis system. The result is similar to the case *a* coupling of electron spin[35, 36].

In the limit of a strongly anisotropic potential, the rotational motion of XY is completely quenched and becomes a librational motion; it now corresponds to a bending vibration of a triatomic molecule. In Figure 3.3 we have assumed that the preferred equilibrium is linear but an analogous energy level scheme can be obtained for a non-linear triatomic molecule in which l is replaced by the symmetric top quantum number K.

A complex formed by coupling together two diatomic molecules produces a similar correlation diagram but the energy levels are far more complex. However, in the limit of negligible anisotropy in the intermolecular potential the problem is now similar to that for two atoms with vibrational energies E_{vR}. The rotational energy levels are given by

$$E = E_{vR} + B_1G_1(G_1 + 1) + B_2G_2(G_2 + 1) \tag{3.3}$$

where B_i is the rotational constant and $G_i\hbar$ is the rotational angular momentum of each diatomic molecule. The energy levels of the dimer of the hydrogen molecule are well described by this expression[37]. As the anisotropy of the potential increases, $\boldsymbol{G}_1$, $\boldsymbol{G}_2$ and $\boldsymbol{R}$ must be coupled together to give $\boldsymbol{J}$. In the Hund's case *a* limit, $\boldsymbol{G}_1$ and $\boldsymbol{G}_2$ must be measured with respect to the rotating axis system and the total Hamiltonian becomes

$$\mathscr{H} = B_1\boldsymbol{G}_1^2 + B_2\boldsymbol{G}_2^2 + (1/2\mu r^2)(\boldsymbol{J} - \boldsymbol{G}_1 - \boldsymbol{G}_2)^2 + p_r^2/2\mu + V(r, \text{orientation}) \tag{3.4}$$

The expressions for the Hamiltonians and energy levels of more complex molecules may be obtained in a similar manner. However, detailed experimental results are not yet available for these molecules.

3.4 EXPERIMENTAL TECHNIQUES FOR THE STUDY OF VAN DER WAALS MOLECULES

Most of the techniques that are suitable for the study of individual molecules have been applied to the study of molecular clusters. In the following sections we shall discuss those methods that have proved successful and we shall indicate what information is available on the structure and properties of van der Waals molecules.

3.4.1 Infrared spectroscopy

In principle all vibrational modes that possess an oscillating electric dipole moment are infrared active. Thus, since van der Waals molecules contain many more vibrational degrees of freedom than the parent monomers, one might expect a very rich i.r. spectrum. Unfortunately this is not so. Those interesting motions which correspond to the stretching or bending of the van der Waals bond are of a very low frequency (typically less than 100 cm^{-1}) and have yet to be observed directly. Instead, most of the work has been concerned with high-frequency vibrations corresponding to bond stretches of the monomer units.

Before continuing, we shall distinguish two types of i.r. spectra, in which the experimental methods differ significantly. First there are those complexes in which at least one of the parent monomers possesses i.r. active vibrations; the analogous vibrations in the complex are also i.r. active. By contrast, homonuclear diatomic molecules possess no i.r. spectrum, but on complex formation some of the vibrational modes become i.r. active.

The first type of spectroscopy, on dimers of i.r. active monomers, is perhaps the most straightforward. The earliest studies[38-40] were on the dimers of the hydrogen halides HF and HCl. These and all subsequent experiments were performed in a static cell under equilibrium conditions where, as explained in Section 3.2.1, the mole fraction of the dimer is enhanced by moving to higher pressures and lower temperatures. In these favourable cases it was sufficient to work near room temperature at pressures in the region of one atmosphere. The spectra observed correspond to the H—X stretching motion which has a frequency on dimer formation only slightly displaced from that of the monomer. Thus the dimer spectrum is usually hidden under the much stronger spectrum of the monomer. However, from a knowledge of the low-pressure spectrum in which the dimer may be considered absent, it is possible to compute the high-pressure monomer spectrum. After subtraction of this from the observed spectrum, we are ideally left with a spectrum due to the dimer. This remaining absorption should have the correct (quadratic) pressure dependence. It is important that the wings of the monomer lines be correctly computed since these too show a quadratic variation with pressure[39]. On the other hand, only lines due to the complex should show a strong temperature dependence.

For diatomic monomers there is a gap in the i.r. spectrum at the fundamental vibrational frequency because of the absence of a 'Q-branch' ($\Delta J = 0$ transitions). This would appear to be a good region to look for the dimer spectrum. However, at high pressure two types of features are produced. First, collisions between the molecules relax the $\Delta J = \pm 1$ selection rule and induced $\Delta J = 0$ transitions are obtained[42]. A more detailed account of collision induced spectra is given later in this section. On top of these broad features there are sharp lines due to bound states of the dimer. For example, with HCl dimer[40, 41] there are three lines in the ratio of 12:7:1. These are attributed to $H^{35}Cl$—$H^{35}Cl$, $H^{35}Cl$—$H^{37}Cl$ and $H^{37}Cl$—$H^{37}Cl$, whose theoretical intensities are in the ratio of 9.4:6.1:1.0. The lines vary quadratically with pressure and show a strong temperature dependence from which an enthalpy of formation of -2.14 ± 0.2 kcal mol^{-1} has been calculated. In a similar way, complexes of HCl with the inert gases have been observed[40, 43, 44]. 'Arrhenius' type plots for the temperature dependence of these lines have shown the enthalpies of formation of Ar—HCl and Xe—HCl to be respectively -1.1 ± 0.2 and -1.6 ± 0.2 kcal mol^{-1}. No rotational analysis was attempted and no structure for these molecules was obtained. However, Neilsen and Gordon[45] have used the i.r. linewidths, line-shifts and band shapes of HCl in the presence of argon to obtain information on the intermolecular interactions. They treat the inelastic collisions between argon and HCl semi-classically and obtain a full angle dependent intermolecular potential.

Hydrogen fluoride dimer has been the subject of many studies[38, 39, 46-48]. Much rotational fine structure has been observed. Herget *et al.*[46] were reluctant to interpret their spectra but they showed that the regularly spaced lines in their bands were compatible with a B rotational constant of 0.29 cm^{-1}. Hines and Wiggins obtained additional weak lines[48] at half the spacing of those of Herget. If these are interpreted as true rotational structure, they obtain a rotational constant of 0.138 cm^{-1}. Recent molecular beam studies[29] have shown that the energy levels are further complicated by an internal tunnelling motion and the i.r. spectrum is consequently more complex than originally appreciated.

There have been a large number of other papers on hydrogen bonded species. We shall just mention the complexes of hydrogen cyanide and acetonitrile with HCl[49] and HF[50, 51]. In all cases fine structure has been observed and rotational constants have been derived.

Recently, Dinerman and Ewing[52, 53] observed the i.r. spectrum of the nitric oxide dimer at a pressure of 270 Torr and at 123 K. Their analysis indicates a *cis*-$(NO)_2$ structure with an N—N bond. The band shapes are consistent with an N—N distance of 1.75 Å and an N—N—O angle of 90°, although 120° also gives a reasonable fit. This may be compared with the x-ray studies[54] on the solid which yield N—N = 2.18 ± 0.06 Å and NNO = $101 \pm 3°$. Although Dinerman and Ewing point out that this is not necessarily inconsistent with their observed spectra, it is interesting to note that a recent *ab initio* SCF calculation[55] gives N—N = 1.74 Å and NNO = 107° and supports the original results. Chemical intuition might suggest that $(NO)_2$ is a very stable entity but the temperature dependence of the spectrum shows that the enthalpy of formation is only -2.45 kcal mol^{-1}.

We shall now return to the subject of homonuclear diatomic molecules. When such a molecule approaches another molecule or atom, its electron distribution is perturbed. In general an electric dipole moment is induced. This oscillates with the vibrational frequency so that the i.r. spectrum of the individual molecules is observable. The full theory of such collision-induced spectra has been worked out by Van Kranendonk and co-workers[56-58]. These spectra depend quadratically upon the pressure and show structure due to translational motion. They may thus be considered as originating from unbound dimers. It should be noted that these induced features are very weak and require either high pressures, greater than an atmosphere, or long path lengths of many metres in order to be observed. The whole subject of the induced spectra of hydrogen has been recently reviewed by Welsh[59].

At low temperatures and high pressures, Welsh and his colleagues have been able to form several stable dimers containing H_2 and D_2. The spectra occur in the same region as the collision induced features. However, the bound state spectra may be obtained by subtracting the calculated translational spectra. The resulting integrated absorption has a quadratic pressure dependence but unlike the translational features shows a strong temperature variation.

The first spectrum to be observed[60] was that of $(H_2)_2$, which required temperatures below 40 K at 1 atm pressure and an optical path length of 13.6 m; this reflects the very low bond strength of the dimer. Subsequently, complexes of hydrogen (or deuterium) with inert gases have been observed. Initial work[61-63] was performed at pressures above 5 atm so that many of the interesting spectral features were blurred out by collision broadening of the lines. However, by using greatly increased path lengths of 165 m, McKellar and Welsh[64-66] were able to work at lower pressures and obtain spectra with substantially improved resolution. Figure 3.4 shows the features due to H_2—Ar which accompany the $S_1(0)$ ($J = 0 \rightarrow 2$) transition of para-hydrogen as the pressure is reduced.

The spectra correspond to those for a nearly isotropic intermolecular potential. They occur in the region of the quadrupole induced vibrational spectrum of hydrogen ($\Delta v = 1$, $\Delta G = 0, \pm 1$) and show fine structure due to changes in the end over end rotational momentum of the complex; this obeys the selection rules $\Delta R = \pm 1, \pm 3$. From the spacing of these lines the rotational constants of the complexes have been obtained and their mean intermolecular distances calculated. The values for the different complexes are: H_2—Ne = 3.64, H_2—Ar = 3.94, H_2—Kr = 4.07 and H_2—Xe = 4.27 Å. The equilibrium distances are somewhat smaller than these. In the complexes containing argon, krypton and xenon there is evidence of a further

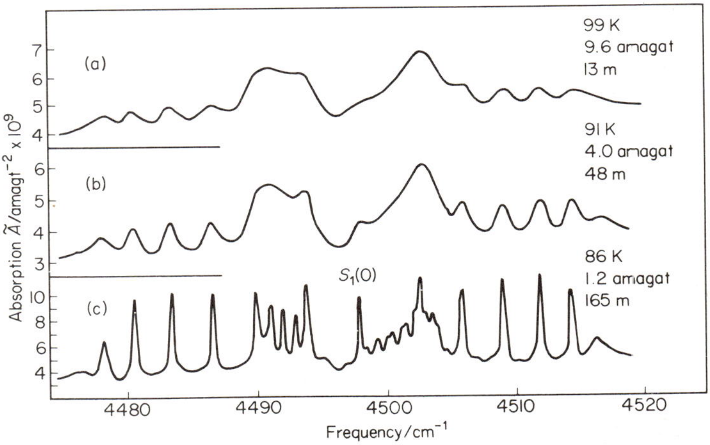

Figure 3.4 The spectrum due to H_2—Ar in the region of the $S_1(0)$ transition of para-H_2 + Ar mixtures as a function of the experimental conditions (From McKellar and Welsh[64], by courtesy of the American Institute of Physics)

small splitting in the fine structure owing to a small anisotropy in the intermolecular potential; this removes the degeneracy of the states with a given *G* and *R* but different *J*. McKellar and Welsh[64] have also observed the spectrum of H_2—N_2.

Ewing and co-workers have observed the spectra of similar dimers containing oxygen. First, Long and Ewing[67,68] observed the i.r. spectrum of $(O_2)_2$ in the region of the forbidden O_2 fundamental. They identified three discrete bands superimposed on a broad collision induced band. Unlike the hydrogen complexes, rotational structure is not well resolved, so consequently it has not been possible to determine the equilibrium structure. However, the separation of the maxima of the observed P and R branches indicates an average intermolecular distance of 4.8 Å. Furthermore, the temperature dependence of these bands at constant density gives the energy of formation as $\Delta E = -0.53 \pm 0.07$ kcal mol^{-1}; this small value is in basic agreement with that obtained from second virial coefficient data[2].

The i.r. spectrum of O_2—Ar has been observed[69] in mixtures of oxygen and argon at 93 K. It occurs in the same region as the oxygen dimer bands whose known structure must be first subtracted. The resulting absorption spectrum shows much fine structure due to the internal rotation or libration of the oxygen molecule. Analysis of the spectrum indicates a T-type equilibrium structure but with a barrier to internal rotation of only 30 cm^{-1}. The separation of the maxima of the unresolved P and R branches indicates that the distance between the argon atom and the centre of mass of the oxygen is approximately 3.5 Å. Similar work[70] on mixtures of argon and nitrogen has provided the spectrum of N_2—Ar. It possesses several features in common with O_2—Ar. The fine structure indicates a similar T-shaped equilibrium configuration with a barrier to internal rotation of the N_2 of only 20 cm^{-1}. The shape of the rotational envelope yields an intermolecular distance of 3.9 Å; this is close to 4.0 Å, the mean of the values for Ar_2 and $(N_2)_2$ obtained from second virial coefficients.

One of the most complex van der Waals molecules to have been studied by i.r. techniques is the CO_2 dimer. Mannik *et al.*[71] studied the region of the i.r. inactive Fermi resonance (ν_1 and $2\nu_2$) transitions at $-80\,^\circ$C. After removal of the collision-induced spectra and weakly allowed bands of the isotopic forms of CO_2, a dimer spectrum results. It is interpreted in terms of a rigid T-structure, which is the stable form for quadrupole–quadrupole interactions.

3.4.2 Raman spectroscopy

By comparison with i.r. spectroscopy there has been very little work using Raman spectroscopy to study van der Waals molecules. This is perhaps surprising since this technique permits the study of inert gas dimers and similar species formed from two identical atoms. To the author's knowledge the only definitive work is that of Morgan and Frommhold[72].

There have been several studies[73] of Raman scattering from inert gas atoms at high pressures, but these are probably due to collision induced effects and may be compared with the translational i.r. spectra of the previous section. However, a theoretical study by Levine[74] showed that at low temperatures spectra due to bound dimers should be observable, provided the pressure is not too high to blur out the features. By using 3 atm of argon and temperatures in the region 103–300 K, Morgan and Frommhold obtained unresolved rotational Raman lines close to the Raleigh line. The profile yields an effective interatomic distance of 4.57 ± 0.5 Å, the weighted average of many vibrational levels.

3.4.3 Ultraviolet spectroscopy

van der Waals molecules, like all other molecules, possess an electronic spectrum and, provided the excited states are not too strongly bound, the transitions occur in the same region as those of the parent monomers. In principle both absorption and emission spectra may be observed. However, because the excited states are usually more strongly bound with potential minima at shorter intermolecular distances than in the ground state, emission occurs at short intermolecular distances where the ground state potential is repulsive. The corresponding spectra usually involve non-bound states of the ground state molecule. Instead it is preferable to observe absorption spectra at high pressures and low temperatures when bound states of the dimer are sufficiently populated.

Much of the work in this subject has been performed by Tanaka and Yoshino on the inert gas dimers. For all the inert gas atoms the first absorption bands occur in the vacuum u.v. Provided the temperature is kept sufficiently low, bands due to the dimer are found to accompany these lines. For He_2 [75], Ar_2 [76], Ne_2 [77,78], Kr_2 [79] and the mixed dimers HeNe [77] and ArKr [79], Tanaka and co-workers observed a number of band systems with vibrational and in most cases rotational structure. Except for Ne_2 it was not possible to make a rotational analysis. Instead the band heads were measured and used to indicate the relative energies of the vibrational levels. Care should be taken using this argument since in these very floppy molecules there is substantial rotation–vibration interaction and the band heads each appear at slightly different separatior s from the band origin. Thus for really accurate results it is necessary to have a complete rotational analysis. However, even with the above approximation significant information has been obtained. It is possible to extrapolate the vibrational levels to the dissociation limit and hence obtain a value of the dissociation energy. Birge–Sponer extrapolation has yielded a dissociation energy D_0° of 76.9 cm^{-1} for Ar_2. For Ne_2, $D_e = 30.2$ cm^{-1} and a rotational analysis has given $B_0 = 0.17$ cm^{-1} or $r_0 = 3.1$ Å. There is, however, a certain amount of doubt about the accuracy of the extrapolation method for calculating the dissociation energy. Tanaka, Yoshino and Freeman[79] discuss the problems associated with this procedure for the case of Kr_2. By assuming an inverse sixth power long-range potential near the dissociation limit, they obtain a well depth of 138.4 ± 1.9 cm^{-1}.

Maitland and Smith[80,81] have re-analysed the absorption spectra of Ne_2 and Ar_2 using the Rydberg–Klein–Rees method to obtain the width of the potential well as a function of its depth. They combine these RKR turning points with the equilibrium and transport properties of the gas to obtain a new potential function. For Ar_2 they obtain a well depth of 99.0 cm^{-1} (ε/k = 142.5 K) at r_e = 3.75 Å; the results for Ne_2 are ε/k = 39.6 K and r_e = 3.07 Å. These are in basic agreement with the molecular beam scattering results.

Xe_2 possesses the strongest van der Waals bond of the inert gas dimers but it is the least well studied in the vacuum u.v. This is primarily because of the very small values of the ground state vibrational and rotational constants (estimated to be ω_e = 17 cm^{-1} and B_e = 0.013 cm^{-1}), which cause poorly resolved, strongly overlapping rotation–vibration bands[82].

Oscillations have been observed in the continuum absorption and emission spectra of the alkali–inert gas molecules[83-85]. It was not possible to obtain information on the ground $X^2\Sigma$ state but it appears that the excited $A^2\Pi$ state of each molecule has a well depth not much greater than kT at room temperature.

Some of the metal atoms, especially those of Group II, form dimers that may be correctly called van der Waals molecules. Early work on the diffuse emission and absorption spectra of Hg_2, Cd_2, Zn_2, Mg_2 and HgAr has been reviewed by Finkelnburg and Peters[86]. More recently, Balfour and Douglas have obtained the well-resolved absorption spectrum of Mg_2 [87-89]. The spectrum displays large rotation–vibration coupling. Their rotational analysis leads to the ground state equilibrium constants ω_e'' = 51.12 cm^{-1} and B_e'' = 0.0929 cm^{-1} (r_e'' = 3.8915 Å) as well as many higher order terms. Extrapolating their vibrational levels yields a dissociation energy, D_0° of 399 ± 5 cm^{-1}. Stwalley[90], who used a calculated long-range potential, found that this value should be altered to 403.7 ± 0.7 cm^{-1}. Similar absorption spectra for Ca_2 yield ω_e'' = 65.0 cm^{-1}, B_e'' = 0.0460 cm^{-1}, r_e'' = 4.28 Å and D_e'' = 940 ± 40 cm^{-1}.

There has been a lot of work on forbidden electronic transitions or simultaneous excitation of two molecules at twice the frequency[92-95]. These features normally depend quadratically on the pressure of the gas but should be considered as the spectra of unbound collision pairs just as in the many i.r. examples. They are induced by the collision when the selection rules for molecular transitions are relaxed. However, recently Long and Ewing[68], working at 87 K, observed a group of features in the visible region on top of the collision induced absorption. These eight fine structure bands possess a quadratic pressure dependence and their intensity decreases rapidly with increase in temperature. They may thus reasonably be assigned to $(O_2)_2$. The progression of bands was assigned to vibrational structure and led to dimer dissociation energies for the ground state ($^3\Sigma_g^- + {}^3\Sigma_g^-$) of 87 cm^{-1} and for the excited state ($^1\Delta_g + {}^1\Delta_g$) of 50 cm^{-1}.

In addition, unresolved absorption bands due to the dimers $(X_2)_2$ of the halogens have been observed. Ogryzlo and Sanctuary[96] noticed an anomalous bump in the absorption profile of Br_2; it showed a quadratic pressure dependence and disappeared rapidly with increasing temperature. The temperature dependence yielded an enthalpy of formation of −2.6 kcal mol^{-1} for $(Br_2)_2$. More precise work by Passchier *et al.*[97] allowed separation of the linear and quadratic pressure dependences of this band and gave ΔH° = −2.27 ± 0.25 kcal mol^{-1}. These results are in remarkable agreement with the value of −2.64 kcal mol^{-1} obtained by Kokovin[98] from *PVT* data. Passchier and

Gregory[99] have performed similar u.v. studies on bands in iodine vapour at 270 nm. Although elevated temperatures were required to obtain significant pressures of iodine, it was possible with difficulty to obtain $\Delta H = -2.9 \pm 0.4$ kcal mol^{-1} at 332°C. Unfortunately no vibrational or rotational structure was observed in this work.

3.4.4 Mass spectrometry

It has been mentioned earlier in Section 3.2.2 that mass spectrometers were used to analyse the products of a supersonic nozzle source. They can, however, provide a general means of monitoring dimer formation. They can also give near quantitative results on dimer concentrations provided some allowance is made for the possibility of fragmentation of the clusters in the ionisation process.

By using a Knudsen effusion source, Leckenby and Robbins[100] have been able to obtain the concentration of dimers as a function of pressure and temperature for a number of molecules. They used gas pressures of up to 200 Torr in order to form significant concentrations of the dimers and they used very small orifices with dimensions less than 2×10^{-4} cm in order to prevent adiabatic cooling from occuring. Under these conditions, dimers of argon, xenon, nitrogen, oxygen, water, carbon dioxide and nitrous oxide were observed. In order to obtain quantitative concentrations of the dimers it was necessary to correct the dimer signal by a factor of two to allow for its larger ionisation cross-section when comparing with the monomer signal. As expected, their results show that the fractional concentration of the dimers increases linearly with source pressure up to the region where continuum flow takes over from molecular flow. Also their results basically support the equilibrium concentrations calculated by Stogryn and Hirschfelder[11], except for carbon dioxide, where the calculated value is about an order of magnitude too large.

The early work of Henkes[23] and Bentley[22] proved the formation of van der Waals molecules in supersonic nozzles. Following this work, Leckenby *et al.*[101] have made detailed studies on the variation of polymer concentrations with source pressure and temperature. They compared gases with similar critical constants, which provide a means of comparing the condensability of gases. Their results show that gases with low values of γ (the ratio of specific heats) form dimers much less readily. In fact SF_6 with $\gamma = 1.093$ produced no detectable polymers.

Argon has been the most frequently studied molecule. Milne and Greene[102] initially studied the equilibrium concentrations of argon polymers formed in supersonic nozzles. For this work they assumed that negligible fragmentation occurs upon ionisation. They argue that, since the potential minima of neutral dimers occur at larger intermolecular distances than those of the corresponding ions, vertical transitions should lead mainly to bound states of the dimer ions. This argument is, however, rather less convincing for higher clusters. Like Leckenby and Robbins[100], these authors assume that the ionisation cross-section varies linearly with the polymer number n. It was also necessary to correct the intensity by another factor of n because of the smaller spreading of the heavier polymers in the nozzle beam. From these results they were able to obtain detailed information on the variation of the mole fraction of clusters with pressure and nozzle diameter; they observe that the mole fraction of dimer formed in the nozzle varies with $p_0^2 d$. By extrapolating their results to zero orifice diameter d (where the mean free path of the gas is greater than the source dimensions), Milne and Greene obtain what are to be considered as equilibrium mole fractions; these are very similar to the results of Leckenby *et al.*

In a later paper, Milne *et al.*[26] used the mass spectrometer to study the kinetics of dimer growth in the free-jet expansion. The analysis of this process is complicated by the fact that the effective temperature of the gas is continuously changing. However, by using Ashkenas and Sherman's expression[103] for the Mach number as a function of distance and a rate constant of the form $k_f = B/T^n$ where $n = 2$ or 3, they were able to obtain a good fit to the observed dimer growth at several pressures and orifice diameters. Strangely, an Arrhenius type function, $k_f = B \exp(\varepsilon/T)$, with a negative activation energy gave a far inferior fit. An alternative approach by Golomb *et al.*[25] to the mechanism of dimer formation used the normal termolecular collision rate constant, $k_f = 3 \times 10^{-33}\ T^{\frac{1}{2}}$ cm^6 molecule^{-2} s^{-1}. They show that the observed Ar_2 abundances are in reasonable agreement with the calculated values. These same authors performed a similar study of the rate of growth of the nitric oxide dimer.

3.4.5 Molecular beam spectroscopy

The technique of molecular beam spectroscopy is fully described in the books of Ramsey[16] and Kusch and Hughes[104]. It is also reviewed in this volume[105]. The method is well suited to the study of van der Waals molecules since the spectroscopy is performed at very high vacua, typically less than 10^{-6} Torr, and the complexes once formed are not destroyed by intermolecular collisions.

Briefly the experiment involves monitoring the intensity of a well collimated beam of molecules after it has passed through two inhomogeneous (deflecting) field regions. Between these is applied an oscillating field. The spectrum appears as a change in the intensity of the beam as the frequency of the oscillating field is changed. The resulting data are normally of very high precision since the linewidths depend primarily on the inverse of the time the molecule spends in the resonance region and are typically only 1–10 kHz.

The source of a molecular beam spectrometer may be a suitable small hole. However, in the field of van der Waals molecules, a supersonic nozzle provides the largest yield of the weakly bound complexes and is almost universally preferred. The detector is usually a mass spectrometer since it permits the study of a definite polymer in what may be a complex mixture.

Recently Klemperer and co-workers[106-108] have observed the spectra of a number of complexes using the molecular beam electric resonance technique. They have studied a number of complexes between argon and some stable diatomic molecules. These include Ar—HCl [106], Ar—HF [107] and Ar—ClF [108]. All these molecules behave like near symmetric prolate tops. Consequently the observed rotational transitions with $\Delta K = 0$ provide values of the mean rotational constant $(B + C)/2$ and its centrifugal distortion D_J. These results also give reliable estimates of the length of the van der Waals bond provided we can assume that the geometry of each diatomic molecule remains unchanged on complex formation.

The nuclear hyperfine structure and Stark effect provide other interesting information. In the $K = 0$ states of these molecules the component of the dipole moment along the a inertial axis can be measured by applying an electric field. Assuming that there is no charge transfer between the two molecules and that the only charge redistribution is due to the polarisation of the argon atom in the field of the approaching diatomic molecule, Novick *et al.*[106] show that the observed dipole moment may be given as a function of known molecular constants:

$$\mu_a = \mu\langle\cos\theta\rangle + a_{\mathrm{Ar}}\{2\mu\langle\cos\theta\rangle/R^3 + \tfrac{3}{2}Q\langle 3\cos^2\theta - 1\rangle/R^4\}$$

where μ is the dipole moment of the isolated diatomic molecule, Q is its quadrupole moment and a_{Ar} is the polarisability of the argon atom. R is the distance between the centres of mass of the two species and θ is the angle between the inertial axis and that of the diatomic molecule. The rotational constant data provide a vibrational mean of $\cos\theta$ in what may be a very floppy molecule.

Similarly, if the diatomic molecule possesses a nucleus with spin $I \geqslant 1$, the quadrupole coupling constant may provide further structural information. Although quadrupole coupling is a second rank tensor interaction it is still the a component which is measured in $K = 0$ states. Assuming the environment of the quadrupole nucleus is unchanged on complex formation, the observed coupling constant is given by

$$(eqQ)_a = \tfrac{1}{2}eqQ\langle 3\cos^2\theta - 1\rangle$$

It is interesting to observe that the effective structures for Ar—HCl vary greatly on isotopic substitution. For the deuterated species, dipole moment and quadrupole data yield $\langle\theta\rangle = 33.5°$. On the other hand, the angles are much larger for the protonated species: arc cos $\langle\cos\theta\rangle = 47.5°$ and arc cos $\langle\cos^2\theta\rangle^{\frac{1}{2}} = 41.5°$. These are consistent with a weak angular potential with a barrier to rotation approximately 5 times the HCl rotational constant.

Kleppner and co-workers[109] have observed the r.f. spectrum of ArK using a magnetic resonance spectrometer. They studied $\Delta m_F = 1$ Zeeman transitions in the $F = 2$ level at low magnetic field. These yield an electron spin–rotation constant $\bar{\gamma}$ of 0.24 MHz.

Molecular beam spectroscopy has also provided much useful information on a number of hydrogen bonded species. The work[29] on $(HF)_2$ and its deuterated forms (see Figure 3.5) has led to significant information on this, one of the simplest of hydrogen bonded systems. The rotational constants have led to a reliable F—F distance of 2.79 Å. The dipole moment and deuterium quadrupole coupling constants suggest a bent structure; the end hydrogen is approximately 60° off axis while the inner-bonded hydrogen is much nearer the axis and may support a near linear hydrogen bond. Recent MO calculations seem to be converging on this result[110-112]. One novel feature observed in the symmetric dimers is a tunnelling motion between the two possible equivalent structures. It involves the breaking and making of the hydrogen bond. This motion manifests itself as a splitting of all rotational levels. HF—DF shows no such motion.

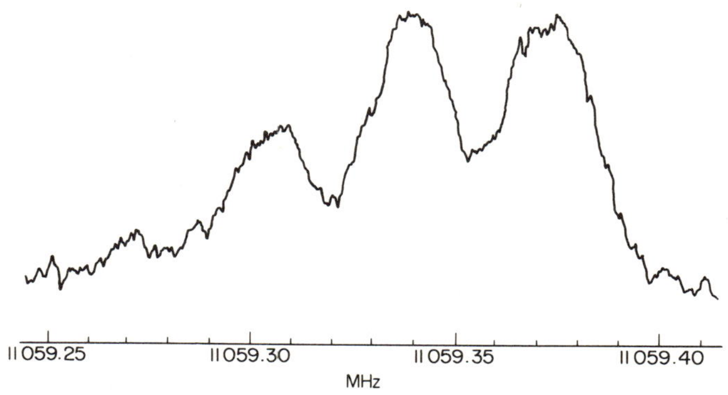

Figure 3.5 Deuterium quadrupole structure of the $J = 1^+$, $M_J = 0$ to $J = 0^-$ transition of $(DF)_2$ at 1082.6 V cm^{-1}. The theoretical statistical weights of the components of the triplet are 3:8:10 for a tunnelling molecule. (From Dyke *et al.*[29], by courtesy of the American Institute of Physics.)

The molecular beam electric resonance spectrum of $(H_2O)_2$ has recently been observed by Dyke and Muenter[113]. Their work lends support to the favoured 'linear hydrogen bond' structure[114, 115].

In addition to the study of resonance spectra, the molecular beam equipment may be used to test the polarity of sample species[116]. For a quadrupole deflecting field, molecules with a positive Stark effect (energy increases in an electric field) are focused towards the axis of the apparatus. The intensity of the beam at the mass spectrometer is seen to increase when the field is increased. Molecules which interact with an electric field only through their polarisability are defocused. There have been a number of such studies on van der Waals molecules[117-120]. We list here some of the more interesting and perhaps unexpected results. The dimer of nitric oxide[117] is polar, indicating an open structure and supporting the i.r. structure[52]. The higher polymers are non-polar. Although HF dimer is polar[29] and its structure has been determined, the higher polymers are non-polar. They are considered to form cyclic structures. Similar work on HCl complexes[117] also shows a polar dimer and non-polar higher polymers. Although BF_3 has a planar and hence non-polar structure, its complexes with inert gases definitely possess a significant dipole moment. This is interpreted as being due to a distortion of the BF_3 framework rather than to charge transfer. It is also possible that the polarity is due to a dipole induced in the argon by the quadrupole moment of the BF_3 [141]. Another amazing result is that for carbon dioxide, which by itself is symmetric and non-polar. Its dimer is polar, in agreement with the locked T-structure obtained from vibrational spectra[71]. The dipole might again be attributed to a quadrupole-induced moment.

In the water system, both the monomer and the dimer refocus strongly[118]; this rules out a symmetric dimer structure. The weak focusing and defocusing of the higher polymers is less easy to interpret. There is, however, some evidence that the heavier polymers have cyclic structures. Nitrogen dioxide is a polar molecule but all its polymers have been shown to be non-polar[119]. This again indicates symmetric and possibly cyclic structures.

3.4.6 Electron diffraction

Diffraction techniques are a useful means of determining the structure of van der Waals molecules since they yield the interatomic distances directly. Rouault, Audit and co-workers[120-124] have observed electron patterns due to a number of polymers of CO_2, argon and xenon. The complexes were formed by the isentropic expansion of the required gas through a supersonic nozzle followed by a skimmer. The resulting beam of molecules is intercepted by a beam of high-energy electrons and the diffraction pattern of the scattered electrons is detected photographically. By a suitable choice of source temperature and pressure, it has been possible to observe dimers, polymers, liquid droplets and crystals; each has its own characteristic scattering pattern. Similar work by Stein and Armstrong[125] on clusters of CO_2 and H_2O containing more than 1000 molecules has yielded diffraction patterns and structures which are the same as in the bulk material.

Some of the most informative results obtained by electron diffraction[124] concern the structure of Ar_2 and Xe_2, which may be present in up to 50% abundance. The diffraction pattern for argon dimer yields a mean interatomic distance of 3.93 ± 0.06 Å. This is the mean of several vibrational levels and agrees well with the value calculated from a Lennard-Jones potential with a minimum at $r_e = 3.82$ Å and using an effective vibrational temperature of 68 K. The corresponding mean bond length for Xe_2 was determined to be 4.41 ± 0.03 Å. The large number of oscillations in the observed interference function showed a gradual decay across the photographic plate. This is a result of the slightly different diffraction patterns from molecules in different vibrational states. Analysis of this decay provided a mean vibrational displacement of 0.11 Å.

Janzen and Bartell[126] have observed the electron diffraction pattern of the condensation products of HF. They interpret their results in terms of a monomer plus a hexamer with an F—F distance of 2.55 Å. It seems likely that there is a range of polymers present in the gas, but their results fit nicely between the value of 2.79 Å for the dimer and 2.49 Å for the chains in the crystal[127].

3.4.7 Other techniques

Bouchiat and co-workers[128-131] have found evidence for alkali–inert gas dimers by observing the rate of relaxation of optically polarised rubidium atoms (in a magnetic field) by inert gas atoms. They have shown that the relaxation is governed by the spin–rotation interaction of an atom pair and is strongly affected by bound van der Waals molecules, which are formed by three-body collisions and exist until destroyed by a subsequent two-body collision. From their study of the relaxation rate as a function of the inert gas pressure and d.c. magnetic field they were able to determine molecular parameters such as the rate of formation of the complex, its lifetime and its spin–rotation constant, $\bar{\gamma}$. In order to obtain $\bar{\gamma}$ they assumed a Lennard-Jones potential for the interaction of the two atoms. This yielded the following values of the mean spin–rotation coupling constants: Rb—Ar = 0.09–0.11, Rb—Kr = 0.647 ± 0.018 and Rb—Xe = 1.4–1.7 MHz. The value for the krypton complex is in good agreement with that calculated by Herman[132]. The result for Rb—Ar is somewhat lower than the value of 0.24 MHz obtained by molecular beam studies; this is probably due to the poorly known interatomic potential. The relaxation method provides the mean contribution to $\bar{\gamma}$ from many rotational and vibrational states. Attempts to overcome this by observing the effect of a magnetic field on the relaxation rate have yet to produce definitive results[131]. It does not appear likely that other structural information will be obtained by this technique.

The methods of crossed-beam scattering at thermal energies have provided extensive information on the interaction potentials between atoms and in some cases molecules. A detailed discussion of the experimental techniques and of the inversion of the data to obtain intermolecular potentials is given in the reviews of Pauly and Toennies[133] and Bernstein and Muckerman[134]. However, in the past few years there have been a number of significant improvements in the accuracy of the potential energy functions. For example, Lee and co-workers have obtained very precise potentials of the inert gas pairs[135-137, 142]. The results of these and other workers are compared and contrasted in the review by Certain and Bruch[4]. An interesting molecular potential is that between two hydrogen molecules, since its anisotropy is expected to be very small. Farrar and Lee[138] assumed a spherically symmetric potential to explain their differential scattering results and obtained ε/k = 34.8 K and r_e = 3.49 Å. A calculation of the stretching vibrations yields just one bound vibrational level only 2.0 cm^{-1} below the continuum; this explains the considerable difficulty[60] in observing spectra from $(H_2)_2$.

Another means of obtaining an intermolecular potential is via the equation of state. The virial expansion is

$$PV/RT = 1 + B(T)/V + C(T)/V^2 + \cdots$$

where the coefficients $B(T)$ and $C(T)$ depend upon both the temperature and the effective intermolecular potential. Using classical statistical mechanics, the second virial coefficient for a spherical potential $V(r)$ is given by

$$B(T) = 2\pi N_0 \int_0^\infty \{1 - \exp[-V(r)/kT]\} r^2 \mathrm{d}r$$

Under certain circumstances quantum mechanical corrections must be added[139]. The measurement of $B(T)$ over a range of temperatures gives information on the potential. For example, if an isotropic Lennard-Jones potential is assumed, it is possible to obtain the well depths and equilibrium intermolecular distances[2]. In a similar way, intermolecular potentials can be obtained from transport properties like viscosities, thermal conductivities and diffusion coefficients[2]. Many of the results have been reviewed by Certain and Bruch[4]. The problem with all these methods is that, for large ranges of the temperature, the results are insensitive to the potential function. The inadequacies of many of the present potentials are emphasised by Rowlinson[140].

3.5 CONCLUSION

At the conclusion of this review, it is evident that the study of van der Waals molecules is still in its infancy, both from the experimental and theoretical points of view. A large number of techniques have been used but as yet have been applied to a few systems. Often each molecule has been studied by just one technique and it is impossible to compare and contrast the results of different experiments. Except for the atom–atom systems, it appears that the i.r. and molecular beam studies are likely to be the most fruitful. These methods have given effective structures and also significant information on the potential function for the bending and stretching of the van der Waals bond. With improvements in experimental techniques it may also prove possible to observe the rotational spectra of dissimilar atom dimers; they possess small but non-zero electric dipole moments[143-145].

The present work indicates that many of the anisotropic intermolecular potentials can be represented by classical electrostatic models, but the data are not sufficiently detailed for firm conclusions to be drawn. It is the author's hope that simple but realistic models for the repulsive potential will soon appear; this will greatly aid the interpretation of the structures of van der Waals molecules. There is also scope for detailed calculations on the complete potential energy surface for the van der Waals bond. This will give not only the equilibrium structures but also insight into the interesting dynamics of the low frequency molecular motions.

Notes Added in Proof

The kinetics of dimer formation during three-body collisions have been studied by molecular dynamics[146]. The mole fraction of stable and metastable dimers are in good agreement with the results of equilibrium statistical mechanics[11].

The i.r. spectra of $(H_2)_2$ has been obtained with greatly improved resolution by McKellar and Welsh[147]. They have also observed the analogous spectra of $(D_2)_2$ and H_2–D_2. Their results indicate an effective intermolecular distance of 4.4 Å in $(H_2)_2$ and 3.9 Å in $(D_2)_2$. McKellar[148] has also observed the i.r. spectrum of HD–Ar. Unlike the similar spectra of H_2–Ar and D_2–Ar, it is necessary to invoke a non-spherical intermolecular potential; this is probably due to the non-coincidence of the centre of rotation of HD with its geometrical centre. The form of the intermolecular potential has been determined by Le Roy and Van Kranendonk[149].

The vacuum-u.v. absorption spectrum of Xe_2 has now been observed with resolved vibrational structure by Tanaka and co-workers[150]. They obtained a spectroscopic value of $195.5\ cm^{-1}$ for the well depth, which is in good agreement with the results of molecular beam scattering experiments. Drummond and Gallagher[151] have observed electronic spectra due to rubidium–inert gas complexes. These yield both ground and excited state potential functions.

References

1. Jeans, J. H. (1904). *The Dynamical Theory of Gases* (Cambridge: University Press)
2. Hirschfelder, J. O., Curtiss, C. F. and Bird, R. B. (1954). *Molecular Theory of Gases and Liquids* (New York: John Wiley)
3. Margenau, H. and Kestner, N. R. (1971). *Theory of Intermolecular Forces*, 2nd Ed. (Oxford: Pergamon Press)
4. Certain, P. R. and Bruch, L. W. (1972). *MTP International Review of Science, Physical Chemistry Series One*, Vol. 1, 113 (W. Byers Brown, editor) (London: Butterworths)
5. Hirschfelder, J. O. (1967). *Advan. Chem. Phys.*, **12**
6. Dalgarno, A. and Davison, W. D. (1966). *Advan. At. Mol. Phys.*, **2,** 1
7. Buckingham, A. D. and Utting, B. D. (1970). *Ann. Rev. Phys. Chem.*, **21,** 287
8. Ewing, G. E. (1972). *Angew. Chem. Int. Ed. Engl.*, **11,** 486
9. Hill, T. L. (1955). *J. Chem. Phys.*, **23,** 617
10. Hill, T. L. (1956). *Statistical Mechanics*, Chap. 5 (New York: McGraw-Hill)
11. Stogryn, D. E. and Hirschfelder, J. O. (1959). *J. Chem. Phys.*, **31,** 1531
12. Buluggiu, E. and Foglia, C. (1967). *Chem. Phys. Lett.*, **1,** 82
13. Konowalow, D. D. and Hirschfelder, J. O. (1961). *Phys. Fluids*, **4,** 629
14. Kantrowitz, A. and Grey, J. (1951). *Rev. Sci. Instr.*, **22,** 328
15. Kistiakowsky, G. B. and Slichter, W. P. (1951). *Rev. Sci. Instr.*, **22,** 333
16. Ramsey, N. F. (1955). *Molecular Beams*, Chap. 2 (London: Oxford University Press)
17. Anderson, J. B., Andres, R. P. and Fenn, J. B. (1966). *Advan. Chem. Phys.*, **10,** 275
18. Gspann, J. and Korting, K. (1973). *J. Chem. Phys.*, **59,** 4726
19. Becker, E. W., Bier, K. and Henkes, W. (1956). *Z. Phys.*, **146,** 333
20. Becker, E. W., Klingelhofer, R. and Lohse, P. (1961). *Z. Naturforsch.*, **16a,** 1259
21. Becker, E. W., Klingelhofer, R. and Lohse, P. (1962). *Z. Naturforsch.*, **17a,** 432
22. Bentley, P. G. (1961). *Nature* (*London*), **190,** 432
23. Henkes, W. (1961). *Z. Naturforsch.*, **16a,** 842
24. Bier, K. and Hagena, O. (1966). *Proc. Symp. Rarefied Gas Dyn. 4th*, **2,** 260
25. Golomb, D., Good, R. E. and Brown, R. F. (1970). *J. Chem. Phys.*, **52,** 1545
26. Milne, T. A., Vandegrift, A. E. and Greene, F. T. (1970). *J. Chem. Phys.*, **52,** 1552
27. Dyke, T. R. (1974). Personal communication
28. Janzen, J. and Bartell, L. S. (1968). *U.S. Atomic Energy Comm., Report* IS–1940
29. Dyke, T. R., Howard, B. J. and Klemperer, W. (1972). *J. Chem. Phys.*, **56,** 2442
30. Kilpatrick, J. E. and Kilpatrick, M. F. (1951). *J. Chem. Phys.*, **19,** 930
31. Cashion, J. K. (1968). *J. Chem. Phys.*, **48,** 94
32. Cashion, J. K. (1966). *J. Chem. Phys.*, **45,** 1656
33. Bratoz, S. and Martin, M. L. (1965). *J. Chem. Phys.*, **42,** 1051
34. Born, M. and Oppenheimer, R. (1927). *Ann. Phys.*, **84,** 457
35. Herzberg, G. (1950). *Spectra of Diatomic Molecules* (New York: Van Nostrand)
36. Carrington, A., Levy, D. H. and Miller, T. A. (1970). *Advan. Chem. Phys.*, **18,** 149
37. Gordon, R. G. and Cashion, J. K. (1966). *J. Chem. Phys.*, **44,** 1190
38. Kuipers, G. A. (1958). *J. Mol. Spectrosc.*, **2,** 75
39. Smith, D. F. (1959). *J. Mol. Spectrosc.*, **3,** 473
40. Rank, D. H., Sitaram, P., Glickman, W. A. and Wiggins, T. A. (1963). *J. Chem. Phys.*, **39,** 2673
41. Rank, D. H., Glickman, W. A. and Wiggins, T. A. (1965). *J. Chem. Phys.*, **43,** 1304
42. Rank, D. H., Rao, B. S. and Wiggins, T. A. (1962). *J. Chem. Phys.*, **39,** 2673
43. Larvor, M., Houdeau, J.-P. and Haeusler, C. (1973). *Compt. Rend. Acad. Sci. Ser. B*, **276,** 421
44. Larvor, M., Houdeau, J.-P. and Haeusler, C. (1973). *Spectrochim. Acta*, **39,** 2673
45. Neilsen, W. B. and Gordon, R. G. (1973). *J. Chem. Phys.*, **58,** 4149

46. Herget, W. F., Gailar, N. M., Lovell, R. J. and Nielsen, A. H. (1960). *J. Opt. Soc. Amer.*, **50,** 1264
47. Huong, P. V. and Couzi, M. (1969). *J. Chim. Phys.*, **66,** 1309
48. Himes, J. L. and Wiggins, T. A. (1971). *J. Mol. Spectrosc.*, **40,** 418
49. Thomas, R. K. and Thompson, H. W. (1970). *Proc. Roy. Soc. A*, **316,** 303
50. Thomas, R. K. (1971). *Proc. Roy. Soc. A*, **325,** 133
51. Couzi, M. and Huong, P. V. (1970). *Compt. Rend. Acad. Sci. Ser. B*, **270,** 832
52. Dinerman, C. E. and Ewing, G. E. (1970). *J. Chem. Phys.*, **53,** 626
53. Dinerman, C. E. and Ewing, G. E. (1971). *J. Chem. Phys.*, **54,** 3660
54. Lipscomb, W. N., Wang, F. E., May, W. R. and Lippert, E. (1961). *Acta Crystallogr.*, **14,** 1100
55. Skancke, P. N. and Boggs, J. E. (1973). *Chem. Phys. Lett.*, **21,** 316
56. Van Kranendonk, J. (1958). *Physica*, **24,** 347
57. Van Kranendonk, J. and Kiss, Z. J. (1959). *Can. J. Phys.*, **37,** 1137
58. Poll, J. D. and Van Kranendonk, J. (1961). *Can. J. Phys.*, **39,** 189
59. Welsh, H. L. (1972). *MTP International Review of Science, Physical Chemistry Series One*, Vol. 3, 33 (D. A. Ramsay, editor) (London: Butterworths)
60. Watanabe, A. and Welsh, H. L. (1964). *Phys. Rev. Lett.*, **13,** 810
61. Kudian, A., Welsh, H. L. and Watanabe, A. (1965). *J. Chem. Phys.*, **43,** 3397
62. Kudian, A., Welsh, H. L. and Watanabe, A. (1967). *J. Chem. Phys.*, **47,** 1553
63. Kudian, A. K. and Welsh, H. L. (1971). *Can. J. Phys.*, **49,** 230
64. McKellar, A. R. W. and Welsh, H. L. (1971). *J. Chem. Phys.*, **55,** 595
65. McKellar, A. R. W. and Welsh, H. L. (1972). *Can. J. Phys.*, **50,** 1458
66. McKellar, A. R. W., Rich, N. and Soots, V. (1970). *Appl. Opt.*, **9,** 222
67. Long, C. A. and Ewing, G. E. (1971). *Chem. Phys. Lett.*, **9,** 225
68. Long, C. A. and Ewing, G. E. (1973). *J. Chem. Phys.*, **58,** 4824
69. Henderson, G. and Ewing, G. E. (1973). *J. Chem. Phys.*, **59,** 2280
70. Henderson, G. and Ewing, G. E. (1974). *Mol. Phys.*, **27,** 903
71. Mannik, L., Stryland, J. C. and Welsh, H. L. (1971). *Can. J. Phys.*, **49,** 3056
72. Morgan, C. E. and Frommhold, L. (1972). *Phys. Rev. Lett.*, **29,** 1053
73. McTague, J. P. and Birnbaum, G. (1971). *Phys. Rev. A*, **3,** 1376
74. Levine, H. B. (1972). *J. Chem. Phys.*, **56,** 2455
75. Tanaka, Y. and Yoshino, K. (1969). *J. Chem. Phys.*, **50,** 3087
76. Tanaka, Y. and Yoshino, K. (1970). *J. Chem. Phys.*, **53,** 2012
77. Tanaka, Y. and Yoshino, K. (1972). *J. Chem. Phys.*, **57,** 2964
78. Tanaka, Y., Yoshino, K. and Freeman, D. E. (1973). *J. Chem. Phys.*, **59,** 564
79. Tanaka, Y., Yoshino, K. and Freeman, D. E. (1973). *J. Chem. Phys.*, **59,** 5160
80. Maitland, G. C. (1973). *Mol. Phys.*, **26,** 513
81. Maitland, G. C. and Smith, E. B. (1970). *Mol. Phys.*, **22,** 861
82. Castex, M. C. and Damany, N. (1974). *Chem. Phys. Lett.*, **24,** 437
83. Jefimenko, O. and Chen, S.-Y. (1957). *J. Chem. Phys.*, **26,** 913
84. Chen, C. L. and Phelps, A. V. (1973). *Phys. Rev. A*, **7,** 470
85. Carrington, C. G., Drummond, D., Gallagher, A. and Phelps, A. V. (1973). *Chem. Phys. Lett.*, **22,** 511
86. Finkelnburg, W. and Peters, Th. (1957). *Handbuch der Physik*, Vol. 28, 173 (Berlin: Springer-Verlag)
87. Balfour, W. J. and Douglas, A. E. (1970). *Can. J. Phys.*, **48,** 901
88. Brett, A. C. and Balfour, W. J. (1971). *J. Chem. Phys.*, **54,** 3240
89. Balfour, W. J. and Whitlock, R. F. (1972). *Can. J. Phys.*, **50,** 1648
90. Stwalley, W. C. (1970). *Chem. Phys. Lett.*, **7,** 600
91. Balfour, W. J. and Whitloch, R. F. (1971). *Chem. Commun.*, 1231
92. Tabisz, G. C., Allin, E. J. and Welsh, H. L. (1969). *Can. J. Phys.*, **47,** 2860
93. Blickensderfer, R. P. and Ewing, G. E. (1969). *J. Chem. Phys.*, **51,** 873
94. Blickensderfer, R. P. and Ewing, G. E. (1969). *J. Chem. Phys.*, **51,** 5284
95. McKellar, A. R. W., Rich, N. H. and Welsh, H. L. (1972). *Can. J. Phys.*, **50,** 1
96. Ogryzlo, E. A. and Sanctuary, B. C. (1965). *J. Phys. Chem.*, **69,** 4422
97. Passchier, A. A., Christian, J. D. and Gregory, N. W. (1967). *J. Phys. Chem.*, **71,** 937
98. Kokovin, G. A. (1965). *Russ. J. Inorg. Chem.*, **10,** 150
99. Passchier, A. A. and Gregory, N. W. (1968). *J. Phys. Chem.*, **72,** 2697

100. Leckenby, R. E. and Robbins, E. J. (1966). *Proc. Roy. Soc. A*, **291**, 389
101. Leckenby, R. E., Robbins, E. J. and Trevalion, P. A. (1964). *Proc. Roy. Soc. A*, **280**, 409
102. Milne, T. A. and Greene, F. T. (1967). *J. Chem. Phys.*, **47**, 4095
103. Ashkenas, H. and Sherman, F. S. (1966). *Proc. Symp. Rarefied Gas Dyn. 4th*, **2**, 84
104. Kusch, P. and Hughes, V. W. (1959). *Handbuch der Physik*, Vol. 37, 52 (Berlin: Springer-Verlag)
105. Muenter, J. S. (1974). *MTP International Review of Science, Physical Chemistry Series Two*, Vol. 2 (A. D. Buckingham, editor) (London: Butterworths)
106. Novick, S. E., Davies, P., Harris, S. J. and Klemperer, W. (1973). *J. Chem. Phys.*, **59**, 2293
107. Harris, S. J., Novick, S. E. and Klemperer, W. (1974). *J. Chem. Phys.*, **60**, 3208
108. Harris, S. J., Novick, S. E. and Klemperer, W. (1974). *J. Chem. Phys.*, **61**, 193
109. Mattison, E. M., Pritchard, D. E. and Kleppner, D. (1974). *Phys. Rev. Lett.*, **32**, 507
110. Kollman, P. A. and Allen, L. C. (1970). *J. Chem. Phys.*, **52**, 5085
111. Del Bene, J. and Pople, J. A. (1971). *J. Chem. Phys.*, **55**, 2296
112. O'Neil, S. V., Schaefer H. F., III, Baskin, C. P. and Bender, C. F. (1974). *J. Chem. Phys.*, **60**, 855
113. Dyke, T. R. and Muenter, J. S. (1974). *J. Chem. Phys.*, to be published
114. Hanskins, D., Moskowitz, J. W. and Stillinger, F. H. (1970). *J. Chem. Phys.*, **53**, 4544
115. Del Bene, J. and Pople, J. A. (1970). *J. Chem. Phys.*, **52**, 4858
116. Buchler, A., Stauffer, L. and Klemperer, W. (1964). *J. Amer. Chem. Soc.*, **86**, 4544
117. Novick, S. E., Davies, P. B., Dyke, T. R. and Klemperer, W. (1973). *J. Amer. Chem. Soc.*, **95**, 8547
118. Dyke, T. R. and Muenter, J. S. (1972). *J. Chem. Phys.*, **57**, 5011
119. Novick, S. E., Howard, B. J. and Klemperer, W. (1972). *J. Chem. Phys.*, **57**, 5619
120. Novick, S. E., Lehn, J. M. and Klemperer, W. (1973). *J. Amer. Chem. Soc.*, **95**, 8189
121. Audit, P. and Rouault, M. (1967). *Entropie*, **18**, 22
122. Raoult, B., Farges, J. and Rouault, M. (1968). *Compt. Rend. Acad. Sci. Ser. B*, **267**, 942
123. Jaegle, A., Duguet, A. and Rouault, M. (1968). *Compt. Rend. Acad. Sci. Ser. B*, **267**, 1081
124. Audit, P. (1969). *J. Phys. (Paris)*, **30**, 192
125. Stein, G. D. and Armstrong, J. A. (1973). *J. Chem. Phys.*, **58**, 1999
126. Janzen, J. and Bartell, L. (1969). *J. Chem. Phys.*, **50**, 3611
127. Atoji, M. and Lipscomb, W. N. (1954). *Acta Crystallogr.*, **7**, 173
128. Bouchiat, C. C., Bouchiat, M. A. and Pottier, L. C. L. (1969). *Phys. Rev.*, **181**, 144
129. Bouchiat, C. C. and Bouchiat, M. A. (1970). *Phys. Rev. A*, **2**, 1274
130. Bouchiat, M. A., Brossel, J. and Pottier, L. C. (1972). *J. Chem. Phys.*, **56**, 3703
131. Bouchiat, M. A. and Pottier, L. C. (1972). *J. Phys. (Paris)*, **33**, 213
132. Herman, R. M. (1964). *Phys. Rev.*, **136**, A1576
133. Pauly, H. and Toennies, J. P. (1965). *Advan. At. Mol. Phys.*, **1**, 195
134. Bernstein, R. B. and Muckerman, J. T. (1967). *Advan. Chem. Phys.*, **12**, 389
135. Siska, P. E., Parson, J. M., Schafer, T. P. and Lee, Y. T. (1971). *J. Chem. Phys.*, **55**, 5762
136. Parson, J. M., Siska, P. E. and Lee, Y. T. (1971). *J. Chem. Phys.*, **56**, 1511
137. Farrar, J. M. and Lee, Y. T. (1972). *J. Chem. Phys.*, **56**, 5801
138. Farrar, J. M. and Lee, Y. T. (1973). *J. Chem. Phys.*, **57**, 5492
139. de Boer, J. (1949). *Rep. Prog. in Phys.*, **12**, 305
140. Rowlinson, J. S. (1966). *Discuss. Faraday Soc.*, **40**, 19
141. Buckingham, A. D. and Pople, J. A. (1955). *Trans. Faraday Soc.*, **51**, 1029
142. Hanley, H. J. M., Barker, J. A., Parson, J. M., Lee, Y. T. and Klein, M. (1972). *Mol. Phys.*, **24**, 11
143. Buckingham, A. D. (1959). *Propriétés optiques et acoustiques des fluides comprimés et actions intermoléculaires*, 57 (Paris: Centre national de la Recherche Scientifique)
144. Byers Brown, W. and Whisnant, D. M. (1973). *Mol. Phys.*, **25**, 1385
145. Matcha, R. L. and Nesbet, R. K. (1967). *Phys. Rev.*, **160**, 72
146. Schieve, W. C. and Harrison, H. W. (1974). *J. Chem. Phys.*, **61**, 700
147. McKellar, A. R. W. and Welsh, H. L. (1974). *Can. J. Phys.*, **52**, 1082
148. McKellar, A. R. W. (1974). *J. Chem. Phys.*, **61**, 4636
149. Le Roy, R. J. and Van Kranendonk, J. (1974). *J. Chem. Phys.*, **61**, 4750
150. Freeman, D. E., Yoshino, K. and Tanaka, Y. (1974). *J. Chem. Phys.*, **61**, 4880
151. Drummond, D. L. and Gallagher, A. (1974). *J. Chem. Phys.*, **60**, 3426

4
Structure Determination by N.M.R. Spectroscopy

K. A. McLAUCHLAN
University of Oxford

4.1	INTRODUCTION	119
4.2	DIAMAGNETIC MOLECULES	121
	4.2.1 *The solid state*	123
	4.2.2 *Partially orientated molecules*	125
	4.2.3 *Relaxation methods*	129
	4.2.4 *The nuclear Overhauser effect*	132
4.3	PARAMAGNETIC MOLECULES	133
	4.3.1 *The lanthanides*	135
	4.3.1.1 *The determination of ion-bound shifts*	136
	4.3.1.2 *The interpretation of ion-bound shifts*	137
	4.3.2 *Relaxation methods*	139
4.4	CONCLUSIONS	141

4.1 INTRODUCTION

This review is concerned with the use of n.m.r. spectroscopy in the determination of the precise, geometric, structures of molecules. Its use in solution to investigate chemical structure through measurements of chemical shifts (or screening constants) and coupling constants is ubiquitous, but a frustrating aspect of n.m.r. has long been the lack of direct correlation between these parameters and molecular geometry. The chemist has come to regard them as scalars but they are derived from tensors and in solution it is the tensor traces that are measured; it will be seen below that their tensor properties complicate interpretation of those types of spectra from which precise structural data are obtained.

Other magnetic interactions may exist within the molecule which are sensitive functions of internuclear distance between the magnetic nuclei. In diamagnetic molecules one such interaction exists, the 'through-space' dipolar interaction of the nuclei. This interaction averages to zero in the normal isotropic motion in a liquid (the trace of the tensor is zero) and the distance information is lost from the normal high-resolution spectrum. However, it has long been used to determine interproton distances in solids and recent developments of ordered fluid media (liquid crystals) allow dipolar interactions to be observed in solution. The effect of dipolar coupling between inequivalent nuclei in solution is simply to affect the magnitude of the observed spin–spin splitting, which for protons it dominates; only if coupling occurs between equivalent nuclei does extra splitting result. However, the long-range nature of the dipolar interaction leads to splittings that could not be resolved in the isotropic case. Even in isotropic solution the dipolar coupling in molecules which contain only protons as magnetic nuclei may be the major source of spin relaxation, and internuclear distances may be obtained from measurement of spin–lattice (T_1) and spin–spin (T_2) relaxation times. The relaxation effects may sometimes be estimated by means of the nuclear Overhauser effect (NOE).

Whereas in diamagnetic molecules dipolar coupling affects the splitting of the resonances rather than their position, the presence of a paramagnetic centre (such as a paramagnetic ion) in a molecule may cause all resonances in its vicinity to shift. This may again be a dipolar effect, and dependent on distance, but one which originates in the magnetism of the paramagnetic centre. This interaction does not average to zero in normal isotropic solution and the structures of molecules in a normal solution may be determined by coordinating on to them a paramagnetic centre. Not all paramagnetic ions cause shifts, however, some merely inducing relaxation in nearby nuclei, but once again the effect on the relaxation time is distance-dependent. One specific technique, proton resonance enhancement (PRE) relies on measuring the 'extra' relaxation which occurs in the presence of a macromolecule.

In principle, therefore, n.m.r. is capable of giving geometric information on the structures of molecules through two interactions, each of which is of a simple dipolar nature; in solution the method is unique. Unfortunately, complexities arise in diamagnetic molecules from the anisotropies of the spin-coupling and screening constant tensors and from the fact that in liquid crystalline solvents the orientation of the molecules is only partial. In paramagnetics, contributions to the observed shift from the dipolar ('pseudo-contact') term may be augmented by a through-bond ('contact') effect whose magnitude needs to be known. Furthermore, the chemical shifts of the nuclei in the complexed molecule are required whereas the observed ones are usually exchange-averaged shifts between ion-bound molecules and free ones. All the relaxation methods depend upon the assumption of local isotropic motion in a liquid and that the dipole relaxation mechanism is dominant.

Notwithstanding these difficulties, n.m.r. is at present used to determine internuclear distances and angular information in a wide variety of compounds, ranging from simple diatomics to enzymes, sometimes yielding whole structures and sometimes only partial ones. It is the purpose of this review to establish the bases for these determinations and to consider possible errors in them. In some of the more widely used techniques it is possible to assess the accuracy of the measurements. Most of the article is devoted to spin-$\frac{1}{2}$ nuclei and a major conclusion is that n.m.r. methods are most suitable for determining geometric information in proton systems; for other nuclei the contributions to the observed shifts and splittings from other than the dipolar effects are not usually negligible. However, it is the positions of protons that are most difficult to locate by other physical methods.

The review is divided into two main parts, the first dealing with diamagnetic molecules and the second with those made paramagnetic by complexing. Each part discusses the basic theory of the interactions and the application of this theory through different experimental techniques. There are so many of these that in a short review it is impossible to provide a complete bibliography of each; it is intended that this review should be considered an essay illustrated with recent examples.

4.2 DIAMAGNETIC MOLECULES

The spin Hamiltonian may be written as the sum of four independent terms, the Zeeman term, $\hat{H}_Z$, the screening constant term, $\hat{H}_S$, the dipolar coupling term, $\hat{H}_D$, and the spin–spin coupling term, $\hat{H}_J$:

$$\hat{H} = \hat{H}_Z + \hat{H}_S + \hat{H}_D + \hat{H}_J \tag{4.1}$$

Of these, geometric information is carried only in the dipolar term:

$$\hat{H}_D = \sum_{i<j} \hat{\boldsymbol{I}}_i \cdot \mathscr{D}_{ij} \cdot \hat{\boldsymbol{I}}_j \tag{4.2}$$

where $\hat{\boldsymbol{I}}_i$ and $\hat{\boldsymbol{I}}_j$ are the operators of the spin vectors of nuclei i and j and $\mathscr{D}_{ij}$ is the dipole coupling tensor with components

$$\mathscr{D}_{ij}^{\alpha\beta} = \frac{\mu_0}{4\pi}\frac{\hbar\,\gamma_i\gamma_j}{2\pi r_{ij}^5}\{r_{ij}^2\,\delta^{\alpha\beta} - 3r_{ij}^{\alpha}\,r_{ij}^{\beta}\} \tag{4.3}$$

Here r_{ij} is the internuclear distance, α and β are the axes of a Cartesian axis system and $\delta^{\alpha\beta} = 1$ if $\alpha = \beta$ and is zero otherwise; μ_0 is the permeability of free space ($\mu_0/4\pi = 10^{-7}$ H m^{-1} = 1 e.m.u.). This tensor is traceless and its magnitude is determined by the relative positions of the nuclei. The Hamiltonian is treated most conveniently by transforming the tensor to spherical polar coordinates. For two nuclei with the magnetic field along the z-axis, $x = r \sin\theta \cos\phi$, $y = r \sin\theta \sin\phi$ and $z = r\cos\theta$ and

$$\hat{H}_D = \frac{\mu_0}{4\pi}\frac{\hbar\gamma_i\gamma_j}{2\pi r_{ij}^3}\{A + B + C + D + E + F\} \tag{4.4}$$

where

$$\begin{aligned}
A &= (1 - 3\cos^2\theta)\hat{I}_{iz}\hat{I}_{jz}\\
B &= -\tfrac{1}{4}(1 - 3\cos^2\theta)[\hat{I}_i^+\,\hat{I}_j^- + \hat{I}_i^-\,\hat{I}_j^+]\\
C &= -\tfrac{3}{2}\sin\theta\cos\theta\exp(-\mathrm{i}\phi)[\hat{I}_{iz}\,\hat{I}_j^+ + \hat{I}_i^+\,\hat{I}_{jz}]\\
D &= -\tfrac{3}{2}\sin\theta\cos\theta\exp(\mathrm{i}\phi)[\hat{I}_{iz}\,\hat{I}_j^- + \hat{I}_i^-\,\hat{I}_{jz}]\\
E &= -\tfrac{3}{4}\sin^2\theta\exp(-2\mathrm{i}\phi)\hat{I}_i^+\,\hat{I}_j^+\\
F &= -\tfrac{3}{4}\sin^2\theta\exp(2\mathrm{i}\phi)\hat{I}_i^-\,\hat{I}_j^-
\end{aligned}$$

where $\hat{I}^{\pm} = \hat{I}_x \pm i\hat{I}_y$. Of these terms only A and B effect the energy of the spin system to first order, whereas C, D, E and F determine the relaxation behaviour and are used to calculate the relaxation time, T_1. Truncating the Hamiltonian, in general

$$\hat{H}_D = \frac{\mu_0}{4\pi}\sum_{i<j} \hbar\gamma_i\gamma_j \frac{(A_{ij} + B_{ij})}{r_{ij}^3} \tag{4.5}$$

For the two nuclei case

$$\hat{H}_D = D_{ij}\{\hat{I}_{iz}\hat{I}_{jz} - \tfrac{1}{4}(\hat{I}_i^+\hat{I}_j^- + \hat{I}_i^-\hat{I}_j^+)\} \tag{4.6}$$

where

$$D_{ij} = (\mu_0/4\pi)\,\hbar\gamma_i\gamma_j(1 - 3\cos^2\theta)(2\pi r_{ij}^3)^{-1}\ \text{Hz} \tag{4.7}$$

Unfortunately for geometric measurements the nuclei are also coupled through the electrons in the bonds:

$$\hat{H}_J = \sum_{i<j} \hat{\boldsymbol{I}}_i\cdot\mathscr{J}_{ij}\cdot\hat{\boldsymbol{I}}_j \tag{4.8}$$

where $\mathscr{J}_{ij}$ is the spin–spin coupling tensor and may have nine independent components, although this number reduces if the nuclei which couple lie on symmetry axes or planes within the molecule. In the simplest case, with both nuclei on the axis of a C_{nv} ($n > 2$) molecule, only two components, $J_{\parallel}$ and $J_{\perp}$, are independent. $\mathscr{J}_{ij}$ differs from $\mathscr{D}_{ij}$ in that its trace is non-zero and its effects on energy are observable in solution.

It can be written conveniently as the sum of an isotropic part, a symmetric anisotropic part with zero trace and a skew symmetric part; the last affects the energy only to second order and will be ignored[1,2]. The anisotropic part has the same transformation properties as the other second-rank tensor $\mathscr{D}_{ij}$. Equation (4.8) may be re-written

$$\hat{H}_J = \sum_{i<j} J_{ij}\hat{\boldsymbol{I}}_i\hat{\boldsymbol{I}}_j + \sum_{i<j} J_{ij}^A\{\hat{I}_{iz}\hat{I}_{jz} - \tfrac{1}{4}(\hat{I}_i^+\hat{I}_j^- + \hat{I}_i^-\hat{I}_j^+)\} \tag{4.9}$$

where J_{ij} and J_{ij}^A are the isotropic and anisotropic coupling constants, respectively. Expanding the scalar product and combining $\hat{H}_D$ and $\hat{H}_J$, in the high-field limit,

$$\hat{H}_I = \sum_{i<j}\{(J_{ij} + J_{ij}^A + D_{ij})\hat{I}_{iz}\hat{I}_{jz} + \tfrac{1}{4}(2J_{ij} - J_{ij}^A - D_{ij})(\hat{I}_i^+\hat{I}_j^- + \hat{I}_i^-\hat{I}_j^+)\} \tag{4.10}$$

This Hamiltonian determines the splittings observed in the magnetic resonance spectra of coupled nuclei, whereas the sum of the Zeeman and screening terms,

$$\hat{H}_Z + \hat{H}_S = -(2\pi)^{-1}\sum_i \gamma_i B_0(1 - \sigma_i - \sigma_i^A)\cdot\boldsymbol{I}_i \tag{4.11}$$

determines the overall positions of the resonances, where σ_i and σ_i^A are the isotropic and symmetric anisotropic parts of the screening constant tensor and B_0 is the applied field. Under the conditions (lack of isotropic motion) which make J_{ij}^A and D_{ij} non-zero, σ_i^A is also non-zero and causes shifts of the resonance positions relative to those in isotropic solution.

For two non-equivalent nuclei, application of the total Hamiltonian implies that there should be two resonances, each split by $(J_{ij} + J_{ij}^A + D_{ij})$ Hz, whereas for equivalent nuclei the splitting of the one resonance is $\frac{3}{2}(J_{ij}^A + D_{ij})$ Hz[2]. The geometric information is carried only in the dipolar coupling which can be extracted if the J values are known; J_{ij} is measurable in isotropic solution but J_{ij}^A must either be measured in an oriented system, or calculated. In the solid state the dipolar interaction is dominant in the coupling of light nuclei (for two protons 1 Å apart D_{ij} is 120.067 kHz), but in liquid crystal media the full contribution is not observed and the J corrections may be significant (see below).

Under the isotropic re-orientation conditions prevalent in normal liquids, $\mathscr{D}_{ij}$, being traceless, averages to zero but the dipolar coupling remains a relaxation mechanism through the terms C, D, E and F in equation (4.4). Other relaxation mechanisms, through the anisotropy of the screening tensor, spin–rotational coupling or scalar coupling, may occur in diamagnetic molecules and their possible effects must be considered. The dipolar contribution to the relaxation for two nuclei, i and j, in a system undergoing isotropic reorientation with a single characteristic correlation time τ_c is given for spin–lattice relaxation by[3]

$$\frac{1}{T_1} = (3/10)(\mu_0/4\pi)^2\, \gamma_i^2 \gamma_j^2 \hbar^2\, \overline{(r_{ij}^{-6})} \left\{\frac{\tau_c}{1 + \omega_0^2\tau_c^2} + \frac{4\tau_c}{1 + 4\omega_0^2\tau_c^2}\right\} \qquad (4.12)$$

where the bar denotes an average over molecular motion and ω_0 is the resonance frequency. In general several nuclei may relax a particular nucleus and the process is not describable by a single relaxation time[4]; this effect is usually small[5] and equation (4.12) can usually be summed over pair-wise interactions[6] by replacing $\overline{(r_{ij}^{-6})}$ by $\sum_{i<j} \overline{(r_{ij}^{-6})}$.

For spin–spin relaxation

$$\frac{1}{T_2} = \frac{3}{20}(\mu_0/4\pi)^2\, \gamma_i^2 \gamma_j^2 \hbar^2 \sum_{i<j} \overline{(r_{ij}^{-6})} \left\{3\tau_c + \frac{5\tau_c}{1 + \omega_0^2\tau_c^2} + \frac{4\tau_c}{1 + 4\omega_0^2\tau_c^2}\right\} \qquad (4.13)$$

These relaxation equations apply to intramolecular relaxation whilst the actual relaxation results both from intramolecular and intermolecular terms:

$$\frac{1}{T_1\,(\text{obs})} = \frac{1}{T_1\,(\text{inter})} + \frac{1}{T_1\,(\text{intra})} \qquad (4.14)$$

Expressions for intermolecular relaxation times can be obtained but the intermolecular contribution is better minimised by working with dilute solutions in aprotic solvents. The average internuclear distance may be extracted from measurements of the intramolecular relaxation times T_1 or T_2, if τ_c is known; methods for obtaining τ_c are discussed below.

4.2.1 The solid state

From the foregoing it would appear that geometric information may be obtained directly from observations of the dipolar splittings in molecules of any complexity provided that single crystals could be obtained. In practice this has been possible until recently only for molecules containing two or three magnetic nuclei, whether equivalent or inequivalent[7-9]. For these simple systems the dipolar couplings can be obtained even for polycrystalline samples although with a consequent increase in linewidth. Even in single crystals the lines are very broad owing to intermolecular dipolar couplings from protons on other molecules in the crystal; in consequence the dipolar fine structure cannot usually be resolved in the spectra of molecules containing more than three magnetic nuclei although the width of the spectrum still depends to a large extent on the intramolecular dipolar contribution. This problem may be overcome either by experiment (by minimising the intermolecular contributions), by analytical methods or by a combination of the two. The analytical approach is considered first.

The lineshapes of magnetic resonance transitions in the solid state are notoriously difficult to calculate; Van Vleck[10] showed, however, that the moments of lines (in particular the second moment) are easily calculable if a normalised lineshape function, $f(B)$, is assumed. The second moment of a line is its mean square width $(\Delta B)^2$ measured from its centre, B_{av}

$$(\Delta B)^2 = \int_0^\infty (B - B_{av})^2 f(B)\, \mathrm{d}B \tag{4.15}$$

For a Gaussian lineshape, and with the field ΔB originating in dipolar interactions of a given nucleus with all other equivalent nuclei (n_1/unit cell),

$$(\Delta B)_1^2 = \tfrac{3}{4}(\mu_0/4\pi)^2\, \gamma_1^2 \hbar^2\, I_1(I_1 + 1)\,(1/n_1) \sum_{i,j} \frac{(1 - 3\cos^2\theta_{ij})^2}{r_{ij}^6} \tag{4.16}$$

A further contribution to the second moment comes from nuclei (n_2/unit cell) not at resonance simultaneously with the others:

$$(\Delta B)_2^2 = \tfrac{1}{3}(\mu_0/4\pi)^2\, \gamma_2^2 \hbar^2\, I_2(I_2 + 1)\,(1/n_2) \sum_{j,k} \frac{(1 - 3\cos^2\theta_{jk})^2}{r_{jk}^6} \tag{4.17}$$

For a polycrystalline sample $(1 - 3\cos^2\theta)^2$ is replaced by its average over all θ, 4/5.

The second moment calculated in this way is compared with experimental values[11, 12] (corrected for modulation effects[13]) to yield geometric information. Errors in measurement owing to truncation of the low intensity wings of spectra [which contribute significantly to $(\Delta B)^2$] have been discussed[14], as have the effects of departure from Gaussian lineshapes[15]. In principle, study of the second moment of a single crystal as a function of the angles between the crystal axes and the applied field should yield the geometries of complex molecules[12, 16] but, in practice, the accuracy of measurement of the second moment limits application to simple problems in crystals of low crystallographic symmetry[17].

The second moment method is reliable only for proton–proton interactions for with nuclei such as ^{19}F [18-21] and ^{31}P [22] the anisotropy of the screening tensor severely affects the lineshape. If the crystal contains a paramagnetic centre, distance information can be obtained from the shifts it causes[23]. In principle a combination of paramagnetic and dipolar effects should allow decipherment of complex structures, but the paramagnetic centre would have to be chosen to satisfy stringent relaxation conditions which might be difficult to meet in the solid.

Intermolecular contributions to the second moment are often of the same order as, or greater than, intramolecular ones. Great effort has therefore gone into minimising intermolecular dipolar interactions to increase the accuracy of measurement of the intramolecular term and to extend the method to more complex molecules. Two physical methods have been used to reduce effective dipolar interactions either by spinning a single crystal about some optimum axis[24] or by effectively spin decoupling by pulse methods[20, 25]; it may be possible to minimise the intermolecular contribution selectively. The former requires rotation at frequencies of the order of the dipole coupling frequencies, which are very high in proton systems, and the method is used more readily to observe sharper resonances from other nuclei. Both techniques require sophisticated equipment extra to the conventional 'wide-line' spectrometer. The simpler chemical methods of enclosing a molecule in a magnetically dilute crystal[26, 27] or clathrate[28] are more general. For example, for SF_6 included in an H_2O clathrate 19% of the second moment is intermolecular in origin, but in a D_2O clathrate only 2.2% is.

The method has been used extensively in the study of interproton distances in water molecules[9, 29, 30], and recent studies include inorganic fluorides[15, 19, 28], HF [31], tetramethylammonium halides[32], methyl chloroform[26], 1,3,5-trichlorobenzene[27], 2,4,6-tribromoaniline[33], phloroglucinol[33], phenanthrene[34] and diethylamine[35]. The emphasis in many of these experiments has been on molecular motion rather than geometry. Bond lengths obtained are typically accurate to $\pm$ 0.01 Å, or better.

Several reviews exist of recent work on both diamagnetic[23, 36–39] and paramagnetic systems[23, 40].

4.2.2 Partially orientated molecules

The problem of broad lines in the solid state observations of dipolar couplings can be overcome in 'partial-orientation' experiments in which molecules, whilst allowed to translate freely so as to average out intermolecular contributions, are constrained to have anisotropic rotational diffusion; the intramolecular couplings alone are observed. Linewidths normal to high-resolution techniques (ca. 0.5 Hz rather than kHz as in the solid state) are obtained and the measurement accuracy is increased. Partial orientation can be caused in several ways[2, 41, 42] but, recently, dissolving solutes in liquid-crystalline solvents has become pre-eminent[43]. In either thermotropic or lyotropic[44, 45] nematic phases the solvent molecules align with the applied magnetic field and constitute an axially symmetric orientation medium in which the solutes are themselves orientated by intermolecular forces. Smectic phases have some application[44, 46, 47] and, since molecular re-orientation in them is slow, enable the angle between the orientation direction and the applied field to be varied; cholesteric phases may be 'unwound' by the field to form nematic phases[48].

In these systems the full dipolar coupling between nuclei is not observed and the dipole coupling constant in equation (4.10), and in the formulae for observed splittings, must be replaced by an average value $\overline{D_{ij}}$. The increased resolution of this technique allows measurement of averaged dipolar couplings between several pairs of nuclei which can be interpreted only in a molecular frame of reference. In the laboratory, measurements are in a frame defined by the applied field and these measurements must be interpreted in the molecular frame, whilst allowing for incomplete motional averaging. Three methods are currently in use for this. One relates the tensor components to a probability function through an expansion in real spherical harmonics[1,49]; another employs spherical coordinate representation and takes advantage of the transformation properties of Wigner matrices[44]. Here the Cartesian coordinate representation of Saupe[50,51] is used because it has the simplest nomenclature. The inter-relations between the motional constants defined in each way, and which can be determined by experiment, have been listed[44].

Providing that the molecule is rigid, the effect of motion on the angle and distance parts of the dipolar term can be averaged separately and in general equation (4.7) becomes

$$\overline{D_{ij}} = -2K_{ij}S_{ij}\langle r_{ij}^{-3}\rangle\tfrac{1}{2}(3\cos^2 a - 1) \tag{4.18}$$

where $K_{ij} = (\mu_0/4\pi)(\hbar\gamma_i\gamma_j/2\pi)$, a is the angle between the constraint direction and B_0 [in a nematic phase $a = 0°$ and $\tfrac{1}{2}(3\cos^2 a - 1) = 1$] and $S_{ij} = \tfrac{1}{2}\langle 3\cos^2\theta_{ij} - 1\rangle$ is the degree of orientation of the axis passing through nuclei i and j. In general an S value is required for each axis and five independent ones suffice to define the orientation of a rigid molecule. They are the elements of a symmetric traceless tensor $\mathscr{S}$ and are given by

$$S_{pq} = \tfrac{1}{2}\langle 3\cos\theta_p\cos\theta_q - \delta_{pq}\rangle \tag{4.19}$$

where p, q denote x, y, z in the molecular frame and θ_x, θ_y, θ_z are the angles between these axes and the direction of the constraint. Relative to any other axis which forms angles a_x, a_y and a_z with the molecule-fixed coordinate system,

$$S = \sum_{p,q}\cos a_p\cos a_q S_{pq} \tag{4.20}$$

If a molecule has a threefold or higher axis of rotation only one S parameter is required to describe its mean orientation; if it has two perpendicular planes of symmetry, two are required (since $S_{xx} + S_{yy} + S_{zz} = 0$); with one plane, three independent S parameters are needed.

For the purposes of this review, two important implications are that the measured dipolar couplings in a rigid molecule are the products of the full coupling and some orientation parameter which must be determined before geometry can be obtained, and that to obtain geometries the number of independent dipolar couplings must exceed the number of S parameters; this usually limits application to molecules of C_{2v} or higher symmetry. Measurement of orientation parameters by independent methods[52] are rare and nearly always only ratios of dipolar couplings, and hence of internuclear distances, can be determined by this method. In simple cases, such as

^{13}C-substituted methyl halides, this ratio allows direct calculation of bond angles[53,54]. The technique cannot yield absolute geometries and it is normal to compare internuclear distance ratios with those determined to similar accuracy in the gas phase using either microwave spectroscopy or electron diffraction, although these techniques measure $\langle r^{-2}\rangle$ and $\langle r\rangle$, respectively, and the problem of how to reduce the measurements of all three techniques to an average structure arises[55-58]. The complete geometry in solution is obtained by using the minimum number of reliable gas phase values to allow its calculation.

Agreement between the structures determined by n.m.r. and other methods is generally excellent[44,59] and it is instructive to consider whether the n.m.r. values are sufficiently reliable to assess whether molecular geometries vary with medium. The procedure usually employed is to calculate the dipolar couplings in an assumed geometry and to compare them with experiment; if the r.m.s. error between the two differs by greater than the experimental error, the structure is rejected. This discrepancy may be typically 4 Hz in average dipolar couplings of up to a few kHz [60,62]. There are many possible errors and limitations to the method:

(i) The experimental errors (couplings are often quoted to ± 0.5 Hz or less) may be assessed unrealistically. Errors as small as this are feasible only if the position of each line in a spectrum is measured absolutely, without reliance on precalibrated charts; the sweep of normal spectrometers is rarely linear over a few kHz and the resolution of Fourier spectrometers is limited by their core size. Additionally, linewidth variations due to non-uniform orientations are often observed.

(ii) The isotropic and anisotropic spin couplings, J and J^A, contribute significantly to the observed splittings and must be determined before values of $\overline{D_{ij}}$ can be obtained. Isotropic couplings are normally in the range -20 to $+20$ Hz for proton–proton couplings over two or three bonds but may be three times this between two fluorines, and directly bonded couplings are even bigger. They change by about 1% (but sometimes more) on changing solvent and should be determined in the liquid-crystalline solvent used[63]. Anisotropic couplings appear to be negligible for couplings of protons, either between themselves[44,64] or with ^{13}C [54,65,66]. Proton–fluorine couplings[64,67] and fluorine–fluorine ones may show appreciable anisotropies[67-69a]. For example, the anisotropy appears to be negligible in pentafluoropyridine[70] but in 1,1-difluoroethylene J^A_{FF} is said[71] to have components -720 ± 39 Hz along the C—C bond, 478 ± 26 Hz perpendicular to the plane and 399 ± 39 Hz, although J^A should be traceless. Such values may be obtained from studies of a given solute/solvent system at different orientation parameters, obtained by varying either temperature or concentration[72].

Partial-orientation experiments constitute the major source of measurement of J^A from observations at a single orientation parameter via calculation of the dipolar interaction for an assumed geometry; unfortunately the values obtained are critically dependent on the assumptions[73].

(iii) It is assumed that molecular motions are so fast that the dipolar couplings are time-independent and that the magnetic interactions may be averaged independently of the geometric terms[44].

(iv) The effect of vibrational averaging of $\langle r^{-3}\rangle$ is ignored in this simple treatment. Considerable attention has been paid to this [44,59,74]. Diehl[75], in considering benzene, concluded that the effect was significant over short distances, as in ^{13}C–H couplings, but negligible for interproton ones. In small ring compounds high vibration amplitudes pertain and averaging of interproton distances is important[60,76-78].

(v) In methyl fluoride the disagreement between calculated and observed dipolar couplings cannot be rationalised in terms of spin–spin coupling anisotropies or of vibrational averaging effects[64,73,74]. It has been suggested that it originates in a distortion of the molecule in its uniaxial environment[79-81] (but see Ref. 82).

(vi) The ratios of dipolar couplings may not be sensitive, within experimental error, to geometry changes[44,68].

In the light of these limitations, reports of small changes in the geometries of molecules with variation in temperature or concentration of the nematic phase[83-85] must be considered unproven. Bond angle variations do seem to occur, however[54]. Small ring molecules show variations in dipolar coupling ratios between lyotropic and thermotropic phases which originate in different amplitudes of vibration[60,76,77]. Occasionally spectra have been observed from molecules, e.g. tetramethylsilane[86] and methane[87,87a], apparently of too high symmetry to undergo orientation. This provides direct evidence for molecular distortion in the nematic phase, which has been observed also in e.s.r. spectra[88]. Overall it appears that medium effects are small and, because of uncertainties in reducing n.m.r., microwave and electron diffraction structures to average ones[89], it is not obvious yet whether they are general. Non-rigid molecules pose extra problems concerning description of their orientation: depending on whether the internal motion is faster than or slower than the re-orientational motion in the medium, one set or two sets of S parameters are required[44,87a,90].

Spectra become very complex if a molecule contains more than about eight magnetic nuclei but this problem can be overcome by deuterium substitution and observation of the proton spectrum whilst the deuterium is decoupled[91-95].

Many reviews of the technique exist[2,44,59,91,92,96-101] and its wide application may be judged from these and the recent publication of the geometries of the following molecules: methanol[102]; methyl halides[54]; methylmercuric halides[103]; ethylene carbonate[76,104,105], monothiocarbonate[76,105], thiocarbonate[104,106], sulphite[60] and imine[107]; acetylenes[108]; formamide[109]; dimethylformamide[48]; acetone[110]; dimethyl sulphoxide[110]; ethanol[111]; trimethylene oxide[61,77,78] and sulphide[61,78]; *cis*-but-2-ene[89]; buta-1,3-diene[62]; butadiene sulphone[112]; cyclobutene[113]; β-propiolactone[114]; furan-2-aldehyde[90]; furan and thiophen-2,5-dialdehyde[115]; 1,4-dioxin[116]; 1,4-dithiin[116,117]; monosubstituted benzenes[83,85,118-120]; disubstituted benzenes[69,71,83]; *o*-xylene[121]; 2,5-dichloro-*p*-xylene[122]; tetrafluorobenzenes[68]; thieno[2,3-*b*]toluene[123]; pentafluoropyridine[70]; a pyridine–bromine complex[124]; 2,5-dimethylpyrazine[125]; γ-picoline[126]; phthalazine[127]; 4,4′-dichlorobiphenyl[128]; norbornadiene[129]; tropone[130,131]; cyclo-octatetraene[93]; π-cyclopentadienylmanganese tricarbonyl[132]; π-cyclobutadienyliron tricarbonyl[133]; tris-methylenemethaneiron carbonyl[134]; phosphorus trifluoride[135]. The metal–hydrogen bond lengths in the compounds $H_3M_3(CO)_9CMe$, where M = Ru [135a] or Os [135b], have been determined.

4.2.3 Relaxation methods

In recent years Fourier transform spectrometers have enabled spin–lattice (T_1)[136] and spin–spin (T_2) relaxation times[137-139] to be determined for each resonance in the high-resolution spectrum of a molecule in isotropic solution. In a T_1 measurement a 180° pulse is applied to invert the magnetisation and the rate of its recovery is monitored by applying a 90° pulse at a series of later times[140, 141]. It is assumed that the recovery is exponential and T_1 is estimated from the equation

$$\ln\{(I_\infty - I_t)/2I_\infty\} = -t/T_1 \tag{4.21}$$

where I_t and I_∞ are the peak intensities at times t and ∞ after the pulse. This assumption is correct for a system of isolated spins or for a set decoupled from other nuclei[142], but cross-relaxation processes cause non-exponential behaviour in coupled systems[143]; this leads to errors of about 10% in assessing T_1 values for protons[144]. Experimental errors of ±10% are commonplace.

In interpreting experimental relaxation times account must be taken of all possible relaxation processes, including contributions from spin–rotation (SR), chemical shift anisotropy (A), scalar coupling (S) and dipolar coupling (Dip); geometric information is carried only in the latter:

$$\frac{1}{T_{1,\text{obs}}} = \frac{1}{T_{1,\text{Dip}}} + \frac{1}{T_{1,\text{SR}}} + \frac{1}{T_{1,\text{A}}} + \frac{1}{T_{1,\text{S}}} \tag{4.22}$$

In either ^{1}H or ^{13}C measurements the chemical shift anisotropy term appears to be negligible at normal field strengths[145-147] whilst the scalar term contributes appreciably only if the nucleus is bonded directly to a quadrupolar one[147]. The spin–rotation contribution is significant in small molecules[148] and may be dominant (as spin–internal rotation) in the relaxation of ^{1}H or ^{13}C in methyl groups[147, 149]. Each contribution to relaxation depends upon the product of an interaction and a motional term, which characterises the modulation of the interaction necessary for relaxation. Different relaxation processes have different correlation times and spin–rotation and dipolar relaxation make their optimum contributions in different time zones; they differ in their temperature dependence[150] and their effects can be separated[147]. Spin–rotation becomes increasingly significant for small molecules at high temperatures but for large molecules with long correlation times the dipolar contribution is dominant. It is with the latter that relaxation methods of distance measurement have their greatest potential.

In aprotic solvents intermolecular contributions to relaxation are negligible in the absence of paramagnetic substances, such as oxygen or trace metals in biological preparations (from buffer solutions).

The accuracy of relaxation structure determinations, either from the experimental or interpretive points of view, falls short of that obtainable from direct measurement of dipolar couplings. However, application is not limited to highly symmetrical small molecules and since the distance information is carried as $\langle r^{-6}\rangle$ [equations (4.12) and (4.13)] an error of ±10% in T_1 causes an error of only $\pm 2\%$ in r. This distance dependence ensures that only those nuclei which are within about 3.5 Å of the observed ones contribute to the relaxation, which often implies a single pairwise interaction and justifies the use of a single correlation time for the interacting nuclei. The sensitive dependence on distance makes relaxation methods (and nuclear Overhauser methods, see below) appropriate for investigating molecular stereochemistry, including the secondary and tertiary structures of proteins.

To obtain $\langle r^{-6} \rangle$ from equations (4.12) and (4.13) the motional factor must be determined. Three methods exist for this: τ_c can be determined from the ratio T_1/T_2 for a resonance[151] or from the frequency-dependence of T_2 [151] or T_1 [151, 152]. The latter is independent of exchange effects and is to be preferred. A plot of the motional term in equation (4.12) as a function of τ_c is given in Figure 4.1 for two different values of ω_0, 270 and 90 MHz. If $\tau_c \leqslant 10^{-10}$ s

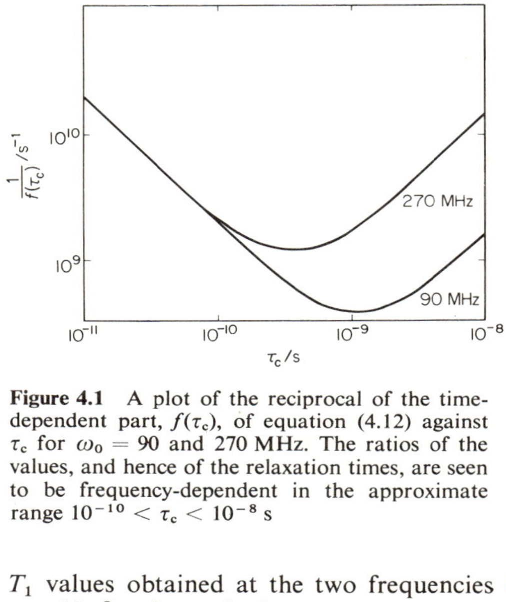

Figure 4.1 A plot of the reciprocal of the time-dependent part, $f(\tau_c)$, of equation (4.12) against τ_c for $\omega_0 = 90$ and 270 MHz. The ratios of the values, and hence of the relaxation times, are seen to be frequency-dependent in the approximate range $10^{-10} < \tau_c < 10^{-8}$ s

the ratio of the T_1 values obtained at the two frequencies is independent of τ_c whilst if $\tau_c \geqslant 10^{-8}$ s the ratio becomes constant at nine; between these two limits it is a sensitive function of τ_c which can be measured. This range encompasses the rotational correlation times of molecules with rigidity imposed by secondary or tertiary structure. However, molecules of widely differing molecular weight, polymyxin (1200), bacitracin (1400), lysozyme (14 000), ribonuclease *A* (14 000) and triose phosphate isomerase (50 500), show similar T_1 ratios, which suggests relaxation mainly results from local segmental motion. In constrast, random-coil polyglycine (6000) shows no frequency-dependence of T_1 [152].

The assumptions inherent in the method, and especially with the use of one correlation time in equation (4.12), have been tested in a study of the phenylalanine aromatic protons in polymyxin B [152]. By studying T_1 ratios at two frequencies at several temperatures a series of τ_c values was obtained and used to calculate $\langle r^{-6} \rangle$: the values were found to be independent of temperature and in agreement with those predicted from crystallographic data.

Figure 4.2 shows the relationship between the relaxation times calculated from crystallographic distances as a function of τ_c and the experimental values of T_1 obtained at the measured τ_c values. Similar good agreement has been found for the phenylalanine aromatic protons and thiazoline methylene ones in bacitracin[153] and for resonances from sulphonamide inhibitors bound to bovine carbonic anhydrase[151].

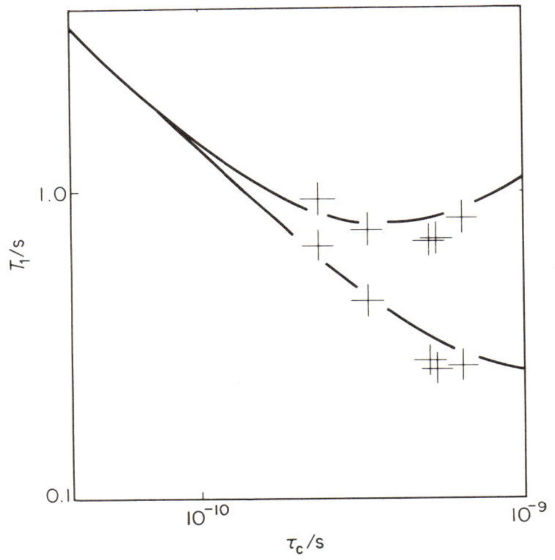

Figure 4.2 A plot of T_1 against τ_c for the phenylalanine protons in polymyxin B. The crosses are experimental values of τ_c found from the T_1 ratios whilst the solid line represents the theoretical dependence of T_1 on τ_c. The upper curve corresponds to 270 MHz measurements and the lower one to 90 MHz

Some protons in proteins, for example the C-2 proton in the histidine residue, are remote from other protons in the same residue and their relaxation reflects the proximity of protons in other side-chains: they act as probes of secondary and tertiary structure. For example, for the isolated dehydroserine ureide proton in viomycin, variation in concentration from 0·04 to 0·2 mol l^{-1} causes $\sum_{i<j} \overline{\langle r_{ij}^{-6} \rangle}$ to vary from 1.87×10^{45} cm^{-6} to 3.70×10^{45} cm^{-6} owing to a conformational change[153].

Unfortunately, contributions to relaxation from anisotropic motions often occur and the analysis is then more complex. The problem has been treated theoretically in general[154] and for the rotations of methyl groups within molecules[155-157]. Relaxation techniques have had little use to date in internuclear distance measurements and cannot be used without a full understanding of relaxation processes. Complexities which occur are demonstrated in recent ^1H studies of pyridine[158] and ^{13}C studies of camphor[148]. Several recent reviews of relaxation processes have been published[159-161], including some specific to ^{13}C [162, 163].

4.2.4 The nuclear Overhauser effect (NOE)

If in a coupled spin system the resonance of one set of spins is saturated in a double resonance experiment whilst the others are observed, the intensities of the latter may be perturbed. This effect, which may be intra- or inter-molecular in origin, is termed the nuclear Overhauser effect and results from redistribution of spin populations between levels via relaxation mechanisms. If the dipolar mechanism is dominant the magnitude of the effect depends upon $\langle r^{-6} \rangle$. It is not a requirement that the nuclei should be scalar coupled, only that they should be sufficiently close for mutual dipolar relaxation. From a geometric point of view, it is the intramolecular effect which is important and this may be obtained by extrapolating to infinite dilution.

The primary measurement is that of intensity change and this is prone to approximately 10% random error besides being sensitive to the precise setting of the irradiation frequency[164].

The theory of the method has been considered in considerable detail by Bell and Saunders[165] and by Noggle and Schirmer[166] and is liable to similar limitations to relaxation methods, chiefly that isotropic motion with one correlation time is assumed. For two nuclei the fractional enhancement, $f_i(j)$, of the intensity of the i spin whilst the j spin is saturated is given by[167]

$$[f_i(j)]^{-1} = 2 + Ar_{ij}^6 \tag{4.23}$$

where A is a constant for different molecules provided that the ratio of relaxation rate by other processes to τ_c is constant. This relationship has been tested for both H–H interactions and CH_3–H ones[167]. In the limit when $r_{ij} \rightarrow 0$, $f_i(j) \rightarrow \frac{1}{2}$ and is independent of r. It should be emphasised that this relationship can be used only for molecules of similar size (hence τ_c).

A more typical case is the three-spin one which yields internuclear distances either from a combination of relaxation and NOE methods or, if it is assumed that all the internuclear vectors reorient with similar correlation times, their ratios can be obtained from pure NOE methods[166]:

$$\left(\frac{r_{ij}}{r_{ik}}\right)^6 = \frac{\gamma_j^3}{\gamma_k^3}\left\{\frac{f_i(k) + f_i(j)f_j(k)}{f_i(j) + f_i(k)f_k(j)}\right\} \tag{4.24}$$

This introduces a feature of all such estimates that all the pair-wise NOE enhancements must be measured.

The technique, which promises to be of greater use in future, has so far been used to elucidate distances in ochotensimine[166, 168] and in 2′,3′-isopropylidene-3,5′-cycloguanosine[164], with experimental errors of roughly 8%, although this depends on r_{ij}; if it is large the fractional enhancements are small. Extension of the method to non-rigid molecules and heteronuclear cases has been considered in detail[166]. The full general theory for a multi-spin system has been given and possible errors discussed; emphasis has been placed on removing paramagnetic impurities and on causing intermolecular effects to be minimal[166]. Recently, intermolecular effects have received much attention[169-171].

The major application of both relaxation and NOE methods of structure determination is likely to be in investigating local structure rather than whole structure; they are already used widely in semi-quantitative fashion to investigate stereochemistry, conformation and secondary and tertiary structure in solution[165, 166, 172, 173].

4.3 PARAMAGNETIC MOLECULES

In the presence of a paramagnetic centre in a molecule the Hamiltonian given in equation (4.1) must be augmented with terms in the electron Zeeman interaction (which does not affect the n.m.r. spectrum and is neglected here) and in the interaction between electron spins and nuclei. There are two of these, a through-bond contact term ($\hat{H}_C$) and a through-space dipolar term. Another dipolar contribution arises in the magnetic anisotropy of the paramagnetic centre; the sum of the two dipolar terms is called the pseudo-contact interaction ($\hat{H}_{Dip}$). It is a tensor with a non-zero trace[174].

In the high-field limit in solution

$$\hat{H}_C + \hat{H}_{Dip} = \sum_i (A_i + A_{Di})\, \hat{\boldsymbol{I}}_{iz} \cdot \hat{\boldsymbol{S}}_z \tag{4.25}$$

where A_i and A_{Di} represent the contact and pseudo-contact hyperfine coupling constants. In principle this total interaction should produce a splitting in the resonance line of each nucleus but the electron relaxation time (τ_S) or the electron exchange time (τ_X) or both are generally much shorter than $(A_i + A_{Di})^{-1}$ and the hyperfine structure is not resolved in the n.m.r. spectrum. If $\langle S_z \rangle$ is the average value of the electron spin component, the nucleus instead experiences an effective field change $2\pi \sum_i (A_i + A_{Di}) \langle S_z \rangle / \gamma_i$ which causes an observable chemical shift provided that $|(A_i + A_{Di})\tau_S|$ or $|(A_i + A_{Di})\tau_X| \ll 1$. In these circumstances the n.m.r. lines in rigid molecules are sharp (3–5 Hz). If, however, τ_S is long no shifts are observed and the paramagnetic centre simply constitutes an efficient relaxation centre.

As with diamagnetic molecules, geometric information may be obtained either from steady-state or relaxation methods, provided that the through-bond and the dipolar effects can be separated. However, the magnetic moments of paramagnetic ions are much greater than those of protons and the effects are observable over greater distances (up to 20 Å). The effects of the two interactions on the observed shifts are simply additive (although of course they may be of opposite sign): if Δ_B is the shift in the fully complexed molecule

$$\Delta_B = \Delta_C + \Delta_{Dip} \tag{4.26}$$

In general,

$$\Delta_C(i) = -\, A_i \hbar \frac{\gamma_e}{\gamma_i} g_e \mu_B \frac{S(S+1)}{3kT} \text{ p.p.m.} \tag{4.27}$$

where g_e is the electron g-factor, μ_B the Bohr magneton and S the total electron spin[174]. For lanthanides, which are the most widely used paramagnetic species, this has the form[175]

$$\Delta_C(i) = -\frac{2\pi\mu_B A_i \hbar J(J+1)\, g(g-1)}{3kT\gamma_i} \text{ p.p.m.} \tag{4.28}$$

where J (in $\hbar$ units) is the resultant electron spin angular momentum and g is the resultant g-factor. For some species the observed shifts from the resonance positions of the closest diamagnetic analogues are almost entirely due to this term[176-178] and in any study of molecular structure care must be taken to consider whether this term contributes. The pseudo-contact contribution is given in general by

$$\Delta_{Dip}(i) = -(1/r_i^3)\{k_1(3\cos^2\theta_i - 1) - k_2 \sin^2\theta_i \cos 2\Omega_i\} \text{ p.p.m.} \tag{4.29}$$

where r_i, θ_i and Ω_i are the spherical polar coordinates of the nucleus from the paramagnetic species taken as a point centre (Figure 4.3). Originally the parameters k_1 and k_2 were considered to arise in the anisotropy of the g-tensor[174] but they are more properly treated in terms of the magnetic

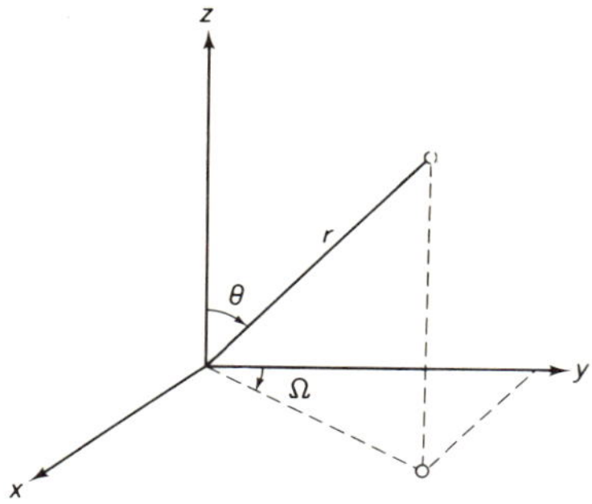

Figure 4.3 The spherical polar coordinate system used to define the position of a magnetic nucleus relative to a paramagnetic ion treated as a point centre. The principal components of the susceptibility tensor (χ_x, χ_y and χ_z) lie along the Cartesian axes which are not necessarily collinear with a similar axis system within the bound molecule

susceptibility tensor[179, 180]. The precise values depend on the components (χ_x, χ_y and χ_z) and on the crystal field coefficient: $k_1 \propto [\chi_z - \frac{1}{2}(\chi_x + \chi_y)]$ and $k_2 \propto (\chi_x - \chi_y)$ [179, 182]. For most lanthanides (the exceptions being Sm^{3+} and Eu^{3+}) this theory, unlike the g-tensor theory, predicts an inverse square temperature dependence, for which there is some experimental evidence[183].

If the susceptibility tensor is cylindrically-symmetric, $\chi_x = \chi_y$ and k_2 is zero and equation (4.29) becomes

$$\Delta_{Dip}(i) = -k_1(3\cos^2\theta_i - 1)/r_i^3 \text{ p.p.m.} \tag{4.30}$$

This is the form of the equation most often used now, with the constant k_1 eliminated by considering ratios of shifts; it is dangerous to calculate k_1 or k_2 since the components of the susceptibility tensor of a given ion may vary in magnitude (and possibly symmetry) with changing environment between molecules. The angular term may be of either sign and so either upfield or downfield shifts can result from a given paramagnetic centre.

Under conditions of slow electron relaxation, the relaxation times T_{1B} and T_{2B} of nuclei in paramagnetic molecules are given by the Solomon–Bloembergen equations[184, 185]

$$\frac{1}{T_{1B}} = \frac{2}{15}(\mu_0/4\pi)^2 \frac{\gamma_i^2 g_e^2 S(S+1)\mu_B^2}{r_i^6}\left\{\frac{3\tau_c}{1+\omega_I^2\tau_c^2} + \frac{7\tau_c}{1+\omega_s^2\tau_c^2}\right\} + \tfrac{2}{3} S(S+1) A^2 \left(\frac{\tau_e}{1+\omega_s^2\tau_e^2}\right) \quad (4.31)$$

and

$$\frac{1}{T_{2B}} = \frac{1}{15}(\mu_0/4\pi)^2 \frac{\gamma_i^2 g_e^2 S(S+1)\mu_B^2}{r_i^6}\left\{4\tau_c + \frac{3\tau_c}{1+\omega_I^2\tau_c^2} + \frac{13\tau_c}{1+\omega_s^2\tau_c^2}\right\} + \tfrac{1}{3} S(S+1) A^2 \left\{\frac{\tau_e}{1+\omega_s^2\tau_e^2} + \tau_e\right\} \quad (4.32)$$

where ω_I and ω_s are the nuclear and electron Larmor precession frequencies and τ_c and τ_e are correlation times for the processes which modulate the dipolar interaction and the hyperfine interaction, respectively:

$$\frac{1}{\tau_c} = \frac{1}{\tau_R} + \frac{1}{\tau_S} + \frac{1}{\tau_B} \quad (4.33)$$

and

$$\frac{1}{\tau_e} = \frac{1}{\tau_S} + \frac{1}{\tau_B} \quad (4.34)$$

where τ_R is the rotational correlation time and τ_B is the residence time of a ligand in the coordination sphere of the ion.

In equations (4.31) and (4.32), which only apply to paramagnetics with isotropic g-values, the first term arises in the pseudo-contact interaction and the latter in the contact one; if the latter is negligible either relaxation time is given by an equation of the type

$$T_{1B}^{-1} = C' r_i^{-6} \quad (4.35)$$

where C' is a constant provided all parts of the molecule move with a single correlation time, an approximation valid in rigid molecules. Distance information can be obtained either from T_1 or T_2 measurements but whilst the latter seem more easily measurable from linewidths the possible effects of chemical exchange must be considered.

Relaxation measurements in paramagnetic molecules have the advantage over diamagnetic ones that there exists one major source of relaxation.

4.3.1 The lanthanides

Paramagnetic centres may often be introduced into diamagnetic molecules simply by coordinating a paramagnetic ion on to them via a nitrogen or oxygen atom. The lanthanide ions are particularly suitable both in their coordination properties and in that many of them possess large magnetic moments. They are added to solutions of the molecule under investigation in the form of complexes, typically the 2,2,6,6-tetramethylheptane-3,5-dionato (thd) or the 1,1,1,2,2,3,3-heptafluoro-7,7-dimethyloctane-4,6-dionato (fod) chelates, which allow further coordination to the lanthanide. From the earliest experiments[186] it was apparent that the chemical shifts observed varied with the amount of complex added, which showed that the situation corresponded to the fast-exchange limit. Geometric information is obtainable only from the shift of the fully-bound form and the initial problem is to extract this limiting shift from the exchange-averaged observations; this is considered below. In a few cases slow exchange has been obtained at low temperature and separate spectra from free and bound ligands observed[187,188].

Initially the choice of lanthanide must be made. In the past this has been largely an empirical decision based on availability and also on the fact that many of the earliest observations used Eu^{3+}, and its use has continued. However, for geometric studies the pseudo-contact contribution should be optimised relative to the contact one. On this criterion it has been shown[189,190] that for protons Yb^{3+} ($\Delta_{Dip}/\Delta_C \approx 1340$) is the most favourable, closely followed by Pr^{3+} (ratio $\approx$ 1000); Eu^{3+} is the least favourable of all (ratio $\approx$ 160). However, the corollary of this is that if the structure is not completely rigid the former pair of ions, giving larger shifts, are more likely to produce motional broadening than is europium. The constant in equation (4.30) is of opposite sign for Pr^{3+} than it is for Eu^{3+} or Yb^{3+} and the former causes shifts of a given proton in an opposite direction to what would be observed with the latter. This can be useful in detecting contact contributions to shifts (see below), which are lowest for protons. Contact contributions in the presence of lanthanides have been detected for ^1H [191-194], ^{13}C [191,192,195-199] and ^{19}F [192-194] nuclei whilst for ^{14}N, ^{15}N [200], ^{17}O [200] and ^{31}P nuclei the presence of lone pairs in their atoms allows coordination to the lanthanides and may produce a predominant contact shift.

Of especial significance is the S ground-state in Gd^{3+}, which has no susceptibility anisotropy and can cause no pseudo-contact shifts. It has a relatively long electron relaxation time (τ_S) and so acts as a source of relaxation.

4.3.1.1 The determination of ion-bound shifts

To obtain the ion-bound shift Δ_B from observed values Δ_{obs} in the fast exchange limit (defined by $\tau_B^{-1} > \Delta_B$), the complex formation equilibria must be considered[201]. Although 1:1 complexes (LS) are often assumed between substrate (S) and ligand (L), 1:2 complexes (LS_2) may also be present[202,203] with two separate equilibria [$L + S \rightleftharpoons LS$ and $LS + S \rightleftharpoons LS_2$]. Following Reuben[190], for the situation of n equivalent and independent binding sites, the intrinsic dissociation constant K is

$$K = \{[S_F](n[L_T] - [S_B])\}/[S_B] \tag{4.36}$$

where the subscripts F, B and T denote free, bound and total concentrations. This may be re-cast in the form

$$\frac{[S_B]}{[S_T]} = f_B = \frac{n[L_T]/[S_T]}{1 + K[S_F]^{-1}} \tag{4.37}$$

where f_B is the fraction of substrate molecules bound to the ions. In the fast-exchange limit $f_B = \Delta_{obs}/\Delta_B$ and, at constant $[S_T]$ and low $[L_T]$, a plot of Δ_{obs} *versus* $[L_T]/[S_T]$ is linear. Its slope is $n\Delta_B/(1 + K[S_T]^{-1})$ (since $[S_F] \approx [S_T]$), from which Δ_B can be obtained if n, K and $[S_T]$ are known. In the past the factor $n/(1 + K[S_T]^{-1})$ has usually been assumed to be unity but this introduces major errors in Δ_B values if 1:2 complexes are formed, and approximately 10% errors in 1:1 complexes. At higher values of $[L_T]$ the curves show appreciable departure from linearity.

Alternatively, if $[L_T]$ is maintained constant and $[S_T]$ varied whilst maintaining $[S_T] \gg [L_T],[S_B]$, equation (4.37) becomes

$$[S_T] = n[L_T]\frac{\Delta_B}{\Delta_{obs}} - K \qquad (4.38)$$

and Δ_B can be obtained from the slope of the plot of $[S_T]$ *versus* $[\Delta_{obs}]^{-1}$, if n is known. It is essential, both for this and other methods[204-206], to determine the stoichiometries of any complexes present[202, 203]; these may vary with solvent for a given lanthanide–substrate system[207].

The slopes and linearities of the plots are affected if any other competitive equilibria occur with alternative ligands, such as solvents, trace amounts of water or impurities in TMS [208]. Analytical procedures for coping with trace impurities[204, 208], and with experimental errors[209], have been given. Even changing from benzene to cyclohexane as solvent may change Δ_B, although Δ_B ratios are unaffected; this seems to imply an effect on the susceptibility tensor[210].

If the stoichiometries are known, the total interpretive and experimental error in the absolute value of the slope may still exceed 10%, but the errors in the slope ratios are less and it is these which yield the geometric information (provided $\Delta_B = \Delta_{Dip}$). The dependence of Δ_{Dip} on $\langle r^{-3} \rangle$ again decreases the error in the derived quantity.

4.3.1.2 The interpretation of ion-bound shifts

The initial problem is to identify whether or not Δ_B contains a contribution from Δ_C. This can be done by using two different lanthanides, which cause shifts in opposite directions[211]: if the dipolar term alone contributes, the ratios of the shifts are independent of k_1 and k_2 which vary in sign and magnitude between the ions. This assumes, of course, that the geometry is invariant to the ion. To a first approximation this is reasonable, the major differences coming in changes in the lengths of Ln—O or Ln—N bonds, which may make a minor contribution to a Ln—^{1}H distance. If this technique is not used the rapid attenuation of Δ_C with distance from the ion in systems which do not contain delocalised electrons can be invoked: information from the closest protons (through bonds) should be neglected but that from others may be used. The occurrence of contact shifts can also be confirmed from comparisons of shift ratios in molecules of known rigid structure.

The value of Δ_B should be corrected for contributions from electron-pair shielding by subtracting from it the shifts observed on complex formation with the diamagnetic ions La^{3+} or Lu^{3+} [211].

Having obtained Δ_{Dip} it must be interpreted in terms of equations (4.29) or (4.30). The coordinates r, θ and Ω are defined relative to the axes of the susceptibility tensor of the paramagnetic ion whose orientation must be determined relative to a molecule-fixed axis system if the geometry of the molecule is to be obtained[212]. It is usually assumed that one axis of the susceptibility tensor lies along the Ln—O or Ln—N bond and there is experimental evidence which suggests that the deviation between the two directions is only a few degrees[213]. The lack of coincidence between the two does not affect the form of equation (4.30) provided that rapid rotation occurs about the bond[214].

In no case to date has the full equation (4.29) been employed in a structure determination in solution; at best equation (4.30) has been used, although a great many publications assume a direct dependence of Δ_{Dip} on $\langle r^{-3} \rangle$. The latter works to a reasonable approximation since the lanthanide is usually some distance from each magnetic nucleus so that the angles θ do not vary greatly, but this procedure cannot yield accurate structures. The question arises whether the assumption of cylindrical symmetry is justified. There are two approaches to this problem. Firstly, the symmetry of the complexes formed may be measured either in solution, by optical rotatory dispersion[215], or in the solid after crystallisation[216-218]. For the assumption to be justified a minimum of C_3 symmetry is required but in no case has this high symmetry been observed. Secondly, an experimental check can be performed for approximate cylindrical symmetry[211]: if two different lanthanides produce shifts whose ratios are independent of the lanthanide the tensor is axially symmetric to a good approximation. Experiments on rigid molecules of known geometry suggest that the assumption is usually reasonable but it cannot be justified absolutely.

Under these conditions further checks on geometry are essential and advantage may be taken of the relaxation caused by Gd^{3+} [219, 220]: with the assumptions listed above, equation (4.35), or its equivalent for T_{2B}, predicts a direct dependence of relaxation rate on $\langle r^{-6} \rangle$, with no angle dependence. This is a very different function of geometry than is $\langle (3 \cos^2 \theta - 1) r^{-3} \rangle$ and if relaxation ratios produce ratios of distances in agreement with those deduced from equation (4.30) then the results may be regarded with considerable confidence. (This conclusion is also subject to the condition that the geometry is unchanged on exchanging lanthanides.)

In calculating geometries the positioning of the lanthanide is usually treated as a variable so that no preconceptions are included of the length of Ln—O and Ln—N bonds, the requirement being consistency between calculated and observed shifts for the magnetic nuclei. This procedure has led to some discrepancies between these bond lengths (which are sensitive only to steric effects at the binding site[221]) and even to absurdly short values[222], but these do not greatly affect the derived geometry of the organic ligand. The calculations can be refined further by making them subject to normal ranges of bond angles and distances and of van der Waals radii in the ligand. The problem is best treated by iteration on a computer, in which the relative orientations of the lanthanide–ligand bond axis and the axis of the susceptibility tensor may be included[211].

Fundamental assumptions in these determinations are those of there being a single complex present which exists as a single geometric isomer, with complexing to only one donor site (although shifts from more than one site are independent[223]), and that the substrate ligand exists in a single conformation.

Whilst the technique abounds with assumptions it is not restricted to simple molecules of high symmetry and constitutes a method of structure determination in solution which may be extended to molecules over a wide range of molecular weights, even to enzymes. (Usually some chemical ruse can be found to attach a lanthanide to any molecule, if necessary via an introduced intermediary group.) Many of the assumptions would be encountered in any determination of the structure of a labile molecule. For example, it is not clear what sort of averaging occurs in any given case over an internal motion, but motion of this sort may lengthen or decrease apparent distances to the lanthanide. It should not be assumed that simply because a given rigid structure satisfies the experimental data that it is necessarily the correct or complete one; the method is unlikely to detect relatively unpopulated conformations.

Very few investigations have satisfied the criteria discussed above and relatively few of the several hundred papers published in the field have even employed equation (4.30) with or without tests for its applicability, although its inexactitude has been demonstrated occasionally[224, 225]. Some of the more reliable applications to molecular structure include investigations of amides and diamides[226], 4-t-butylcyclohexanol[227, 228], adamantone[221] and adamantol[228, 229], camphor[221] and norcamphor[221], rigid bicyclic ethers[230, 231], quinoline[232], isoquinoline[232] and benzoquinoline[232], pyridine[199], borneol[213, 229, 233, 234] and norborneol[213, 235], norbornylamine[231] α,β-unsaturated carboxylic acids[236], γ-picoline[199, 237] and γ-picoline *N*-oxide[237], 1,10-phenanthroline[225], α,α'-bipyridyl[225] and nucleotides[220]. A quantitative discussion of agreement between experimental and calculated shifts has been given[235].

Lanthanides are widely used in n.m.r. as 'shift reagents' to facilitate analysis of complex spectra and may be employed in difference spectroscopy[238], which can lead to considerable gain in measurement accuracy of shift effects in large molecules. Several reviews exist which cover the full range of applications in some detail[190, 239, 240-244]; the application of paramagnetic probes in biochemistry has been considered in a recent book[245].

4.3.2 Relaxation methods

If the correlation times can be determined, values of $\langle r^{-6} \rangle$ can be obtained directly from relaxation measurements using equations (4.31) and (4.32), but once again exchange conditions and corrections to the observations are important. The paramagnetic contributions to the relaxation times $T_{i,\mathrm{p}}$ (where $i = 1$ or 2) are given by

$$\frac{1}{T_{i,\mathrm{p}}} = \frac{1}{T_{i,\mathrm{obs}}} - \frac{1}{T_{i,\mathrm{dia}}} \tag{4.39}$$

where $T_{i,\mathrm{obs}}$ and $T_{i,\mathrm{dia}}$ are the measured relaxation times in the presence and absence of paramagnetic ions, respectively. In the fast exchange limit, and if relaxation of ligands in the outer solvation sphere is neglected, these are related to the relaxation time $T_{i,\mathrm{B}}$ in the bound molecule by

$$\frac{1}{pT_{i,\mathrm{p}}} = \frac{q}{T_{i,\mathrm{B}} + \tau_{\mathrm{B}}} \tag{4.40}$$

where p is the ratio of the concentration of the ion to that of the ligand and q is the coordination number. This equation applies in many aqueous solutions of the ions Mn^{2+} and Gd^{3+}; Mn^{2+} is particularly useful in that it can often replace Mg^{2+} in biological systems without destroying their activity. $T_{i,\mathrm{B}}$ can be extracted from equation (4.40) if the ligand exchange rate is known. In aqueous Mn^{2+} solutions, for example, the rate of water exchange, q/τ_{B}, is rapid compared with the relaxation rate $q/T_{1,\mathrm{B}}$ and also for this ion $\omega_s^2\tau_e^2$ and $\omega_s^2\tau_c^2 \gg 1$. A combination of equations (4.40) and (4.31) then gives[246]

$$\frac{1}{pT_{1,\mathrm{p}}} = \frac{q}{T_{1,\mathrm{B}}} = \frac{2}{15}\left(\frac{\mu_0}{4\pi}\right)^2 \frac{\gamma_I^2 g_e^2 \mu_{\mathrm{B}}^2 S(S+1)}{r^6} \left\{\frac{3\tau_c}{1+\omega_I^2\tau_c^2}\right\} \tag{4.41}$$

It is emphasised that this expression stems from a careful consideration of exchange rates and correlation times for this particular ion in aqueous solutions; use of any other ion requires equally careful experimental justification.

Similarly it has been shown for Gd^{3+} that the contact contribution to $T_{2,\mathrm{B}}$ can be neglected[247] and

$$\frac{1}{T_{2,\mathrm{B}}} = \frac{1}{15}\left(\frac{\mu_0}{4\pi}\right)^2 \frac{\gamma_I^2 g_e^2 \mu_{\mathrm{B}}^2 S(S+1)}{r^6} \left\{4\tau_c + \frac{3\tau_c}{1+\omega_I^2\tau_c^2} + \frac{13\tau_c}{1+\omega_s^2\tau_c^2}\right\} \tag{4.42}$$

This equation is also assumed to apply to Mn^{2+}.

To evaluate $\langle r^{-6}\rangle$, values of the coordination number q and τ_c are required. The former is usually obtained by analogy from model compounds and the latter can be obtained from studies of the frequency dependence of relaxation[248-250], its temperature dependence[251] or from T_1/T_2 ratios[252].

In the particular case of an ion binding to a macromolecule, use may be made of the proton resonance enhancement (PRE) phenomenon[253], which is the increase in relaxation rate by a factor ε when the ion is bound to the macromolecule. This results because the slower tumbling rate of the macromolecule is closer to the Larmor precession frequency than is the motion of a small molecule. If the relaxation time in the presence of a macromolecule is $T_{i,\mathrm{p}}^*$

$$\varepsilon = \frac{1/T_{i,\mathrm{p}}^*}{1/T_{i,\mathrm{p}}} = \frac{q^* f(\tau_c)^*}{q f(\tau_c)} \tag{4.43}$$

where q^* is the coordination number in the presence of the macromolecule and $f(\tau_c)$ is the time-dependent part of the relaxation equations. Thus if the proton resonance enhancement is measured, $f(\tau_c)^*$ can be calculated for the macromolecule if $f(\tau_c)$ has been determined by one of the above methods and if q and q^* are correctly guessed.

Relaxation measurements of the distances from Mn^{2+} to other magnetic nuclei (^{1}H, ^{13}C or ^{31}P) have been made, for example, in Mn^{II} complexes with AMP [254], phosphorylase *b*/AMP [254], pyruvate carboxylase[250], pyruvate kinase[250, 255], carbonic anhydrase/sulphacetamide[252], phosphofructokinase[256] and bovine erythrocyte superoxide dismutase[257]. Similar studies with Gd^{III} have been reported for the β-methyl-*N*-acetyl glucosamine/lysozyme system[258,259].

Inspection of this list shows that the method is giving information on distances from nuclei on substrate molecules bound to enzymes which contain paramagnetic ions near their active sites. For the bigger enzymes, for example phosphofructokinase, such investigations are beyond the scope of present x-ray technology in the solid state. The distances obtained are typically in the 4–12 Å region, with quoted errors of about ±0.4 Å; these are assessed mainly in terms of experimental observations and the uncertainty of q^*, and may be unrealistically low.

This brief summary serves to show the detail in which the dynamics of the solutions must be known before $\langle r^{-6}\rangle$ can be obtained. Detailed expositions of theory and practice are to be found in a number of recent papers[258-260], reviews[246, 247, 253, 261] and a book[245]. Emphasis has been placed here on the use of metal ions as paramagnetic probes but stable free radicals ('spin labels') may be used also[262].

4.4 CONCLUSIONS

The methods of structure determination discussed in this review are ones which are capable of yielding complete structures by n.m.r. methods, but there also exists the possibility of combining n.m.r. and other physical methods. For example, the structure of collagen, which is too irregular a molecule to allow a full x-ray study at high resolution, has been obtained by combining x-ray and deuterium n.m.r. techniques[263]. This depended on the orientation-dependence of nuclear quadrupole couplings, a subject neglected here since it has been little used in structure determination.

Magnetic resonance is the only method available for determining structures of molecules in solution. In small rigid molecules of high symmetry the partial-orientation method is of comparable accuracy for interproton distances with the conventional gas-phase or solid-state methods; for other nuclei the accuracy is more difficult to assess. In non-rigid molecules new problems of the effects of internal motions on apparent geometries of molecules are encountered in all the methods discussed above. The methods which depend on relaxation, the nuclear Overhauser effect and the addition of paramagnetic probes, are applicable to molecules of any size and symmetry. All of them are encompassed by assumptions, approximations and guesswork but they yield entirely new information about the dissolved molecule, where structure is most important to the synthetic chemist and the biochemist. At high molecular weights they even extend the range of molecules whose structures can be wholly or partially determined by physical methods in any phase. However, it is difficult at present to assess the acuracy of the distance determinations, although there seems to exist a high degree of self-consistency.

If this review has persuaded the reader not only of the potential of magnetic resonance methods for studying molecular structure but also of their limitations, and of the need for an absolute understanding of both theory and the magnetic or exchange conditions pertaining in solution before a structure determination is attempted, then it will have served its purpose.

References

1. Pople, J. A. and Buckingham, A. D. (1963). *Trans. Faraday Soc.*, **59,** 2421
2. Buckingham, A. D. and McLauchlan, K. A. (1967). *Progr. NMR Spectrosc.*, **2,** 63
3. Bloembergen, N., Purcell, E. M. and Pound, R. V. (1948). *Phys. Rev.*, **73,** 679
4. Hubbard, P. S. (1970). *J. Chem. Phys.*, **52,** 563
5. Kattawar, G. W. and Eisner, M. (1961). *Phys. Rev.*, **126,** 1054
6. Gutowsky, H. S. and Woessner, D. E. (1956). *Phys. Rev.*, **104,** 843
7. Andrew, E. R. and Bersohn, R. (1950). *J. Chem. Phys.*, **18,** 159
8. Andrew, E. R. and Bersohn, R. (1952). *J. Chem. Phys.*, **20,** 924
9. Haland, K. and Pedersen, B. (1968). *J. Chem. Phys.*, **49,** 3194
10. Van Vleck, J. H. (1948). *Phys. Rev.*, **74,** 1168
11. Powles, J. G. (1958). *J. Appl. Phys.*, **9,** 299
12. McCall, D. W. and Hamming, R. W. (1959). *Acta Crystallogr.*, **12,** 81
13. Andrew, E. R. (1953). *Phys. Rev.*, **91,** 425
14. Hughes, D. G. (1973). *J. Magn. Resonance*, **11,** 83
15. Farnes, R. E., Parker, G. W. and Memory, J. D. (1970). *Phys. Rev.*, **31,** 4228
16. McCall, D. W. and Hamming, R. W. (1963). *Acta Crystallogr.*, **16,** 1071
17. Lundin, A. G., Falaleev, O. V., Sergeev, N. A., Gurevich, A. S. and Falaleeva, L. G. (1970). *Colloque Ampère*, **16,** 1071
18. Carolan, J. L. (1971). *Chem. Phys. Lett.*, **12,** 389
19. Hindermann, D. K. and Falconer, W. E. (1970). *J. Chem. Phys.*, **52,** 6198
20. Mehrig, M., Griffin, R. G. and Waugh, J. S. (1971). *J. Chem. Phys.*, **55,** 746
21. O'Reilly, D. E., Petersen, E. M., El Saffar, Z. M. and Scheie, C. E. (1971). *Chem. Phys. Lett.*, **8,** 470
22. Gibby, M. G., Pines, A., Rhim, W. K. and Waugh, J. S. (1972). *J. Chem. Phys.*, **56,** 991
23. Weiss, A. (1972). *Angew. Chem. Int. Ed. Engl.*, **11,** 607
24. Andrew, E. R. (1971). *Progr. NMR Spectrosc.*, **8,** 1
25. Mansfield, P. (1971). *Progr. NMR Spectrosc.*, **8,** 41
26. McIntyre, H. M., Cobb, T. M. and Johnson, C. S. (1970). *Chem. Phys. Lett.*, **4,** 585
27. El Moghazi, M., Ernst, R. R. and Gunthard, H. H. (1970). *J. Magn. Resonance*, **3,** 480
28. Garg, S. K. and Davidson, D. W. (1972). *Chem. Phys. Lett.*, **13,** 73
29. Pedersen, B. (1964). *Proton Magnetic Resonance in Hydrates* (Oslo: University Press)
30. Dereppe, J. M., Touissaint, A. and Van Meersche, M. (1973). *J. Chim. Phys., Physicochim. Biol.*, **70,** 146
31. Habuda, S. P. and Gagarinsky, Y. U. (1971). *Acta Crystallogr.*, **B27,** 1667
32. Mahajan, M. and Rao, B. D. (1972). *J. Phys. Chem. Solids*, **33,** 2191
33. Agarwal, S. C., Mirza, P., Agarwal, V. D. and Gupta, R. C. (1972). *Indian J. Pure Appl. Phys.*, **10,** 602
34. Banerjee, A. K., Agrawal, V. D. and Gupta, R. C. (1972). *Indian J. Pure Appl. Phys.*, **10,** 538
35. Allen, P. S., Khanzada, A. W. K. and McDowell, C. A. (1973). *Molec. Phys.*, **25,** 1273
36. Finer, E. G. (1972). *Nucl. Magn. Resonance*, **1,** 273
37. Derbyshire, W. (1973). *Nucl. Magn. Resonance*, **2,** 323
38. Andrew, E. R. (1972). *Magn. Resonance Rev.*, **1,** 33
39. Kosfeld, R. (1972). *NMR: Basic Principles and Progress*, **4,** 181 (P. Diehl, editor) (Berlin: Springer)
40. Petrov, M. P. and Turov, E. A. (1971). *Appl. Spectrosc. Rev.*, **5,** 265
41. Hilbers, C. W. and MacLean, C. (1972). *NMR: Basic Principles and Progress*, **7,** 1 (P. Diehl, editor) (Berlin: Springer)
42. Biemond, J. and MacLean, C. (1973). *Molec. Phys.*, **26,** 409

43. Englert, G. and Saupe, A. (1962). *Z. Naturforsch.*, **19a,** 172
44. Bulthuis, J., Hilbers, C. W. and MacLean, C. (1972). *MTP International Review of Science, Physical Chemistry Series One*, Vol. 4, *Magnetic Resonance* (C. A. McDowell, editor), 201 (London: Butterworths)
45. Long, R. C. and Goldstein, J. H. (1973). *Molec. Cryst. Liq. Cryst.*, **23,** 137
46. Luz, Z. and Meiboom, S. (1973). *J. Chem. Phys.*, **59,** 275
47. Luz, Z. and Meiboom, S. (1973). *Molec. Cryst. Liq. Cryst.*, **22,** 143
48. Samulski, E. T. and Berendsen, H. J. C. (1972). *J. Chem. Phys.*, **56,** 3920
49. Snyder, L. C. (1965). *J. Chem. Phys.*, **43,** 4041
50. Saupe, A. (1964). *Z. Naturforsch.*, **19a,** 161
51. Saupe, A. (1972). *Molec. Cryst. Liq. Cryst.*, **16,** 87
52. Chapman, G. E., Long, E. M. and McLauchlan, K. A. (1969). *Molec. Phys.*, **17,** 189
53. Englert, G. and Saupe, A. (1966). *Molec. Cryst. Liq. Cryst.*, **1,** 503
54. Bhattacharyya, P. K. and Dailey, B. P. (1973). *Molec. Phys.*, **26,** 1379
55. Herschbach, D. R. and Laurie, V. W. (1962). *J. Chem. Phys.*, **37,** 1668
56. Herschbach, D. R. and Laurie, V. W. (1962). *J. Chem. Phys.*, **37,** 1687
57. Kuchitsu, K. (1968). *J. Chem. Phys.*, **49,** 4456
58. Whiffen, D. H. (1971). *Chem. Brit.*, **7,** 57
59. Diehl, P. and Khetrapal, C. L. (1969). *NMR: Basic Principles and Progress*, (P. Diehl, editor) (Berlin: Springer)
60. Raza, M. A. and Reeves, L. W. (1972). *Can. J. Chem.*, **50,** 2370
61. Khetrapal, C. L., Kunwar, A. C. and Saupe, A. (1973). *Molec. Phys.*, **25,** 1405
62. Segre, A. L. and Castellano, S. (1972). *J. Magn. Resonance*, **7,** 5
63. Bulthuis, J. and MacLean, C. (1971). *J. Magn. Resonance*, **4,** 148
64. Krugh, T. R. and Bernheim, R. A. (1970). *J. Chem. Phys.*, **52,** 4242
65. Barfield, M. (1970). *Chem. Phys. Lett.*, **4,** 518
66. Nakasuji, H., Kato, H., Moroshima, I. and Yonezawa, T. (1970). *Chem. Phys. Lett.*, **4,** 607
67. Haigh, C. W. and Sykes, S. (1973). *Chem. Phys. Lett.*, **19,** 571
67a. den Otter, G. J., MacLean, C., Haigh, C. W. and Sykes, S. (1974). *J. Chem. Soc. Chem. Commun.*, 24
68. Gerritsen, J., Koopmans, G., Rollema, H. S. and MacLean, C. (1972). *J. Magn. Resonance*, **8,** 20
69. den Otter, G. J., Gerritsen, J. and MacLean, C. (1973). *J. Molec. Struct.*, **16,** 379
69a. den Otter, G. J., Heijser, W. and MacLean, C. (1974). *J. Magn. Resonance*, **13,** 11
70. Emsley, J. W., Lindon, J. C. and Salman, S. R. (1972). *J. Chem. Soc. Faraday Trans. II*, **68,** 1343
71. Gerritsen, J. and MacLean, C. (1971). *Molec. Cryst. Liq. Cryst.*, **12,** 97
72. Gerritsen, J. and MacLean, C. (1971). *Molec. Cryst. Liq. Cryst.*, **7,** 57
73. Ditchfield, R. and Snyder, L. C. (1972). *J. Chem. Phys.*, **56,** 5823
74. Bulthuis, J. and MacLean, C. (1970). *Chem. Phys. Lett.*, **7,** 242
75. Diehl, P. and Niederberger, W. (1973). *J. Magn. Resonance*, **9,** 495
76. Raza, M. A. and Reeves, L. W. (1972). *J. Magn. Resonance*, **8,** 222
77. Cole, K. C. and Gilson, D. F. R. (1972). *J. Chem. Phys.*, **56,** 4363
78. Khetrapal, C. L., Kunwar, A. C. and Saupe, A. (1973). *Molec. Phys.*, **25,** 1405
79. Lucas, N. J. D. (1971). *Molec. Phys.*, **22,** 147
80. Lucas, N. J. D. (1971). *Molec. Phys.*, **22,** 239
81. Buckingham, A. D., Burnell, E. E. and de Lange, C. A. (1969). *Molec. Phys.*, **16,** 299
82. Gerritsen, J. and MacLean, C. (1971). *J. Magn. Resonance*, **5,** 44
83. Veracini, C. A., Bucci, P. and Barili, P. L. (1972). *Molec. Phys.*, **23,** 59
84. Barili, P. L. and Veracini, C. A. (1971). *Chem. Phys. Lett.*, **8,** 299
85. Canet, D., Haloui, E. and Nery, H. (1973). *J. Magn. Resonance*, **10,** 121
86. Snyder, L. C. and Meiboom, S. (1966). *J. Chem. Phys.*, **44,** 4057
87. Ader, R. and Loewenstein, A. (1972). *Molec. Phys.*, **24,** 455
87a. Buckingham, A. D. (1974). *5th Int. Symp. Magn. Resonance, Bombay*
88. Pedulli, G. F., Zannoni, C. and Alberti, A. (1973). *J. Magn. Resonance*, **10,** 372
89. Burnell, E. E. and Diehl, P. (1973). *Org. Magn. Resonance*, **5,** 137
90. Barili, P. L., Lunazzi, L. and Veracini, C. A. (1972). *Molec. Phys.*, **24,** 673
91. Meiboom, S., Hewitt, R. C. and Snyder, L. C. (1972). *Pure Appl. Chem.*, **32,** 251
92. Meiboom, S. and Snyder, L. C. (1971). *Accounts Chem. Res.*, **4,** 81

93. Luz, Z. and Meiboom, S. (1973). *J. Chem. Phys.*, **59,** 1077
94. Snyder, L. C. and Meiboom, S. (1973). *J. Chem. Phys.*, **58,** 5096
95. Hewitt, R. C., Meiboom, S. and Snyder, L. C. (1973). *J. Chem. Phys.*, **58,** 5089
96. Luckhurst, G. R. (1968). *Quart. Rev. Chem. Soc.*, **22,** 179
97. Diehl, P. and Henricks, P. M. (1972). *Nucl. Magn. Resonance*, **1,** 321
98. Bernheim, R. A. and Lavery, B. J. (1967). *J. Colloid Interface Sci.*, **26,** 291
99. Saupe, A. (1970). *Magnetic Resonance* (C. K. Coogan, N. S. Ham, S. N. Stuart, J. R. Pilbrow and G. V. H. Wilson, editors) (New York: Plenum Press)
100. Luckhurst, G. R. (1973). *Molec. Cryst. Liq. Cryst.*, **21,** 125
101. Meiboom, S. and Snyder, L. C. (1968). *Science*, **162,** 1337
102. Bhattacharyya, P. K. and Dailey, B. P. (1973). *J. Chem. Phys.*, **59,** 3737
103. Khetrapal, C. L. and Saupe, A. (1973). *Molec. Cryst. Liq. Cryst.*, **19,** 195
104. Swinton, P. and Gatti, G. (1972). *J. Magn. Resonance*, **8,** 293
105. Barton, S. A., Raza, M. A. and Reeves, L. W. (1973). *J. Magn. Resonance*, **9,** 45
106. Raza, M. A. and Reeves, L. W. (1972). *J. Chem. Phys.*, **57,** 821
107. Gazzard, I. J. (1973). *Molec. Phys.*, **25,** 469
108. Spiesecke, H. (1970). *Liq. Cryst. Ordered Fluids. Proc. Amer. Chem. Soc., Symp.*, 123
109. Reeves, L. W., Riveros, J. M., Spragg, R. A. and Varin, J. A. (1973). *Molec. Phys.*, **25,** 9
110. Zauer, M. and Azman, A. (1973). *J. Magn. Resonance*, **11,** 105
111. Emsley, J. W., Lindon, J. C. and Tabony, J. (1973). *Molec. Phys.*, **26,** 1485
112. Khetrapal, C. L., Kunwar, A. C. and Saupe, A. (1972). *J. Magn. Resonance*, **7,** 18
113. Herrig, W. and Guenther, H. (1972). *J. Magn. Resonance*, **8,** 284
114. Raza, M. A. and Reeves, L. W. (1972). *Molec. Phys.*, **23,** 1007
115. Huckerby, T. N. (1971). *Tetrahedron Lett.*, 3497
116. Russell, J. (1972). *Org. Magn. Resonance*, **4,** 433
117. Long, R. C. and Goldstein, J. H. (1971). *J. Molec. Spectrosc.*, **40,** 632
118. Degelaen, J., Diehl, P. and Niederberger, W. (1972). *Org. Magn. Resonance*, **4,** 721
119. Haloui, E. and Canet, D. (1972). *Compt. Rend. Acad. Sci., Ser. C*, **275,** 447
120. Gerritsen, J. and MacLean, C. (1972). *Rec. Trav. Chim. Pays-Bas*, **91,** 1393
121. Burnell, E. E. and Diehl, P. (1972). *Molec. Phys.*, **24,** 489
122. Canet, D. and Price, R. (1973). *J. Magn. Resonance*, **9,** 35
123. Boicelli, C. A., Mangini, A., Lunazzi, L. and Tiecco, M. (1972). *J. Chem. Soc., Perkin Trans. II*, 599
124. Veracini, C. A., Longeri, M. and Barili, P. L. (1973). *Chem. Phys. Lett.*, **19,** 592
125. Canet, D. (1973). *Compt. Rend. Acad. Sci. Ser. C*, **276,** 315
126. Khetrapal, C. L. and Saupe, A. (1973). *J. Magn. Resonance*, **9,** 275
127. Khetrapal, C. L., Saupe, A. and Kunwar, A. C. (1972). *Molec. Cryst. Liq. Cryst.*, **17,** 121
128. Niederberger, W., Diehl, P. and Lunazzi, L. (1973). *Molec. Phys.*, **26,** 571
129. Burnell, E. E. and Diehl, P. (1972). *Can. J. Chem.*, **50,** 3566
130. Emsley, J. W. and Lindon, J. C. (1973). *Molec. Phys.*, **25,** 641
131. Veracini, C. A. and Pietra, F. (1972). *J. Chem. Soc. Chem. Commun.*, 1262
132. Bailey, D., Buckingham, A. D., McIvor, M. C. and Rest, A. J. (1973). *Molec. Phys.*, **25,** 479
133. Bailey, D., Buckingham, A. D. and Rest, A. J. (1973). *Molec. Phys.*, **26,** 233
134. Buckingham, A. D., Rest, A. J. and Yesinowski, J. P. (1973). *Molec. Phys.*, **25,** 1457
135. Zumbulyadis, N. and Dailey, B. P. (1973). *Molec. Phys.*, **26,** 777
135a. Buckingham, A. D., Yesinowski, J. P., Canty, A. J. and Rest, A. J. (1973). *J. Amer. Chem. Soc.*, **95,** 2732
135b. Yesinowski, J. P. and Bailey, D. (1974). *J. Organomet. Chem.*, **65,** C27
136. Freeman, R. and Hill, H. D. W. (1970). *J. Chem. Phys.*, **53,** 4103
137. McLauchlin, A. C., McDonald, G. G. and Leigh, J. S. (1973). *J. Magn. Resonance*, **11,** 107
138. Freeman, R. and Hill, H. D. W. (1971). *J. Chem. Phys.*, **55,** 1985
139. Haeberlen, U., Spiesse, H. W. and Schweitzer, D. (1972). *J. Magn. Resonance*, **6,** 39
140. Carr, H. Y. and Purcell, E. M. (1954). *Phys. Rev.*, **94,** 630
141. Wold, R. L., Waugh, J. S., Klein, M. P. and Phelps, D. E. (1968). *J. Chem. Phys.*, **48,** 3831
142. Alger, T. D., Collins, S. W. and Grant, D. M. (1971). *J. Chem. Phys.*, **54,** 2820

143. Buchner, W. (1973). *J. Magn. Resonance*, **12,** 82
144. Campbell, I. D. and Freeman, R. (1973). *J. Magn. Resonance*, **11,** 143
145. Blichorski, J. S. (1972). *Z. Naturforsch.*, **27a,** 1456
146. Levy, G. C., Cargioli, J. D. and Anet, F. A. L. (1973). *J. Amer. Chem. Soc.*, **95,** 1527
147. Ferrar, T. C., Druck, S. J., Shoup, R. R. and Becker, E. D. (1972). *J. Amer. Chem. Soc.*, **94,** 699
148. Grandjean, J., Laszlo, P. and Price, R. (1973). *Molec. Phys.*, **25,** 695
149. Burke, T. E. and Chan, S. I. (1970). *J. Magn. Resonance*, **2,** 170
150. Hubbard, P. S. (1963). *Phys. Rev.*, **131,** 1155
151. Navon, G. and Lanir, A. (1972). *J. Magn. Resonance*, **8,** 144
152. Coates, H. B., McLauchlan, K. A., Campbell, I. D. and McColl, C. E. (1973). *Biochim. Biophys. Acta*, **310,** 1
153. Coates, H. B. and McLauchlan, K. A. unpublished work; Coates, H. B. (1973). *D.Phil. thesis*, Oxford
154. Levine, Y. K., Partington, P. and Roberts, G. C. K. (1973). *Molec. Phys.*, **25,** 497
155. Buchner, W. (1973). *J. Magn. Resonance*, **11,** 46
156. Versmold, H. (1973). *J. Chem. Phys.*, **58,** 5649
157. Woessner, D. E. and Snowden, B. S. (1972). *Adv. Molec. Relaxation Processes*, **3,** 181
158. Tomchuk, E., Czubryt, J. J., Bick, E. and Chatterjee, N. (1973). *J. Magn. Resonance*, **12,** 20
159. Boden, N. (1973). *Nucl. Magn. Resonance*, **2,** 112
160. Dwek, R. A. (1973). *NMR in Biochemistry* (Oxford: University Press)
161. Klose, G. (1972). *Math-Naturwiss. Reihe*, **21,** 752
162. Hyerla, J. R. and Grant, D. M. (1972). *MTP International Review of Science, Physical Chemistry Series One*, Vol. 4, *Magnetic Resonance* (C. A. McDowell, editor), 1 (London: Butterworths)
163. Levy, G. C. (1973). *Accounts Chem. Res.*, **6,** 161
164. Schirmer, R. E., Noggle, J. H., Davis, J. P. and Hart, P. A. (1970). *J. Amer. Chem. Soc.*, **92,** 3266
165. Bell, R. A. and Saunders, J. K. (1973). *Topics in Stereochemistry*, **7,** 1
166. Noggle, J. H. and Schirmer, R. E. (1971). *Nuclear Overhauser Effect* (New York: Academic Press)
167. Saunders, J. K. and Bell, R. A. (1970). *Can. J. Chem.*, **48,** 1114
168. Bell, R. A. and Saunders, J. K. (1968). *Can. J. Chem.*, **46,** 3421
169. Krishna, N. R. and Gordon, S. L. (1973). *J. Chem. Phys.*, **58,** 5687
170. Miller, D. P., Ternai, B. and Maciel, G. E. (1973). *J. Amer. Chem. Soc.*, **95,** 1336
171. Barbiou, V. and Gabe, I. (1970). *Colloque Ampère*, **16,** 312
172. Bachers, G. E. and Schaefer, T. (1971). *Chem. Rev.*, **71,** 617
173. Shaw, D. (1973). *Nucl. Magn. Resonance*, **2,** 269
174. McConnell, H. M. and Robertson, R. E. (1958). *J. Chem. Phys.*, **29,** 1361
175. Reuben, J. and Fiat, D. (1969). *J. Chem. Phys.*, **51,** 4909
176. Eaton, D. R. (1973). *Can. J. Chem.*, **51,** 2632
177. de Boer, E. and Van Willingen, H. (1967). *Progr. NMR Spectrosc.*, **2,** 111
178. Eaton, D. R. and Phillips, W. D. (1965). *Adv. Magn. Resonance*, **1,** 103
179. Bleaney, B. (1972). *J. Magn. Resonance*, **8,** 91
180. Golding, R. M. and Pyykko, P. (1973). *Molec. Phys.*, **26,** 1389
181. Horrocks, W. D. and Sipe, J. P. (1971). *J. Amer. Chem. Soc.*, **93,** 6300
182. La Mar, G. N., Horrocks, W. D. and Allen, L. C. (1964). *J. Chem. Phys.*, **41,** 2126
183. Grotens, A. M., Backhus, J. J. M. and de Boer, E. (1973). *Tetrahedron Lett.*, 1465
184. Bloembergen, N. and Morgan, L. O. (1961). *J. Chem. Phys.*, **34,** 842
185. Solomon, I. (1955). *Phys. Rev.*, **99,** 559
186. Hinckley, C. C. (1969). *J. Amer. Chem. Soc.*, **91,** 5760
187. Evans, D. F. and Wyatt, M. (1972). *J. Chem. Soc. Chem. Commun.*, 312
188. Grotens, A. M., Backhus, J. J. M., Pypers, F. M. and de Boer, E. (1973). *Tetrahedron Lett.*, 1467
189. Reuben, J. (1973). *J. Magn. Resonance*, **11,** 103
190. Reuben, J. (1973). *Progr. NMR Spectrosc.*, **9,** 1
191. Gansow, O. A., Loeffler, P. A., Davis, R. E., Willcott, M. R. and Lenkinski, R. E. (1973). *J. Amer. Chem. Soc.*, **95,** 3389
192. Kainosho, M. and Ajisaka, K. (1973). *Yuki Gosei Kagaku Kyokai Shi.*, **31,** 126

193. Kainosho, M., Ajisaka, K. and Tori, K. (1972). *Chem. Lett.*, **11,** 1061
194. Tori, K., Yoshimura, Y., Kainosho, M and Ajisaka, K. (1973). *Tetrahedron Lett.*, 1573
195. Chalmers, A. A. and Pachler, K. G. R. (1972). *Tetrahedron Lett.*, 4033
196. Gansow, O. A., Loeffler, P. A., Davis, R. E., Willcot, M. R. and Lenkinski, R. E. (1973). *J. Amer. Chem. Soc.*, **95,** 3389
197. Gansow, O. A., Loeffler, P. A., Davis, R. E., Willcott, M. R. and Lenkinski, R. E. (1973). *J. Amer. Chem. Soc.*, **95,** 3390
198. Hawkes, G. E., Marzin, C., Johns, S. R. and Roberts, J. D. (1973). *J. Amer. Chem. Soc.*, **95,** 1661
199. Hirayama, M., Edagawa, E. and Hanyu, Y. (1972). *J. Chem. Soc. Chem. Commun.*, 1343
200. Golding, R. M. and Halton, M. P. (1972). *Aust. J. Chem.*, **25,** 2577
201. Goldberg, L. and Ritchey, W. M. (1972). *Spectrosc. Lett.*, **5,** 201
202. ApSimon, J. W., Beierbeck, H. and Fruchier, A. (1973). *J. Amer. Chem. Soc.*, **95,** 939
203. Shapiro, B. L. and Johnston, M. D. (1972). *J. Amer. Chem. Soc.*, **94,** 8185
204. ApSimon, J. W., Beierbeck, H. and Fruchier, A. (1972). *Can. J. Chem.*, **50,** 2725
205. Williams, D. E. (1972). *Tetrahedron Lett.*, 1345
206. Bouquant, J. and Chuche, J. (1972). *Tetrahedron Lett.*, 2337
207. Gibb, V. G., Armitage, I. M., Hall, L. D. and Marshall, A. G. (1972). *J. Amer. Chem. Soc.*, **94,** 8919
208. Shapiro, B. L., Shapiro, M. J., Godwin, A. D. and Johnston, M. D. (1972). *J. Magn. Resonance*, **8,** 402
209. ApSimon, J. W. and Beierbeck, H. (1972). *J. Chem. Soc. Chem. Commun.*, 172
210. Bouquant, J. and Chuche, J. (1973). *Tetrahedron Lett.*, 493
211. Bleaney, B., Dobson, C. M., Levine, B., Martin, R. B., Williams, R. J. P. and Xavier, A. V. (1972). *J. Chem. Soc. Chem. Commun.*, 791
212. Honeybourne, C. L. (1972). *Tetrahedron Lett.*, 1095
213. Hawkes, G. E., Leibfritz, D., Roberts, D. W. and Roberts, J. D. (1973). *J. Amer. Chem. Soc.*, **95,** 1659
214. Briggs, J. M., Moss, G., Randall, E. W. and Sales, K. D. (1972). *J. Chem. Soc. Chem. Commun.*, 1180
215. Andersen, N. H., Bottino, B. J. and Smith, S. E. (1972). *J. Chem. Soc. Chem. Commun.*, 1193
216. Cramer, R. E. and Seff, K. (1972). *J. Chem. Soc. Chem. Commun.*, 400
217. Horrocks, W. D., Sipe, J. P. and Luber, J. R. (1971). *J. Amer. Chem. Soc.*, **93,** 5258
218. Uebel, J. J. and Wing, R. M. (1972). *J. Amer. Chem. Soc.*, **94,** 8910
219. La Mar, G. N. and Faller, J. W. (1973). *J. Amer. Chem. Soc.*, **95,** 3817
220. Barry, C. D., North, A. C. T., Glasel, J. A., Williams, R. J. P. and Xavier, A. V. (1971). *Nature* (*London*), **232,** 236
221. De Marco, P. U., Cerimele, C. R., Crane, R. W. and Thakkar, A. L. (1972). *Tetrahedron Lett.*, 3539
222. Mazzvichi, P. H., Ammon, H. L. and Jameson, C. W. (1973). *Tetrahedron Lett.*, 573
223. Grotens, A. M., de Boer, E. and Smid, J. (1973). *Tetrahedron Lett.*, 1471
224. Beaute, C., Cornuel, S., Lelandais, D., Thouai, N. and Wolkowski, Z. W. (1972). *Tetrahedron Lett.*, 1099
225. Bhacca, N. S., Selbin, J. and Wander, J. D. (1972). *J. Amer. Chem. Soc.*, **94,** 8719
226. Montaudo, G. and Finocchiaro, P. (1972). *J. Org. Chem.*, **37,** 3434
227. Wing, R. M., Early, T. A. and Uebel, J. J. (1972). *Tetrahedron Lett.*, 4153
228. Farid, S., Ateya, A. and Maggio, M. (1971). *Chem. Commun.*, 1285
229. Goodisman, J. and Matthews, R. S. (1972). *J. Chem. Soc. Chem. Commun.*, 127
230. Caple, R. and Kuo, S. C. (1971). *Tetrahedron Lett.*, 4413
231. Caple, R., Harriss, D. K. and Kuo, S. C. (1973). *J. Org. Chem.*, **38,** 381
232. Huber, H. and Pascual, C. (1971). *Helv. Chim. Acta*, **54,** 913
233. Briggs, J. M., Hart, F. A. and Moss, G. P. (1970). *Chem. Commun.*, 1506
234. Briggs, J. M., Hart, F. A., Moss, G. P. and Randall, E. W. (1971). *Chem. Commun.*, 364
235. Willcott, M. R., Lenkinski, R. E. and Davis, R. E. (1972). *J. Amer. Chem. Soc.*, **94,** 1742
236. Cedar, O. and Beijer, B. (1972). *Acta Chem. Scand.*, **26,** 2977

237. Horrocks, W. D. and Sipe, J. P. (1971). *J. Amer. Chem. Soc.*, **93,** 6800
238. Campbell, I. D., Dobson, C. M., Williams, R. J. P. and Xavier, A. V. (1973). *J. Magn. Resonance*, **11,** 172
239. Cockerill, A. F., Davies, G. L. O., Harden, R. C. and Rackham, D. M. (1973). *Chem. Rev.*, **73,** 553
240. Sanders, J. K. M. and Williams, D. H. (1972). *Nature* (*London*), **240,** 385
241. Mayo, B. C. (1973). *Chem. Soc. Rev.*, **2,** 49
242. (1973). *NMR Shift Reagents* (R. E. Sievers, editor) (New York: Academic Press)
243. Foreman, M. I. (1973). *Nucl. Magn. Resonance*, **2,** 378
244. von Ammon, R. and Fischer, R. D. (1972). *Angew. Chem. Int. Ed. Engl.*, **11,** 675
245. Dwek, R. A. (1973). *NMR in Biochemistry*, (Oxford: University Press)
246. Mildvan, A. S. and Engle, J. L. (1972). *Methods in Enzymology*, **26C,** 654
247. Dwek, R. A. (1972). *Adv. Molec. Relaxation Processes*, **4,** 1
248. Reuben, J. and Cohn, M. (1970). *J. Biol. Chem.*, **245,** 6539
249. Peacocke, A. R., Richards, R. E. and Sheard, B. (1969). *Molec. Phys.*, **16,** 177
250. Fung, C. H., Mildvan, A. S., Allerhand, A., Komoroski, R. and Scrutton, M. C. (1973). *Biochem.*, **12,** 620
251. Villafranca, J. J. and Mildvan, A. S. (1971). *J. Biol. Chem.*, **246,** 772
252. Lanir, A. and Navon, G. (1972). *Biochem.*, **11,** 3536
253. Mildvan, A. S. and Cohn, M. (1970). *Adv. Enzymology*, **33,** 1
254. Bennick, A., Campbell, I. D., Dwek, R. A., Radda, G. K. and Salmon, A. J. (1971). *Nature* (*London*), 190
255. Nowak, T. and Mildvan, A. S. (1972). *Biochem.*, **11,** 2819
256. Jones, R., Dwek, R. A. and Walker, I. O. (1972). *Eur. J. Biochem.*, **28,** 74
257. Gaber, B. P., Brown, R. D., Koenig, S. H. and Fee, J. A. (1972). *Biochim. Biophys. Acta*, **271,** 1
258. Morallee, K. G., Niebohr, E., Rossotti, F. J. C., Williams, R. J. P. and Xavier, A. V. (1970). *Chem. Commun.*, 1132
259. Dwek, R. A., Richards, R. E., Morallee, K. G., Niebohr, E., Williams, R. J. P. and Xavier, A. V. (1971). *Eur. J. Biochem.*, **21,** 204
260. Dwek, R. A., Radda, G. K., Richards, R. E. and Salmon, A. G. (1972). *Eur. J. Biochem.*, **29,** 509
261. Cohn, M. and Reuben, J. (1971). *Accounts Chem. Res.*, **4,** 214
262. Cohn, M. (1970). *Quart. Rev. Biochem.*, **3,** 61
263. Campbell, I. D., Chapman, G. E. and McLauchlan, K. A. (1970). *Nature* (*London*), **225,** 639

5
Electric Dipole Polarisabilities of Atoms and Molecules

M. P. BOGAARD and B. J. ORR
University of New South Wales

5.1 INTRODUCTION 150

5.2 PERTURBATION THEORY OF POLARISABILITIES 152
5.2.1 *First-order polarisabilities,* $\boldsymbol{\alpha}$ 152
5.2.2 *Higher-order polarisabilities,* $\boldsymbol{\beta}$, γ, . . . 156
5.2.3 *Effects of molecular symmetry* 160

5.3 PHENOMENA INVOLVING $\boldsymbol{\alpha}$ 162
5.3.1 *Refractive index and dielectric constant* 162
5.3.1.1 *Refractive index* 162
5.3.1.2 *Dispersion* 163
5.3.1.3 *Dielectric constant* 165
5.3.2 *Rayleigh and Raman scattering* 166
5.3.3 *The quadratic Stark effect* 170

5.4 NON-LINEAR OPTICAL PHENOMENA 172
5.4.1 *Second harmonic generation* 172
5.4.1.1 *Coherent second harmonic generation* 172
5.4.1.2 *Hyper-Rayleigh and hyper-Raman scattering* 172
5.4.2 *Third harmonic generation* 174
5.4.3 *Electro-optic effects* 175
5.4.3.1 *Electric field-induced second harmonic generation* 175
5.4.3.2 *Electro-optic Kerr effect* 177
5.4.4 *Miscellaneous non-linear optical phenomena* 179

5.5 INTERACTING ATOMS AND MOLECULES 180
5.5.1 *Virial coefficients of electrical and optical properties* 180
5.5.2 *Vibrational and rotational dependence of* $\boldsymbol{\alpha}$ 181
5.5.3 *Intermolecular forces* 182

5.6 THEORETICAL ESTIMATES 183
5.6.1 *Static polarisability* 183

5.6.2 *Dynamic polarisability* 185
5.6.3 *Hyperpolarisabilities* 185
5.6.4 *Bond and atomic additivity schemes* 185
5.6.5 *Interacting atoms* 186

5.7 CONCLUSIONS AND COMPARISON OF DATA 188

5.8 ACKNOWLEDGEMENTS 190

5.1 INTRODUCTION

Many important physical phenomena depend on the electric dipole moments induced in molecules, and atoms, by externally applied electric fields. Provided that the electric field strength $\boldsymbol{E}$ at the molecule is not too large, the electric dipole moment $\boldsymbol{\mu}$ may be expanded in powers of $\boldsymbol{E}$:

$$\boldsymbol{\mu}=\boldsymbol{\mu}^{(0)}+\boldsymbol{\alpha}\boldsymbol{E}+\kappa\boldsymbol{\beta}\boldsymbol{E}^2+\lambda\boldsymbol{\gamma}\boldsymbol{E}^3+\ldots \qquad (5.1)$$

where $\boldsymbol{\mu}^{(0)}$ is the (permanent) electric dipole moment in the absence of the field $\boldsymbol{E}$. The coefficients $\boldsymbol{\alpha}$, $\boldsymbol{\beta}$, γ, . . . are tensor quantities, which allow the induced moment to be in a direction other than that of the applied field. The *polarisability* $\boldsymbol{\alpha}$ is a tensor of rank 2 and the higher-order polarisabilities (*hyperpolarisabilities*) $\boldsymbol{\beta}$, γ, . . . are of rank 3, 4, . . . , respectively. The quantities κ, λ, . . . are numerical factors which depend on the combination of electric field frequencies involved in each non-linear process.*

We are concerned in this article with the theoretical description and measurement of electric dipole polarisabilities. As a result, equation (5.1) need not contain terms proportional to powers of the electric field gradients or of any associated magnetic fields. This is not to imply that such additional terms are unimportant; the phenomena which they describe include, for example, optical activity and the Faraday effect[1]. We shall also confine attention primarily to phenomena involving isolated atoms and molecules, i.e. to effects observed in media of low density. Finally, a distinction is to be made between resonant and non-resonant phenomena involving polarisabilities. By the former we mean any effect in which the frequency of oscillation of the applied electric field (for example, the frequency of an electromagnetic wave) is close to a molecular resonance frequency. Our treatment of resonant phenomena is superficial, despite the interest which they currently attract, particularly with the advent of tunable lasers.

A few simple examples illustrate the occurrence of polarisability-based phenomena. A light beam of frequency ω incident on a gas produces an electric field $\boldsymbol{E}_0 \cos \omega t$ at the molecules of the gas. The electric dipole moment $\boldsymbol{\mu}$ of each molecule then contains a component, proportional to $\boldsymbol{\alpha}$, which oscillates at frequency ω. This oscillating electric dipole moment emits

*There exists in the literature considerable variation in the choice of constants κ, λ, . . . ; the convention prescribed in this review (Section 5.2.2) is a particularly convenient one, as it allows a direct comparison of hyperpolarisabilities derived from different non-linear processes. For a static electric field $\boldsymbol{E}$, our convention yields $\kappa=\frac{1}{2}$, $\lambda=\frac{1}{6}$.

radiation, forming a spherical scattered wave from each molecule. The spherical waves from all the molecules interfere to produce net scattered radiation of two types: coherent and incoherent. Constructive interference of the net scattered radiation with the transmitted electromagnetic wave produces a resultant wave which, for positive $\boldsymbol{\alpha}$, is retarded in phase relative to that of the incident wave. This is the phenomenon of *refraction*, consisting of forward scattering of an intensity comparable with that of the incident wave. The process is said to be *coherent*, in view of the phase relationship between the incident and resultant waves. For a medium of uniformly distributed molecules (for example, a large perfect crystal), no scattering other than in the forward direction occurs, owing to destructive interference in other directions. However, for a random medium such as a fluid, scattering is observed in other directions as well. The molecular motion in the fluid produces random phase relations between the spherical waves, whose mutual interference results in incoherent scattering, known as *Rayleigh scattering*.

If the molecules of the medium lack a centre of inversion, their hyperpolarisability $\boldsymbol{\beta}$ is non-zero and the term $\boldsymbol{\beta}\boldsymbol{E}_0^2 \cos^2 \omega t$ produces a component of $\boldsymbol{\mu}$ oscillating at frequency 2ω. Under suitable conditions, both coherent (*second harmonic generation*) and incoherent (*hyper-Rayleigh*) scattering at frequency 2ω may be observed.

Some phenomena depend on the energy W of the molecule in an electric field $\boldsymbol{E}$. Using the relation,

$$W = -\int \boldsymbol{\mu}\cdot \mathrm{d}\boldsymbol{E} \tag{5.2}$$

equation (5.1) yields, for a static electric field:

$$W = W^{(0)} - \boldsymbol{\mu}^{(0)}\boldsymbol{E} - \tfrac{1}{2}\boldsymbol{\alpha}\boldsymbol{E}^2 - \tfrac{1}{6}\boldsymbol{\beta}\boldsymbol{E}^3 - \tfrac{1}{24}\boldsymbol{\gamma}\boldsymbol{E}^4 - \dots \tag{5.3}$$

where $W^{(0)}$ is the energy in the absence of the field. An illustration of the term in $\boldsymbol{\alpha}$ is the deflection of an S-state atom, for which $\boldsymbol{\mu}^0 = 0$, in a non-uniform electric field. This is a result of the force

$$\boldsymbol{F} = -\partial W/\partial \boldsymbol{r} = +\boldsymbol{\alpha}\boldsymbol{E}(\partial \boldsymbol{E}/\partial \boldsymbol{r}) \tag{5.4}$$

which acts on the atom. Another example, of a quantum mechanical nature, is the quadratic Stark effect which may be observed in atomic and molecular spectra.

A further example involving the energy W arises in calculation of the macroscopic dielectric polarisation $\boldsymbol{P}$ for a gas:

$$\boldsymbol{P} = (\varepsilon - 1)\varepsilon_0 \boldsymbol{E} \tag{5.5}$$

where ε is the dielectric constant of the gas and ε_0 the permittivity of a vacuum. The polarisation $\boldsymbol{P}$ is proportional to the mean electric dipole moment per molecule

$$\bar{\boldsymbol{\mu}} = \int \boldsymbol{\mu}(\tau)\exp[-W(\tau)/kT]\mathrm{d}\tau \Big/ \int \exp[-W(\tau)/kT]\mathrm{d}\tau \tag{5.6}$$

where $\boldsymbol{\mu}(\tau)$ is the electric dipole moment of a molecule with orientation τ, and $W(\tau)$ the corresponding energy, in the presence of the electric field $\boldsymbol{E}$. Contributions to $\bar{\boldsymbol{\mu}}$ which are linear in $\boldsymbol{E}$ can arise from two sources: one is the linear dependence on $\boldsymbol{E}$ of $\boldsymbol{\mu}(\tau)$ itself, namely the term $\boldsymbol{\alpha}\boldsymbol{E}$ of equation (5.1); the other is the linear dependence on $\boldsymbol{E}$ of the energy of orientation $W(\tau)$, namely the term $-\boldsymbol{\mu}^{(0)}\mathbf{E}$ in equation (5.3). The result of the classical statistical average [equation (5.6)] is

$$\bar{\mu} = (a + \mu^2/3kT)\boldsymbol{E} \tag{5.7}$$

where a is the isotropic part of $\boldsymbol{a}$ and μ the magnitude of the moment $\boldsymbol{\mu}^{(0)}$ This example illustrates how an observable of a given order in the applied fields may have contributions both from the molecular quantity which is being averaged and from field dependence of the molecular energy. The latter terms always contain inverse powers of kT and arise from partial orientation of the molecules. Other phenomena in which these two types of contribution are important include the electro-optic and optical Kerr effects and d.c.-induced second-harmonic generation in fluids.

For electric field strengths normally attainable in the laboratory ($\sim 10^7$ V m^{-1}), the series (5.1) and (5.3) converge rapidly; typical orders of magnitude in SI units* are: $\boldsymbol{\mu}$, 10^{-30} C m; $\boldsymbol{\alpha}$, 10^{-40} C^2 m^2 J^{-1}; $\boldsymbol{\beta}$, 10^{-50} C^3 m^3 J^{-2}; γ, 10^{-60} C^4 m^4 J^{-3}. In view of the rapid convergence which is normally found in equation (5.1), considerable subtlety must be applied in the experimental determination of the higher-order polarisabilities $\boldsymbol{\beta}$, γ, . . .

It is our aim in this review to cover the range of phenomena in which atomic and molecular polarisabilities are involved and to outline the experimental and theoretical methods used for their evaluation. It is not intended to provide a comprehensive coverage of the literature, which is extensive, but rather to give the reader an insight into the structure of the field and an awareness of its growth points. A number of reviews which may serve as introduction exist and are listed below. Reference to more specialised articles is given in the relevant sections of this review.

Author(s) and year	*Subject of review*	*Ref.*
Buckingham (1967)	Theory of molecular polarisabilities	1
Buckingham (1971)	Electric moments of molecules	2
Buckingham and Orr (1967)	Molecular hyperpolarisabilities	3
Le Fèvre (1965)	Refractivity and polarisability	4
Franken and Ward (1963)	Non-linear optics	5
Bloembergen (1965)	Non-linear optics	6
Terhune *et al.* (1966)	Non-linear optics	7
Kielich (1972)	General survey (540 references)	8

5.2 PERTURBATION THEORY OF POLARISABILITIES

5.2.1 First-order polarisabilities, $\boldsymbol{\alpha}$

The theory of molecular polarisability may be formulated with the aid of quantum mechanical perturbation theory. The most general problem in our context is that of a polarisable molecule in interaction with an oscillating

*Much of the literature is in electrostatic units, to which conversion may be made readily. For $\boldsymbol{\mu}$, 1 C m$=2.9979\times10^{11}$ e.s.u.$=0.299\ 79\times10\ ^{30}$D. Electrostatic units for $\boldsymbol{a}$, $\boldsymbol{\beta}$, γ, . . . may be obtained by multiplying the conversion factor for $\boldsymbol{\mu}$ by $(2.9979\times10^4)^n$, where $n=1, 2, 3, \ldots$ respectively. Hence for $\boldsymbol{a}$, 1 C^2 m^2 J$^{-1}=0.898\ 76\times10^{16}$ e.s.u. (or cm^3).

electric field $\boldsymbol{E}(t)$. This may be represented by a molecule in an electromagnetic field, for which a semi-classical description (molecule quantised, radiation not quantised) is adequate in many cases. Formal treatments of this problem, yielding the results which follow, are presented in a number of quantum mechanical texts[9–13]. Alternative approaches, particularly applicable to numerical computation of polarisabilities, have also been reviewed[13,14]. This section is confined to the perturbation-theoretical description of the first-order polarisability $\boldsymbol{\alpha}$. Higher-order polarisabilities are considered in Section 5.2.2.

It is well established[15,16] that the Hamiltonian $H'(t)$ describing the interaction of a molecule and an electromagnetic field may be written in terms of the electric and magnetic multipole operators of the molecule, and that, for wavelengths much larger than molecular dimensions, only the leading (electric dipole) term need be considered:

$$H'(t) = -\boldsymbol{\mu}\cdot\boldsymbol{E}_0 \cos \omega t \tag{5.8}$$

Here $\boldsymbol{\mu}$ is the electric dipole operator of the molecule and an appropriate phase choice has made $\boldsymbol{E}(t)$ real. The assumptions inherent in our use of the electric dipole approximation prove equivalent to our earlier restriction of attention to electric dipole polarisabilities. Use of time-dependent perturbation theory shows how solutions $|\Psi(t)\rangle$ of the time-dependent Schrödinger equation for the system evolve in the presence of the perturbation (5.8). This enables evaluation of the expression,

$$\begin{aligned}\mu_\xi^{(mn)}(t) &= \langle \Psi_m(t)|\mu_\xi|\Psi_n(t)\rangle \\ &= \exp(\mathrm{i}\omega_{mn}t)\langle m|\mu_\xi|n\rangle \\ &+ (2\hbar)^{-1}\sum_{\pm}\exp[\mathrm{i}(\omega_{mn}\pm\omega)t]\sum_k\left[\frac{\langle m|\mu_\xi|k\rangle\langle k|\mu_\eta|n\rangle}{(\Omega_{kn}\pm\omega)}\right. \\ &\left.\pm\frac{\langle m|\mu_\eta|k\rangle\langle k|\mu_\xi|n\rangle}{(\Omega_{km}^{*}\mp\omega)}\right]E_{o\eta} + \ldots\end{aligned} \tag{5.9}$$

which is the electric transition dipole between quantum states m and n of the molecule. The notation used above is as follows: $\omega_{mn} = \omega_m - \omega_n$, where ω_m and ω_n are energy eigenvalues (in angular frequency units) of the initially unperturbed molecule; $\Omega_{kn} = \omega_{kn} - \frac{1}{2}\mathrm{i}\Gamma_{kn}$ and is complex in order to provide for radiative damping in resonant situations[17]; $\sum_{\pm}$ denotes a summation over ensuing $\pm$ alternatives; $\sum_k$ denotes summation over a complete set of intermediate eigenstates k. Cartesian tensor notation, employing the Einstein summation convention†, has also been introduced at this stage. Equation (5.9) is the master equation for our description of a number of phenomena, including Raman ($m \neq n$) and Rayleigh ($m = n$) scattering in both resonant and non-resonant cases.

The source of Rayleigh scattering in the non-resonant case may be obtained from equation (5.9) by setting $m = n$ and omitting radiative damping, to yield

†Repeated Greek suffixes imply summation over cartesian x, y and z components. For example, $A_\xi B_{\xi\eta} = A_x B_{x\eta} + A_y B_{y\eta} + A_z B_{z\eta}$ ($\eta = x$, y or z); $C_\xi D_\xi = \boldsymbol{C}\cdot\boldsymbol{D}$.

$$\mu_{\xi}^{(nn)}(t)=\mu_{\xi}^{(n)}+a_{\xi\eta}^{(n)}(-\omega;\omega)E_{0\eta}\cos\omega t+\ldots \tag{5.10}$$

Here $\mu_{\xi}^{(n)}=\langle n|\mu_{\xi}|n\rangle$ is the permanent electric dipole moment for the molecule in quantum state n. The optical polarisability tensor is given by

$$a_{\xi\eta}^{(n)}(-\omega;\omega)=\sum_{k}\frac{2\omega_{kn}}{\hbar(\omega_{kn}^{2}-\omega^{2})}\langle n|\mu_{\xi}|k\rangle\langle k|\mu_{\eta}|n\rangle \tag{5.11}$$

Use has been made of the fact that the product of electric dipole matrix elements in (5.11) is invariant to interchange of ξ and η, for a non-degenerate state n in the absence of a static magnetic field[1].

On taking the limit of equations (5.10) and (5.11) as the frequency ω approaches zero, we obtain the result for perturbation by a static electric field E_{η}:

$$\mu_{\xi}^{(nn)}(\text{static})=\mu_{\xi}^{(n)}+a_{\xi\eta}^{(n)}(0;0)E_{\eta}+\ldots \tag{5.12}$$

$$a_{\xi\eta}^{(n)}(0;0)=\sum_{k}{}'\frac{2}{\hbar\omega_{kn}}\langle n|\mu_{\xi}|k\rangle\langle k|\mu_{\eta}|n\rangle \tag{5.13}$$

where $\sum_{k}{}'$ denotes a summation over intermediate states k from which the initial state n has been excluded.

If the optical frequency ω approaches a molecular transition frequency ω_{kn}, provision must be made for resonance effects. Retention of radiative damping in equation (5.9), with $m=n$, yields the result

$$\begin{aligned}\mu_{\xi}^{(nn)}(t)=\mu_{\xi}^{(n)}&+\text{Re}\,[a_{\xi\eta}^{(n)}(-\omega;\omega)]E_{0\eta}\cos\omega t\\&+\text{Im}\,[a_{\xi\eta}^{(n)}(-\omega;\omega)]E_{0\eta}\sin\omega t+\ldots\end{aligned} \tag{5.14}$$

$$\text{Re}\,[a_{\xi\eta}^{(n)}(-\omega;\omega)]=\sum_{k}\frac{2\omega_{kn}(\omega_{kn}^{2}-\omega^{2}+\frac{1}{4}\Gamma_{kn}^{2})}{\hbar(\omega_{kn}^{2}-\omega^{2}+\frac{1}{4}\Gamma_{kn}^{2})^{2}+\hbar\omega^{2}\Gamma_{kn}^{2}}\langle n|\mu_{\xi}|k\rangle\langle k|\mu_{\eta}|n\rangle \tag{5.15}$$

$$\text{Im}\,[a_{\xi\eta}^{(n)}(-\omega;\omega)]=\sum_{k}\frac{2\omega_{kn}\omega\Gamma_{kn}}{\hbar(\omega_{kn}^{2}-\omega^{2}+\frac{1}{4}\Gamma_{kn}^{2})^{2}+\hbar\omega^{2}\Gamma_{kn}^{2}}\langle n|\mu_{\xi}|k\rangle\langle k|\mu_{\eta}|n\rangle \tag{5.16}$$

where Re and Im denote real and imaginary parts, respectively. In addition to the contribution due to Re ($\boldsymbol{\alpha}$), which oscillates in phase with the radiation field $\boldsymbol{E}(t)$, a second contribution, which involves the tensor Im ($\boldsymbol{\alpha}$) and is out of phase with $\boldsymbol{E}(t)$, appears in the resonant case. The former (in-phase) contribution is responsible for so-called *anomalous dispersion* of refractive index and the latter may be identified with *absorption* of radiation at resonance. The frequency dependence of Re ($\boldsymbol{\alpha}$) and Im ($\boldsymbol{\alpha}$) in the vicinity of a resonance frequency ω_{kn} are illustrated in Figure 5.1.

The above cases are examples of elastic processes, in which the frequencies of incident and scattered radiation are identical. The origin of inelastic (Raman) scattering may be established by considering the real part of equation (5.9), with $m\neq n$. In the non-resonant case,

$$\begin{aligned}\mathrm{Re}\,[\mu_\xi^{(mn)}(t)] &= \cos(\omega_{mn}t+\varphi_{mn})\langle m|\mu_\xi|n\rangle \\ &+(2\hbar)^{-1}\cos[(\omega_{mn}+\omega)t+\varphi_{mn}]\sum_k\left[\frac{\langle m|\mu_\xi|k\rangle\langle k|\mu_\eta|n\rangle}{\omega_{kn}+\omega}\right. \\ &\left.+\frac{\langle m|\mu_\eta|k\rangle\langle k|\mu_\xi|n\rangle}{\omega_{km}-\omega}\right]E_{0\eta} \\ &+(2\hbar)^{-1}\cos[(\omega_{mn}-\omega)t+\varphi_{mn}]\sum_k\left[\frac{\langle m|\mu_\xi|k\rangle\langle k|\mu_\eta|n\rangle}{\omega_{kn}-\omega}\right. \\ &\left.+\frac{\langle m|\mu_\eta|k\rangle\langle k|\mu_\xi|n\rangle}{\omega_{km}+\omega}\right]E_{0\eta}+\ldots\end{aligned} \tag{5.17}$$

where $\langle m|\boldsymbol{\mu}|n\rangle$ and $\langle m|\boldsymbol{\mu}|k\rangle\langle k|\boldsymbol{\mu}|n\rangle$ have been made real by extracting a complex phase factor $\exp(\mathrm{i}\varphi_{mn})$. The first term in equation (5.17), oscillating at frequency ω_{mn}, corresponds to *spontaneous emission* $m\rightarrow n$. The remaining two

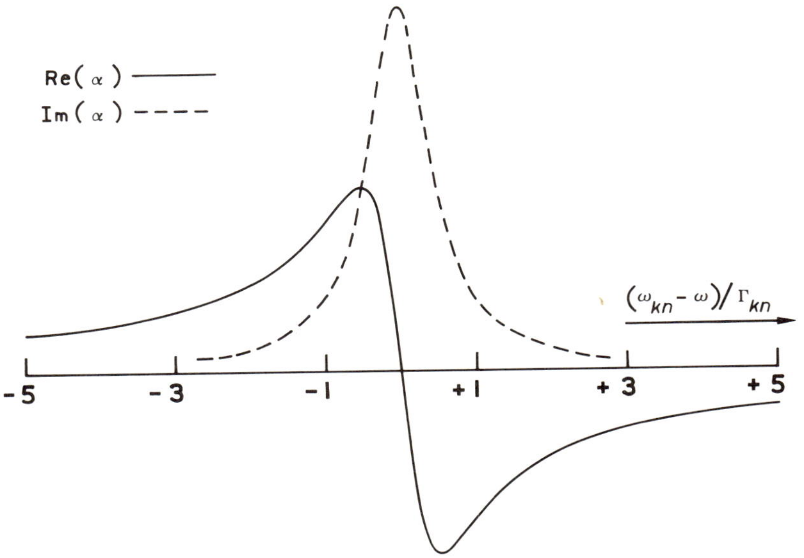

Figure 5.1

terms, having frequencies $(\omega+\omega_{mn})$ and $(\omega-\omega_{mn})$, correspond respectively to *anti-Stokes* and *Stokes* Raman processes (if $\omega_{mn}>0$). These Raman processes may be described in terms of transition polarisabilities, of form

$$\alpha_{\xi\eta}^{(mn)}(\omega_{mn}\pm\omega;\mp\omega)=\sum_k\left[\frac{\langle m|\mu_\xi|k\rangle\langle k|\mu_\eta|n\rangle}{\hbar(\omega_{kn}\pm\omega)}+\frac{\langle m|\mu_\eta|k\rangle\langle k|\mu_\xi|n\rangle}{\hbar(\omega_{kn}\mp\omega)}\right] \tag{5.18}$$

where the upper and lower signs refer to anti-Stokes and Stokes processes, respectively.

5.2.2 Higher-order polarisabilities, β, γ, . . .

The perturbation theory may be extended to higher orders to yield expressions for the various hyperpolarisabilities which contribute to non-linear optic and other phenomena. Such an extension was first formulated in the pioneering paper of Göppert-Mayer[15], which preceded by thirty years the experimental realisation of non-linear optics. A number of detailed semi-classical treatments of higher-order polarisabilities exist, including those by Bloembergen *et al.*[18] and Ward[19]. The latter work employs diagrammatic techniques to minimise the labour of evaluating non-linear optical polarisabilities. Useful treatments of higher-order perturbation theory for time-independent[20] and time-dependent[13] processes have been presented by Dalgarno. A recent treatment of non-linear optical polarisation[17] pays particular attention to resonant situations.

From the treatment of Orr and Ward[17] the results of higher-order polarisability theory for a non-resonant situation may be summarised as follows. A perturbation Hamiltonian $H'(t)$ may be defined generally to be of form:

$$H'(t) = H^0 + \tfrac{1}{2}\sum_j [H'^{\omega_j}\exp(-i\omega_j t) + H'^{-\omega_j}\exp(+i\omega_j t)] \tag{5.19}$$

where $H'^{\omega_j} = -\boldsymbol{\mu}\cdot\boldsymbol{E}_0^{\omega_j}$ and $H'^{-\omega_j} = (H'^{\omega_j})^*$. The perturbation $H'(t)$ induces an electric dipole moment $\boldsymbol{\mu}(t)$ in the molecule. This may be resolved into Fourier components

$$\boldsymbol{\mu}(t) = \boldsymbol{\mu}^0 + \tfrac{1}{2}\sum_\sigma [\boldsymbol{\mu}^{\omega_\sigma}\exp(-i\omega_\sigma t) + \boldsymbol{\mu}^{-\omega_\sigma}\exp(+i\omega_\sigma t)], \tag{5.20}$$

where $\boldsymbol{\mu}^{-\omega_\sigma} = (\boldsymbol{\mu}^{\omega_\sigma})^*$. The non-linear optical process of interest can then be defined by stating the frequencies involved and the dependence of the induced dipole on the perturbation; for example, for a third-order process

$$\omega_\sigma = \omega_1 + \omega_2 + \omega_3 \tag{5.21a}$$

$$\boldsymbol{\mu}^{\omega_\sigma} \propto H'^{\omega_1} \times H'^{\omega_2} \times H'^{\omega_3} \tag{5.21b}$$

with specific values for the field frequencies ω_1, ω_2, ω_3. The detailed form of relations such as equation (5.21b) is found[17], for a molecule in the quantum state n, to be

For a first-order non-resonant process ($\omega_\sigma = \omega$):

$$\boldsymbol{\mu}^{\omega} = 2(-\hbar)^{-1} I_{-\omega,\omega} \sum_k{}' \frac{\langle n|\boldsymbol{\mu}|k\rangle\langle k|H'^{\omega}|n\rangle}{\omega_{kn} - \omega} \tag{5.22}$$

For a second-order non-resonant process ($\omega_\sigma = \omega_1 + \omega_2$):

$$\boldsymbol{\mu}^{\omega_\sigma} = 3K(-\omega_\sigma;\omega_1,\omega_2)(-\hbar)^{-2} I_{-\sigma,1,2} \times \sum_{k,l}{}' \frac{\langle n|\boldsymbol{\mu}|l\rangle\langle l|\overline{H'^{\omega_2}}|k\rangle\langle k|H'^{\omega_1}|n\rangle}{(\omega_{ln} - \omega_\sigma)(\omega_{kn} - \omega_1)} \tag{5.23}$$

For a third-order non-resonant process ($\omega_\sigma = \omega_1 + \omega_2 + \omega_3$):

$$\mu^{\omega_\sigma}=4K(-\omega_\sigma;\omega_1,\omega_2,\omega_3)(-\hbar)^{-3}I_{-\sigma,1,2,3}$$
$$\times\left[\sum_{k,l,m}{}'\frac{\langle n|\boldsymbol{\mu}|m\rangle\langle m|\overline{H'^{\omega_3}}|l\rangle\langle l|\overline{H'^{\omega_2}}|k\rangle\langle k|H'^{\omega_1}|n\rangle}{(\omega_{mn}-\omega_\sigma)(\omega_{ln}-\omega_1-\omega_2)(\omega_{kn}-\omega_1)}\right. \tag{5.24}$$
$$\left.-\sum_{k,l}{}'\frac{\langle n|\boldsymbol{\mu}|l\rangle\langle l|H'^{\omega_3}|n\rangle\langle n|H'^{\omega_2}|k\rangle\langle k|H'^{\omega_1}|n\rangle}{(\omega_{ln}-\omega_\sigma)(\omega_{kn}-\omega_1)(\omega_{kn}+\omega_2)}\right]$$

The notation employed needs explanation. The symbols I denote permutation operators: for example, $I_{-\sigma,1,2,3}$ denotes the average of all terms generated by simultaneously permuting the frequencies $-\omega_\sigma$, ω_1, ω_2, ω_3 and the corresponding operators $\boldsymbol{\mu}$, H'^{ω_1}, H'^{ω_2}, H'^{ω_3}. The notation Σ' again denotes a summation from which the initial state n is excluded. The barred notation $\langle l|\overline{H'^{\omega_j}}|k\rangle$ is shorthand for the expression $(\langle l|H'^{\omega_j}|k\rangle-\langle n|H'^{\omega_j}|n\rangle)$. The numerical coefficients $K(-\omega_\sigma;\omega_1,\omega_2)$ and $K(-\omega_\sigma;\omega_1,\omega_2,\omega_3)$ depend on the non-linear optical process of interest. Their evaluation has been discussed elsewhere[17] and selected results are summarised in Table 5.1, which also specifies representative optical phenomena for a given combination of frequencies.

Table 5.1 Representative non-linear optical processes, with corresponding resultant (ω_σ) and incident (ω_1, ω_2 . . . .) frequencies and numerical factors K($-\omega_\sigma$; ω_1, ω_2, . . .)*

Non-linear optical process	$-\omega_\sigma$	ω_1	ω_2	ω_3	K
First static hyperpolarisability	0	0	0	–	1
Second harmonic generation (SHG)	-2ω	ω	ω	–	1/2
Electro-optic Pockels effect	$-\omega$	0	ω	–	2
Optical rectification	0	ω	$-\omega$	–	1/2
Two-wave mixing	$-(\omega_1+\omega_2)$	ω_1	ω_2	–	1
Second static hyperpolarisability	0	0	0	0	1
Third harmonic generation (THG)	-3ω	ω	ω	ω	1/4
Intensity-dependent refractive index	$-\omega$	ω	ω	$-\omega$	3/4
Optical Kerr effect	$-\omega_1$	ω_1	ω_2	$-\omega_2$	3/2
D.C.-induced optical rectification	0	0	ω	$-\omega$	3/8
D.C.-induced SHG (ESHG)	-2ω	0	ω	ω	3/2
Electro-optic Kerr effect	$-\omega$	ω	0	0	3
Three-wave mixing	$-\omega_\sigma$	ω_1	ω_1	ω_2	3/4
D.C.-induced two-wave mixing	$-(\omega_1+\omega_2)$	0	ω_1	ω_2	3
Three-wave mixing	$-\omega_\sigma$	ω_1	ω_2	ω_3	3/2

*A more extensive tabulation of factors K is given in Ref. 17.

As an optical frequency ω_j approaches a molecular transition frequency ω_{kn}, a resonant situation materialises and it becomes necessary to use a more complicated formulation than that shown in equations (5.22)–(5.24). The relevant formulae, which include complex energy denominators to take account of damping, are given by Orr and Ward[17]. The derivation in the resonant case requires caution to avoid secular and resonant divergences; this problem has also been discussed recently by Langhoff, Epstein and Karplus[14].

The expressions in equations (5.22)–(5.24) enable the non-linear optical polarisation $\boldsymbol{\mu}(t)$ to be evaluated, after substitution into equation (5.20). Additional information is required, however, before hyperpolarisabilities $\boldsymbol{\beta}$ and γ can be specified unambiguously. The definition prescribed in the present article is such that (a) static hyperpolarisabilities $\boldsymbol{\beta}$ (0;0,0), γ (0;0,0,0) are the coefficients in a Taylor expansion in the static electric field $\boldsymbol{E}^0$; and (b) optical hyperpolarisabilities converge to the static hyperpolarisability of corresponding order as all frequencies tend to zero. Hence

$$\mu_\xi(\text{static}) = \mu_\xi + a_{\xi\eta}(0;0)E^0_\eta + \tfrac{1}{2}\beta_{\xi\eta\zeta}(0;0,0)E^0_\eta E^0_\zeta + \tfrac{1}{6}\gamma_{\xi\eta\zeta\varphi}(0;0,0,0)E^0_\eta E^0_\zeta E^0_\varphi + \ldots \tag{5.25}$$

where, for example, $\beta_{\xi\eta\zeta}(0;0,0) = \left[\partial^2[\mu_\xi(\text{static})]/\partial E^0_\eta \partial E^0_\zeta\right]_{E^0=0}$.

This definition was introduced by Buckingham and Pople[21] and has been used extensively in the chemical literature[1,3,8,22].

Additional application of condition (b) yields, for general second- and third-order processes with ω_σ non- zero:

$$\mu^{\pm\omega_\sigma}_\xi = \tfrac{1}{2}K(-\omega_\sigma;\omega_1,\omega_2)\beta_{\xi\eta\zeta}(\mp\omega_\sigma;\pm\omega_1,\pm\omega_2)E^{\pm\omega_1}_{0\,\eta}E^{\pm\omega_2}_{0\zeta} \tag{5.26}$$

$$\mu^{\pm\omega_\sigma}_\xi = \tfrac{1}{6}K(-\omega_\sigma;\omega_1,\omega_2,\omega_3)\gamma_{\xi\eta\zeta\varphi}(\mp\omega_\sigma;\pm\omega_1,\pm\omega_2,\pm\omega_3)E^{\pm\omega_1}_{0\,\eta}E^{\pm\omega_2}_{0\zeta}E^{\pm\omega_3}_{0\,\varphi} \tag{5.27}$$

Special definitions are required for optical rectification processes[23]:

$$\mu^0_\xi = \tfrac{1}{2}K(0;\omega,-\omega)\sum_{\pm}\beta_{\xi\eta\zeta}(0;\pm\omega,\mp\omega)E^{\pm\omega}_{0\eta}E^{\mp\omega}_{0\zeta} \tag{5.28}$$

$$\mu^0_\xi = \tfrac{1}{6}K(0;\omega_1,\omega_2,-\omega_\sigma)\sum_{\pm}\gamma_{\xi\eta\zeta\varphi}(0;\pm\omega_1,\pm\omega_2,\mp\omega_\sigma)E^{\pm\omega_1}_{0\eta}E^{\pm\omega_2}_{0\zeta}E^{\mp\omega_\sigma}_{0\varphi} \tag{5.29}$$

where $\omega_\sigma = \omega_1 + \omega_2$.

A number of *symmetry properties* are inherent in the hyperpolarisability definitions presented above. They may be summarised as follows:

(a) The hyperpolarisabilities are invariant to interchange of field frequency labels together with coordinate suffixes, e.g. $\beta_{\xi\eta\zeta}(-\omega_\sigma;\omega_1,\omega_2) = \beta_{\xi\zeta\eta}(-\omega_\sigma;\omega_2,\omega_1)$. It should be noted that the interchange of field and polarisation frequency labels is not covered by this symmetry rule.

(b) Complex conjugation is equivalent to changing the signs of all frequency labels, e.g. $\beta^*_{\xi\eta\zeta}(-\omega_\sigma;\omega_1,\omega_2) = \beta_{\xi\eta\zeta}(\omega_\sigma;-\omega_1,-\omega_2)$.

(c) Permutation symmetry[24]: If damping is negligible, the hyperpolarisabilities are invariant to interchange of field and polarisation frequency labels, together with interchange of corresponding suffixes, e.g. $\beta_{\xi\eta\zeta}(-\omega_\sigma;\omega_1,\omega_2) \approx \beta_{\eta\xi\zeta}(\omega_1;-\omega_\sigma,\omega_2)$.

(d) Kleinman symmetry[5,25]: If the dominant contributions to the hyperpolarisabilities are from transitions whose frequencies are remote from all relevant field and polarisation frequencies, the hyperpolarisabilities are approximately invariant to interchange of polarisation and field coordinate suffixes, e.g. $\beta_{\xi\eta\zeta}(-\omega_\sigma;\omega_1,\omega_2) \approx \beta_{\eta\xi\zeta}(-\omega_\sigma;\omega_1,\omega_2)$.

(e) In the limit of negligible dispersion (e.g. $\partial\boldsymbol{\beta}/\partial\omega_\sigma$, $\partial\boldsymbol{\beta}/\partial\omega_1$, $\partial\boldsymbol{\beta}/\partial\omega_2$ negligible at all frequencies) all hyperpolarisabilities of a given order are equal, e.g. $\beta_{\xi\eta\zeta}(-\omega_\sigma;\omega_1,\omega_2) \approx \beta_{\xi\eta\zeta}(0;0,0)$.

In viewing the above definitions and symmetry properties it should be realised that the literature contains a number of conflicting definitions of higher-order polarisabilities. Our convention is consistent with most of the chemical literature[1,3,8,22] and corresponds closely to that adopted for non-linear optical susceptibilities by Ward[17,23,26], whose definitions exclude the numerical factors of $\frac{1}{2}$ and $\frac{1}{6}$ in equations (5.25)–(5.29). Early workers in non-linear optics[5,6,18] chose to define all non-linear susceptibilities χ by the relation,

$$\mu^{\omega_\sigma}=\chi(-\omega_\sigma;\omega_1,\omega_2,\ \ldots)E_0^{\omega_1}E_0^{\omega_2}\ \ldots \tag{5.30}$$

that is, without numerical factors involving $K(-\omega_\sigma;\omega_1,\omega_2,\ \ldots\ )$. The advantage of simplicity for an individual non-linear process in this definition must be weighed against the disadvantage that susceptibilities for different processes do not converge in the limit of negligible dispersion. This complicates comparisons: for example, non-resonant susceptibilities for optical rectification and the linear electro-optic effect are related[27], in the limit $\omega\to 0$, by

$$\chi_{\xi\eta\zeta}(0;\omega,-\omega)=\tfrac{1}{4}\chi_{\zeta\eta\xi}(-\omega;\omega,0) \tag{5.31}$$

whereas with our definition there is no factor of $\frac{1}{4}$ in a comparison of the corresponding $\boldsymbol{\beta}$ components. Other currently accepted definitions are those of Terhune and Maker[28] and Pershan[29].

It is relevant at this point to illustrate the extraction of hyperpolarisability formulae, by means of the above definitions, from equations (5.23) and (5.24). In the case of second-harmonic generation ($\omega_\sigma=2\omega;\omega_1=\omega_2=\omega$), substitution into equation (5.23) yields

$$\mu_\xi^{2\omega}=\tfrac{1}{4}\hbar^{-2}E_{0\eta}^{\omega}E_{0\zeta}^{\omega}X_{\xi\eta\xi}^{(n)} \tag{5.32}$$

$$X_{\xi\eta\zeta}^{(n)}=S_{\eta\zeta}\sum_{k,l}{}'\left[\frac{\langle n|\mu_\xi|l\rangle\langle l|\overline{\mu_\eta}|k\rangle\langle k|\mu_\zeta|n\rangle}{(\omega_{ln}-2\omega)(\omega_{kn}-\omega)}\right.$$
$$\left.+\frac{\langle n|\mu_\eta|l\rangle\langle l|\overline{\mu_\xi}|k\rangle\langle k|\mu_\zeta|n\rangle}{(\omega_{ln}+\omega)(\omega_{kn}-\omega)}+\frac{\langle n|\mu_\eta|l\rangle\langle l|\overline{\mu_\zeta}|k\rangle\langle k|\eta_\xi|n\rangle}{(\omega_{ln}+\omega)(\omega_{kn}+2\omega)}\right] \tag{5.33}$$

where $\langle l|\overline{\mu_\xi}|k\rangle=(\langle l|\mu_\xi|k\rangle-\langle n|\mu_\xi|n\rangle)$ and $S_{\eta\zeta}\ \ldots$ specifies summation over all permutations of cartesian suffixes η, ζ, . . .

Equation (5.26) yields

$$\mu_\xi^{\pm 2\omega}=\tfrac{1}{4}\beta_{\xi\eta\zeta}(\mp 2\omega;\pm\omega,\pm\omega)E_{0\eta}^{\pm\omega}E_{0\zeta}^{\pm\omega} \tag{5.34}$$

from which $\boldsymbol{\beta}$ may be identified:

$$\beta_{\xi\eta\zeta}(-2\omega;\omega,\omega)=\hbar^{-2}X_{\xi\eta\zeta}^{(n)} \tag{5.35}$$

Similar manipulations in the case of static fields produce the result

$$\beta_{\xi\eta\zeta}(0;0,0)=\hbar^{-2}S_{\xi\eta\zeta}\sum_{k,l}{}'\frac{\langle n|\mu_\xi|l\rangle\langle l|\overline{\mu_\eta}|k\rangle\langle k|\mu_\zeta|n\rangle}{\omega_{ln}\omega_{kn}} \tag{5.36}$$

It may readily be confirmed that $\boldsymbol{\beta}(-2\omega;\omega,\omega)$ reduces to $\boldsymbol{\beta}(0;0,0)$ as ω tends to zero, in accord with part (b) of the definition. The static second hyperpolarisability may be shown, in like fashion, to be

$$\gamma_{\xi\eta\zeta\varphi}(0;0,0,0)=\hbar^{-2}S_{\xi\eta\zeta\varphi}\left[\sum_{k,l,m}{}' \frac{\langle n|\mu_\xi|m\rangle\langle m|\bar{\mu}_\eta|l\rangle\langle l|\bar{\mu}_\zeta|k\rangle\langle k|\mu_\varphi|n\rangle}{\omega_{mn}\omega_{ln}\omega_{kn}} - \sum_{k,l}{}' \frac{\langle n|\mu_\xi|l\rangle\langle l|\mu_\eta|n\rangle\langle n|\mu_\zeta|k\rangle\langle k|\mu_\varphi|n\rangle}{\omega_{ln}\omega_{kn}^2}\right]. \tag{5.37}$$

Hyperpolarisabilities may also be defined for *inelastic* non-linear optical processes, of which *hyper-Raman scattering*[22] is a good example. This phenomenon consists of incoherent scattering at frequencies $(2\omega\pm\omega_{mn})$ as a result of excitation by two photons of frequency ω. The transition hyperpolarisability $\boldsymbol{\beta}^{(mn)}(\omega_{mn}\pm2\omega;\mp\omega,\mp\omega)$ which describes such a process is analogous to the transition polarisability $\boldsymbol{\alpha}^{(mn)}(\omega_{mn}\pm\omega;\mp\omega)$ which is responsible for the conventional Raman effect [equation (5.18)]. A density matrix treatment of hyper-Raman scattering has been given by Long and Stanton[30,31].

5.2.3 Effects of molecular symmetry

In terms of the symmetry properties outlined above, complete specification of a polarisability tensor of rank n may require as many as 3^n independent elements. This number of independent elements may be substantially reduced, however, if symmetry due to molecular geometry is taken into account. For example, the number of independent elements of $\boldsymbol{\beta}(-\omega_\sigma;\omega_1,\omega_2)$ is reduced from 27 to 1 in the case of a tetrahedral molecule. The non-zero elements[3,8] and the number of independent elements[1,8] of the polarisability and the static hyperpolarisabilities $\boldsymbol{\beta}(0;0,0)$ and $\gamma(0;0,0,0)$ have been tabulated for many molecular symmetry groups. An extensive compilation of the symmetry properties of the transition tensors $\boldsymbol{\mu}^{(mn)}$, $\boldsymbol{\alpha}^{(mn)}$ and $\boldsymbol{\beta}^{(mn)}$ has also been presented[22] in the context of selection rules for infrared, Raman and hyper-Raman spectra. It was assumed[22,31] in this compilation, however, that $\boldsymbol{\beta}(-2\omega;\omega,\omega)$ is symmetric in all suffixes; the validity of this assumption has been discussed by Bersohn *et al.*[32].

The effect of molecular symmetry on optical hyperpolarisabilities does not correspond to that for the static case, since dispersion effects destroy the (Kleinman) symmetry with respect to interchange of all suffixes. The numbers of independent elements, required to specify the tensors for various optical processes and selected molecular symmetry groups, are listed in Table 5.2. The tabulation is based on the group-theoretical procedure of Jahn[33]; the symbol V denotes the representation of a polar vector and $[V^n]$ its nth order symmetrical product. The results in Table 5.2 apply to non-resonant situations where the permutation symmetry property (c) of the previous section is applicable. For situations where resonance occurs or where only the real portion of the tensor is relevant, the transformation properties of the tensor sometimes differ from those given in Table 5.2 and the appropriate representation is then given in footnotes to the table.

For most molecular symmetries, a given experiment is capable of yielding only a particular combination of tensor elements. For example, in the case of the polarisability $\boldsymbol{\alpha}$ it is convenient to define a mean a (determinable from refractive index measurements) and an anisotropy κ (observable in depolarisation of Rayleigh scattering):

$$a=\tfrac{1}{3}a_{\xi\xi};\ \kappa^2=(a_{\xi\eta}a_{\xi\eta}-3a^2)/6a^2 \qquad (5.38)$$

where the convention that repeated Greek suffixes imply summation over x, y, z has again been used. For a molecule with a threefold or higher rotational axis (labelled z),

$$a=\tfrac{1}{3}(a_{\parallel}+2a_{\perp});\ \kappa=(a_{\parallel}-a_{\perp})/3a \qquad (5.39)$$

where $a_{\parallel}=a_{zz}$ and $a_{\perp}=a_{xx}=a_{yy}$ are the two independent elements of $\boldsymbol{\alpha}$. Observable combinations may also be defined for the various hyperpolarisabilities, but the number of independent tensor elements often vastly

Table 5.2 The number of independent tensor elements for selected optical processes and molecular symmetries

Tensor	*Representation*	*Molecular symmetry group* C_{2v}	D_{2h}	C_{3v}	D_{3h}	$C_{\infty v}$	$D_{\infty h}$	T_d	O_h	Sphere
$\boldsymbol{\mu}^{(0)}$	V	1	0	1	0	1	0	0	0	0
$\boldsymbol{\alpha}(0;0)$	$[V^2]$	3	3	2	2	2	2	1	1	1
$\boldsymbol{\alpha}(-\omega;\omega)$	$[V^2]$*	3	3	2	2	2	2	1	1	1
$\boldsymbol{\beta}(0;0,0)$	$[V^3]$	3	0	3	1	2	0	1	0	0
$\boldsymbol{\beta}(-2\omega;\omega,\omega)$	$V[V^2]$	5	0	4	1	3	0	1	0	0
$\boldsymbol{\beta}(-\omega;\omega,0)$	$[V^2]V$†	5	0	4	1	3	0	1	0	0
$\boldsymbol{\beta}(0;\omega,-\omega)$	V^3‡	7	0	5	1	4	0	1	0	0
$\boldsymbol{\beta}(-\omega_\sigma;\omega_1,\omega_2)$	V^3	7	0	5	1	4	0	1	0	0
$\gamma(0;0,0,0)$	$[V^4]$	6	6	4	3	3	3	2	2	1
$\gamma(-3\omega;\omega,\omega,\omega)$	$V[V^3]$	9	9	6	4	4	4	2	2	1
$\gamma(-\omega;\omega,\omega,-\omega)$	$[[V^2]^2]$§	9	9	6	5	5	5	3	3	2
$\gamma(-\omega_1;\omega_1,\omega_2,-\omega_2)$	$[V^2]V^2$‖	15	15	10	7	7	7	3	3	2
$\gamma(0;0,\omega,-\omega)$	$[V^2]V^2$¶	15	15	10	7	7	7	3	3	2
$\gamma(-2\omega;0,\omega,\omega)$	$V^2[V^2]$	15	15	10	7	7	7	3	3	2
$\gamma(-\omega;\omega,0,0)$	$[V^2][V^2]$**	12	12	8	6	6	6	3	3	2
$\gamma(-\omega_\sigma;\omega_1,\omega_1,\omega_2)$	$V[V^2]V$	15	15	10	7	7	7	3	3	2
$\gamma(-\omega_\sigma;0,\omega_1,\omega_2)$	V^4	21	21	14	10	10	10	4	4	3
$\gamma(-\omega_\sigma;\omega_1,\omega_2,\omega_3)$	V^4	21	21	14	10	10	10	4	4	3

* $\boldsymbol{\alpha}_{res} \sim V^2$, which gives same number of elements for groups listed.
† $\boldsymbol{\beta}_{res} \sim V^3$
‡ Re $(\boldsymbol{\beta}) \sim V[V^2]$
§ $\gamma_{res} \sim V[V^2]V$
‖ $\gamma_{res} \sim V^4$
¶ Re $(\gamma) \sim [V^2]^2$
** $\gamma_{res} \sim V^2[V^2]$

exceeds the number of accessible experiments. Two simple examples are the mean first and second static hyperpolarisabilities[3]:

$$\beta^0=\tfrac{3}{5}\beta_{\xi\xi z}(0;0,0) \qquad (5.40a)$$

$$\gamma^0 = \tfrac{1}{5}\gamma_{\xi\xi\eta\eta}(0;0,0,0) \tag{5.40b}$$

where z corresponds to the direction of the permanent electric dipole moment of the molecule.

It will prove convenient and efficient to adopt a more compact notation for the dynamic and static polarisabilities, as follows:

$$\boldsymbol{\alpha} = \boldsymbol{\alpha}(-\omega;\omega) \tag{5.41a}$$

$$\boldsymbol{\alpha}^0 = \boldsymbol{\alpha}(0;0) \tag{5.41b}$$

5.3 PHENOMENA INVOLVING $\boldsymbol{\alpha}$

5.3.1 Refractive index and dielectric constant

5.3.1.1 Refractive index

The first-order contribution to the electric dipole moment induced in a molecule by an electric field involves the polarisability $\boldsymbol{\alpha}$; it is not surprising therefore that the more easily determined electromagnetic properties of matter such as the refractive index n and the dielectric constant ε are determined by $\boldsymbol{\alpha}$.

The relation of the *refractive index* to the polarisability, the Lorentz–Lorenz relation

$$\frac{n^2-1}{n^2+2} = \frac{Na}{3\varepsilon_0 V_m} \tag{5.42}$$

was first derived almost simultaneously by Lorentz[34] and Lorenz[35]. N is Avogadro's number and V_m is the molar volume; a is the mean optical polarisability for the frequency ω at which n is measured. At low densities n approaches unity and (5.42) becomes

$$(n-1) = Na/(2\varepsilon_0 V_m) \tag{5.43}$$

The refractive index of a medium is the ratio of c, the velocity of light in vacuum, to v, the velocity of light in the medium; for a transparent substance, $(n-1)$ observed with visible light is normally a positive quantity and this implies a slowing down of the light wave. A rigorous derivation[36,37] of (5.42) is complicated. However, a facile derivation of (5.43) by Kauzmann[38] gives a clear insight into the phenomenon and is of sufficient interest for an outline to be presented here.

Consider a plane wave, the incident wave, with electric vector

$$\boldsymbol{E} = \boldsymbol{E}_0 \cos \omega(t - z/c) \tag{5.44}$$

propagating in the z-direction with velocity c, incident on a thin slab of the medium containing N/V_m molecules per unit volume. The slab thickness Δz is chosen to be much less than the wavelength λ of the light. The induced molecular dipole moment is

$$\boldsymbol{\mu} = a\boldsymbol{E}_0 \cos \omega(t - z/c) \tag{5.45}$$

where it is assumed for convenience that $\boldsymbol{\alpha}$ is isotropic. (For anisotropically polarisable molecules the theory involves an isotropic average over molecular orientations, so that only the mean polarisability a enters the final formula.) Each molecule produces a scattered spherical wave with electric vector of magnitude

$$E_s = -(a\omega^2/4\pi\varepsilon_0 cr)\sin\theta E_0 \cos\omega(t-z/c) \tag{5.46}$$

where θ is the angle between $\boldsymbol{E}_0$ and the vector from the molecule to the point, distant r, at which $\boldsymbol{E}_s$ is measured. The negative sign indicates that, for positive a, the scattered wave is phase retarded by π with respect to the incident wave. At a distance (much greater than λ) from the layer the spherical waves combine to form a plane wave moving in the forward direction. However, for a given point on the plane wave front the spherical waves do not all arrive at the same time; this results[38] in an advance in phase by $\pi/2$ of the resultant plane wave, E_r, with respect to the spherical wave, E_s:

$$\boldsymbol{E}_r = -\delta\Delta z \boldsymbol{E}_0 \sin\omega(t-z/c) \tag{5.47}$$

where $\delta = Na\omega/2\varepsilon_0 cV_m$. The total electric field due to the scattered waves becomes

$$\boldsymbol{E} = \boldsymbol{E}_0[\cos\omega(t-z/c) - \delta\Delta z \sin\omega(t-z/c)] \tag{5.48}$$

On defining a time τ such that $\tan\omega\tau = \delta\Delta z$, we obtain

$$\boldsymbol{E} = \boldsymbol{E}_0 \cos\omega(t+\tau-z/c)/\cos(\omega\tau) \approx \boldsymbol{E}_0 \cos\omega(t+\tau-z/c) \tag{5.49}$$

if $\delta\Delta z$ is small. The light wave thus appears as though it has been delayed by a time τ in its passage through the layer. Recalling that v is the velocity of light in the medium,

$$\tau = \Delta z/v - \Delta z/c = \Delta z(n-1)/c \tag{5.50}$$

since $n = c/v$. Use of the relation $\tan\omega\tau \approx \omega\tau$ produces equation (5.43). The phase relationship of incident and scattered waves is clearly of paramount importance in forward scattering and it will be mentioned again in Section 5.4, in connection with second and third harmonic generation.

According to equation (5.42) a determination of the optical polarisability requires a measurement of the refractive index n and molar volume V_m of a gas. Interferometric measurements[39,40] allow $(n-1)$ to be determined to a few parts in 10^4. Great care must be taken in the determination of V_m, for this is often the factor which limits the accuracy of a. Due allowance must therefore be made for the non-ideality of the gases studied.

5.3.1.2 Dispersion

On combining equations (5.11), (5.38) and (5.43) we obtain

$$n-1 = \frac{N}{3\varepsilon_0 V_m}\sum_k \frac{\omega_{kg}}{\hbar(\omega_{kg}^2-\omega^2)}\langle g|\mu_\xi|k\rangle\langle k|\mu_\xi|g\rangle \tag{5.51}$$

$$n-1=(N/V_{\mathrm{m}})\sum_{k}C_k/(\omega_{kg}^2-\omega^2) \qquad (5.52)$$

where $|g\rangle$ is the ground state of the molecules under consideration. The constants C_k are related to the oscillator strengths f_k for the transitions $|k\rangle\leftarrow|g\rangle$: $C_k=(e^2/9m\varepsilon_0)f_k$, where e and m are respectively the electronic charge and mass. Equation (5.51) describes the variation of refractive index with frequency — a phenomenon known as *dispersion*. In the visible part of the spectrum the refractive index for transparent substances usually increases with increasing frequency (decreasing wavelength); this is termed normal dispersion. Such behaviour implies that the dominant terms in (5.51) have ω_{kg} values corresponding to ultraviolet transitions. The observed dispersion data can usually be adequately represented by retaining one (or at most two) terms in (5.52) using empirically determined 'effective' values for C_k and ω_{kg}. Many of the data in the literature[41] are presented in this manner. Other empirical dispersion equations are often used[4]. Near an absorption band for the transition $|k\rangle\leftarrow|g\rangle$ the kth term in (5.51) becomes resonant and the dispersion curve resembles the solid curve of Figure 5.1, superimposed on a rising background due to the other terms. At frequencies just higher than ω_{kg} the refractive index decreases with increasing frequency; this is termed, for historical reasons, anomalous dispersion. At very high frequencies such that $\omega>\omega_{kg}$ for the important k, equation (5.51) predicts n to be less than unity. For x-rays, refractive indices less than unity are found.

If the constants in (5.52) are derived empirically from refractivity data in the visible region, extrapolation to $\omega=0$ does not in general yield an accurate estimate of the static polarisability a^0. The sum over states k in (5.51) includes terms for rotationally and vibrationally excited levels in the ground electronic state of the molecule. For molecules with allowed transitions in the i.r. spectrum these terms are non-zero; equation (5.52) therefore extrapolates correctly to a^0 only for the noble gases and for homonuclear diatomic molecules. The contribution to the sum in equation (5.52) from vibrationally excited states k may be obtained from measurement of the refractive index near the appropriate i.r. absorption bands[42]. Usually only the vibrational fundamentals have oscillator strengths f_k, and hence C_k values, sufficiently large to make appreciable contributions[43,44] to a. Alternatively a measurement of the integrated absorption intensity yields f_k directly[45,46]. [That the absorption intensity and refractive index are related is obvious from equations (5.15) and (5.16), for the same constants appear in them. The general relation between absorption and dispersion is expressed by the Kramers–Kronig relationship[42,46,47].] The contribution to a^2 of the 'vibrational' terms in (5.51) is usually less than 10%; uncertainties in these terms therefore often have only a minor effect on estimates of a^0. The contribution to the refractive index from the rotational states in the sum over k in equation (5.52) becomes important only for far-i.r. and microwave wavelengths. These contributions are not usually considered to contribute to the molecular polarisability, as our discussion of the dielectric constant shows.

Equation (5.42) is approximately valid for dense fluids,[37,48] and polarisabilities derived from liquid data[4] are usually in close agreement with those derived from measurements on gases. Vogel and co-workers[49] have reported values of the molar refraction.

$$R=\frac{(n^2-1)}{(n^2+2)}V_{\mathrm{m}} \approx \frac{Na}{3\varepsilon_0} \tag{5.53}$$

measured at several visible wavelengths, for a wide range of organic and inorganic liquids.

5.3.1.3 *Dielectric constant*

The application of a static electric field E_Z to a gas results in an induced electric moment. The polarisation P_Z is the induced moment per unit volume and

$$P_Z=\varepsilon_0(\varepsilon-1)E_Z \tag{5.54}$$

where ε is the static dielectric constant. P_Z may be expanded in powers of the field strength,

$$P_Z=P_Z^0+A_{ZZ}E_Z+\ \ldots . \tag{5.55}$$

and, because the permanent polarisation P_Z^0 is zero for an isotropic fluid, the dielectric constant is seen to be proportional to A_{ZZ}, the bulk polarisability of the gas. If the gas is composed of molecules with average polarisability $\overline{a_{ZZ}}$, then, at low densities where local field corrections[48] are negligible, $A_{ZZ}=(N/V_{\mathrm{m}})\overline{a_{ZZ}}$. For a Boltzmann distribution of molecular energies W_n,

$$\varepsilon-1=(N/\varepsilon_0 V_{\mathrm{m}})\overline{a_{ZZ}}=(N\varepsilon_0 V_{\mathrm{m}})\sum_n a_{ZZ}^{(n)} \exp(-W_n/kT)/\sum_n \exp(-W_n/kT) \tag{5.56}$$

Measurements of the dielectric constant are generally made not with static fields but with slowly varying fields of frequency 10^3–10^6 Hz. For gases these frequencies are much less than any molecular transition frequencies and the polarisability $a_{ZZ}^{(n)}$ can be evaluated by the use of equation (5.13) for the static polarisability, summing over the rotational, vibrational and electronic states of the molecule. The sum over the rotational states may be explicitly evaluated[2,50] in terms of μ, the ground state molecular dipole moment. For temperatures such that the rotational energy level separation is much less than kT this results in

$$(\varepsilon-1)=(N/\varepsilon_0 V_{\mathrm{m}})[a^0+\mu^2/3kT] \tag{5.57}$$

where a^0 is the mean static polarisability given by equation (5.13); the sum in this equation is now over only the vibrational and electronic states. The term in $\mu^2/3kT$ arises from a distortion of the rotational states by the field and its contribution to the polarisation is known as the *orientation polarisation*, because classically[51] the application of the electric field results in a non-uniform distribution of the directions in which the permanent dipoles point. The term in a^0 arises from a distortion of the internal states of the molecule and its contribution to P_Z is known as the *distortion polarisation*. If the contributions to a^0 from the vibrational and electronic terms in (5.13) are calculated separately their contributions to P_Z are known as the *atomic* and *electronic polarisations*, respectively[52–54].

Equation (5.57), first derived by Debye[51], has been extensively used for measuring molecular dipole moments[52–54]; here its importance lies in the presence of the term in a^0. Consider first non-dipolar molecules; for these, (5.57) reduces to

$$\varepsilon - 1 = (N/\varepsilon_0 V_m)a^0 \tag{5.58}$$

which is equivalent to the Clausius–Mossotti relation,

$$\frac{(\varepsilon-1)}{(\varepsilon+2)} = (N/3\varepsilon_0 V_m)a^0 \tag{5.59}$$

for gases at low densities where ε is close to unity. Evidently a measurement of $(\varepsilon-1)V_m$ yields a^0 directly. For dipolar molecules a measurement of $(\varepsilon-1)V_m$ at a number of temperatures enables a separation of the two terms of (5.57). However, because the temperature range available for measurement is limited and a^0 typically contributes only 10% to $(\varepsilon-1)$ for dipolar gases, accurate measurements are required. If a measurement of $(\varepsilon-1)V_m$ is available for only one temperature, a^0 may still be obtained if μ is known from other measurements such as, for example, the Stark effect[55].

Experimental techniques for measuring the dielectric constant have been reviewed[56]; accuracies of 1 part in 10^6 can be obtained. Recent results by Cole[57], Sutton[58] and co-workers indicate that use of (5.57) can yield a^0 with an accuracy of a few percent for dipolar molecules. For non-dipolar molecules accuracies are typically an order of magnitude better. A selection of a^0 values from various sources[57,59–69] is listed in Table 5.4 in Section 5.7. Teachout and Pack[178] have presented a compilation of a^0 values for neutral atoms in their ground states.

5.3.2 Rayleigh and Raman scattering

A light beam incident on a medium consisting of randomly distributed molecules produces scattered light not only in the forward direction but in other directions as well. The early classical theory given by Rayleigh[71] showed the scattered light to have the same frequency ω as the incident light and the scattered intensity to be proportional to ω^4. Furthermore, it was predicted that for spherical molecules the scattered light would be plane polarised perpendicular to the plane formed by the incident and scattered propagation directions. Investigation of the nature of the scattered light by Strutt[72] showed it to be depolarised, a result which was attributed to molecular anisotropy.

A classical derivation[73] of the depolarisation ratio, ρ, is relatively simple. Consider a light beam with electric vector $E_X \cos \omega(t-z/c)$ along the X direction and propagating along the Z axis; such a beam is polarised in a plane parallel to the XZ plane. If this light beam is incident on a molecule situated at the origin, the electric field E_η along the η direction ($\eta = x$, y or z) of an axis system fixed in the molecule induces a dipole moment

$$\mu_\xi = a_{\xi\eta}E_\eta = a_{\xi\eta}X_\eta E_X \cos \omega(t-z/c) \tag{5.60}$$

where X_η is the direction cosine between the η axis and the X axis. At a point on the Y axis, sufficiently far from the origin that the spherical wave scattered from the molecule approximates to a plane wave, we place an analyser which transmits light polarised with electric vector parallel to X. If I is the intensity (the power per unit area) of the incident beam and S_X the scattered power per unit solid angle,

$$S_X=(4\pi\varepsilon_0)^{-2}\,(\omega/c)^4\,(a_{\xi\eta}X_\xi X_\eta)^2 I \tag{5.61a}$$

On rotating the analyser so that light polarised parallel to the YZ plane is transmitted, we obtain

$$S_Z=(4\pi\varepsilon_0)^{-2}\,(\omega/c)^4\,(a_{\xi\eta}Z_\xi X_\eta)^2 I \tag{5.61b}$$

For a random distribution of molecules, such as is found in gases at low pressure, the phase of the scattered waves from different molecules is uncorrelated in all but forward scattering. The total scattered intensity from an ensemble of molecules is therefore the sum of the contributions from each molecule. If the number of molecules is large, the mean value of S_X and S_Z for each molecule is obtained from equations (5.61) by averaging[74] over all orientations of the scattering molecule. This yields

$$\langle S_X\rangle=(4\pi\varepsilon_0)^{-2}(\omega/c)^4\frac{1}{15}[2a_{\xi\eta}a_{\xi\eta}+a_{\xi\xi}a_{\eta\eta}]I$$

$$=(4\pi\varepsilon_0)^{-2}(\omega/c)^4\frac{a^2}{5}[4\kappa^2+5]I \tag{5.62a}$$

$$\langle S_Z\rangle=(4\pi\varepsilon_0)^{-2}(\omega/c)^4\frac{1}{30}[3a_{\xi\eta}a_{\xi\eta}-a_{\xi\xi}a_{\eta\eta}]I$$

$$=(4\pi\varepsilon_0)^{-2}(\omega/c)^4\frac{3a^2}{5}\kappa^2 I \tag{5.62b}$$

where a and κ are defined as in equation (5.38). The depolarisation ratio, ρ, of the scattered light is defined as

$$\rho=\langle S_Z\rangle/\langle S_X\rangle \tag{5.63}$$

and from equations (5.62):

$$\rho=3\kappa^2/(4\kappa^2+5) \tag{5.64}$$

Equation (5.64) holds for incident light polarised perpendicular to the plane containing the incident and scattered beams. For molecules with a threefold or higher rotational axis of symmetry, $\kappa=(a_\parallel-a_\perp)/3a$ [equation (5.39)] and a measurement of ρ and a (from the refractive index of the gas) determines the magnitude of $(a_\parallel-a_\perp)$; its sign can often be assigned by use of bond polarisability models.

The classical derivation of the depolarisation ratio given above ignores the fact that not all the scattered radiation occurs at the frequency ω of the incident light. In addition to scattered light at the frequency ω (Rayleigh scattering) there will be present light scattered at frequencies $\omega\pm\omega_{mn}$ (Raman scattering). The frequencies ω_{mn} correspond to the difference in energy between the molecular states m and n, and molecules giving rise to scattering at the frequency $\omega \pm \omega_{mn}$ undergo the transition $m \leftrightarrow n$. If ω is much less than the lowest electronic transition frequency, the transition occurs between a pair of vibrational or rotational states. Scattering involving changes in vibrational states usually yields light at frequencies sufficiently different from ω for it to be easily eliminated with a filter. Light from scattering involving changes in rotational states only cannot usually be removed in this way; it is depolarised[75] and makes an important contribution to the measured depolarisation ratio. However, Bridge and Buckingham[73] have shown that, to a good approximation, the mean depolarisation ratio of the light scattered at frequencies corresponding to rotational transitions (the rotational Raman lines) taken together with the light scattered at the unchanged frequency (the Rayleigh line) is equal to the classical depolarisation ratio of equation (5.64). The approximations leading to this result have been discussed in detail[73]; for light molecules some corrections are necessary.

Although the advent of the laser as a light source has made possible measurements of the depolarisation ratio at well defined frequencies and with good accuracy, there are few recent studies[73,76]. Bridge and Buckingham[73] have discussed the experimental technique in detail; for molecules with values of $\kappa \approx 0.1$ an accuracy of $\sim 1\%$ in κ is attained. If equation (5.64) is used to deduce κ from the depolarisation ratio, ρ, it is important that the light from all rotational Raman lines of appreciable intensity is included when ρ is measured; some data in the literature are subject to great uncertainty because, in attempts to exclude vibrational Raman lines, filters with a bandwidth sufficiently narrow to exclude some rotational lines have been used. For light molecules it is sometimes possible to isolate the Rayleigh line and measure its depolarisation ratio[73,76].

The absolute intensity of Rayleigh scattering is given by equations (5.62) after multiplying by the number of molecules in the scattering volume, i.e. in that volume of the gas both illuminated by the incident beam and 'seen' by the detector of the scattered light. Accurate measurement of the scattering volume is very difficult; recent results by Geindre *et al.*[77] quote an uncertainty of $\pm 20\%$. The relative accuracy, for differing gases and a fixed but unknown scattering volume, is much better. The scattered intensity is determined by a^2 and κ^2; since the polarisability a can be obtained more easily, and much more precisely, from measurements of the refractive index and κ^2 may be obtained from the depolarisation ratio, it seems that studies of absolute Rayleigh intensities for low density gases are not profitable.

The polarisability components $a_{\parallel}$ and $a_{\perp}$, obtained from depolarisation ratios and refractive index measurements with the aid of equations (5.39), (5.42) and (5.64), are of course frequency dependent. The frequency dependence is, in general, described by equation (5.13) and, by analogy with equation (5.52), we can write

$$a_{\parallel} = \sum_k C_{\parallel k}/(\omega_{kg}^2 - \omega^2) \tag{5.65a}$$

$$a_{\perp} = \sum_k C_{\perp k}/(\omega_{kg}^2 - \omega^2) \tag{5.65b}$$

where the sums over k are for transitions, electronic or vibrational, with parallel and perpendicular polarisation, respectively. For frequencies ω much less than the first electronic transition frequency the contribution of the electronic terms in equation (5.65) could, like that for the mean polarisability, be represented by empirically determined 'effective' constants. For molecular hydrogen the depolarisation ratio has been measured at 632.8 nm; Victor and Dalgarno[78] have used this, refractive index dispersion and related data to determine twenty constants for equation (5.65a) and thirteen for equation (5.65b). These equations very satisfactorily reproduce the dispersion of a and $(a_{\parallel}-a_{\perp})$ for wavelengths $\lambda=\infty$ (static polarisabilities) down to $\lambda=122$ nm. The same constants may be used to calculate a number of other properties of hydrogen[78]. The application of equations (5.65) to estimate $a_{\parallel}^{0}$ and $a_{\perp}^{0}$ for molecules other than hydrogen is hampered by a lack of data on the frequency dependence of the depolarisation ratio. For molecules with i.r. transitions the 'vibrational' terms (see Section 5.3.1.2) in (5.65a) and (5.65b) can be estimated from i.r. dispersion or absorption data provided the assignment, parallel or perpendicular polarisation, of the absorption bands is known. The static polarisability anisotropy, $(a_{\parallel}^{0}-a_{\perp}^{0})$, of a number of molecules has been estimated in this manner[79] from their optical polarisability anisotropy.

The intensity and depolarisation of Raman scattered light[75,80,81,85] corresponding to the transition $m\leftrightarrow n$ is given by the equations governing Rayleigh scattering if the polarisability, $\boldsymbol{\alpha}$, in these equations is replaced by the transition polarisability, $\boldsymbol{\alpha}^{(mn)}$ of equation (5.18). Placzek[75] has shown that, for molecules in electronically non-degenerate ground states, the expression for the transition polarisability can be replaced, to a good approximation, by the ground state polarisability of the molecule for fixed nuclei, and by the effect of the nuclear motions on the polarisability. This approximation constitutes the polarisability theory of Raman scattering and is subject to the *proviso* that the frequency ω of the incident light is much less than the lowest electronic transition frequency ω_{E} and that $(\omega_{\mathrm{E}}-\omega)$ is much greater than ω_{mn}, the energy difference between states m and n. For rotational Raman scattering by a rigid molecule the polarisability theory allows[73,75] the expressions for the intensity and depolarisation of the scattered light to be written in terms of a^2 and κ^2. The contribution of the rotational Raman transitions to the depolarisation ratio of 'Rayleigh' scattered light has been mentioned above. For vibrational Raman scattering the effect of the nuclear configuration on the polarisability is described by expanding $\boldsymbol{\alpha}$ in a Taylor series in Q_i, the normal coordinates of the molecule[75,80,82]:

$$\boldsymbol{\alpha}^{(mn)}=(\boldsymbol{\alpha})_{\mathrm{e}}^{(mn)}+\sum_{i}[(\partial\boldsymbol{\alpha}/\partial Q_i)_{\mathrm{e}}Q_i]^{mn}+\ldots \tag{5.66}$$

where Q_i corresponds to the ith normal mode of the molecular vibrations and $(\;)_{\mathrm{e}}$ denotes evaluation at the equilibrium nuclear configuration. With $m=n$, the first term in equation (5.66) is the main contributor to Rayleigh scattering; the second term, with $m\neq n$, governs vibrational Raman scattering. The transition $m\leftrightarrow n$ is allowed if, for the appropriate normal mode, $(\partial\boldsymbol{\alpha}/\partial Q)_{\mathrm{e}}$ is non-zero; for harmonic oscillators there is an additional criterion: states m and n must differ by a single quantum in the normal mode i. The symmetry properties of $\boldsymbol{\alpha}$ determining the selection rules for vibrational Raman scattering have been tabulated in numerous places[75,82]. The depolarisation ratio ρ^{mn} for the vibrational Raman transition $m \leftrightarrow n$ [see equation (5.64)],

$$\rho^{mn} = 3(\kappa^{mn})^2/[4(\kappa^{mn})^2 + 5] \tag{5.67}$$

is of considerable diagnostic value in determining the symmetry species of the normal mode excited in the transition. If n is the ground vibrational state of the molecule and m a state in which one quantum of a totally symmetric vibration has been excited, $0 < \rho^{mn} < 3/4$ and the Raman band is said to be polarised; if m is a state in which a non-totally symmetric or a degenerate vibration is excited, $\rho^{mn} = 3/4$ and the band is said to be depolarised.

Measurement of the absolute intensity[83,84] and depolarisation ratio[81] of a vibrational Raman band allows estimates of the transition polarisability, $(a^{(mn)})^2$, and of $(\kappa^{mn})^2$. Within the polarisability approximation, and usually also the harmonic oscillator approximation, $(a^{(mn)})^2$ and $(\kappa^{mn})^2$ can yield[83], for molecules of sufficient symmetry, individual components of the $\partial\boldsymbol{\alpha}/\partial Q$ tensor. The measurement of absolute Raman intensities[83,84] is subject to all the difficulties associated with the measurement of absolute Rayleigh scattered intensities. In addition, allowance must be made for the frequency dependence of the monochromator and detector efficiency. Problems concerned with measurement of the scattering volume are usually avoided by measuring intensities relative either to the Rayleigh intensity[84], whose absolute intensity is assumed to be governed by equations (5.62), or to the $J = 1 \rightarrow 3$ rotational Raman transition of hydrogen[83], whose intensity is determined by $a^2\kappa^2$. The measurement of the depolarisation ratio of vibrational Raman bands is accomplished in the manner used for Rayleigh scattering; the polarisation dependence of the monochromator transmission must be allowed for.

The frequency dependence of the transition polarisability, $\boldsymbol{\alpha}^{(mn)}$, is given in equation (5.18). Within the confines of the polarisability theory of Raman scattering it is also given by the frequency dependence of the polarisability, $\boldsymbol{\alpha}$. The variation of Raman scattering with ω, the frequency of the incident light, is determined by the dispersion of $\boldsymbol{\alpha}^{(mn)}$, or $\boldsymbol{\alpha}$, and by the additional $(\omega \pm \omega_{mn})^4$ dependence of the scattered intensity[85]. As ω approaches the first electronic transition frequency ω_E, the scattered intensity rises very rapidly and we encounter resonance Raman scattering[86,87]. Although the polarisability theory of Raman scattering predicts a resonance as $\omega \rightarrow \omega_E$, it fails to account for the details of the vibrational selection rules, especially with regard to the number of overtones commonly observed in resonance Raman scattering; the transition polarisability $\boldsymbol{\alpha}^{(mn)}$ must therefore be used.

5.3.3 The quadratic Stark effect

The application of an external electric field, E, to an atom or molecule results in a perturbation which shifts the energy levels and which may also partially lift the degeneracy of these levels. The electric field therefore causes a shift or splitting of the spectral lines arising from transitions between those levels. This is the Stark effect[55]. If we except hydrogen-like atoms and symmetric-top rotors, for which the Stark effect is first-order in the applied field[55], the effect is quadratic in E. The Stark effect is readily observed in the pure rotational spectrum of molecules and in the electronic spectra of atoms. Let W_J be the energy of a linear molecule with rotational angular momentum $\boldsymbol{J}$ and in the absence of the electric field. In the presence of the field the energy becomes[55]

$$W_{JM}(E) = W_J - \tfrac{1}{2}a^0E^2 + \frac{J(J+1)-3M^2}{(2J-1)(2J+3)}\left[\frac{\mu^2}{2BJ(J+1)} - \tfrac{1}{3}(a_{\parallel}^0 - a_{\perp}^0)\right]E^2 \quad (5.68a)$$

for $J \geqslant 1$; for $J=0$:

$$W_{00}(E) = W_0 - \tfrac{1}{2}a^0E^2 - \tfrac{1}{6}\mu^2E^2/B \quad (5.68b)$$

In these equations, B is the rotational constant and M is the component of $\boldsymbol{J}$ in the direction of the electric field. The term in a^0E^2 causes a change in energy which is the same for all rotational levels; its effect is not readily observable. The expression in the square brackets of equation (5.68a) is, for most dipolar molecules, dominated by the term in μ^2. Molecular dipole moments may be obtained with high accuracy from observations of Stark splittings of microwave rotational spectra by ignoring in equations (5.68) the terms in a^0. However, the difference in the J-dependence of the μ^2 and $(a_{\parallel}^0 - a_{\perp}^0)$ terms allows their separation[55,88]. The relative importance of the $(a_{\parallel}^0 - a_{\perp}^0)$ term increases with J and high-J transitions are therefore preferred for the determination of the polarisability anisotropy. High field strengths (up to 50 kV cm^{-1}) [88] are necessary in order that sufficient accuracy in the Stark shifts may be obtained. The static polarisability of a number of molecules has been obtained by this method[88,89].

Atoms have no permanent dipole moment and the quadratic Stark shift for a given atomic energy level is determined by a^0 and $(a_{\parallel}^0 - a_{\perp}^0)$. For a transition between two electronic states the polarisability of each state must be considered when computing the Stark shifts. Level-crossing spectroscopy has been used to measure the Stark shifts for a number of resonance lines of various alkali atoms[90,91]. The results have yielded the difference in a^0 for the ground $ns\ ^2S_{1/2}$ state and selected $np\ ^2P_{3/2}$ states (n is the principal quantum number of the valence electron); the polarisability anisotropy of the $np\ ^2P_{3/2}$ states was also obtained. Both a^0 and $(a_{\parallel}^0 - a_{\perp}^0)$ increase rapidly with n.

The quadratic Stark shift of the rotational energy levels of non-dipolar molecules is also determined by a^0 and $(a_{\parallel}^0 - a_{\perp}^0)$. Although these molecules display no pure rotational spectrum, the Stark shifts may be obtained from molecular beam measurements. Since the determination of molecular properties from molecular beam resonance spectroscopy is reviewed elsewhere in this volume[92] we need not elaborate on this method of determining polarisability anisotropies. An important example is the determination of $(a_{\parallel}^0 - a_{\perp}^0)$ for the hydrogen molecule[93]. This quantity is used for the calibration of absolute Raman scattering intensities[82] (see Section 5.3.2). The mean polarisability for atoms or non-dipolar molecules may be obtained by studying the deflection of molecular beams when passed through an inhomogenous field.

5.4 NON-LINEAR OPTICAL PHENOMENA

5.4.1 Second harmonic generation

5.4.1.1 Coherent second harmonic generation

Optical second harmonic generation (SHG) was first achieved in 1961 by Franken *et al.*[94], when a pulsed ruby laser beam (λ=694.3 nm) was focused into a quartz crystal to produce a tiny output of light at 347.1 nm. Conversion efficiencies have since been increased enormously, to the extent that SHG is now a widely exploited means of producing intense coherent light at shorter wavelengths[28,95,96]. The phenomenon is a coherent scattering effect, analogous to that responsible for refractive index, and arises from interaction of the electric field of the laser beam with the non-linear polarisability $\boldsymbol{\beta}(-2\omega;\omega,\omega)$ of the material. In contrast to conventional refraction, however, the incident and scattered waves differ in frequency and usually propagate with different velocities in the material, as a result of the natural dispersion of refractive index between frequencies ω and 2ω. Whereas the incident wave and the scattered second harmonic wave are in phase as they leave a given scattering centre in the medium, their difference in velocity causes an increasing phase difference as they propagate through the medium. The net effect is that the SHG intensity from a slab of non-linear optical material varies periodically as a function of the thickness of the slab. This behaviour may be described in terms of a characteristic *coherence length* l_c of the material; the SHG intensity will be at a maximum if the slab thickness of the material is l_c or odd integral multiples thereof. It may be shown[5,6] that

$$l_c=\lambda_1/4(n_1-n_2) \tag{5.69}$$

where λ_1 is the fundamental wavelength and n_1, n_2 the fundamental and SH refractive indexes. For crystals and red incident light, l_c is typically 10 μm.

The limitations on SHG output imposed by coherence length effects may be overcome by *index matching* techniques, which minimise (n_1-n_2) and thereby increase l_c. This can be achieved in certain birefringent crystals by orienting the crystal with respect to the direction of propagation of the laser beam, such that the extraordinary refractive index at one frequency equals the ordinary refractive index at the other.

The symmetry conditions for generation of *coherent* second harmonic radiation require that the scattering material as a whole should lack a centre of inversion. This can only be achieved in 21 of the 32 possible crystal classes. Isotropic materials such as fluids are rigorously excluded. Coherent SHG is therefore of only passing relevance to the determination of $\boldsymbol{\beta}(-2\omega;\omega,\omega)$ for *isolated* molecules.

5.4.1.2 Hyper-Rayleigh and hyper-Raman scattering

SHG in fluids can nevertheless be observed as *incoherent* scattering. We refer to this phenomenon as *hyper-Rayleigh scattering*, since it bears the same relation to coherent SHG as does Rayleigh scattering to refraction. The symmetry criterion for hyper-Rayleigh scattering is that the molecules of the scattering medium should lack a centre of inversion, so that at least one element of the molecular hyperpolarisability tensor $\boldsymbol{\beta}(-2\omega;\omega,\omega)$ is non-zero (see Section 5.2.3).

The theory of hyper-Rayleigh scattering has been dealt with by Bersohn, Pao and Frisch[32] and others[3,22,97–99]. For a fluid of non-interacting molecules, five independent quadratic combinations of $\boldsymbol{\beta}(-2\omega;\omega,\omega)$ can in principle be obtained from hyper-Rayleigh observables. The assumption that Kleinman symmetry in all tensor suffixes is applicable[22,32] reduces the number of independent combinations from five to two. Molecular symmetry may also reduce this number. For example, for tetrahedral (T_d) symmetry only a single element, $\beta_{xyz}(-2\omega;\omega,\omega)$, is needed to represent hyper-Rayleigh scattering. The depolarisation ratio for hyper-Rayleigh scattering observed with linearly polarised light should, for T_d symmetry, be 2/3 and independent of β_{xyz}. For isolated molecules of general symmetry, the depolarisation ratio is

$$\rho(2\omega)=\frac{4\beta_{\xi\eta\zeta}\beta_{\xi\eta\zeta}-\beta_{\xi\eta\eta}\beta_{\xi\zeta\zeta}}{6\beta_{\xi\eta\zeta}\beta_{\xi\eta\zeta}+9\beta_{\xi\eta\eta}\beta_{\xi\zeta\zeta}} \tag{5.70}$$

where Kleinman symmetry is assumed[3,22]; the corresponding result* in the absence of Kleinman symmetry has been derived[32].

Expressions[22,32] for absolute hyper-Rayleigh scattering intensities, which are more difficult to measure accurately than depolarisation ratios, involve a quadratic dependence on elements of $\boldsymbol{\beta}(-2\omega;\ \omega,\ \omega)$ and proportionality to ω^4 and the square of incident intensity. In dense fluids, hyper-Rayleigh scattering is sensitive to orientational correlation between neighbouring molecules and to hyperpolarisability contributions of form $\gamma(-2\omega;\ \omega,\ \omega,\ 0)\times \boldsymbol{F}_{\mathrm{loc}}$, where $\boldsymbol{F}_{\mathrm{loc}}$ is the instantaneous local electric field at the scattering molecule. The influence of such effects on total scattered intensities and depolarisation ratios[32,98,99,100] and hyper-Rayleigh linewidths[99,100] has been treated theoretically, and related to dielectric relaxation and depolarised Rayleigh scattering[99].

Associated with the essentially elastic hyper-Rayleigh scattering at frequency 2ω is an inelastic spectrum due to *hyper-Raman scattering* at frequencies $(2\omega\pm\omega_{mn})$, where ω_{mn} corresponds typically to vibrational energy intervals of the molecule. The effect has been thoroughly reviewed by Long[31,101]. A density matrix treatment[30,31] shows how hyper-Raman scattering originates in the off-diagonal hyperpolarisability $\boldsymbol{\beta}^{(mn)}(\omega_{mn}\pm 2\omega;\ \mp\omega,\ \mp\omega)$; this can be expanded in the normal coordinates Q_i of the molecule:

$$\boldsymbol{\beta}^{(mn)}(\omega_{mn}\pm 2\omega;\mp\omega,\mp\omega)=(\boldsymbol{\beta})^{(mn)}_{\mathrm{e}}+\sum_i[(\partial\boldsymbol{\beta}/\partial Q_i)_{\mathrm{e}}Q_i]^{mn}+\ \ldots\ . \tag{5.71}$$

where $(\ )_{\mathrm{e}}$ denotes evaluation at the equilibrium molecular geometry. The first term in equation (5.71) gives rise to hyper-Rayleigh scattering, with $m=n$. The second governs the hyper-Raman effect, such that a transition $m\leftrightarrow n$ is allowed if states m and n differ by a single quantum of one of the normal modes i and if $(\partial\boldsymbol{\beta}/\partial Q_i)_{\mathrm{e}}$ is non-zero. The symmetry properties of

*The third term in equation (74) of Ref. 32 is in error and should read:

$$-\frac{2}{35}\beta_{zzz}(\beta_{xzx}+\beta_{xxz}).$$

$\beta^{(mn)}$ which determine hyper-Raman selection rules have been tabulated[22,31], assuming that Kleinman symmetry is applicable[179]. They predict that all i.r.-active vibrational modes are also active in the hyper-Raman spectrum and that certain i.r.- and Raman-inactive modes (e.g. the a_u twisting mode of C_2H_4) are allowed in the hyper-Raman spectrum. A further point is that a hyper-Raman spectrum need not necessarily be accompanied by a hyper-Rayleigh peak; in contrast, the totally symmetric nature of the mean polarisability α for any molecule ensures that Rayleigh scattering is never symmetry-forbidden. A theory of hyper-Raman polarisation effects[103] shows how vibrations of different symmetries may be distinguished, using circularly polarised light. Rotational selection rules have also been derived.

Experimental work on hyper-Rayleigh and hyper-Raman scattering has been performed almost exclusively by Maker and co-workers[99,102,104]. Use of a pulsed ruby laser and multichannel photon counting detection enable signals amounting to less than one photo-event per ten laser flashes to be discerned[104]. Results for a number of liquids[99,102] and gases[104] have been reported. Absolute hyper-Rayleigh intensity measurements on liquid CCl_4 and gaseous CH_4 yield estimates of $|\beta_{xyz}|$ for the two tetrahedral molecules[99,100,102]; the interpretation of results is complicated by co-operative scattering effects. An early discrepancy[102] in the observed depolarisation ratio $\rho(2\omega)$ for liquid CCl_4 has been attributed[99] to polarised background scattering. Hyper-Raman spectra reported for liquids such as H_2O [102] and CH_3OH [101], and the gases[104] CH_4, C_2H_6 and C_2H_4 confirm the above selection rules. Hyper-Rayleigh scattering from liquids composed of centrosymmetric molecules has also been observed[135] and depolarisation ratios related to a non-linear anisotropy factor involving elements of $\gamma(-2\omega; \omega, \omega, 0)$.

5.4.2 Third harmonic generation

Optical third harmonic generation (THG), which has been observed in a range of crystals, glasses, liquids[105] and gases[106], arises from the non-linear polarisability $\gamma(-3\omega; \omega, \omega, \omega)$ of the material. As in its second harmonic analogue, complications associated with coherence length arise, owing to dispersion of refractive index between frequencies ω and 3ω. In contrast to SHG, however, coherent THG is not restricted to non-centrosymmetric materials, and has been observed in crystals, such as calcite, and in isotropic fluids.

Coherent THG in various gases[106] has yielded estimates of elements of $\gamma(-3\omega; \omega, \omega, \omega)$ for individual atoms and molecules; these are of higher reliability than elements of $\beta(-2\omega; \omega, \omega)$ obtained from SHG (hyper-Rayleigh) experiments. The technique of Ward and New[106] employs a pulsed ruby laser beam (694.3 nm) brought to a focus ahead of a gas cell fitted with an entrance window which absorbs third harmonic radiation (231.4 nm). After traversing the gas cell the laser beam is absorbed in a $NiSO_4$ filter solution, which transmits the weak forward-scattered third harmonic radiation to a monochromator and photomultiplier. The choice of laser beam configuration in the gas cell is the result of an analysis[106] showing that no net THG results from focusing into a homogenous medium of infinite extent. The technique provides a differential measurement of the THG properties of the gas in the cell, relative to those of the absorbing entrance window. Reliable estimates of the parameter

$$\gamma^{\mathrm{THG}}=\tfrac{1}{5}\gamma_{\xi\xi\eta\eta}(-3\omega;\ \omega,\ \omega,\ \omega) \tag{5.72}$$

have been determined for the inert gases, H_2 CO_2, and N_2 [106]. THG in Cs vapour, enhanced by a two-photon resonance, has also been observed[107].

Optical THG may also be observed in index-matched gas mixtures. The index-matching condition, that refractive indexes n_1 and n_3 for frequencies ω and 3ω are equal, may be attained by mixing two gases, one exhibiting normal dispersion (n decreasing as wavelength increases) and the other with anomalous dispersion of refractive index, at an appropriate ratio of partial pressures. For alkali metal vapours[108] the incident wavelength may readily be selected in the visible or near i.r. such that the anomalous dispersion curve is centred between the fundamental and third harmonic wavelengths; admixture of a suitable inert buffer gas, such as xenon, which shows normal dispersion at both wavelengths, enables the index-matching condition to be established. In the absence of index-matching, THG occurs effectively only over the short coherence length l_c of the vapour, where

$$l_c=\lambda_1/6(n_1-n_2) \tag{5.73}$$

Addition of buffer gas in index-matching proportions allows the coherence length to exceed the path length l through the vapour and increases the THG output power by a factor of order $(l/l_c)^2$. The effectiveness of such THG processes may also depend[109] on resonant enhancement of the hyperpolarisability $\gamma(-3\omega;\ \omega,\ \omega,\ \omega)$ and on high u.v. transparency. Index-matching of Rb vapour with Xe provides a 33-fold enhancement of THG output[110], confirming theoretical predictions. Similar experiments for Cd in Ar[111] have produced coherent vacuum u.v. picosecond pulses at 177 and 118 nm, with a peak power of 7 kW at the former wavelength.

5.4.3 Electro-optic effects

A number of phenomena involving hyperpolarisabilities had been discovered, in experiments employing strong static electric and magnetic fields, well in advance of the laser era[5]. In this section we consider several non-linear optical processes induced by static uniform electric fields.

The *linear electro-optic effect* (also known as the *Pockels effect*, after its discoverer[112]) arises from the hyperpolarisability $\boldsymbol{\beta}(-\omega;\ \omega,\ 0)$ and manifests itself as a birefringence proportional to the applied static electric field strength. The effect is restricted to non-centrosymmetric crystals and is useful for optical switching and modulation applications.

5.4.3.1 Electric field-induced second harmonic generation

The SHG analogue of the Pockels effect is *d.c. electric field-induced SHG* (ESHG), which arises from the hyperpolarisability $\gamma(-2\omega;\ \omega,\ \omega,\ 0)$. The process was first demonstrated in a crystal of calcite[105]; calcite is centrosymmetric and does not display normal SHG. Application of the static electric field destroys the centre of inversion of the crystal, enabling SHG to take place. Index-matching conditions are critical for optimum output.

Since ESHG can be observed in isotropic media, experiments in gases are possible[113–116]. It may be shown that, if the static field $\boldsymbol{E}^0$ and the optical field (amplitude $\boldsymbol{E}^{\omega}$) are both polarised in the space-fixed Z direction, a Z-polarised electric dipole is induced at frequency 2ω:

$$\overline{\mu_Z^{2\omega}}=\tfrac{1}{4}\left[\frac{\mu}{3kT}\beta_{\parallel}^{\mathrm{SHG}}+\gamma_{\parallel}^{\mathrm{ESHG}}\right](E_Z^{\omega})^2E_Z^0 \tag{5.74}$$

where the barred notation indicates an isotropic average over molecular orientations and

$$\beta_{\parallel}^{\mathrm{SHG}}=\tfrac{2}{5}\beta_{\xi\xi z}(-2\omega;\ \omega,\ \omega)+\tfrac{1}{5}\beta_{z\xi\xi}(-2\omega;\ \omega,\ \omega) \tag{5.75}$$

$$\gamma_{\parallel}^{\mathrm{ESHG}}=\tfrac{2}{15}\gamma_{\xi\xi\eta\eta}(-2\omega;\ \omega,\ \omega,\ 0)+\tfrac{1}{15}\gamma_{\xi\eta\eta\xi}(-2\omega;\ \omega,\ \omega,\ 0) \tag{5.76}$$

z denotes the molecular-fixed electric dipole axis. Similarly, for $\boldsymbol{E}^{\omega}$ perpendicular to $\boldsymbol{E}^0$:

$$\overline{\mu_Z^{2\omega}}=\tfrac{1}{12}\left[\frac{\mu}{3kT}\beta_{\perp}^{\mathrm{SHG}}+\gamma_{\perp}^{\mathrm{ESHG}}\right](E_X^{\omega})^2E_Z^0 \tag{5.77}$$

$$\beta_{\perp}^{\mathrm{SHG}}=\tfrac{6}{5}\beta_{z\xi\xi}(-2\omega;\ \omega,\ \omega)-\tfrac{3}{5}\beta_{\xi\xi z}(-2\omega;\ \omega,\ \omega) \tag{5.78}$$

$$\gamma_{\perp}^{\mathrm{ESHG}}=\tfrac{2}{5}\gamma_{\xi\eta\eta\xi}(-2\omega;\ \omega,\ \omega,\ 0)-\tfrac{1}{5}\gamma_{\xi\xi\eta\eta}(-2\omega;\ \omega,\ \omega,\ 0) \tag{5.79}$$

The induced moments $\overline{\mu_Z^{2\omega}}$ give rise to scattering at the second harmonic frequency, interference of which in the forward direction is characterised by a coherence length l_c as defined in equation (5.69). The terms involving β and γ in equations (5.74) and (5.77) arise respectively from orientation and distortion mechanisms; the former is similar to that contributing to dielectric polarisation (Sections 5.1 and 5.3.1) and is therefore inversely proportional to temperature T. For non-dipolar molecules the orientation term is zero, since both μ and $\beta_{\parallel \text{ or } \perp}^{\mathrm{SHG}}$ (which transforms as μ) are zero. For dipolar molecules the distortion and orientation terms can be separated by studies of temperature dependence. In contrast to THG experiments in gases, from which only a single parameter [γ^{THG}, equation (5.72)] is obtainable, four independent parameters ($\beta_{\parallel}^{\mathrm{SHG}}$, $\beta_{\perp}^{\mathrm{SHG}}$, $\gamma_{\parallel}^{\mathrm{ESHG}}$, $\gamma_{\perp}^{\mathrm{ESHG}}$) may be extracted from ESHG studies. In the limit where Kleinman symmetry becomes applicable, $\beta_{\parallel}^{\mathrm{SHG}}$ is identical to $\beta_{\perp}^{\mathrm{SHG}}$ and $\gamma_{\parallel}^{\mathrm{ESHG}}$ to $\gamma_{\perp}^{\mathrm{ESHG}}$.

Gas-phase ESHG experiments have been reported by the groups of Mayer[113] and Ward[114–116,177]. In the apparatus of Finn and Ward[114], a ruby-laser beam is focused in the gas of interest and midway between a pair of cylindrical electrodes, arranged transverse to the beam. The electrodes establish the field $\boldsymbol{E}^0$ in the focal region and the resulting second harmonic light passes through a filter and monochromator to a photomultiplier. The experiments have yielded hyperpolarisability estimates for the inert gases[114] and a number of simple molecules[113,115]. In extracting estimates of $\beta_{\parallel}^{\mathrm{SHG}}$ for dipolar molecules, allowance has been made for the minor contribution of $\gamma_{\parallel}^{\mathrm{ESHG}}$, either by equating $\gamma_{\parallel}^{\mathrm{ESHG}}$ to a closely related parameter $\gamma_{\parallel}^{\mathrm{TWM}}$ from three-wave mixing experiments (Section 5.4.4)[113] or by applying a bond additivity approximation[115]. Recent studies[177] of the temperature dependence of ESHG have enabled the contributions of $\beta_{\|}^{\mathrm{SHG}}$ and $\gamma_{\|}^{\mathrm{ESHG}}$ to be separated more rigorously. ESHG polarisation experiments[116] indicate that Kleinman symmetry applies to within experimental uncertainty for inert gas atoms.

5.4.3.2 Electro-optic Kerr effect

A third electro-optic effect of great usefulness in obtaining polarisability data is the *electro-optic Kerr effect*, which was first observed in 1875 [117]. The effect consists of birefringence induced in a material by a static electric field and is quadratically dependent on the electric field strength. In contrast to the linear electro-optic (Pockels) effect, the Kerr effect is not restricted to non-centrosymmetric materials. A recent review by Le Fevre and Le Fevre[118] covers techniques and applications and cites earlier review articles. The Kerr birefringence induced in a medium is the difference between the refractive indexes, $n_{\|}$ and $n_{\perp}$, for light linearly polarised parallel and perpendicular, respectively, to the static electric field. The birefringence may be measured by passing light, which is linearly polarised at 45° to the electric field, through the medium and observing the phase retardation φ of the elliptically polarised emergent light:

$$\varphi = 2\pi l(n_{\|} - n_{\perp})/\lambda \text{ radians} \tag{5.80}$$

where l is the length of the birefringent medium and λ the optical wavelength *in vacuo*. The retardation may be measured by use of a suitable optical compensator or by a null method using a second calibrated Kerr cell[118]; an apparatus of high sensitivity designed for gas phase studies has been described by Buckingham *et al.*[119]. Kerr effect experiments rely on optical retardation determination for their accuracy, rather than on intensity measurements as is the case in ESHG, THG and other non-linear optical experiments. The Kerr effect therefore provides better absolute accuracy than other techniques, which generally need to be scaled by relative intensity measurements to some standard reference material[106,113–115].

For gases, the molar Kerr constant ${}_{\mathrm{m}}K$, which is related to individual molecular observables, is defined as

$${}_{\mathrm{m}}K = 6n(n^2+2)^{-2}(\varepsilon+2)^{-2}[(n_{\|}-n_{\perp})V_{\mathrm{m}}E^{-2}]_{E\to 0} = A_{\mathrm{K}} + B_{\mathrm{K}}V_{\mathrm{m}}^{-1} + \ldots \tag{5.81}$$

where ε is the dielectric constant, n the isotropic refractive index, V_{m} the molar volume and E the uniform electric field strength. The first and second Kerr virial coefficients are represented by A_{K} and B_{K}; they arise respectively from contributions by individual molecules and by pairs of molecules. The theory of Buckingham and Pople[120] shows that, for a molecule of general symmetry,

$$A_{\mathrm{K}} = \frac{N}{81\varepsilon_0}\left\{\gamma^{\mathrm{K}} + (kT)^{-1}\left[\tfrac{2}{3}\mu\beta^{\mathrm{K}} + \tfrac{3}{10}(a_{\xi\eta}a_{\xi\eta}^{0} - 3aa^{0})\right] + \tfrac{3}{10}(kT)^{-2}\mu^2(a_{zz} - a)\right\} \tag{5.82}$$

where ε_0 is the permittivity of a vacuum, $\boldsymbol{\alpha}$ and $\boldsymbol{\alpha}^0$ are the optical and static polarisabilities, respectively, and z denotes the molecule-fixed direction of the electric dipole moment μ. The Kerr hyperpolarisability parameters β^K and γ^K are defined by

$$\beta^K = \tfrac{3}{10}[3\beta_{\xi z \xi}(-\omega;\ \omega, 0) - \beta_{\xi\xi z}(-\omega;\ \omega, 0)] \tag{5.83}$$

$$\gamma^K = \tfrac{1}{10}[3\gamma_{\xi\eta\xi\eta}(-\omega;\ \omega, 0, 0) - \gamma_{\xi\xi\eta\eta}(-\omega;\ \omega, 0, 0)] \tag{5.84}$$

Of the four terms in equation (5.82), three arise from orientation mechanisms. The term involving γ^K is associated entirely with distortion of the molecule by the electric field. Equation (5.82) simplifies in a number of special cases. For non-dipolar molecules, only the terms involving γ^K and $\boldsymbol{\alpha}\boldsymbol{\alpha}^0$ survive. If dispersion in the polarisability anisotropy parameter κ^2 [equation (5.38)] is negligible, the $\boldsymbol{\alpha}\boldsymbol{\alpha}^0$ term can be more compactly written as $(N/45\varepsilon_0)aa^0\kappa^2$, which enables comparison to be made with Rayleigh depolarisation measurements[79]. For molecules possessing a threefold or higher rotational axis,

$$A_K = \frac{N}{81\varepsilon_0}\left\{\gamma^K + (kT)^{-1}\left[\tfrac{2}{3}\mu\beta^K + \tfrac{1}{5}(a_\parallel - a_\perp)(a_\parallel^0 - a_\perp^0)\right] + \tfrac{1}{5}(kT)^{-2}\mu^2(a_\parallel - a_\perp)\right\} \tag{5.85}$$

Finally, if the molecule is isotropically polarisable, only the term in γ^K is non-zero, so that A_K is directly proportional to γ^K. For molecules with moderately large electric dipole moments ($\sim 3\times 10^{-30}$ C m ≈ 1 D), the relative magnitudes of the various contributions to A_K are typically in the order:

$$|\mu^2\boldsymbol{\alpha}\ \text{term}| > |\mu\beta^K\ \text{term}| > |\boldsymbol{\alpha}\boldsymbol{\alpha}^0\ \text{term}| > |\gamma^K\ \text{term}|$$

There are, however, striking exceptions to the above rule. For example, in any molecule of form CH_2X_2 with bond angles close to tetrahedral, the distribution of polarisability anisotropy within the molecule may cause $a_{zz} - a$ to be nearly zero, making the $\mu^2\boldsymbol{\alpha}$ term unusually small[3].

The density dependence of ${}_mK$ deserves some emphasis. Equation (5.82) describes the contribution to ${}_mK$ only from non-interacting molecules of a gas. Extrapolation of ${}_mK$ to zero density is therefore essential in accurate work.

Extensive temperature and gas density studies of the Kerr effect in a number of gases and vapours have been performed by Buckingham and co-workers, so that data exist from which quantities such as γ^K, β^K and $(a_{zz} - a)$ may be extracted. For atoms and effectively spherical molecules, A_K gives[119,121] a direct measure of γ^K. For non-dipolar anisotropic molecules, the γ^K and $\boldsymbol{\alpha}\boldsymbol{\alpha}^0$ terms can be separated[79,123,124] from a linear plot of A_K against T^{-1}. In H_2 gas it is necessary[123] to take account of quantum corrections to the $\boldsymbol{\alpha}\boldsymbol{\alpha}^0$ orientation term, in view of the large rotational constant of H_2. For dipolar molecules, where all four contributions to A_K appear, it has been found possible to estimate the small contribution of γ^K and to plot $(A_K - N\gamma^K/81\varepsilon_0)T$ linearly against T^{-1}; this yields[122] estimates of β^K and $(a_{zz} - a)$.

Polarisability data for individual molecules can also be obtained from Kerr effect studies of solutes at high dilution in non-dipolar solvents. Techniques of this type have been developed and applied to a wide range of molecules by Le Fèvre and co-workers[118]. Attempts have also been made to extract values of γ^K from Kerr effect data for pure liquids[125]; in a combined Kerr effect and light scattering study[126], estimates of γ^K for a series of liquid n-alkanes were found to extrapolate linearly against chain-length to the values of γ^K for CH_4 as a gas.

5.4.4 Miscellaneous non-linear optical phenomena

The non-linear optical processes which have been discovered during the past decade are extremely diverse and in many cases beyond the defined scope of this review. They include: *multiphoton absorption*[127,128], which may be shown to have its origins in the imaginary parts of optical hyperpolarisabilities $\boldsymbol{\beta}$, γ, . . . in resonant situations; *intensity-dependent refractive index*[105], which arises from the hyperpolarisability $\gamma(-\omega; \omega, \omega, -\omega)$ and is intrinsically related to the phenomenon of *self-focusing* of intense laser beams in solids and liquids[7]; the closely related *optical Kerr effect*[129,130], in which a material is perturbed by one laser beam and the resulting birefringence monitored by a second beam; *optical parametric conversion*[7,96], in which one or more incident ('pump') waves are converted to two output waves at 'idler' and 'signal' frequencies; the *stimulated Raman effect*[7,28,131], which may be regarded as a resonant four-wave parametric conversion process; other *resonant parametric conversion processes*, which promise to provide coherent tunable light sources throughout the infrared[132] and ultraviolet[107] regions; coherent *three-wave mixing* experiments[105,113,133], in which a forward-scattered resultant wave is produced via the hyperpolarisability $\gamma(-\omega-\Delta; \omega, \omega, -\omega+\Delta)$; *resonant optical harmonic conversion* processes of orders higher than the third, which may provide useful sources of coherent vacuum ultraviolet and soft x-radiation[134]; *optical rectification*[27], which involves the hyperpolarisability $\boldsymbol{\beta}(0; \omega, -\omega)$ and is coupled to SHG in non-centrosymmetric crystals; *dielectric saturation* phenomena[8], including electrostrictive and electrocaloric effects.

All the above phenomena depend on molecular polarisabilities and hyperpolarisabilities, and a number have already been classified in Table 5.1. In only a few instances, however, has it been possible to perform experiments in the gas phase and to extract quantitative data for individual molecules. *Three-wave mixing* experiments, first performed in liquids[105] and later in gases[113,133], have been particularly useful in this regard. Radiation of frequency ω from a pulsed ruby laser is focused with coherent Raman-shifted radiation of frequency $(\omega-\Delta)$ into the sample and the forward-scattered resultant frequency $(\omega+\Delta)$ detected. As for THG, and in contrast to electro-optic Kerr and ESHG experiments, three-wave mixing (TWM) has no contribution from molecular orientation processes; the TWM intensity is therefore related to the parameter

$$\gamma_{\|}^{\mathrm{TWM}} = \tfrac{2}{15}\gamma_{\xi\xi\eta\eta}(-\omega-\Delta; \omega, \omega, -\omega+\Delta) + \tfrac{1}{15}\gamma_{\xi\eta\eta\xi}(-\omega-\Delta; \omega, \omega, -\omega+\Delta) \qquad (5.86)$$

where the waves of frequency ω and $(\omega-\Delta)$ have been assumed to be parallel in polarisation. Values of this parameter have been reported[113,133] for a number of gases and vapours, using H_2 gas to produce the laser Raman shift Δ. Measurements of the optical Kerr effect and of the intensity-dependent refractive index in gases are also capable in principle of yielding useful estimates of components of γ. Experiments so far have concentrated on liquids[105,130], however. The interpretation of either phenomenon would be complicated by a molecular orientation contribution[129], inversely proportional to kT and involving the product of two optical polarisabilities $\boldsymbol{\alpha}$.

A comparison of experimentally determined β and γ hyperpolarisabilities for a range of atoms and molecules is made in Section 5.7.

5.5 INTERACTING ATOMS AND MOLECULES

Although our primary concern is with the polarisabilities of isolated atoms and molecules, interactions between molecules must be considered in analysing bulk properties for gases of finite density to yield results for isolated molecules. Certain intramolecular effects, notably the dependence of molecular properties on vibrational and rotational state, are also considered in this section.

5.5.1 Virial coefficients of electrical and optical properties

The initial deviation from isolated-molecule behaviour may be examined systematically by means of a virial equation[136]. The observed value, Q, of a measurable property of the gas is expanded in inverse powers of the molar volume V_m:

$$Q = A_Q + B_Q V_m^{-1} + C_Q V_m^{-2} + \dots \qquad (5.87)$$

The coefficients A_Q, B_Q, C_Q, . . . , known as first, second, third, . . . virial coefficients of property Q, are functions of temperature but not of density. From equation (5.87), A_Q is the limiting value of Q at zero density; this pertains to a set of independent molecules, so that A_Q is proportional to $\bar{q}$, the mean contribution to Q from an isolated molecule. At higher densities some of the molecules are, at any instant, interacting in pairs, giving rise to a second virial coefficient B_Q. This may be evaluated from the relation,

$$B_Q = \operatorname*{Limit}_{V_m \to \infty} (Q - A_Q) V_m \qquad (5.88)$$

or by fitting observed values of Q to equation (5.87). Higher terms in equation (5.87) arise from multiple interactions between molecules. Effects due to second virial coefficients have been observed in measurements of refractive index[40,137–139a], dielectric constant[57–59,65–67], Rayleigh depolarisation ratio[140–142] and the Kerr effect[79,121–122a]. These studies indicate that, at s.t.p., the contribution due to B_Q comprises 1 % or less of Q. An exception to this generalisation is the Kerr effect where, as an extreme example[122], the molar Kerr constant for CH_2F_2 at 303 K changes sign at a pressure of 0.38 MPa (~3.8 atm), indicating that $B_K V_m^{-1}$ is of comparable magnitude to A_K.

It is relevant to discuss in more detail the results for the polarisability of pairs of atoms and 'spherical' molecules, in view of current interest[143–147]. Expansion of the Clausius–Mossotti function in powers of V_m^{-1} yields

$$\frac{\varepsilon-1}{\varepsilon+2}V_m = A_\varepsilon + B_\varepsilon V_m^{-1} + C_\varepsilon V_m^{-2} + \ldots \tag{5.89}$$

which defines the dielectric virial coefficients A_ε, B_ε; Sutter[139] has reviewed the subject. From equation (5.59),

$$A_\varepsilon = N a^0 / 3\varepsilon_0 \tag{5.90}$$

The expression for B_ε is[148]

$$B_\varepsilon = \frac{4\pi N^2}{3\varepsilon_0}\int_0^\infty [\tfrac{1}{2}a_{12}^0(r) - a^0] \exp[-W_{12}(r)/kT] r^2 \mathrm{d}r \tag{5.91}$$

where r is the distance between a pair of atoms labelled 1 and 2. For rare gas atoms the interatomic potential $W_{12}(r)$ is well represented by a Lennard-Jones 6–12 potential and a^0 may be obtained from A_ε. The (classical) Silberstein model[143,149] for the pair polarisability $a_{12}^0(r)$ yields a positive value of B_ε. Experimental estimates[59] of B_ε are negative for He and Ne and positive for Ar and Kr. The inadequacies of the Silberstein model have led theoreticians to *ab initio* calculations of the helium pair polarisability[144–147] (see Section 5.6.5). The polarisability anisotropy $(a_{\parallel} - a_{\perp})_{12}$ for a pair of atoms may be determined[121] experimentally from the second Kerr virial coefficient B_K and from pressure-induced depolarised light scattering[140–142]. Calculations[121,122] of B_K using the Silberstein model yield results which are too large, often by a factor of two, again indicating the deficiencies[143] of the model.

5.5.2 Vibrational and rotational dependence of $\boldsymbol{\alpha}$

The expectation value $\langle q \rangle$ of a molecular property q depends not only on the electronic state of the molecule but also on its vibrational and rotational state. This is well demonstrated experimentally by molecular electric dipole moments, for which Gordy and Cook[70] quote a number of examples. Similar effects on the polarisability $\boldsymbol{\alpha}$ have been observed only for molecular hydrogen[62,150–152]. No such effects appear to have been reported for the hyperpolarisabilities $\boldsymbol{\beta}$ and γ.

The polarisability $a_{v,J}$ of a diatomic molecule in the rovibrational state with quantum numbers v,J may be expressed[153] in terms of the derivatives of a with respect to internuclear distance r:

$$a = a_e + a'_e\xi + \tfrac{1}{2}a''_e\xi^2 + \ldots \tag{5.92}$$

where the displacement parameter $\xi = (r - r_e)/r_e$. It follows[153] that, for an anharmonic oscillator,

$$a_{v,J} = a_e + (v+\tfrac{1}{2})(B_e/\omega_e)(a''_e - 3aa'_e) + 4J(J+1)(B_e/\omega_e)^2 a'_e \tag{5.93}$$

where B_e, ω_e are the usual rotational and vibrational constants and a the cubic anharmonic constant. The vibrational contribution to $a_{v,J}$ is given by the second term of (5.93); the term also largely determines the effects of isotopic substitution, owing to its dependence on (B_e/ω_e). The third term of (5.93) describes the effect of centrifugal distortion which, for low J, is less significant than the vibrational contribution, since normally $\omega_e \gg B_e$.

Most experimental determinations of molecular polarisabilities depend on measurements of bulk properties. At typical temperatures a large fraction of the molecules may be in excited rotational states and a smaller, but often appreciable, fraction in excited vibrational states. The bulk properties are then related to the polarisabilities of individual molecules by means of an appropriate statistical average. Since the fraction of molecules in a given state varies with temperature, the average molecular polarisability may be expected to be temperature dependent, owing to the v,J-dependence of $\boldsymbol{\alpha}$. Limited studies of the temperature dependence of the dielectric polarisation for CH_4 [66], CF_4 [66] and C_2H_6 [67] suggest only that such effects, if any, are small.

For molecular hydrogen, it is possible to prepare samples where one or at most two states are significantly populated. Low-temperature studies of the dielectric polarisation[62,150] and refractive index[151] indicate a 0.15% difference in the polarisabilities of $J=1$ and $J=0$ states. The observed 1.26% polarisability difference[62,150] for H_2 and D_2 yields estimates[153] of a'_e and a''_e in close agreement with *ab initio* theory[152] Molecular beam studies[93] have yielded $(a^0_{\parallel}-a^0_{\perp})$ for the $J=1$ state of H_2 and light-scattering results[73] have yielded $(a_{\parallel}-a_{\perp})$ for the $J=0$ state of H_2. After extrapolation to zero frequency[78], the anisotropies for the $J=0$ and $J=1$ states agree to within the 2% experimental error of the light scattering result.

5.5.3. Intermolecular forces

The subject of intermolecular forces is vast[154]; here we make brief mention of the part played by polarisabilities in determining long-range intermolecular forces. The forces acting between two molecules, so far apart that electron exchange may be neglected, may be obtained[1,155] by using perturbation theory to calculate the interaction between them. The perturbation to the Hamiltonian of the non-interacting molecules is the coulombic interaction between the charges on molecules 1 and 2. Using standard perturbation theory, the energy is expressed in ascending powers of the perturbation; the perturbation energy, $U(r)$, may be further expanded in inverse powers of the separation r between the molecules. The first-order perturbed energy is called the electrostatic energy; it arises[1] from the interaction of the permanent electric multipole moments of the molecules. The second-order energy may be divided into two parts: the induction energy and the dispersion energy. The induction energy, which arises from the interaction of permanent and induced electric multipole moments, can be expressed in terms of the field and field gradients produced by the multipole moments of one molecule and the static electric dipole, quadrupole, . . . polarisabilities of the other[1]. The dispersion energy varies as r^{-6} for large r; its presence does not depend on either molecule having a permanent electric multipole. It may be expressed approximately in terms of the static polarisability of the molecules[1,155] or exactly in terms of the dynamic polarisability[154]. Victor and Dalgarno[78] have used their results for the dispersion of $a_{\parallel}$ and $a_{\perp}$ to calculate the long-range dispersion forces between two hydrogen molecules. Similar results are available[154] for the rare gases, N_2 and O_2.

5.6 THEORETICAL ESTIMATES

The calculation of polarisabilities is currently a field of intense activity. We present in this section a brief survey of some of this field. In view of the experimental bias of this review, our selection of papers is based on their relevance to experimental results or problems, rather than on their theoretical merit.

Three basic approaches may be taken to the calculation of polarisabilities: the 'sum-over-states' approach, which uses the perturbation formulae for $\boldsymbol{\alpha}$, $\boldsymbol{\beta}$, . . . given in Section 5.2 and is quantitatively useful only when a small number of terms contribute significantly to the infinite sums; variational techniques, which calculate perturbed wavefunctions to order n in the perturbation $\boldsymbol{\mu} \cdot \boldsymbol{E}$ for a finite value of E and enable the energy $W(E)$ to be calculated correct to order $(2n+1)$, and thence the polarisabilities $\boldsymbol{\alpha}^0$, $\boldsymbol{\beta}^0$, . . . [see equation (5.3)]; a third technique[13,14] in which the effect of the perturbation is described by a series of differential equations, soluble by numerical, variational or analytical means and applicable to both time-independent and time-dependent problems within the Hartree–Fock approximation. Finally, a semi-empirical means of estimating polarisabilities, based on the bond additivity approximation, has been widely applied and is discussed in Section 5.6.4.

5.6.1 Static polarisability

The static polarisability of various states of the alkali metal atoms has been calculated[90,91,156] by the sum-over-states method, using Coulomb-approximation[157] matrix elements. Some authors[90,91] neglect the continuum contribution to the sum despite its ready availability[156]. Its importance may be illustrated by atomic hydrogen, for which the contribution of np states to the 1s ground state polarisability is: 2p, 66%; 3p, 9%; 4p, 3%; 5p, 1%; continuum p states, 19%. For molecular hydrogen, $a_{\parallel}^0$ and $a_{\perp}^0$ have been obtained[78,158] from direct summation over states with matrix elements obtained empirically[78] or from *ab initio* calculations[158]. Very accurate variation calculations of a^0 have been carried out for He in its ground state[159–160a] and in its 2 $^3S_1$ and 2 $^1S_0$ excited states[159]. Kołos and Wolniewicz[152] have produced a definitive calculation of $a_{\parallel}^0$ and $a_{\perp}^0$ for molecular hydrogen. Having calculated the components of $\boldsymbol{\alpha}^0$ at various internuclear distances they performed vibrational and rotational averaging to obtain $a_{\parallel}^0$ and $a_{\perp}^0$ for various rovibrational states. The results for He and H_2 are compared with experiment in Table 5.3. Calculations on other small diatomic molecules[162,163,166] do not display the same quality of agreement, often being in error by 100%.

Table 5.3 Comparison of theoretical and experimental polarisabilities

Molecule	*Property*	*Theory*	*Experiment*	*Comments*
He	$10^{40}\ \alpha^0/C^2\ m^2\ J^{-1}$	0.228044 [160]	0.229 [59]	
	$10^{60}\ \gamma^0/C^4\ m^4\ J^{-3}$	0.002688 [160]	—	
		0.00267 [170]		
	$10^{60}\ \gamma^K/C^4\ m^4\ J^{-3}$	0.00275	0.0033 [121]	$\lambda = 632.8$ nm. Dispersion corrections[165] applied to theoretical result
H_2	$10^{40}\ \alpha^0/C^2\ m^2\ J^{-1}$	0.8952 [152]	0.895_3 [61]	Rotational average for 293 K
			0.896_7 [59]	
	$10^{40}\ (\alpha^0_{\parallel} - \alpha^0_{\perp})/C^2\ m^2\ J^{-1}$	0.3350 [152]	0.3356 [93]	For $v = 0$, $J = 1$
	$10^2\ (\alpha^0_{J=1} - \alpha^0_{J=0})/\alpha^0_{J=0}$	0.19 [152]	0.20 [62]	
D_2	$10^2\ (\alpha^0_{H_2} - \alpha^0_{D_2})/\alpha^0_{H_2}$	1.27 [152]	1.26 [62]	

5.6.2 Dynamic polarisability

Accurate calculations of dynamic polarisability have been made only for He [164,165] and H_2 [78,158], yielding agreement with experiment to $\pm 1\,\%$ or better. The empirically determined sum-over-states result for H_2 [78] generally agrees with experiment[73,76] to within $\pm 0.1\,\%$.

5.6.3 Hyperpolarisabilities

The Hartree–Fock perturbation method has been used to calculate β_{xxz} and β_{zzz} for a number of dipolar diatomic[167] and polyatomic[168] molecules. Finite-field perturbation theory has been used[169]. Comparison with experiment can be made for only one molecule: for CH_3F, estimates of 10^{50} β^K from theory[168] and experiment[122] are -0.059 and (-0.19 ± 0.1) C^3 m^3 J^{-2}, respectively.

Sum-over-states calculations have yielded estimates of γ^K for Na and Li [155] and of γ^{THG} for Cs [107] in the vicinity of optical resonances, where rapid convergence of the summations is attainable. An attempted calculation[155] of γ^0 for Li failed to converge, although the corresponding calculation of a^0 converged rapidly. A very elaborate calculation has been made[160] of γ^0 for He; results are compared with experiment in Table 5.3. Sitz and Yaris[165] have used time-dependent perturbation theory to calculate γ^K, γ^{THG} and γ^{ESHG} for He; this calculation of γ^{THG} has been used[106] as a standard relative to which observed values of γ^{THG} for other gases may be scaled. A more accurate calculation of γ^{THG} for He has recently been reported by Klingbeil[170]. The finite-field method has yielded[171] estimates of γ^0 for He, Ne and Ar; allowing for the difficulty of the calculations, agreement with experiment[121] is reasonable.

5.6.4 Bond and atomic additivity schemes

It has long been well established that the mean polarisability a, and hence the molar refraction R [equation (5.53)], is an approximately additive scalar property of the bonds and functional groups of which a molecule is constituted. For example, Vogel and co-workers[49] have established useful additivity schemes for R and a in a wide range of molecular systems. The topic has been reviewed by Le Fèvre[4]. Deviations from strict additivity are generally attributed to constitutive effects which reflect the environments of the various bonds within the molecule.

The anisotropic part of the polarisability tensor $\boldsymbol{\alpha}$ is known to obey approximate tensor additivity rules[4]. The matrix formulation of bond polarisability additivity of Smith and Mortensen[172] is useful in this context. Additivity rules of the above type have been applied successfully to the interpretation of Rayleigh scattering anisotropies κ^2 and of electro-optic Kerr effect data, in many cases treating aspects of the molecular geometry as the unknown to be determined[4].

It is attractive to propose bond additivity schemes for certain hyper-polarisability parameters[3], notably β and γ for various experiments. Since β

Table 5.4 Summary of polarisability data*

Molecule	$10^{30}\mu$/[a] C m	$10^{40}\alpha^0$/[b] $C^2\ m^2\ J^{-1}$	$10^{40}\alpha$/[c] $C^2\ m^2\ J^{-1}$	$10^{40}(\alpha_\parallel - \alpha_\perp)$/[d] $C^2\ m^2\ J^{-1}$	$10^{50}\beta^K$/[e] $C^3\ m^3\ J^{-2}$
He		0.229 [59]	0.230 [60]		
Ne		0.441 [59]	0.443 [60]		
Ar		1.827 [59]	1.850 [60]		
Kr		2.764 [59]	2.80 [60]		
Xe		4.47 [60]	4.57 [60]		
H_2		0.895_3 [61]	0.916 [78]	0.352	
		0.896_7 [59]		0.335_6 [l]	
D_2		0.885 [61,62]	0.905 [41,78]	0.334	
				0.324_6 [l]	
N_2		1.935_7 [61]	1.967 [41]	0.77_5	
O_2		1.757_6 [61]	1.777 [41]	1.22	
CO	0.37	2.20 [63]	2.200 [41]	0.59_2	—
NO	0.51	—	1.74 [41]	0.93_9	—
N_2O	0.55	3.37 [64]	3.318 [41]	3.28	—
				3.58 [m]	
CO_2		3.241_6 [61]	2.933 [41]	2.34	
CS_2		9.7 [63]	9.54 [41]	10.5_0 [v]	
OCS	2.39	5.8 [63]	5.79 [41]	4.58 [v]	—
				5.2 [m]	
SO_2	5.4	4.76 [63]	4.326 [40,41]	2.27	—
SF_6		7.268 [65]	5.03 [41]	0.00 [73]	
Methane		2.885 [66]	2.901 [41]		
Ethane		4.9_4 [63]	5.01 [41]	0.86_4	
Ethene		4.733 [67]	4.70 [39,41]	1.81	
Ethyne		4.3_7 [63]	3.88 [41]	2.07 [n]	
Benzene		11.8 [63]	11.56 [41]	−6.24	
Cyclopropane		6.3 [63]	6.28 [41]	-0.89_8	
CH_3F	6.19	3.30 [57]	2.90 [40,41]		−0.19 [122]
CH_2F_2	6.54	—	(3.04) [k]	(0.50) [k]	-0.04_1 [122]
CHF_3	5.51	3.97 [57]	3.119 [41]	-0.27_0	+0.27 [122]
CF_4		4.270 [66]	3.172 [40,41]		
CH_3Cl	5.94	5.25 [57]	5.04 [40,41]	1.72	—
CH_2Cl_2	5.40	8.8 [63]	7.41 [41]	3.07	(−3) [p]
$CHCl_3$	3.47 [69]	10.6 [o]	9.47 [41]	−2.98	—
CCl_4		12.5 [63]	11.69 [41]	0.00 [73]	

* Alphabetical superscripts refer to footnotes, numerical superscripts to literature references. A dash (—) indicates an entry for which a non-zero value should be found but for which no reliable data are yet available.

a Electric dipole moments taken from appropriate compilations[2,70], except where otherwise stated.

b Mean static polarisabilities determined from dielectric constant data.

c Mean dynamic polarisabilities evaluated at an optical wavelength of 632.8 nm.

d Polarisability anisotropies determined from Rayleigh scattering data (λ=632.8 nm)[73], unless otherwise stated. For non-axially symmetric molecules, the quantity reported is $10^{40} 3\alpha|\kappa|$ [see equation (5.38)].

e First Kerr hyperpolarisability defined by equation (5.83), λ=632.8 nm.

f SHG hyperpolarisability defined by equation (5.75), λ=694.3 nm.

g Second Kerr hyperpolarisability defined by equation (5.84), λ=632.8 nm.

Table 5.4 (contd) **Summary of polarisability data***

$10^{50}\beta_{\parallel}^{SHG}$/[f] $C^3\ m^3\ J^{-2}$	$10^{60}\gamma^{K}$/[g] $C^4\ m^4\ J^{-3}$	$10^{60}\gamma_{\parallel}^{TWM}$/[h] $C^4\ m^4\ J^{-3}$	$10^{60}\gamma_{\parallel}^{ESHG}$/[i] $C^4\ m^4\ J^{-3}$	$10^{60}\gamma_{\parallel}^{THG}$/[j] $C^4\ m^4\ J^{-3}$
	0.0033 [121]	0.004 [133]	(0.0028) [q,w]	(0.0030) [j,w]
	0.0063 [121]	–	0.0078 [q,114]	0.006_6
	0.073 [121]	(0.073) [r]	0.088 [q,114]	0.09_4
	0.17 [121]	–	0.22 [q,114]	0.29
	0.48 [121]	–	0.60 [q,114]	0.7_3
	0.035 [s,123]	0.07 [u,113]	0.05 [r,113]	0.06
	0.03_0 [s,123]	0.06 [133]	–	–
	0.09 [79]	0.06_5 [133]	–	0.08
	–	0.06 [133]	–	–
–	–	0.08_5 [133]	–	–
–	–	0.20 [133]	–	–
–	–	–	–	–
	0.56 [79]	0.09_5 [133]	–	0.12
	7.1 [124]	0.83 [113]	–	–
–	–	–	–	–
–	–	–	–	–
	0.15 [121]	0.12 [113]	0.12 [r,113]	–
		0.12 [133]		
	0.180 [132]	0.20 [113]	0.189 [q,177]	–
		0.14 [133]	0.19 [r,113]	
	0.24 [79]	0.29 [133]	–	–
	−0.04 [79]	–	–	–
	1.3 [79]	–	–	–
	0.8 [124]	–	–	–
	0.5 [79]	–	–	–
−0.19 [q,177]	–	–	0.19 [q,177]	–
−0.14 [q,177]	–	–	0.12 [q,177]	–
−0.086 [q,177]	–	0.11 [113]	0.108 [q,177]	–
−0.055 [r,113]				
	0.093 [122]	–	0.069 [q,177]	–
–	–	–	–	–
–	–	–	–	–
+0.29 [r,113]	–	0.73 [113]	–	–
	1.23 [t]	1.0_2 [113]	1.02 [r,113]	–

h Three-wave mixing (TWM) hyperpolarisability defined by equation (5.86), $\lambda=694.3$ nm, $\Delta=4150\ cm^{-1}$ (from H_2 gas). All values are normalised with respect to γ^K for Ar [121].

i ESHG hyperpolarisability defined by equation (5.76), $\lambda=694.3$ nm.

j THG hyperpolarisability defined by equation (5.72), $\lambda=694.3$ nm. All values are from Ref. 106, normalised with respect to a theoretical value for He [165].

k $\alpha(CH_2F_2)$ estimated approximately as the mean of $\alpha(CH_4)$ and $\alpha(CF_4)$ [122].

l Value of 10^{40} $(\alpha_{\parallel}^{0}-\alpha_{\perp}^{0})$ obtained by molecular beam method[93].

m Value of $10^{40}(\alpha_{\parallel}^{0}-\alpha_{\perp}^{0})$ obtained by microwave spectroscopy[88].

n Compare with value of $10^{40}(\alpha_{\parallel}^{0}-\alpha_{\perp}^{0})$ for [^{2}H]ethyne of $2.0_6\ C^2\ m^2\ J^{-1}$, from microwave spectroscopy[89].

o Calculated from dielectric data[68], using $\mu=3.47$ C m [69].

p Approximate estimate from measurements in solution[3].

transforms as the electric dipole moment μ, a vector additivity scheme, similar to that for μ, might be expected. For the isotropic parameter γ a scalar additivity scheme analogous to that for a should apply. Very few experimental results are as yet available to test the validity of such schemes. For the series of fluorinated methanes CH_3F, CH_2F_2, CHF_3, vector additivity predicts that both μ and β are in the ratios 1 : 1.15 : 1. Experiments indicate that the corresponding ratios are 1 : 1.06 : 0.89, 1 : 0.22 : −1.42 and 1 : 0.74 : 0.44 for μ [70], β^{K} [122], and β^{SHG} [177], respectively. These results do not encourage confidence that reliable additivity schemes for β can be developed. The prospects for γ seem brighter: the series of fluorinated methanes conform[177] to a scalar additivity scheme for $\gamma_{\parallel}^{ESHG}$ to within 15%; however, such a scheme for γ^{K} applied to various hydrocarbons[79,124] yields values of γ^{K} for a C—C bond which vary erratically with molecular environment.

5.6.5 Interacting atoms

The failure of the Silberstein model in predicting components of the pair polarisability tensor $\boldsymbol{\alpha}_{12}(r)$ for the rare gas atoms (Section 5.5.1) has prompted several more rigorous calculations[144–147] for He atom pairs. In contrast to the Silberstein model, all of these calculations predict a decrease in $a_{12}(r)$ at short range and two[146,147] yield a negative value of the second dielectric virial coefficient B_{ε}, as observed experimentally[59].

5.7 CONCLUSIONS AND COMPARISON OF DATA

In concluding this review, we present a tabular summary (Table 5.4) of polarisabilities and hyperpolarisabilities for a selection of atoms and molecules. The data have been experimentally determined in the gas phase and have been critically assessed before inclusion in the table. The table and its footnotes need little explanation but some general remarks are appropriate. It is evident that values of a^0, a and $(a_{\parallel} - a_{\perp})$ are known for almost all the molecules listed; they conform to trends expected on the basis of molecular structure and of dispersion and vibrational contributions to $\boldsymbol{\alpha}$. The compilation of β and γ values is much less complete, however, and a number of apparent discrepancies are in evidence. The various values of γ for the rare gases are internally consistent, showing the general increase in moving from γ^{K} to γ^{THG} expected from simple dispersion considerations. The results for Ne are irregular in this regard; this might possibly be related to anomalies found[173] in average excitation energies for Ne. A number of severe inconsistencies exist amongst the various hyperpolarisabilities of several molecules, notably CO_2, CS_2, CH_2F_2 and CHF_3. The apparent discrepancies in the small body of β hyperpolarisability data are particularly serious and await satisfactory explanation; it is perhaps significant that in both methods of determination (Kerr effect and ESHG), β is obtained only after subtracting contributions from $\boldsymbol{\alpha}$ and γ. The Kerr effect should offer the highest absolute accuracy in determining β and γ, in view of its reliance on measurements of phase retardation, rather than scattered intensity, and of the relative ease of extrapolating Kerr effect data to zero density. However, the newer techniques (TWM, ESHG, THG) should offer comparable accuracy once normalisation against an appropriate reference standard (for example, He or Ar) has been performed.

Looking to the future, there is clearly a need for further careful work on the determination of β and γ hyperpolarisabilities, both in extending the range of molecules studied and in accounting for the apparent discrepancies which currently exist. In this context, we express the hope that various authors will adopt a universal system of notation and definition for non-linear optical polarisabilities, or at least clearly state their chosen conventions. There is also a need to build up a body of accurate data on polarisability derivatives from absolute Raman intensity measurements; such data are relevant to Raman-based methods of detecting atmospheric pollutants. A field awaiting fuller exploitation by molecular spectroscopists is that of hyper-Raman scattering, which with its novel selection rules should provide a useful supplement to i.r. and Raman spectra. Such developments should also produce more extensive hyper-Rayleigh scattering data, from which further components of the hyperpolarisability $\boldsymbol{\beta}$ should be obtainable once intermolecular correlations have been dealt with. A further experiment, which we have not mentioned, is that of d.c. electric field-induced Rayleigh scattering; a theoretical treatment[174] predicts that for tetrahedral molecules $\beta_{xyz}(-\omega; 0, \omega)$ should be obtainable, but experiments on small molecules to date[175] have proved unsuccessful. In this review we have adopted the attitude that, with few exceptions, theoretical estimates of polarisabilities and hyperpolarisabilities are of more use in testing theoretical procedures than in precisely predicting molecular observables. The current high rate of production of computed polarisabilities and, to an increasing degree, hyperpolarisabilities seems likely to continue unabated in the future.

Finally, we remind the reader of a number of important areas which might have been included in a more extensive review. Our restriction of attention to atoms and molecules in media of low density is intended to promote a simple but thorough understanding of polarisability-based physical processes. However, this effectively by-passes the vast body of experiment and (frequently very complicated) theory on phenomena involving polarisabilities in dense media. Furthermore, if interactions with electric field gradients and magnetic fields had been considered, a host of other polarisability-based phenomena would have emerged[1], including optical rotatory power and dispersion, magnetic susceptibility, the Faraday effect, the Cotton–Mouton effect, quadrupolar birefringence, and other processes involving electric quadrupole and magnetic dipole moments. Perhaps our most regrettable omission has

---

q Normalised with respect to a theoretical value of $\gamma_{\parallel}^{\mathrm{ESHG}}$ for He [165].

r Normalised with respect to γ^{K} for Ar [121].

s Extraction of γ^{K} from Kerr effect data for H_2 and D_2 includes allowance for quantisation of molecular rotation[123].

t Bogaard, M. P., Buckingham, A. D. and Ritchie, G. L. D., unpublished results.

u Raman shift $\Delta = 2900\ \mathrm{cm}^{-1}$ (from CH_4 gas) for TWM in H_2.

v Bogaard, M. P., Buckingham, A. D., Pierens, R. K. and White, A. H., unpublished results.

w If the new theoretical value[170] of γ^{THG} for He is accepted, values of γ^{THG} and γ^{ESHG} should be increased by 1.2%.

been a detailed coverage of resonant optical processes involving polarisabilities. The field which in its early days gave rise to the now anachronistic term 'anomalous dispersion' has regained prominence in this present era of tunable lasers. Many fascinating experiments in this area have been reported within the past five years, including those bearing on the intrinsic differences between resonance fluorescence and resonant Raman scattering[87,176], resonant enhancement of non-linear optical processes[107,109] and development of sources of coherent tunable radiation throughout the electromagnetic spectrum[132,134]. The next five years promise to be even more productive.

Acknowledgements

We are much indebted to Professor A. D. Buckingham, for his support and encouragement, and to Professor J. F. Ward and Dr. R. S. Watts, for a number of useful communications.

References

1. Buckingham, A. D. (1967). *Advances in Chemical Physics*, Vol. 12, 107 (J. O. Hirschfelder, editor) (New York: Interscience)
2. Buckingham, A. D. (1970). *Physical Chemistry: An Advanced Treatise*, Vol. 4, 349 (H. Eyring, D. Henderson and W. Jost, editors) (New York: Academic Press)
3. Buckingham, A. D. and Orr, B. J. (1967). *Quart. Rev.*, **21,** 195
4. Le Fèvre, R. J. W. (1965). *Advances in Physical Organic Chemistry*, Vol. 3, 1 (V. Gold, editor) (New York: Academic Press)
5. Franken, P. A. and Ward, J. F. (1963). *Rev. Mod. Phys.*, **35,** 23
6. Bloembergen, N. (1965). *Nonlinear Optics* (New York: Benjamin)
7. Minck, R. W., Terhune, R. W. and Wang, C. C. (1966). *Appl. Opt.*, **5,** 1595
8. Kielich, S. (1972). *Dielectric and Related Molecular Processes*, Vol. 1, 192 (M. Davies, editor) (London: Chemical Society)
9. Eyring, H., Walter, J. and Kimball, G. E. (1944). *Quantum Chemistry* (New York: Wiley)
10. Born, M. and Huang, K. (1954). *Dynamical Theory of Crystal Lattices*, Chapter 4 (Oxford: Clarendon Press)
11. Bates, D. R. (1961). *Quantum Theory. I. Elements*, 251 (D. R. Bates, editor) (New York: Academic Press)
12. Davydov, A. S. (1965). *Quantum Mechanics*, Chapter 9 (Oxford: Pergamon Press)
13. Dalgarno, A. (1966). *Perturbation Theory and its Applications to Quantum Mechanics*, 145 (C. H. Wilcox, editor) (New York: Wiley)
14. Langhoff, P. W., Epstein, S. T. and Karplus, M. (1972). *Rev. Mod. Phys.*, **44,** 602
15. Göppert-Mayer, M. (1931). *Ann. Phys.*, **9,** 273
16. Barron, L. D. and Gray, C. G. (1973). *J. Phys. A*, **6,** 59
17. Orr, B. J. and Ward, J. F. (1971). *Molec. Phys.*, **20,** 513
18. Armstrong, J. A., Bloembergen, N., Ducuing, J. and Pershan, P. S. (1962). *Phys. Rev.*, **127,** 1918
19. Ward, J. F. (1965). *Rev. Mod. Phys.*, **37,** 1
20. Dalgarno, A. (1961). *Quantum Theory. I. Elements*, 171 (D. R. Bates, editor) (New York: Academic Press)
21. Buckingham, A. D. and Pople, J. A. (1955). *Proc. Phys. Soc. A*, **68,** 905
22. Cyvin, S. J., Rauch, J. E. and Decius, J. C. (1965). *J. Chem. Phys.*, **43,** 4083
23. Ward, J. F. (1969). Personal communication
24. Pershan, P. S. (1963). *Phys. Rev.*, **130,** 919
25. Kleinman, D. A. (1962). *Phys. Rev.*, **126,** 1977
26. Ward, J. F. and New, G. H. C. (1969). *Phys. Rev.*, **185,** 57

27. Ward, J. F. and Franken, P. A. (1964). *Phys. Rev.*, **133,** A183
28. Terhune, R. W. and Maker, P. D. (1968). *Lasers — A Series of Advances*, Vol. 2, 295 (A. K. Levine, editor) (New York: Marcel Dekker)
29. Pershan, P. S. (1966). *Progress in Optics*, Vol. 5, 85 (New York: Interscience)
30. Long, D. A. and Stanton, L. (1970). *Proc. Roy. Soc. A*, **318,** 441
31. Long, D. A. (1971). *Essays in Structural Chemistry*, 18 (A. J. Downs, D. A. Long and L. A. K. Staveley, editors) (London: Macmillan)
32. Bersohn, R., Pao, Y.-H. and Frisch, H. L. (1966). *J. Chem. Phys.*, **45,** 3184
33. Jahn, H. A. (1949). *Acta Crystallogr.*, **2,** 30
34 Lorentz, H. A. (1880). *Ann. Phys.*, **9,** 641
35. Lorenz, L. (1880). *Ann. Phys.*, **11,** 70
36. Rosenfeld, L. (1951). *Theory of Electrons* (New York: Interscience)
37. Mazur, P. (1958). *Advances in Chemical Physics*, Vol. 1, 309 (I. Prigogine, editor) (New York: Interscience)
38. Kauzmann, W. (1957). *Quantum Chemistry* (New York: Academic Press)
39. Ashton, H. M. and Halberstadt, E. S. (1958). *Proc. Roy. Soc. A*, **245,** 373
40. Blythe, H. R., Lambert, J. D., Petter, P. J. and Spoel, H. (1960). *Proc. Roy. Soc. A*, **255,** 427
41. Landolt-Bornstein (1962). *Zahlenwerte und Funktionen*, Band II, Teil 8 (Berlin: Springer)
42. Jaffe, J. H. (1961). *Advances in Spectroscopy*, Vol. 2 (H. W. Thompson, editor) (New York: Interscience)
43. Whiffen, D. H. (1958). *Trans. Faraday Soc.*, **54,** 327
44. Illinger, K. H. (1961). *J. Chem. Phys.*, **35,** 409
45. Overend, J. (1963). *Infra-Red Spectroscopy and Molecular Structure*, 345 (M. Davies, editor) (Amsterdam: Elsevier)
46. Haigh, J. and Sutton, L. E. (1972). *MTP International Review of Science, Physical Chemistry Series One, Molecular Structure and Properties*, Vol. 2, 1 (G. Allen, editor) (London: Butterworths)
47. Landau, L. D. and Lifshitz, E. M. (1960). *Electrodynamics of Continuous Media*, 259 (Oxford: Pergamon)
48. Buckingham, A. D. (1972). *MTP International Review of Science, Physical Chemistry Series One, Molecular Structure and Properties*, Vol. 2, 241 (G. Allen, editor) (London: Butterworths)
49. Vogel, A. I. *et al.*, a series of 52 papers entitled *Physical properties and chemical constitution*, published in *J. Chem. Soc.* between 1934 and 1966. The last paper in the series is: (1966). *J. Chem. Soc. B*, 1080
50. Van Vleck, J. H. (1932). *The Theory of Electric and Magnetic Susceptibilities* (New York: McGraw-Hill)
51. Debye, P. (1929). *Polar Molecules* (New York: Chemical Catalogue Co.)
52. Le Fèvre, R. J. W. (1953). *Dipole Moments* (London: Methuen)
53. Smith, J. W. (1955). *Electric Dipole Moments* (London: Butterworths)
54. Smyth, C. P. (1955). *Dielectric Behaviour and Structure* (New York: McGraw-Hill)
55. Buckingham, A. D. (1972). *MTP International Review of Science, Physical Chemistry Series One, Spectroscopy*, Vol. 3, 73 (D. A. Ramsay, editor) (London: Butterworths)
56. Hill, N. E., Vaughan, W. E., Price, A. H. and Davies, M. (1969). *Dielectric Properties and Molecular Behaviour*, Chapter 2 (London: Van Nostrand-Reinhold)
57. Sutter, H. and Cole, R. H. (1970). *J. Chem. Phys.*, **52,** 132
58. Barnes, A. N. M., Turner, D. J. and Sutton, L. E. (1971). *Trans. Faraday Soc.*, **67,** 2902
59. Orcutt, R. H. and Cole, R. H. (1967). *J. Chem. Phys.*, **46,** 697
60. Dalgarno, A. and Kingston, A. E. (1960). *Proc. Roy. Soc. A*, **259,** 424
61. Newell, A. C. and Baird, R. C. (1965). *J. Appl. Phys.*, **36,** 3751
62. Knaap, H. F. P., Hermans, L. J. F., Jonkman, R. M. and Beenakker, J. J. M. (1965). *Disc. Faraday Soc.*, **40,** 135
63. Landolt-Börnstein (1951). *Zahlenwerte und Funktionen*, Band I, Teil 3; (1959). *ibid.*, Band II, Teil 6 (Berlin: Springer)
64. Kirouac, S. and Bose, T. K. (1973). *J. Chem. Phys.*, **59,** 3043
65. Nelson, R. D. and Cole, R. H. (1971). *J. Chem. Phys.*, **54,** 4033
66. Bose, T. K., Sochanski, J. S. and Cole, R. H. (1972). *J. Chem. Phys.*, **57,** 3592

67. Bose, T. K. and Cole, R. H. (1971). *J. Chem. Phys.*, **54,** 3829
68. Buckingham, A. D. and Raab, R. E. (1961). *J. Chem. Soc.*, 5511
69. Rheinhart, P. B. (1970). *J. Chem. Phys.*, **53,** 1418
70. Nelson, R. D., Lide, D. R. and Maryott, A. A. (1967). *Selected Values of Dipole Moments for Molecules in the Gas Phase*, NSRDS-NBS10 (Washington: U.S. Dept. of Commerce); Gordy, W. and Cook, R. L. (1970). *Microwave Molecular Spectra*, Chapter 10 (New York: Wiley)
71. Lord Rayleigh (1899). *Phil. Mag.*, **47,** 375
72. Strutt, R. J. (1918). *Proc. Roy. Soc. A*, **94,** 453
73. Bridge, N. J. and Buckingham, A. D. (1966). *Proc. Roy. Soc. A*, **295,** 334
74. Andrews, A. L. and Buckingham, A. D. (1960). *Molec. Phys.*, **3,** 183
75. Placzek, G. (1934). *Handbuch der Radiologie*, Vol. 6, Part 2, 205 (E. Marx, editor) (Leipzig: Akademische Verlag-gesellschaft); English translation: Werbin, A. (1959). UCRL-Trans.-526(L) (Washington: U.S. Dept. of Commerce)
76. Rowell, R. L., Aval, G. M. and Barrett, J. J. (1971). *J. Chem. Phys.*, **54,** 1960
77. Geindre, J.-P., Gauthier, J.-C., and Delpech, J.-F. (1973). *Phys. Lett.*, **44A,** 149
78. Victor, G. A. and Dalgarno, A. (1969). *J. Chem. Phys.*, **50,** 2535
79. Buckingham, A. D., Bogaard, M. P., Dunmur, D. A., Hobbs, C. P. and Orr, B. J. (1970). *Trans. Faraday Soc.*, **66,** 1548
80. Stoicheff, B. P. (1959). *Advances in Spectroscopy*, Vol. I, 91 (H. W. Thompson, editor) (New York: Interscience)
81. Tobin, M. C. (1971). *Laser Raman Spectroscopy* (New York: Interscience)
82. Herzberg, G. (1945). *Infrared and Raman Spectra of Polyatomic Molecules* (Princeton: Van Nostrand)
83. Golden, D. M. and Crawford, B. (1962). *J. Chem. Phys.*, **36,** 1654; Yoshino, T. and Bernstein, H. J. (1958). *J. Mol. Spectrosc.*, **2,** 213
84. Fenner, W. R., Hyatt, H. A., Kellam, J. M. and Porto, S. P. S. (1973). *J. Opt. Soc. Amer.*, **63,** 73
85. Barnett, G. P. and Albrecht, A. C. (1970). *Raman Spectroscopy*, Vol. 2, 207 (H. A. Szymanski, editor) (New York: Plenum)
86. Behringer, J. (1967). *Raman Spectroscopy*, Vol. 1, 168 (H. A. Szymanski, editor) (New York: Plenum)
87. Holzer, W., Murphy, W. F. and Bernstein, H. J. (1970). *J. Chem. Phys.*, **52,** 399
88. Scharpen, L. H., Muenter, J. S. and Laurie, V. W. (1970). *J. Chem. Phys.*, **53,** 2513
89. Muenter, J. S. and Laurie, V. W. (1964). *J. Amer. Chem. Soc.*, **86,** 3901
90. Khadjavi, A. and Lurio, A. (1968). *Phys. Rev.*, **167,** 128
91. Schmieder, R. W., Lurio, A. and Happer, W. (1971). *Phys. Rev. A*, **3,** 1209
92. Muenter, J. S. (1974). *International Review of Science, Physical Chemistry Series Two, Molecular Structure and Properties*, Chap. 2 (A. D. Buckingham, editor) (London: Butterworths)
93. MacAdam, K. B. and Ramsay, N. F. (1972). *Phys. Rev. A*, **6,** 898
94. Franken, P. A., Hill, A. E., Peters, C. W. and Weinreich, G. (1961). *Phys. Rev. Lett.*, **7,** 118
95. Kielich, S. (1970). *Opto-electronics*, **2,** 125
96. Giordmaine, J. A. (1969). *Physics Today*, **22,** No. 9, 39
97. Li, Y.-Y. (1964). *Acta Phys. Sinica*, **20,** 164
98. Kielich, S. (1968). *Acta Phys. Polon.*, **33,** 89; and cited references
99. Maker, P. D. (1970). *Phys. Rev. A*, **1,** 923
100. Weinberg, D. L. (1967). *J. Chem. Phys.*, **47,** 1307
101. Long, D. A. (1971). *Chem. Brit.*, **7,** 108
102. Terhune, R. W., Maker, P. D. and Savage, C. M. (1965). *Phys. Rev. Lett.*, **14,** 681
103. Stanton, L. (1972). *Molec. Phys.*, **23,** 601
104. Maker, P. D. (1966). *Physics of Quantum Electronics*, 60 (P. L. Kelley, B. Lax and P. E. Tannenwald, editors) (New York: McGraw-Hill); Verdieck, J. F., Peterson, S. H., Savage, C. M. and Maker, P. D. (1970). *Chem. Phys. Lett.*, **7,** 219; Savage, C. M. and Maker, P. D. (1971). *Appl. Opt.*, **10,** 965
105. Maker, P. D. and Terhune, R. W. (1965). *Phys. Rev.*, **137,** A801
106. Ward, J. F. and New, G. H. C. (1969). *Phys. Rev.*, **185,** 57
107. Leung, K. M., Ward, J. F. and Orr, B. J. (1974). *Phys. Rev. A*, **9,** 2440; Leung, K. M. (1972). *Ph.D. dissertation*, University of Michigan

108. Harris, S. E. and Miles, R. B. (1971). *Appl. Phys. Lett.*, **19,** 385
109. Miles, R. B. and Harris, S. E. (1973). *IEEE J. Quantum Electron.*, **9,** 470
110. Young, J. F., Bjorklund, G. C., Kung, A. H., Miles, R. B. and Harris, S. E. (1971). *Phys. Rev. Lett.*, **27,** 1551
111. Kung, A. H., Young, J. F., Bjorklund, G. C. and Harris, S. E. (1972). *Phys. Rev. Lett.*, **29,** 985
112. Pockels, F. (1893). *Abhandl. Ges. Wiss. Göttingen*, **39,** 1
113. Hauchecorne, G., Kerhervé, F. and Mayer, G. (1971). *J. Phys. (Paris)*, **32,** 47; Kielich, S. (1969). *IEEE J. Quantum Electron.*, **QE5,** 562
114. Finn, R. S. and Ward, J. F. (1971). *Phys. Rev. Lett.*, **26,** 285
115. Finn, R. S. and Ward, J. F. (1974). *J. Chem. Phys.*, **60,** 454
116. Bigio, I. J. and Ward, J. F. (1974). *Phys. Rev. A*, **9,** 35
117. Kerr, J. (1875). *Phil. Mag.*, **50,** 337, 446
118. Le Fèvre, C. G. and Le Fèvre, R. J. W. (1972). *Techniques of Chemistry*, Vol. I, Part IIIC, 399 (A. Weissberger, editor) (New York: Wiley)
119. Boyle, L. L., Buckingham, A. D., Disch, R. L. and Dunmur, D. A. (1966). *J. Chem. Phys.*, **45,** 1318
120. Buckingham, A. D. and Pople, J. A. (1955). *Proc. Phys. Soc. A*, **68,** 905
121. Buckingham, A. D. and Dunmur, D. A. (1968). *Trans. Faraday Soc.*, **64,** 1776
122. Buckingham, A. D. and Orr, B. J. (1969). *Trans. Faraday Soc.*, **65,** 673
122a. Schaefer, D. W., Sears, R. E. J. and Waugh, J. S. (1970). *J. Chem. Phys.*, **53,** 2127
123. Buckingham, A. D. and Orr, B. J. (1968). *Proc. Roy. Soc. A*, **305,** 259
124. Bogaard, M. P., Buckingham, A. D. and Ritchie, G. L. D. (1970). *Molec. Phys.*, **18,** 575
125. Hellwarth, R. W. (1970). *J. Chem. Phys.*, **52,** 2128
126. Champion, J. V., Meeten, G. H. and Whittle, C. D. (1970). *Trans. Faraday Soc.*, **66,** 2671
127. Kaiser, W. and Garrett, C. G. B. (1961). *Phys. Rev. Lett.*, **7,** 229; Abella, I. D. (1962). *ibid.*, **9,** 453; Singh, S. and Bradley, L. T. (1964). *ibid.*, **12,** 612
128. Peticolas, W. L. (1967). *Ann. Rev. Phys. Chem.*, **18,** 233
129. Buckingham, A. D. (1956). *Proc. Phys. Soc. B*, **69,** 344
130. Mayer, G. and Gires, F. (1964). *Compt. Rend. Acad. Sci.*, **258,** 2039; Paillette, M. (1966). *ibid.*, **262,** 264; Paillette, M. (1969). *Ann. Phys.*, **4,** 671
131. Bloembergen, N. (1967). *Amer. J. Phys.*, **35,** 989
132. Sorokin, P. P., Wynne, J. J. and Lankard, J. R. (1973). *Appl. Phys. Lett.*, **22,** 342; Wynne, J. J. (1973). *Bull. Amer. Phys. Soc.*, **18,** 534
133. Rado, W. G. (1967). *Appl. Phys. Lett.*, **11,** 123
134. Harris, S. E. (1973). *Phys. Rev. Lett.*, **31,** 341
135. Kielich, S., Lalanne, J. R. and Martin, F. B. (1971). *Phys. Rev. Lett.*, **26,** 1295
136. Buckingham, A. D. and Pople, J. A. (1956). *Disc. Faraday Soc.*, **22,** 17
137. Everett, D. H. and Munn, R. J. (1963). *Trans. Faraday Soc.*, **59,** 2486
138. Beaume, R. and Coulon, R. (1967). *Compt. Rend. Acad. Sci. B*, **265,** 309
139. Sutter, H. (1972). *Dielectric and Related Molecular Processes*, Vol. 1, 65 (M. Davies, editor) (London: Chemical Society)
139a. Buckingham, A. D. and Graham, C. (1974). *Proc. Roy. Soc. A*, **336,** 275
140. Thibeau, M., Tabisz, G. C., Oksengorn, B. and Vodar, B. (1970). *J. Quant. Spectrosc. Radiat. Transfer*, **10,** 839
141. McTague, J. P. and Birnbaum, G. (1971). *Phys. Rev. A*, **3,** 1376
142. Volterra, V., Bucaro, J. A. and Litovitz, T. A. (1971). *Phys. Rev. Lett.*, **26, 55**
143. Buckingham, A. D., Martin, P. H. and Watts, R. S. (1973). *Chem. Phys. Lett.*, **21,** 186
144. Lim, T. K., Linder, B. and Kromhout, R. A. (1970). *J. Chem. Phys.*, **52,** 3831
145. Buckingham, A. D. and Watts, R. S. (1973). *Molec. Phys.*, **26,** 7
146. O'Brien, E. F., Gutschick, V. P., McKoy, V. and McTague, J. P. (1973). *Phys. Rev. A*, **8,** 690
147. Fortune, P. J., Certain, P. R. and Bruch, L. W. (1975). *Chem. Phys. Lett.*, **27,** 233
148. Buckingham, A. D. (1956). *Trans. Faraday Soc.*, **52,** 1035
149. Silberstein, L. (1917). *Phil. Mag.*, **33,** 92 and 251
150. Hermans, L. J. F., de Groot, J. J., Knaap, H. F. P. and Beenakker, J. J. M. (1965). *Physica*, **31,** 1567

151. Diller, D. E. (1968). *J. Chem. Phys.*, **49,** 3096
152. Kołos, W. and Wolniewicz, L. (1967). *J. Chem. Phys.*, **46,** 1426
153. Buckingham, A. D. (1965). *Disc. Faraday Soc.*, **40,** 171
154. Certain, P. R. and Bruch, L. W. (1972). *MTP International Review of Science, Physical Chemistry Series One, Theoretical Chemistry*, Vol. 1, 113 (W. Byers Brown, editor) (London: Butterworths)
155. Buckingham, A. D. (1965). *Disc. Faraday Soc.*, **40,** 232
156. Bogaard, M. P., Buckingham, A. D. and Orr, B. J. (1967). *Molec. Phys.*, **13,** 533
157. Bates, D. R. and Damgaard, A. (1949). *Phil. Trans. Roy. Soc. A*, **242,** 101
158. Dalgarno, A., Ford, A. L. and Brown, J. C., (1971). *Phys. Rev. Lett.*, **27,** 1033
159. Chung, K. T. and Hurst, R. P. (1966). *Phys. Rev.*, **152,** 35
160. Buckingham, A. D. and Hibbard, P. G. (1968). *Symp. Faraday Soc.*, **2,** 41
160a. Weinhold, F. (1972). *Proc. Roy. Soc. A*, **327,** 209
161. Thomas, M. A. and Humberston, J. W. (1972). *J. Phys. B*, **5,** L229
162. Kolker, H. J. and Karplus, M. J. (1963). *J. Chem. Phys.*, **39,** 2011
163. O'Hare, J. M. and Hurst, R. P. (1967). *J. Chem. Phys.*, **46,** 2356
164. Chan, Y. M. and Dalgarno, A. (1965). *Proc. Phys. Soc.*, **85,** 227
165. Sitz, P. and Yaris, R. (1968). *J. Chem. Phys.*, **49,** 3546
166. Stevens, R. M. and Lipscomb, W. N. (1964). *J. Chem. Phys.*, **40,** 2238; *ibid.*, **41,** 184, 3710
167. O'Hare, J. M. and Hurst, R. P. (1967). *J. Chem. Phys.*, **46,** 2536
168. Arrighini, G. P., Maestro, M. and Moccia, R. (1968). *Symp. Faraday Soc.*, **2,** 48
169. McLean, A. D. and Yoshimine, M. (1967). *J. Chem. Phys.*, **46,** 3682
170. Klingbeil, R. (1973). *Phys. Rev. A*, **7,** 48
171. Sitter, R. E. and Hurst, R. P. (1972). *Phys. Rev. A*, **5,** 5
172. Smith, R. P. and Mortensen, E. M. (1960). *J. Chem. Phys.*, **32,** 502, 508
173. Buckingham, A. D. and Jamieson, M. J. (1971). *Molec. Phys.*, **22,** 117
174. Andrews, A. L. and Buckingham, A. D. (1960). *Molec. Phys.*, **3,** 183
175. Lalanne, J. R. and Bothorel, P. (1966). *Compt. Rend. Acad. Sci.*, **263,** 693; Lalanne, J. R. (1967). *ibid.*, **265,** 1181
176. Fouche, D. G. and Chang, R. K. (1972). *Phys. Rev. Lett.*, **29,** 536; St. Peters, R. L., Silverstein, S. D., Lapp, M. and Penney, C. M. (1973). *ibid.*, **30,** 191; Williams, P. F., Rousseau, D. L. and Dworetsky, S. H. (1974). *ibid.*, **32,** 196
177. Bigio, I. J. and Ward, J. F. (1974). Unpublished results; Bigio, I. J. (1974). *Ph.D. dissertation*, University of Michigan
178. Teachout, R. R. and Pack, R. T. (1971). *Atomic Data*, **3,** 195
179. Selection rules without the assumption of Kleinman symmetry are given by Christie, J. H. and Lockwood, D. J. (1971). *J. Chem. Phys.*, **54,** 1141
180. Stanton, L. (1973). *J. Raman Spectrosc.*, **1,** 53

6
Bonding Features in Magnetochemical Models

M. GERLOCH
University of Cambridge

6.1 INTRODUCTION 195

6.2 A MODEL FOR DISTORTION 203
6.2.1 *Many-electron systems* 211
6.2.2 *Magnetic susceptibilities* 215
6.2.3 *Very low temperatures* 219

6.3 MAGNETIC EXCHANGE 221
6.3.1 *μ-Oxo-bis[pentamminechromium(III)]* 223
6.3.2 *Exchange in systems with orbital angular momentum* 227
6.3.3 *Dinuclear octahedral cobalt(II) complexes* 232

ACKNOWLEDGEMENTS 236

6.1 INTRODUCTION

When Pauling introduced magnetism into chemistry it was a time of great theoretical advance in bonding theory when structure and order were being made. His borrowings from physics were bold, and general, and were applied to idealised systems. Detail and accuracy then would have robbed inorganic chemists of a spectacularly successful tool used ever since. With the establishment, acceptance and sometimes disenchantment with Pauling's approach to bonding came the clarification, detailing and refinement of the subject he invented — magnetochemistry.

Pauling demonstrated how it may be possible to distinguish formal oxidation states, limited coordination numbers and coordination geometry and/or bond type in transition metal complexes[1]. His concept of directed, hybrid bonds led him to suppose a total quenching of orbital angular momentum and a consequent simplification of Van Vleck's relation for magnetic moment,

$$\mu_{\text{eff}}=\sqrt{L(L+1)+4S(S+1)} \tag{6.1}$$

A dependence of moment on spin alone led naturally to Pauling's famous equation:

$$\mu_{\text{so}}=\sqrt{n(n+2)} \tag{6.2}$$

from which it followed that magnetic moments were a direct measure of the number n of unpaired electrons in the system. Any two or more structures differing in their numbers of unpaired electrons could be distinguished by a simple Gouy powder susceptibility measurement. In conjunction with other techniques, these ideas are still in use today, some 40 years on. Of course, much is wrong with the spin-only formula. Provided we don't ask too many questions, especially of the 'wrong' sort, however, most forgotten factors tend to cancel and the model works.

An early major improvement in the theory of magnetic moments recognised that large deviations from the spin-only formula need not reflect greater ionic bonding but that not all orbital angular momentum need be quenched even in the presence of strong covalent bonding. A clear distinction between octahedral and tetrahedral nickel(II) compounds, for example, may then be made, a differentiation not possible simply from the unpaired-electron count. This sort of refinement arises directly from symmetry considerations and accordingly has wide applicability. However, the usual problem of group theory furnishing only an 'on–off' result means that distinctions based on formal orbital degeneracies are frequently blurred. Considerable overlap of classes often occurs. Spin–orbit coupling can raise crystal-field degeneracies and lower moments or, indeed, admix degenerate terms into formal orbitally non-degenerate ground terms. An attempt to sharpen classifications based on formal orbital angular momenta produced measurements of magnetic moments over a temperature range. This was exploited most successfully by Nyholm and his school[2].

Having gone to the trouble of measuring magnetic susceptibilities through the liquid-nitrogen temperature range 80–300 K, however, the magnetochemist was naturally disappointed to find that, while his overall goal of the determination of gross structure or bonding type may have been achieved, detailed agreement between theory and experiment was usually lacking. Two refinements were introduced to help rectify this situation at this time. One, distortion[3], acknowledged that idealised cubic or planar geometries are uncommon and the other, orbital reduction factors[4], made the first explicit recognition within crystal-field magnetochemistry that bonds are covalent. Figgis made the first systematic study of susceptibilities using a model which included the effects of spin–orbit coupling, distortion and 'covalency' (orbital reduction) factors simultaneously[5]. He, together with Lewis and their students[6–8], exploited this model with the result that the experimental magnetic moments of some 60 diverse transition metal complexes over the liquid-nitrogen temperature range were almost all fitted virtually exactly by variation of the model parameters. This was an exciting achievement in magnetochemistry for, not only did it represent the first time general quantitative interpretation of magnetic moments was made, but also the parameter values required to fit experiment promised a rich haul to match Pauling's early qualitative treatment of nearly 30 years earlier. Now, it seemed, magnetochemistry could quantify distortions from idealised geometries and, perhaps more exciting, measure covalency. Indeed, Orgel referred[9] to the orbital reduction factor *k* as a 'covalency factor'. Several reviews written at that time displayed an enthusiastic and confident appreciation of the subject's promise.

The essence of the models used by Figgis may be illustrated by his calculations for ions possessing the formal orbital triplet ground term $^2T_{2g}$, e.g. octahedral titanium(III) or tetrahedral copper(II) complexes[5]. The primary cubic crystal-field splits the ^2D free-ion term leaving a thermally-occupied $^2T_{2g}$ term with an unoccupied ^2E_g term several thousands of wavenumbers higher in energy and whose existence is neglected — seemingly for fair reasons. The effects of spin–orbit coupling and a small axial distortion from cubic symmetry are taken as equally important peturbations of the ground term. Diagonalisation of the six $^2T_{2g}$ basis functions under the effective peturbation

$$\mathcal{H}' = V_{\text{axial}} + \lambda \boldsymbol{L}\cdot\boldsymbol{S} \tag{6.3}$$

yields three Kramers' doublets as in Figure 6.1.

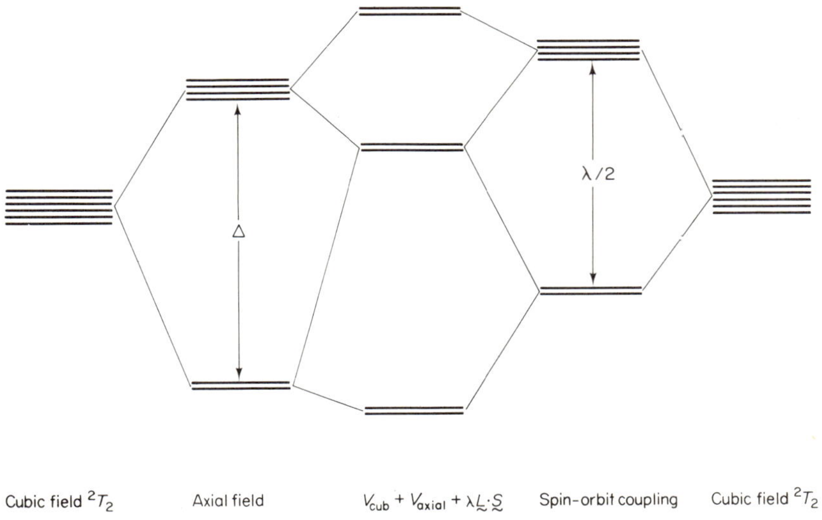

Figure 6.1 Simultaneous perturbation of $^2T_{2g}$ term by an axial ligand-field and spin–orbit coupling.

Calculated magnetic susceptibilities follow from a Boltzmann population of these three doublets whose relative energies depend upon magnitudes of distortion (Δ) and spin–orbit coupling (λ). For comparison with earlier work by Kotani, Figgis chose to show average magnetic moments $\bar{\mu}_{\text{eff}}$ as functions of kT/λ and $v=\Delta/\lambda$. Figure 6.2a shows some examples. Susceptibilities were calculated for directions parallel and perpendicular to the axis of distortion, but only their averages, $\bar{\chi}=\frac{1}{3}(\chi_{\|}+2\chi_{\perp})$, were published at that time owing to a general lack of anisotropy data. The magnetic properties were computed using the operator

$$\mu_\alpha=(kL_\alpha+2S_\alpha)\mu_B \tag{6.4}$$

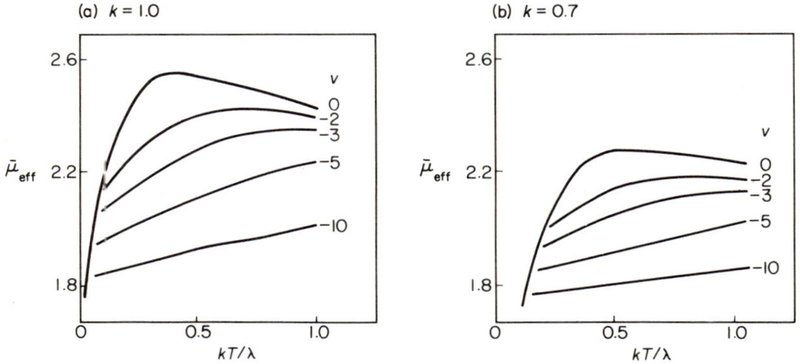

Figure 6.2 Average magnetic moments of ions with a $^2T_{2g}$ ground term as functions of axial distortion and spin–orbit coupling.

for each direction a, where k is Stevens' orbital-reduction factor[4]. At that time, k was regarded as expressing the reduced orbital angular momentum to be expected for a molecular orbital. Thus,

$$k=\frac{\langle\psi_a|L_z|\psi_b\rangle}{\langle d_a|L_z|d_b\rangle} \tag{6.5}$$

for example, where d_a is a pure metal orbital transforming like the molecular orbital ψ_a which in reality contains d_a admixed. Figure 6.2b illustrates Figgis' calculation for a k value less than unity, a situation supposedly representative of more bond covalency than that shown in Figure 6.2a.

The model thus furnished families of curves showing average magnetic moments as functions of the three parameters (in addition to temperature, of course) Δ, λ, k. The calculations initially required a small computer program, but in use this is not necessary since values of $\bar{\mu}_{eff}$ as functions of Δ, k, λ (or v, λ, k an equivalent set) have been published. Use of the model is thus quite straightforward. The idea is that the deviations of k and λ from 1.0 or λ_0 (the free-ion value), respectively, should correlate with covalency while the magnitude and sign of Δ should indicate the size and sense of an axial distortion from ideal cubic geometry.

The simplicity and empiricism of the distortion 'operator' V_{axial} has both convenience and limitations. An orbital triplet term splits into an orbital singlet and orbital doublet in an axial field whether quantised parallel to a three- or four-fold axis of the cube. The functions may be identified immediately by symmetry and, if used as bases, the matrix of V_{axial} is diagonal with elements proportional to the axial splitting Δ. For example, in tetragonal symmetry, we have

$$\langle d_{xz}|V_{\text{axial}}|d_{xz}\rangle = \langle d_{yz}|V_{\text{axial}}|d_{yz}\rangle = \Delta/3$$

and $$\langle d_{xy}|V_{\text{axial}}|d_{xy}\rangle = -2\Delta/3 \tag{6.6}$$

as the only non-zero elements. The facility with which the result is achieved, however, is offset by the complete lack of correlation this model can have with the quantitative nature of the distortion or even its sign. We cannot know, for example, whether a 20% axial bond lengthening, say, in ML_6 gives rise to a greater distortion than chemical substitution to ML_4X_2. An empirical approach is all that remains here. Further, such empiricism becomes useless if an even lower symmetry is appropriate, for then an orbital triplet term like $^2T_{2g}$ is split into three orbital singlets and we would require to parameterise this by two quantities, Δ_1 and Δ_2. This is always the problem, of course, if we parametrise matrix elements, for although these correspond to the system's observables, we generally deny ourselves a theoretical model to relate different properties. However, if any rhombic distortion in a molecule is small as considered from chemical criteria, we should obtain a reasonable result from the earlier model. But is this so? It is worthwhile outlining one example[10] where lower symmetry is crucial for magnetism as it illustrates an important quality of susceptibility. The molecule *trans*-dimesitylbis(diethylphenylphosphine)cobalt(II) is a 4-coordinate low-spin d^7 complex (Figure 6.3). The metal and ligand donor atoms lie in a plane while further

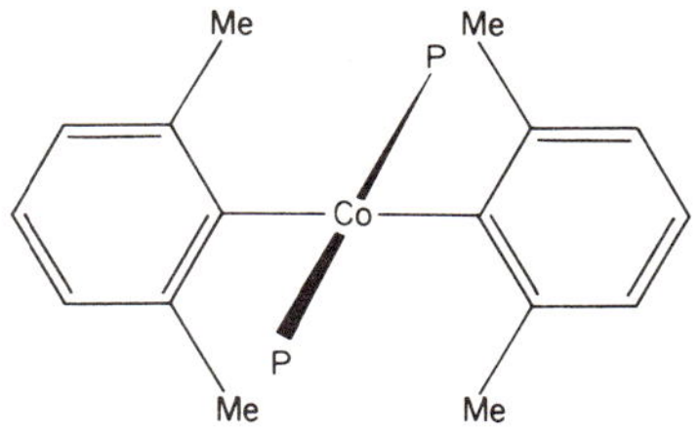

Figure 6.3 Coordination in $Co(mesityl)_2(PEt_2Ph)_2$.

coordination to the possible fifth and sixth sites of an octahedron is prevented by the blocking action of the α-methyl groups of the mesityls[11]. Chemically speaking, it seems a reasonable first approximation to describe the coordination as square planar, even though phosphines and mesityl groups have different ligand field effects. Experimentally, the molecular g tensor is rhombic, although two g values are similar and very different from the third: $g_1=1.96$, $g_2=1.74$, $g_3=3.72$. The dominant magnetic direction (g_3) lies parallel to the P—Co—P axis, however; g_1 is parallel to mesityl–mesityl and g_2 to the initially-presumed approximate fourfold axis. One might suppose the g-values indicate 'field strengths' which are approximately equal for mesityls and nothing (bonding to the methyl groups is not envisaged) and much less than for phosphine. The energy level diagram eventually deduced[10] from these and other data, however, is shown in Figure 6.4 and the points which we note here are as follows:

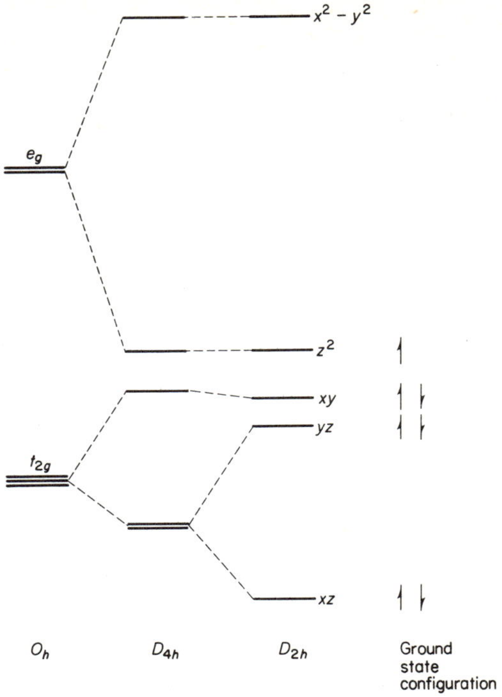

Figure 6.4 Proposed orbital energy diagram for $Co(mesityl)_2(PEt_2Ph)_2$.

(a) Chemical commonsense is vindicated in that the predominant distortion from octahedral is tetragonal. Superimposed on this is a smaller rhombic field expressing the different nature of mesityl and phosphine.

(b) The magnetic susceptibilities and *g*-values are dominated not by the whole splitting diagram but by the ground state $(xz)^2(yz)^2(xy)^2(z^2)^1$ and one particularly low-lying excited state $(xz)^2(yz)^1(xy)^2(z^2)^2$. The details are not important for the present discussion, but this result emphasises a point, well known to those who were first concerned with the subject but one tending to be lost in a concentrated search for chemical or bonding information, namely, magnetic moments and particularly magnetic anisotropies are properties mainly of populated states, giving information about excited states only in second order. More vividly, though less accurately, the *g*-values in $Co(mesityl)_2(phosphine)_2$ relate to the electrons in two orbitals (z^2 and yz) giving little information on the remaining ones which dominate a description of the absorption spectrum and reflect the overall chemical geometry.

However, excited states, several thousands of wavenumbers distant from ground, need not be insignificant. The above example concerned a case dominated by a low-lying ground state. In the absence of such dominance, higher states can be important. Consider the approximately tetrahedral ion $CuCl_4^{2-}$. Figgis' model sought to interpret the magnetic susceptibilities of such an ion within the framework of the $^2T_{2g}$ ground term as basis. Fitting the average moments, by using curves like those in Figure 6.2, gave parameter values $k \approx 1.0$, $\lambda \approx -800$ cm^{-1} and $\Delta \approx +4000$–5000 cm^{-1}. The sign of the Δ value indicated an orbital singlet ground term. The polarised single-crystal electronic spectrum[12] of this complex ion has been assigned as shown in Figure 6.5, an interpretation in accord with the results from the magnetism

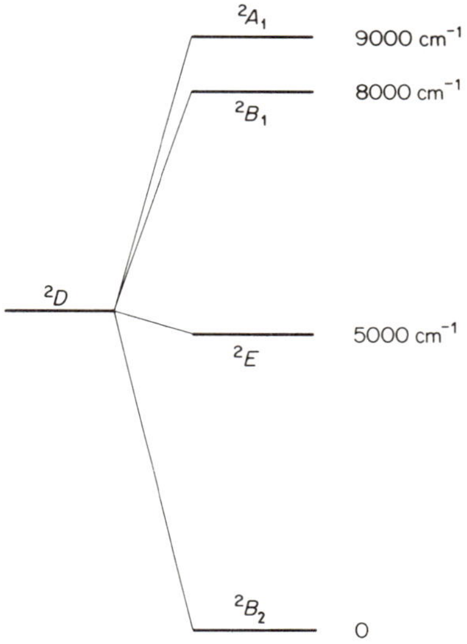

Figure 6.5 Term energy diagram for $CuCl_4^{2-}$ in Cs_2CuCl_4.

and, incidentally, in agreement with the flattened nature of the tetrahedron. However, the parameter values which defined the magnetic fit also show the molecular susceptibility parallel to the distortion axis as greater than that perpendicular, by some 15%. Experiment gives the opposite result[13], by a similar magnitude. When calculations are made within the enlarged basis of the ^2D term, however, contributions arising mostly from the excited $^2B_1$ and $^2A_1$ terms, several thousand wavenumbers above ground and admixed by spin–orbit coupling, alter the interpretation considerably[14]. The net result is to bring theory into agreement with all experimental results — geometry, spectrum, average moments and anisotropies — somewhat different k and λ values now being appropriate. The study demonstrated the much greater sensitivity of magnetic anisotropies to the theoretical model and parameter values than of average moments or, put the other way around, showed how average susceptibility measurements can be too easily fitted by inadequate models. As usual in the theory of magnetism, Van Vleck had previously made the same general point, though in a different way. In his classic book[15], he demonstrates that mean susceptibilities are only slightly affected by crystal-field splittings which are small compared with the thermal energy. On the other hand, of course, thermal distributions of molecules amongst a manifold of closely spaced levels are the essence of anisotropies. A typical behaviour of $\Delta\mu$ with respect to an axial distortion is shown in Figure 6.6 which illustrates how anisotropies may be largest for slight distortion. The octahedron $Fe(CN)_6^{3-}$ is distorted only slightly, by about 150 cm^{-1}, and yet displays an anisotropy approximately equal to its average susceptibility[16]. The factors (distortion or inequivalence of ligands, for example) which give rise to small ground-state splittings may well be small or even negligible so far as 'average' electronic absorption spectral studies are concerned: certainly so, for chemical or geometric classification of molecules. Nevertheless, the consequences of these small effects for magnetic moments and particularly anisotropies can be considerable. It is essentially for these reasons that simple, 'rule-of-thumb' approaches to magnetism can be misleading.

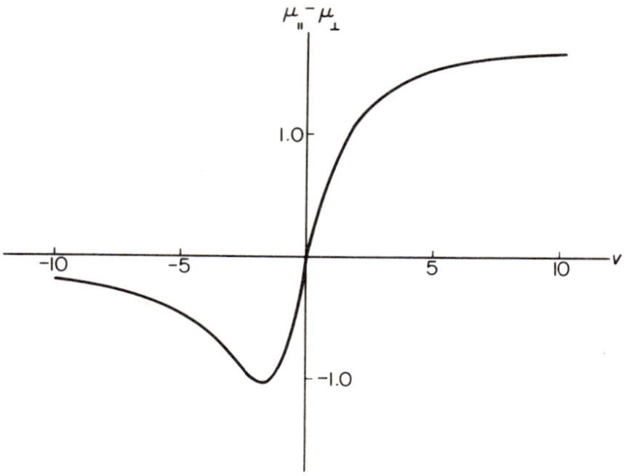

Figure 6.6 Anisotropy ($\Delta\mu=\mu_{\parallel}-\mu_{\perp}$) of ions with $^2T_{2g}$ ground terms as a function of axial distortion.

Consider another example[17] of the dangers of using a too restricted basis set of functions. Ferrous Tutton salt, $(NH_4)_2SO_4 \cdot FeSO_4 \cdot 12H_2O$, and ferrous fluorosilicate, $FeSiF_6 \cdot 6H_2O$, both contain approximately octahedral $Fe(H_2O)_6^{2+}$ ions. The average moments of these compounds at 300 K are ca. 5.2 and ca. 5.5 Bohr magnetons, respectively. It seems obvious that the reason for this difference must lie in the detailed geometries of the ion. Yet, fitting the moments to an axial distortion/spin–orbit model, using the ground cubic term $^5T_{2g}$ as basis, yielded orbital reduction factors k of ca. 1.0 and ca. 0.7, respectively. With interpretations of k then current, but since revised[18], such k values meant a considerable (30%?) increase in covalency in the iron–water bonds of the crystal containing fluorosilicate anions, compared with the one containing ammonia and sulphate ions! In due course the explanation put forward to remove this problem involved a model spanning the whole ^5D term as basis, but with excited state admixing taking place as much via the crystal-field distortion as by spin–orbit coupling. The *symmetry* of the distortion is an important ingredient here, being trigonal for the fluorosilicate and tetragonal (or, at least, assumed so) for the Tutton salt. Details are unimportant for the present discussion, except perhaps one. Figure 6.7 shows how even average moments at room temperature, in trigonally distorted octahedral iron(II) complexes, can be extremely sensitive to the distortion angle (α in Figure 6.7 is the angle subtended by any Fe–ligand bond and the threefold axis).

These and several similar studies led Nyholm to publish a review of magnetochemistry in 1967 which was, to say the least, tinged with caution and pessimism[19]. Several years earlier, Nyholm had been largely responsible for reminding inorganic chemists of Pauling's magnetic criterion, of extending and sharpening these ideas and of firmly establishing magnetic measurements as an almost automatic procedure in a coordination chemist's routine. The models that Figgis developed gave the further promise of quantitative

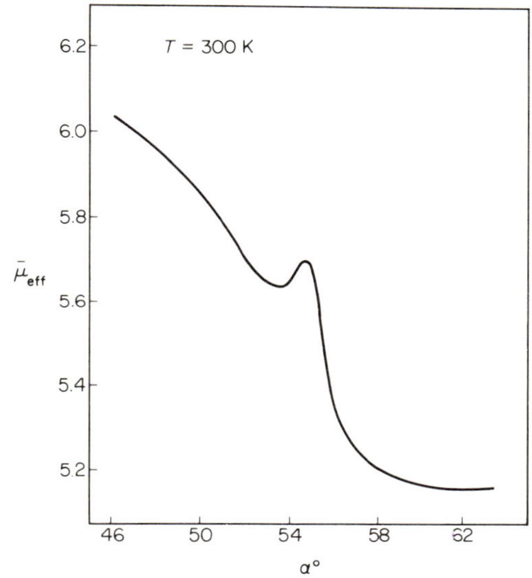

Figure 6.7 Average moments for trigonally-distorted, octahedral iron(II) ions.

indications from magnetism about bonding in chemistry. And yet, when that 'one experiment too many' — anisotropy — was performed or when mean susceptibilities were pursued in great detail, it became apparent that all was far from well. Although a great deal was being learnt of the reasons for various 'discrepancies' and of how models might be better established, it was undeniable that a direct utility for magnetism, beyond that of Pauling, was questionable. If the problem is 'how many unpaired electrons?' or 'is there some orbital contribution or not?', an answer is easy to find. If we wish to distinguish a 5-coordinate complex from a distorted octahedron, or a covalent bond from a nearly ionic one, the chances are slim. After so much work by so many people, this seemed a great pity, but inescapable. A principal function of this article is to rekindle the old hopes and aims with the outline of a model which is generally applicable, which relates to chemical bonding better than any earlier models and which appears to work.

6.2 A MODEL FOR DISTORTION

A calculation of magnetic properties of transition metal complexes requires a knowledge of eigenvalues and eigenvectors of ground and several excited states at a level of detail currently impossible to achieve by *ab initio* Hartree–Fock or other methods. These limitations are even more acute in semi-empirical and approximate molecular orbital methods like CNDO calculations. The most successful and widely used methods to establish energy levels in d and f electron systems are essentially perturbation approaches. A particular difficulty with recent attempts to interpret the magnetic properties of distorted systems has been the definition of parameters and their relation to chemically and structurally *identifiable* features in molecules.

Perhaps the simplest approach is that used first by Bleaney[3] and since by Figgis[5], Swalen[20] and others[13]. As illustrated in the discussion of the magnetic properties of 'tetrahedral' copper(II) ions, it is characterised by the parameterisation of matrix elements, as in equation (6.6), the perturbation assumption being implicit in the description of the model wavefunctions as pure metal d orbitals whose angular parts are unaffected by bonding. The advantages of this approach are its great simplicity and the direct parameterisation of observable quantities. The disadvantages are the large number of parameters required in systems more complex than d^1 or d^9, and that the parameterisation of results (matrix elements) rather than causes (operators) robs the parameters of any clear interpretation in chemical terms and particularly of transferability or comparability from molecule to molecule.

Many of the problems arising from these deficiencies are resolved, at least in part, by parameterisation of the ligand-field operator[21]. Again within a perturbation approach, first- or higher-order eigenvalues but zeroth-order eigenvectors are derived from a pure d orbital set of metal functions operated upon by a one-electron operator V which expresses the 'ligand-field potential' in which the metal electrons find themselves. There are several variations, but the technique is exemplified by the crystal-field point-charge method. In the first place, assuming a one-electron potential, the operator V, expressing the influence of the ligand environment on the central metal ion, may be expanded as a superposition of spherical harmonics centred on the metal:

$$V = \sum_i^{\infty} R_i Y_i \tag{6.7}$$

The series terminates rapidly in practice, as may be seen from symmetry and group theoretical arguments. If attention is limited to matrix elements $\langle\psi|V|\psi'\rangle$ for d electrons, terms in V will only be operative if of even parity (k even) and no term can be of order greater than four, as follows from the vector triangle rule. Thus, for a system with no molecular symmetry, the most extensive expansion of V in equation (6.7) is:

$$\begin{aligned} V = {} & R_0 a Y_0{}^0 + R_2(b Y_2{}^1 + c Y_2{}^0 + d Y_2{}^{-1}) \\ & + R_4(e Y_4{}^4 + f Y_4{}^3 + g Y_4{}^2 + h Y_4{}^1 + p Y_4{}^0 + q Y_4{}^{-1} + r Y_4{}^{-2} + s Y_4{}^{-3} \\ & + t Y_4{}^{-4}) \end{aligned} \tag{6.8}$$

The maximum number of expansion coefficients — parameters operative for splitting of any one d^n configuration — is thus 12. In molecules of high symmetry this number is greatly reduced. The octahedral potential, for example, is

$$V_{\text{oct}} \propto Y_4^0 + \sqrt{\frac{5}{14}}(Y_4^4 + Y_4^{-4}) \tag{6.9}$$

in tetragonal quantisation, leaving only one parameter as an overall scaling factor. It must be emphasised that this expression for V_{oct} follows only from the assumption of a one-electron operator and is otherwise rigorously correct for any model of bonding. There is no restriction to a pure crystal-field assumption of the sort first made by Bethe.

The single scaling factor in octahedral symmetry emerges on computation of matrix elements, *viz.*

$$\langle\psi|\,V\,|\psi'\rangle = \sum_i \langle R(d)\,Y_l^m|\,R_k(i)\,Y_k^q(i)\,|R(d)\,Y_l^{m'}\rangle \tag{6.10}$$

and in octahedral symmetry $k=4$ only, leaving R_k as a common radial factor in the expansion of V. Factorisation of the radial and angular parts gives for each term in (6.10):

$$\langle R(d)|\,R_k\,|R(d)\rangle\langle Y_l^m|\,Y_k^q\,|Y_l^{m'}\rangle$$

The angular parts are evaluated completely, following on the assumption that the angular forms of the metal functions are known. The radial parts of the metal orbitals and of the potential are not assumed and the radial integral $\langle R(d)|\,R_k\,|R(d)\rangle$ is parameterised as proportional to Dq or Δ_{oct}.

Molecules with lower symmetry than cubic require more parameters. An octahedron elongated parallel to a fourfold axis, for example with symmetry D_{4h}, defines the potential

$$V_{D_{4h}} = a\,Y_2^0 + b\,Y_4^0 + c(Y_4^4 + Y_4^{-4}) \tag{6.11}$$

Here matrix elements [equation (6.10)] devolve into two types, involving second- and fourth-order radial integrals associated with harmonics Y_2 and Y_4. There are several ways of tackling this sort of situation. The most direct, commonly used in various physics schools, is to regard a, b and c in equation (6.11) as parameters. Symmetry leaves these three unknowns in the potential expansion so this process is the most simple and correct procedure to adopt. However, in our view, a process in which (say) spectral properties are fitted to a potential parameterised in this way gains nothing over a simple statement of the spectral frequencies. The coefficients are not related to geometry and chemistry in any discernible way and the approach is purely phenomenological. The same criticism is partly true of the most common form of parameterisation in D_{4h} symmetry in which we write

$$V_{D_{4h}} = a\,Y_2^0 + d\,Y_4^0 + V_{\text{oct}} \tag{6.12}$$

and expresses matrix elements in terms of Ds, associated with the second-order term, and Dt and Dq with the fourth-order terms. Certainly it is true that a three-parameter model (Ds, Dt, Dq) properly expresses the three degrees of freedom left in this symmetry, but there is little else to be said in its favour except, perhaps, for the convenience of certain spectral splittings being functions of only one parameter at a time, at least in first order.

The point-charge model offers more chance of establishing a significant relationship between structure and bonding on the one hand and ligand-field theory on the other. Representing ligands as point charges (or point dipoles) endows the model with more known structural features than mere recourse to symmetry. The multipole expansion of $1/r_{ij}$, the inverse distance between metal electron and ligand,

$$\frac{1}{r_{ij}}=\sum_{k}^{\infty}\sum_{q}^{k}\frac{4\pi}{2k+1}\cdot\frac{r_{<}^{\ k}}{r_{>}^{\ k+1}}\cdot Y_k^q(i)\cdot Y_k^{q*}(j) \tag{6.13}$$

specifies the radial part of the operator in equation (6.7) or of the matrix elements in equations (6.9) and (6.10) and leads to an approximate† definition of Dq as

$$Dq=\frac{1}{6}ze^2\langle R(d)|\frac{r^4}{a^5}|R(d)\rangle \tag{6.14}$$

where a is the bond length and ze an effective nuclear charge. An analogous definition for a second-order radial parameter Cp is

$$Cp=\frac{2}{7}ze^2\langle R(d)|\frac{r^2}{a^3}|R(d)\rangle \tag{6.15}$$

the numbers 1/6 and 2/7 appearing only for reasons of historical and computational convenience.

The detailed coordination of the molecule is now implicit in the potential expansion coefficients [a, b and c of equation (6.11)]. There follow from the definitions in equations (6.14) and (6.15) the relationships

$$Ds=Cp(\text{basal})-Cp(\text{axial});\ Dt=\frac{4}{7}[Dq(\text{basal})-Dq(\text{axial})] \tag{6.16}$$

from which we see that Ds and Dt represent differences between radial parameters for axial and equatorial ligands in second- and fourth-orders, respectively. An immediate penalty of this procedure, however, is the use of four, rather than three, symmetry-required parameters. The hope is that this may be offset by closer relationships between Cp and Dq and bonding concepts, together with comparisons between parameters from different systems being chemically reasonable and informative. The utility of the parameters Cp and Dq in D_{4h} and other symmetries has been discussed in detail recently and is not repeated here[21]. The present short review of the point-charge model is intended to present the background against which our more recent model may be viewed.

There are two main problems associated with the point-charge approach. The first concerns the detailed interpretation of Cp and Dq. While a simple point-charge calculation suggests Cp values should be several times larger than Dq, this may not be the case in practice and, anyway, a thorough examination of a purely electrostatic model, with the inclusion of Kleiner's correction for the interaction of metal electrons with the positively charged ligand nuclei, theoretically implies far smaller Dq values than are observed experimentally. This latter objection only concerns the ultimate significance and interpretation of parameter values, however, and does not vitiate their use in an empirical way. A more serious difficulty, associated with the practical

†See Ref. 21, p.133

question of over-parameterisation, concerns the factorisation of radial and angular properties of the model. Bearing in mind the case of trigonally distorted octahedral ions as in the introduction, the potential in (say) D_{3d} symmetry is given as

$$V_{D_{3d}} \propto a Y_2{}^0 + b Y_4{}^0 + c(Y_4{}^3 - Y_4{}^{-3}) \qquad (6.17)$$

Again, this may be parameterised in the matrix elements by $D\sigma$, $D\tau$ and Dq in a manner analogous to Ds, Dt and Dq for D_{4h} symmetry. We chose to use the parameters Cp and Dq, associated with second- and fourth-order radial parameters as before, which has the great advantage of offering comparability with similar quantities derived from molecules of other symmetry. The third parameter is the angle a in Figure 6.8, being the angle subtended by the three-fold axis and any bond. The angle a ($=\cos^{-1}\sqrt{\frac{1}{3}} \approx 54.74°$ in octahedral

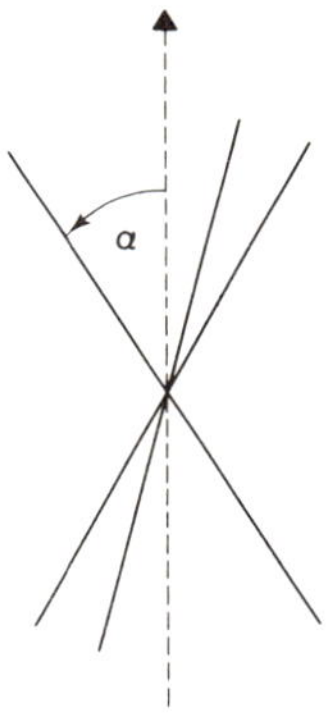

Figure 6.8

symmetry) is to be regarded as a parameter of the system rather than a quantity fixed by an x-ray structure analysis. We insist on this very strongly as the unreality of regarding a multiatom ligand as a point charge must be taken up not only in Cp and Dq as radial, scaling factors but also in the 'angle of maximum ligand influence' not necessarily being directed exactly towards the donor atom. A similar view is taken of the angle β in C_{4v} 5-coordinate systems, as in Figure 6.9.

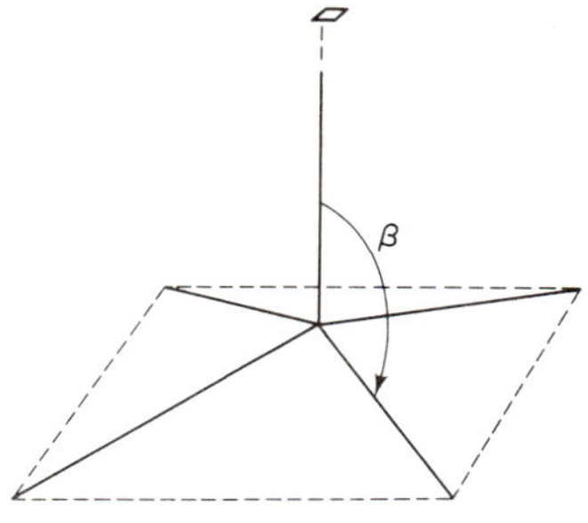

Figure 6.9

It has become abundantly clear from studies of trigonally distorted octahedral iron(II) [17] and cobalt(II) [22] d electron systems, similar f electron lanthanide[23] complexes and some C_{4v} 5-coordinate cobalt(II) and nickel(II) molecules[21], that anisotropies or mean susceptibilities or both are frequently extremely sensitive to the angle of distortion, as expressed say by α or β in Figures 6.8 and 6.9. In some cases the point-charge model has indicated that a $\frac{1}{4}°$ change can affect the observed magnetism measurably. If this is so, and we have reservations as will be discussed, the use of doped crystals in e.s.r. studies to throw light on a concentrated system could be a serious mistake. It is frequently necessary to study very dilute solid solutions in order to prevent e.s.r. signals being washed out or averaged by exchange broadening and other effects. It is very unlikely that the geometry of the paramagnetic ions in a dilute system agrees to within $\frac{1}{4}°$ of that in the pure substance and the implication of the point-charge results cited above is that such experiments may only yield information about the dilute system. But there are reservations, as we said.

There is additionally a major overall practical limitation of the point-charge approach we have outlined. Almost all molecules studied to date have, or are assumed to have, axial symmetry. Few molecules fall into this class and, in the light of our comments about the extreme sensitivity to distortion, models which average in-plane ligand fields as 'perpendicular' in less-than-ideal circumstances may be badly at fault. Furthermore, most transition metal complexes of any chemical interest possess little or no symmetry in any approximation. Were we to adopt a point-charge approach similar to that above, we would have to parameterise all angles in the coordination sphere, resulting in a model as over-parameterised as that given by the potential in equation (6.8). It is important to recognise that the factorisation of radial (Cp and Dq) and angular parameters is only approximate. Interpretations of these parameters are blurred and this is the origin of our reservations about the earlier models' great sensitivity to angular distortions. The fact remains, however, that within a point-charge model, such angles must be varied for fit. It is clear that a different approach is normally essential for low-symmetry systems. Such an approach is the angular overlap model of Schäffer and Jørgensen[24].

The angular overlap model (AOM) has been reviewed in varying degrees of formality[21,24–27] so we confine ourselves here to an outline designed to discuss its use in the present context of magnetic susceptibilities. The model may be seen as a first-order perturbation approach to covalent bonding. With the central assumption of *weak* covalency, the perturbation energies of the central metal d (or f) orbitals may be taken as proportional to the squares of relevant metal–ligand overlap integrals. The AOM is thus based on covalency, in contrast to the crystal-field model resting on electrostatic interactions. However, in common with crystal field theory, the AOM assumes that the effects of the several ligands bonded to the central metal are additive, an assumption which is equivalent to the neglect of ligand–ligand overlap. This assumption is not essential to either model but is usually made with both.

The power of AOM follows from the classification of bonding interactions between any ligand and the metal with respect to the metal–donor atom axis. Consider a σ bonding interaction as in Figure 6.10a in which the orbitals are mutually directed for maximum overlap, or in 6.10b involving identical orbitals and bond lengths but where there is an angular displacement. If we write the overlap integral in Figure 6.10a as S^*_{ML}, then that in Figure 6.10b may be written as $S^*_{ML} \cdot A$ where A is a function of θ and dependent *only* on the *angular* form of the *metal* function. Clearly there is symmetry in the argument, but by choosing to classify orbitals on the ligand as σ, π, δ, etc. with respect to the M—L vector, the variation of overlap integral with orientation of the

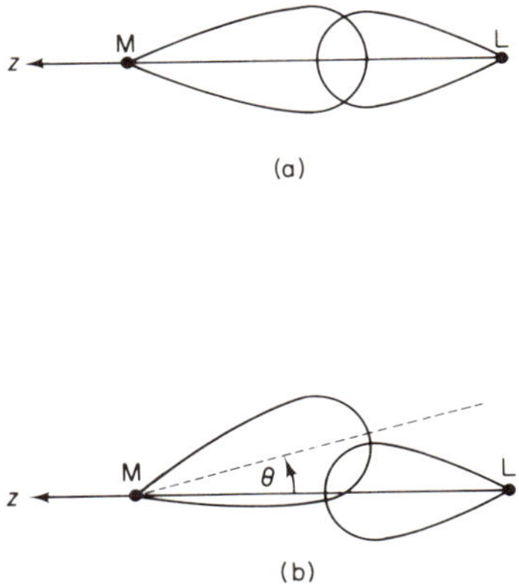

Figure 6.10

metal orbital is expressed solely in terms of metal functions. Thus, in common with the point-charge model, for example, the angular forms of the d wavefunction need be assumed but not those of the ligand orbitals. For example, consider only σ-bonding in an ML_6 complex of any symmetry but in which all bond lengths are equal. Interactions between metal and ligand orbitals are then written as proportional to angular factors like A, the overlap integral S^*_{ML} — common to all bonds — being sequestered into an overall scaling factor. The relative energies of the antibonding metal orbitals are now determined by the angular factors, the *angular overlap integrals*, whose magnitudes and signs depend solely upon the angular properties of the metal functions and their relative orientations with respect to the M—L bonds. Parallel remarks apply to the π-bonding but with a different scaling factor now being appropriate. Formally we write the antibonding energy E^* of a given d orbital as:

$$E^* = k(S^*_{ML} \cdot F^l_\lambda)^2 = e_\lambda F^{l2}_\lambda \tag{6.18}$$

Here k is a proportionality constant dependent on factors including the relative valence orbital ionisation energies of metal and ligand. S^*_{ML} is the maximal overlap integral for antibonding interaction between metal and ligand functions of a given type and given bond length. F^l_λ is the angular overlap integral previously called A. The superscript labels the l quantum number of the metal functions while the subscript classifies the bonding symmetry with respect to the metal–ligand axis. For example, σ-bonding in transition metal complexes requires F^d_σ; π-bonding requires F^d_π.

More generally, recognition of inequivalent π-bonding effects in two directions perpendicular to the M—L axis involves the more comprehensive angular factors D_λ^l, where λ now represents σ-, π_x- and π_y-bonding, for example. The D_λ^l form a matrix expressing the rotation of a set of d wavefunctions, say, into the ligand axis frame, there being one D matrix per ligand. Consider ligand and metal axis frames as in Figure 6.11. The D_λ^d matrix is most readily, and theoretically conveniently, constructed in terms of the Eulerian rotations

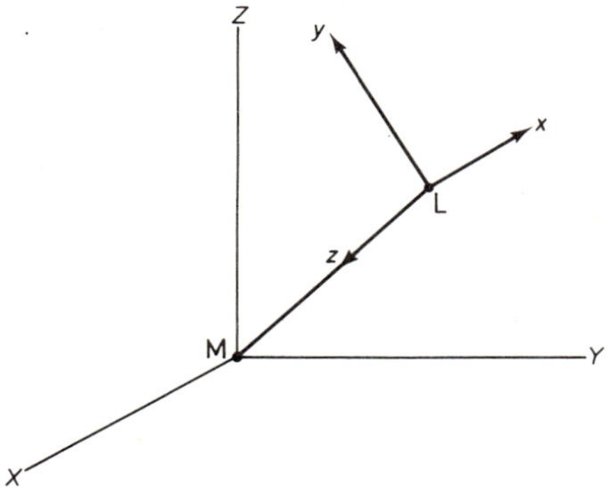

Figure 6.11 Ligand (x, y, z), and metal (X, Y, Z) coordinate frames.

required to bring the two frames into coincidence. In practice it is often more convenient to express the mutual orientation of these frames in terms of direction cosines as in matrix (6.19):

$$\begin{array}{c|ccc} & x & y & z \\ \hline X & a_1 & \beta_1 & \gamma_1 \\ Y & a_2 & \beta_2 & \gamma_2 \\ Z & a_3 & \beta_3 & \gamma_3 \end{array} \tag{6.19}$$

$$\begin{bmatrix} z^2 \\ xz \\ yz \\ xy \\ x^2-y^2 \end{bmatrix} = \begin{bmatrix} (3\gamma^2{}_3-1)/2 & \sqrt{3}\alpha_3\gamma_3 & \sqrt{3}\beta_3\gamma_3 & \sqrt{3}\alpha_3\beta_3 & \sqrt{3}(\alpha^2{}_3-\beta^2{}_3)/2 \\ \sqrt{3}\gamma_1\gamma_3 & \alpha_1\gamma_3+\alpha_3\gamma_1 & \beta_1\gamma_3+\beta_3\gamma_1 & \alpha_1\beta_3+\alpha_3\beta_1 & \alpha_1\alpha_3-\beta_1\beta_3 \\ \sqrt{3}\gamma_2\gamma_3 & \alpha_2\gamma_3+\alpha_3\gamma_2 & \beta_2\gamma_3+\beta_3\gamma_2 & \alpha_2\beta_3+\alpha_3\beta_2 & \alpha_2\alpha_3-\beta_2\beta_3 \\ \sqrt{3}\gamma_1\gamma_2 & \alpha_1\gamma_2+\alpha_2\gamma_1 & \beta_1\gamma_2+\beta_2\gamma_1 & \alpha_1\beta_2+\alpha_2\beta_1 & \alpha_1\alpha_2-\beta_1\beta_2 \\ \sqrt{3}(\gamma_1^2-\gamma_2^2)/2 & \alpha_1\gamma_1-\alpha_2\gamma_2 & \beta_1\gamma_1-\beta_2\gamma_2 & \alpha_1\beta_1-\alpha_2\beta_2 & (\alpha_1^2-\alpha_2^2+\beta_2^2-\beta_1^2)/2 \end{bmatrix} \begin{bmatrix} z'^2 \\ x'z' \\ y'z' \\ x'y' \\ x'^2- \end{bmatrix}$$
(6.

from which the D_λ^d matrix (6.20) may be derived. The relations (6.20) may be interpreted as follows, For the ligand orientation (6.19), σ-bonding involves $[(3\gamma_3^2—1)/2]S^*_{\mathrm{ML}}$ with d_{z^2}, $(\sqrt{3}\gamma_1\gamma_3)S^*_{\mathrm{ML}}$ with d_{xz}, $(\sqrt{3}\gamma_2\gamma_3)S^*_{\mathrm{ML}}$ with d_{yz}, $(\sqrt{3}\gamma_1\gamma_2)S^*_{\mathrm{ML}}$ with d_{xy}, and $[\sqrt{3}(\gamma_1^2-\gamma_2^2)/2]S^*_{\mathrm{ML}}$ with $d_{x^2-y^2}$.

The general matrix element of covalent interaction with ligands between real d orbitals is given by (6.21):

$$\langle u|V|v\rangle = \sum_{j=1}^{\text{ligands}} \sum_t D^d_{ut}(j) D^d_{vt}(j) e_t(j) \tag{6.21}$$

where u and v are d orbitals transforming as u and v (e.g. d_{z^2}, d_{xy}, etc.) and t is the bonding mode (e.g. σ, π_x, π_y, etc.). Taking stock: the covalent interaction between metal and ligand orbitals is separated into δ, π_x, π_y and δ types (the latter usually ignored) and, for each, antibonding matrix elements (6.21) are computed by summing independent contributions from all ligands. Each symmetry-independent ligand has scaling factors e_λ, associated with each type of bonding. In the simple case of O_h symmetry the difference between the antibonding energies of d_{z^2} and d_{xy}, namely $10Dq$, emerges in AOM as $3e_\sigma - 4e_\pi$. Here the equivalence of all ligands means there is only one e_σ and one e_π parameter. An immediate objection to the AOM is apparent: it requires two parameters in O_h where symmetry only needs one, Dq. However, in lower symmetries the degree of parameterisation of AOM and crystal-field models can be similar. For example, a trigonally distorted octahedral D_{3d} molecule ML_6 requires Dq, Cp, θ in the crystal-field framework but only e_σ and e_π in AOM; a tetragonal-pyramidal 5-coordinate C_{4v} system has Dq, Dt, Ds or Cp(axial), Dq(axial), Cp(basal), Dq(basal), and β in crystal-field approaches but e_σ(axial), e_π(axial), e_σ(basal) and e_π(basal) in AOM. Angles need not be parameterised in AOM. The precise orientations of ligands are built into the calculation (via the D^l_λ matrices) and features like asymmetrical π-bonding, which necessitate the angular parameterisation in crystal-field theory, are dealt with explicitly in the e_λ parameters. Magnetic anisotropy in a hexa-aquo near-octahedral complex must be treated in the point-charge model by allowing variation of small angular distortions and hence by artificially sequestering the problems of non-axially symmetrical water ligands into the angle a (Figure 6.8). In the AOM approach, the problem may be treated by including π-bonding effects perpendicular to the water planes but not parallel to them (if the waters are 'trigonally coordinated'), for example. There are many cases in which artificially high symmetry has been supposed in crystal-field models, *faute de mieux*, which can be handled in AOM without undue increase in the number of parameters.

In short, AOM is an ideal model for treating low- or zero-symmetry situations. Above all, the most important appeal of AOM is the way the parameters e_σ and e_π, in referring explicitly to σ- and π-bonding modes, directly relate to chemists' visceral attachment to well-tried and understood features of chemical bonding. One can have little feel for magnitudes and signs of Ds and Dt, for example, while a large π-bonding contribution (as witnessed by large e_π values) for ammonia ligands is immediately seen as unacceptable. Conversely, the degree of parameterisation in a particular situation can be reduced by neglect of those parameters *obviously* associated with 'disallowed' bonding modes.

6.2.1 Many-electron systems

We turn now to some of the practical aspects of computation using the angular overlap model. Matrix elements of the *total* potential V are expressed in a basis of *real* orbitals: this basis is not essential but is natural and convenient. Many-electron systems are most naturally and simply solved in the crystal-field formalism within a complex basis of free ion functions: again for convenience rather than necessity. Further, if the spin–orbit perturbation is to be included, as it must, the natural basis for quantisation is $|J, M_J\rangle$. Harnung and Schäffer[28] have described a tensor operator approach to the

AOM using real orbitals as bases. It appears to gain little over the well-established Racah treatment and is apparently not directly suited to the spin–orbit problem; it is, however, the first comprehensive treatment of the many-electron problem using bases natural to the AOM. We describe our calculational techniques within the more usual Racah formalism[29–31].

Consider first the one-electron problem. The AOM gives matrix elements in the real orbital basis for the total potential as in (6.21) while the crystal-field model gives matrix elements in $|l, m_l\rangle$ quantisation for any one part of the potential as in equation (6.23):

$$V=\sum_j R_j c_j Y_J \tag{6.22}$$

$$\langle l, m_l | Y^q | l, m_l' \rangle = (-1)^{m_l} \sqrt{\frac{2k+1}{4\pi}} (2l+1) \begin{pmatrix} l & k & l \\ -m_l & q & m_l' \end{pmatrix} \begin{pmatrix} l & k & l \\ 0 & 0 & 0 \end{pmatrix} \tag{6.23}$$

The superposition for the total potential (6.22) is then:

$$\langle l, m_l | V | l, m_l' \rangle = \sum_j R_j c_j (-1)^{m_l} {}_J \sqrt{\frac{2k_j+1}{4\pi}} (2l+1) \begin{pmatrix} l & k & l \\ -m_l & q & m_l' \end{pmatrix}_j \begin{pmatrix} l & k & l \\ 0 & 0 & 0 \end{pmatrix}_j \tag{6.24}$$

It is perhaps in these equations that the clearest comparison between AOM and crystal-field theory is to be found. In AOM, the total matrix element is a sum over ligands and bonding modes: in crystal-field theory the sum is over spherical harmonics in what is, after all, an arbitrary expansion (6.22). It cannot be expected, therefore, to see a parallel between the e_λ parameters of AOM and the Rc parameters of crystal-field theory: this is also shown in the way the degree of parameterisation — the number of degrees of freedom — in the two models need not coincide.

Crystal-field theory conventionally follows (6.23) in many-electron systems using the relationship[23] (6.25) from tensor operator theory.

$$\langle lLSJM_J | Y_k^q | lL'SJ'M_J' \rangle = (-1)^{J+J'+L+l+M_J-k-S} \sqrt{\frac{2k+1}{4\pi}} \sqrt{(2J+1)(2J'+1)}$$
$$\times \begin{pmatrix} J & k & J' \\ -M_J & q & M_J' \end{pmatrix} \begin{Bmatrix} L & L' & k \\ J' & J & S \end{Bmatrix} (2l+1) \begin{pmatrix} l & k & l \\ 0 & 0 & 0 \end{pmatrix} \langle lLS \| U^k \| lL'S \rangle \tag{6.25}$$

$|J,M_J\rangle$ quantisation is used, as stated above, for convenience in writing down matrix elements of spin–orbit coupling. In our programme we use (6.25); it is well established and very powerful. The scaling factors $R_j c_j$ to be used with (6.25) are the same as in (6.24) and (6.22), of course, for they represent the potential not the basis. Therefore we may use (6.21) and (6.24) to establish a relationship between the e_λ and the $R_j c_j$ which is then used in the 'master equation' (6.25). In short, we use Racah's tensor operator theory to handle the many-electron problem and relate the symmetry-defined or phenomenological potential (6.22) to the angular overlap model at the one-electron level.

The first stage is to write $R_j c_j$ in terms of matrix elements in $|l, m_l\rangle$ quantisation. A trivial example is:

$$\langle 2 | V | -1 \rangle \propto \begin{pmatrix} 2 & 4 & 2 \\ -2 & 3 & -1 \end{pmatrix} \begin{pmatrix} 2 & 4 & 2 \\ 0 & 0 & 0 \end{pmatrix} R_4 c_{43} \tag{6.26}$$

as in this case the matrix element is non-zero only for the Y_4^3 term in the potential. In other cases simultaneous equations in two or three unknowns need to be solved finally to establish the results in Table 6.1. These results are quite general, of course, and express relationships valid in any symmetry; the table may be completed using the relationship

$$c_{kq}=(-1)^q c^*_{k-q} \tag{6.27}$$

a reminder that the matrix elements are generally complex.

Table 6.1 General relationships between spherical harmonic multipliers and complex matrix elements

$$R_0c_{00}=\tfrac{2}{5}\sqrt{\pi}[2\langle 2|V|2\rangle+2\langle 1|V|1\rangle+\langle 0|V|0\rangle]$$

$$R_2c_{20}=-\sqrt{\frac{4\pi}{5}}[2\langle 2|V|2\rangle-\langle 1|V|1\rangle-\langle 0|V|0\rangle]$$

$$R_2c_{21}=\sqrt{\frac{4\pi}{5}}[\sqrt{6}\langle 2|V|1\rangle+\langle 1|V|0\rangle]$$

$$R_2c_{22}=-\sqrt{\frac{4\pi}{5}}[2\langle 2|V|0\rangle+\sqrt{\tfrac{3}{2}}\langle 1|V|-1\rangle]$$

$$R_4c_{40}=\tfrac{2}{5}\sqrt{\pi}[\langle 2|V|2\rangle-4\langle 1|V|1\rangle+3\langle 0|V|0\rangle]$$

$$R_4c_{41}=2\sqrt{\frac{2\pi}{5}}[\sqrt{3}\langle 1|V|0\rangle-\sqrt{\tfrac{1}{2}}\langle 2|V|1\rangle]$$

$$R_4c_{42}=2\sqrt{\frac{2\pi}{5}}[\sqrt{\tfrac{3}{2}}\langle 2|V|0\rangle-\langle 1|V|-1\rangle]$$

$$R_4c_{43}=-\sqrt{\frac{28\pi}{5}}\langle 2|V|-1\rangle$$

$$R_4c_{44}=\sqrt{\frac{14\pi}{5}}\langle 2|V|-2\rangle$$

The second stage is to express the relationship between matrix elements in the complex $|l, m_l\rangle$ basis and those in the real basis. The bases are related as in (6.28):

$$
\begin{aligned}
|z^2\rangle &= |Y_2^0\rangle \\
|xz\rangle &= \frac{1}{\sqrt{2}}(Y_2^{-1}-Y_2^1) \\
|yz\rangle &= \frac{i}{\sqrt{2}}(Y_2^{-1}+Y_2^1) \\
|xy\rangle &= \frac{i}{\sqrt{2}}(Y_2^{-2}-Y_2^2) \\
|x^2-y^2\rangle &= \frac{1}{\sqrt{2}}(Y_2^{-2}+Y_2^2)
\end{aligned}
\tag{6.28}
$$

Table 6.2 General relationships between spherical harmonic multipliers and real matrix elements

(a) Real parts of $R_j c_j$

$$R_0c_{00}=\tfrac{2}{5}\sqrt{\pi}[\langle x^2-y^2|V|x^2-y^2\rangle+\langle xy|V|xy\rangle+\langle xz|V|xz\rangle+\langle yz|V|yz\rangle+\langle z^2|V|z^2\rangle]$$

$$R_2c_{20}=-\sqrt{\frac{\pi}{5}}[2\langle x^2-y^2|V|x^2-y^2\rangle+2\langle xy|V|xy\rangle-\langle xz|V|xz\rangle-\langle yz|V|yz\rangle -2\langle z^2|V|z^2\rangle]$$

$$R_2c_{21}=-\sqrt{\frac{4\pi}{5}}\left[\frac{\sqrt{6}}{2}\langle yz|V|xy\rangle+\frac{\sqrt{6}}{2}\langle xz|V|x^2-y^2\rangle+\frac{1}{\sqrt{2}}\langle xz|V|z^2\rangle\right]$$

$$R_2c_{22}=-\sqrt{\frac{4\pi}{5}}[\sqrt{2}\langle x^2-y^2|V|z^2\rangle+\sqrt{\tfrac{3}{8}}\langle yz|V|yz\rangle-\sqrt{\tfrac{3}{8}}\langle xz|V|xz\rangle]$$

$$R_4c_{40}=\sqrt{\frac{\pi}{5}}[\langle x^2-y^2|V|x^2-y^2\rangle+\langle xy|V|xy\rangle-4\langle xz|V|xz\rangle-4\langle yz|V|yz\rangle +6\langle z^2|V|z^2\rangle]$$

$$R_4c_{41}=2\sqrt{\frac{2\pi}{5}}\left[-\sqrt{\frac{3}{2}}\langle xz|V|z^2\rangle+\frac{1}{2\sqrt{2}}\langle yz|V|xy\rangle-\frac{1}{2\sqrt{2}}\langle xz|V|x^2-y^2\rangle\right]$$

$$R_4c_{42}=2\sqrt{\frac{2\pi}{5}}\left[\frac{\sqrt{3}}{2}\langle x^2-y^2|V|z^2\rangle+\tfrac{1}{2}\langle xz|V|xz\rangle-\tfrac{1}{2}\langle yz|V|yz\rangle\right]$$

$$R_4c_{43}=\sqrt{\frac{7\pi}{5}}[\langle yz|V|xy\rangle-\langle xz|V|x^2-y^2\rangle]$$

$$R_4c_{44}=\sqrt{\frac{7\pi}{10}}[\langle x^2-y^2|V|x^2-y^2\rangle-\langle xy|V|xy\rangle]$$

(b) Imaginary parts of $R_j c_j$

$$R_0c_{00}=R_2c_{20}=R_4c_{40}=0$$

$$R_2c_{21}=\sqrt{\frac{4\pi}{5}}\left[-\sqrt{\frac{3}{2}}\langle yz|V|x^2-y^2\rangle+\sqrt{\frac{3}{2}}\langle xz|V|xy\rangle+\frac{1}{\sqrt{2}}\langle yz|V|z^2\rangle\right]$$

$$R_2c_{22}=-\sqrt{\frac{4\pi}{5}}\left[-\sqrt{2}\langle xy|V|z^2\rangle+\sqrt{\frac{3}{2}}\langle xz|V|yz\rangle\right]$$

$$R_4c_{41}=2\sqrt{\frac{2\pi}{5}}\left[\sqrt{\frac{3}{2}}\langle yz|V|z^2\rangle+\frac{1}{2\sqrt{2}}\langle yz|V|x^2-y^2\rangle-\frac{1}{2\sqrt{2}}\langle xz|V|xy\rangle\right]$$

$$R_4c_{42}=2\sqrt{\frac{2\pi}{5}}\left[\frac{-\sqrt{3}}{2}\langle xy|V|z^2\rangle-\langle xz|V|yz\rangle\right]$$

$$R_4c_{43}=\sqrt{\frac{7\pi}{5}}[\langle xz|V|xy\rangle+\langle yz|V|x^2-y^2\rangle]$$

$$R_4c_{44}=-\sqrt{\frac{14\pi}{5}}\langle x^2-y^2|V|xy\rangle$$

It is then straightforward to deduce transformations like (6.29):

$$\text{Real part } \langle 1|V|-1\rangle=-\tfrac{1}{2}[\langle xz|V|xz\rangle-\langle yz|V|yz\rangle]$$

$$\text{Imaginary part } \langle 1|V|-1\rangle=\langle xz|V|yz\rangle \qquad (6.29)$$

from which the coefficients $R_j c_j$ may be written in terms of matrix elements in the real basis to yield a second, generally valid Table 6.2. These tables are given in full for the sake of their general utility.

The final stage is to construct a matrix relating the $R_j c_j$ in Table 6.2 with the e_λ in (6.21). The result is now special, being relevant only to a particular geometry, number of symmetry-independent ligands, and bonding modes. Having once established the relationship between $R_j c_j$ and e_λ in our computer programme, the many-electron calculation then proceeds along conventional lines using equation (6.25).

To summarise, let us consider a particular molecule — $Co(py)_4(NCS)_2$. Our programme only requires crystal coordinates of sufficient atoms to define ligand axis sets — Figure 6.12 — in order to construct matrices (6.19) and (6.20) automatically. In addition, a suitable free-ion basis of terms and/or states is given; for example, in susceptibility calculations here, we limit

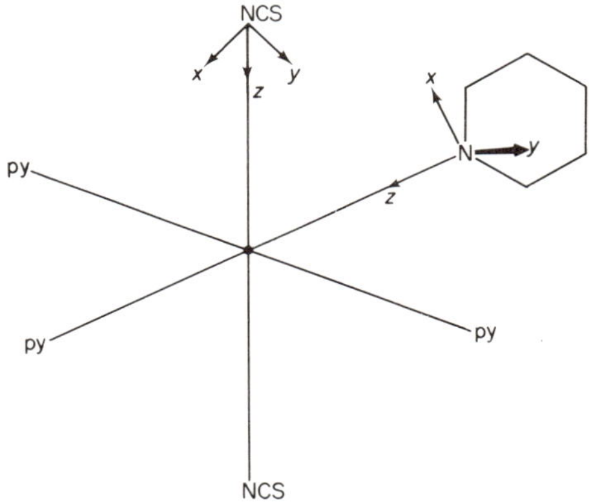

Figure 6.12 Choice of ligand coordinate frames in $M(py)_4(NCS)_2$.

ourselves to the ^4P and ^4F terms. Appropriate reduced matrix elements [of (6.25) and the corresponding spin–orbit expression] are also required input data. Thereafter, the calculation proceeds as described, constructing a matrix involving the AOM parameters e_σ(py), $e_{\pi y}$(py), e_σ(Cl), $e_{\pi x}=e_{\pi y}$(Cl). Spin–orbit and inter-electron repulsion parameters are also included. Eigenvalues and eigenvectors result from diagonalisation of the generally complex matrix. The aim stated at the beginning of Section 6.2, namely the construction of eigenvalues and their associated wavefunctions for a molecule of any symmetry with any metal function basis, is now achieved. All that remains is to calculate the magnetic susceptibilities which follow from these. This turns out to be a non-trivial problem.

6.2.2 Magnetic susceptibilities

Van Vleck's equation for the molar magnetic susceptibility $\chi_m \left\{ = \left[\dfrac{B}{\mu_0 H} - 1 \right] V_m, \right.$ where μ_0 is the permeability of free space ($\mu_0/4\pi = 10^{-7}$H m^{-1} = 1 e.m.u.) and V_m is the molar volume$\left.\right\}$ may be written

$$\chi_{\rm m}=\mu_0 N_{\rm A}\frac{\sum_i (E_i^{\rm I2}/kT-2E_i^{\rm II})\exp(-E_i^0/kT)}{\sum_i \exp(-E_i^0/kT)} \tag{6.30}$$

where $N_{\rm A}$ is Avogadro's number, $E_i^{\rm I}=\langle\psi_i|\mu_H|\psi_i\rangle$ is the first-order magnetic moment in the direction of the field, and E_i^{II} is the second-order term

$$\sum_{j>m}\frac{\langle\psi_i|\mu_H|\psi_j\rangle\langle\psi_j|\mu_H|\psi_i\rangle}{E_i^0-E_j^0}$$

for wavefunctions ψ_i (m-fold degenerate) and ψ_j with energies E_i^0 and E_j^0. The magnetic moment operator is taken as

$$\hat{\mu}_\alpha=(kL_\alpha+2S_\alpha)\mu_{\rm B} \tag{6.31}$$

where μ_B is the Bohr magneton, k is Stevens' orbital reduction factor and a represents a given direction of field B in the magnetic Hamiltonian operator

$$\mathscr{H}'=-\mu_\alpha B_\alpha \tag{6.32}$$

In an axially symmetric molecule, for example, when $a=z$, we calculate $\chi_{\parallel}$ and when $a = x$ or y, $\chi_{\perp}$. However, in cases of low symmetry, we do not know *a priori* the directions of the principal susceptibilities.

Magnetic susceptibility is a second-rank tensor quantity expressing the relation between magnetisation, a vector of effect, and magnetic field, a vector of action:

$$M_\alpha=\chi_{\alpha\beta}B_\beta\mu_0^{-1} \tag{6.33}$$

For a particular orientation of the coordinate frame, the tensor $\chi_{\alpha\beta}$ is diagonal and we have:

$$M_x=\chi_{xx}B_x\mu_0^{-1}$$
$$M_y=\chi_{yy}B_y\mu_0^{-1}$$
$$M_z=\chi_{zz}B_z\mu_0^{-1} \tag{6.34}$$

The susceptibilities $\chi_{\alpha\alpha}$ in this frame are the principal susceptibilities of the system, corresponding to the cases when the resulting magnetisation is parallel to the imposed field. Calculation of the principal susceptibilities simply requires the use of (6.30) for $a=x$, y or z in (6.33), *provided* the principal directions x, y, z are known. In low symmetry when they are not, we must proceed as follows.

The defining equation for the molar magnetisation in direction a is

$$M_\alpha=-N_{\rm A}\sum_i\frac{\partial E_i}{\partial B_\alpha}\exp(-E_i/kT)/\sum_i \exp(-E_i/kT) \tag{6.35}$$

The perturbation expansion of E_i, the energy of the ith level in the magnetic field, is

$$E_i=E_i^0+\langle i|\mu_\alpha|i\rangle B_\alpha+\sum_{j>m}\frac{\langle i|\mu_\alpha|j\rangle\langle j|\mu_\beta|i\rangle B_\alpha B_\beta}{E_i^0-E_j^0} \tag{6.36}$$

We use standard tensor notation so that repeated Greek suffices imply summation ($a=x$, y, z). Hence

$$\frac{\partial E_i}{\partial B_\alpha}=\langle i|\mu_\alpha|i\rangle+\sum_{j>m}\frac{[\langle i|\mu_\alpha|j\rangle\langle j|\mu_\beta|i\rangle+\langle i|\mu_\beta|j\rangle\langle j|\mu_\alpha|i\rangle]B_\beta}{E_i^0-E_j^0} \tag{6.37}$$

Expanding $\exp(-E_i/kT)$,

$$\exp(-E_i/kT)=\exp(-E_i^0/kT)(1-\langle i|\mu_\beta|i\rangle B_\beta/kT+\ldots) \tag{6.38}$$

and retaining only terms linear in B as in (6.34) and remembering the lack of residual magnetisation [all of which is standard procedure in the proof of Van Vleck's equation (6.30)], we obtain:

$$M_\alpha=N_A\sum_i\left(\langle i|\mu_\alpha|i\rangle\langle i|\mu_\beta|i\rangle B_\beta/kT-\sum_{j>m}\frac{[\langle i|\mu_\alpha|j\rangle\langle j|\mu_\beta|i\rangle+\langle i|\mu_\beta|j\rangle\langle j|\mu_\alpha|i\rangle]}{E_i^0-E_j^0}\right)$$
$$\times\exp(-E_i^0/kT)/\sum_i\exp(-E_i^0/kT) \tag{6.39}$$

Whence,

$$\chi_{\alpha\beta}=N_A\sum_i\left(\langle i|\mu_\alpha|i\rangle\langle i|\mu_\beta|i\rangle/kT-\sum_{j>m}\frac{[\langle i|\mu_\alpha|j\rangle\langle j|\mu_\beta|i\rangle+\langle i|\mu_\beta|j\rangle\langle j|\mu_\alpha|i\rangle]}{E_i^0-E_j^0}\right)$$
$$\times\exp(-E_i^0/kT)/\sum_i\exp(-E_i^0/kT) \tag{6.40}$$

Note that, for diagonal elements, equation (6.40) reduces to expressions as given by Van Vleck; for example:

$$\chi_{xx}=N_A\sum_i\frac{\left(\langle i|\mu_x|i\rangle^2/kT-2\sum_{j>m}\dfrac{\langle i|\mu_x|j\rangle\langle j|\mu_x|i\rangle}{E_i^0-E_j^0}\right)\exp(-E_i^0/kT)}{\sum_i\exp(-E_i^0/kT)} \tag{6.41}$$

Although the above proof of (6.39) assumes E_i to be diagonal in $\mu_\alpha B_\alpha$ for each column of $\chi_{\alpha\beta}$, this condition need not hold, for it is simple to show that

$$\sum_i\langle i|\mu_\alpha|i\rangle\langle i|\mu_\beta|i\rangle \text{ in an } \alpha\text{-diagonal basis}$$
$$=\sum_i\sum_j\langle i|\mu_\alpha|j\rangle\langle j|\mu_\beta|i\rangle \text{ in a general basis} \tag{6.42}$$

for levels i and j belonging to a degenerate set. This is the computation procedure we normally follow. This and other computational aids are not discussed in full here. The essence of this approach, however, is to compute all six independent $\chi_{\alpha\beta}$ components in a molecule-fixed frame and then to diagonalise the susceptibility tensor to yield the principal susceptibilities and their orientations in the molecule.

The final stage is a calculation of *crystal* susceptibilities, anisotropies and other observables. In all earlier work, such as that cited in the introduction, the usual procedure was to transform crystal susceptibility measurements into the molecular frame and to compare these with predictions of the model at a molecular level. Such transformations require a knowledge of the orientation of the molecular magnetic ellipsoid. As this is unknown *a priori* in very low symmetry situations, we must compare calculation and observation at the crystal level. In the monoclinic crystal class, for example, the orientation of only one principal crystal susceptibility is fixed by symmetry: $\chi_3 = \chi_b$ lies parallel to b. χ_1 and χ_2 lie arbitrarily in the ac plane as in Figure 6.13.

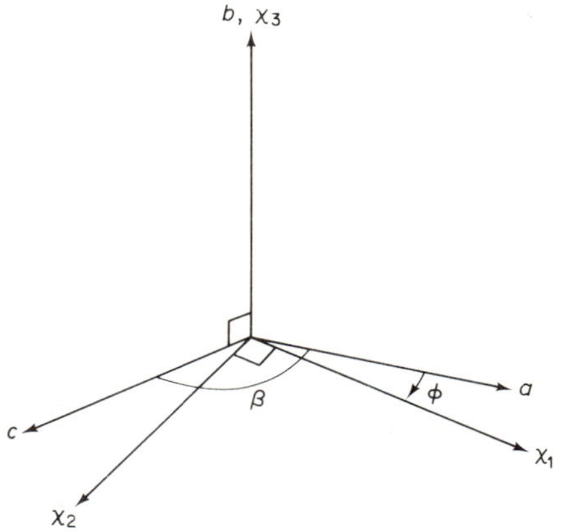

Figure 6.13 Principal crystal susceptibilities in monoclinic crystals.

With the convention[32] $\chi_2 > \chi_1$, we define φ as the angle between χ_1 and a measured from a towards c. This angle may be determined experimentally, as discussed elsewhere[32], and a requirement of the quantum mechanical model is to reproduce it, as well as the magnitudes of χ_1, χ_2 and χ_3.

When calculating molecular from crystal susceptibilities, the sort of approximation previously made for a molecule like *trans*-$Co(py)_4(NCS)_2$ is one of essential fourfold symmetry, or at least that one principal molecular magnetic axis lies parallel to the NCS—NCS direction. Such crystal-to-molecule transformations in monoclinic systems are over-determined in that the information regarding the orientations of the principal molecular magnetic axes is contained both in the assumed direction cosines and in the angle φ. A common occurrence in such transformations is to find inconsistent results when calculations are performed using different parts of the available information. These inconsistencies can be excellent indications of the incorrectness of the sort of symmetry approximations we describe. The fact that the pyridine ligands are not oriented in such a way as to maintain fourfold symmetry automatically and *importantly* reduces the symmetry from axial. Features of this sort are virtually impossible to handle within the framework of the point-charge crystal-field model. The AOM treats the situation simply, the low symmetry manifests itself in the D^l_λ matrices, and the treatment of susceptibilities above allows a rhombic susceptibility tensor to be calculated and, in monoclinic crystals, the angle φ. Despite all the complications following angular displacements, which have large effects upon susceptibilities and anisotropies and which cannot be satisfactorily averaged in a crystal-field model, the AOM merely retains radial parameters which relate directly to σ- and π-bonding with their attendant appeal to chemical intuition. In short, the AOM allows the chemically trivial complications of geometry to be factored out from the chemically interesting features of bonding. Preliminary results for $Co(py)_4(NCS)_2$ show very little π-bonding with the pyridine ligands in order to reproduce the observed magnetism. This is hardly an unexpected result but it goes to illustrate the practicality of the approach. The detailed techniques are now available for study of the magnetism of almost any paramagnetic monomer, and this situation is new. Not the least benefit of this advance is the almost unlimited range of systems now open to uniformly rigorous study, compared with the earlier restriction to molecules with axial symmetry.

6.2.3 Very low temperatures

Increasingly, efforts are being made to measure magnetic properties in the liquid-helium range, down to (say) 1.6 K. It has long been anticipated that complications will arise at lower temperatures resulting from very small magnetic interactions between paramagnetic centres. With luck these may be avoided if bulky ligands are present to maintain effective magnetic dilution. There remains one problem, however, which could be a dominant factor in all systems as base temperatures are approached. This is the phenomenon of *magnetic saturation*. When Zeeman energies are no longer very small compared with kT, the truncated series expansion of the Boltzmann exponentials in (6.38), a fundamental factor in the derivation of Van Vleck's equation, is no longer valid. Retention of the exponential intact leads, in simple first-order Zeeman calculations, to Brillouin functions involving hyperbolic trigonometric functions. For example, Van Vleck[15] gives an expression for the molar magnetisation of a system with specified g and J values, as in equation (6.43):

$$M=\mu_0 N_A J g \mu_B D_J$$

$$D_J=\frac{2J+1}{2J}\coth\left(\frac{2Jy+y}{2J}\right)-\frac{1}{2J}\coth\left(\frac{y}{2J}\right)$$

$$y=\frac{Jg\mu_B B}{kT} \tag{6.43}$$

Ferric alum, $Fe_2(SO_4)_3(NH_4)_2SO_4 \cdot 24H_2O$, is known to obey Curie's law down to less than 1 K. On the other hand, at very low temperatures and high magnetic fields, saturation effects are important: substitution of $g=2$, $J=\frac{5}{2}$, $B=5$ T, $T=10$ K in (6.43) gives a magnetisation only $\approx 60\%$ of the value derived from Curie's law. It is clear that saturation effects generally become important for all systems at such fields (now readily attainable in the laboratory) and low temperatures. It could become as significant as any effects resulting from distortion or even exchange. Certainly, under these conditions, the use of Van Vleck's equation (6.30), or of its generalised form (6.40), is inappropriate.

Suppose we were to retain more terms in the expansion of the exponentials and then follow through the derivation above. The equivalent expression for the magnetisation M_α [equation (6.39)] would now be non-linear in the field and some describe this situation as giving rise to field-dependent susceptibilities. A more general expansion of magnetisation as a power series in the field is now appropriate:

$$M_\alpha = \chi_{\alpha\beta} B_\beta + \Omega_{\alpha\beta\gamma\delta} B_\beta B_\gamma B_\delta + \ldots \tag{6.44}$$

where $\Omega_{\alpha\beta\gamma\delta}$ is a fourth-rank 'hypersusceptibility', and so on. It is not simple to establish analytical expressions for Ω, etc., corresponding to that for χ in equation (6.40). Instead we shall calculate the magnetisation. After all, that is the quantity always measured: susceptibility tensors are merely mathematical, unobservable, quantities which relate the vectors of action (field) and effect (magnetisation). At higher temperature, it happens that magnetisation is linear in field strength and so susceptibility χ is a convenient 'coefficient' to record. Calculation of M_α in equation (6.44) requires not only the non-expansion of the exponentials in the Bolzmann partition functions, but also a recognition of the limitations of the perturbation expansion we use in equation (6.36), say, when Zeeman energies may be comparable with very small ligand-field effects. Accordingly, we adopt the following procedure.

Suppose we wish to calculate the magnetisation in direction $\boldsymbol{a}$ whose direction cosines with respect to the molecule-fixed axes are a_j ($j=x, y, z$). We may diagonalise the complete basis [say $^4P+^4F$ for cobalt(II) systems], either before or after diagonalisation in the ligand-field as convenient, with the Hamiltonian:

$$\mathscr{H}' = \sum_{i<k} \frac{e^2}{r_{ik}} + V_{\mathrm{AOM}} + \sum_i \zeta l_i \cdot s_i + \mu_j \sum_j^3 a_j \mu_j B_j \tag{6.45}$$

Let the resulting eigenvectors be called ψ, with eigenvalues E. In using the defining equation for magnetisation (6.35) we require the derivatives $(\partial E_i/\partial B_\alpha)$. These we obtain via the Hellmann–Feynman theorem, from which we have:

$$\frac{\partial E_i}{\partial B_\alpha} = \mu_B \langle \psi_i | \sum_j^3 a_j u_j | \psi_i \rangle \tag{6.46}$$

In terms of computer programmes, details of which are not given here, most of the quantities required to set up the matrix for equations (6.45) and (6.46) have already been calculated: in practice, the process is therefore not too lengthy. However, the whole procedure in (6.45) and (6.46) must be repeated for each change in the orbital reduction factor k and for each field direction a. Obviously this technique is only useful where Van Vleck's equation is invalid.

There emerges one final and most interesting point about saturation effects. In so far as the higher rank tensors in equation (6.44) are important, the susceptibility is dependent not only upon field strength but also field direction. Higher temperature, principal susceptibility directions need not be retained as saturation begins to affect the system. Many measurements may be made to study the variation of $\Omega_{\alpha\beta\gamma\delta} B_\gamma B_\delta$, say, with field direction. Such measurements, even more than the more usual examination of magnetisation curves (i.e. *M versus B*), should provide more independent experimental information relating to the same quantum mechanical condition. In magnetochemistry, well-known for the high degree of parameterisation, such extra data would be most welcome. Thus very low-temperature, high-field magnetic measurements should provide a rich area for future work.

The techniques, models and philosophy applied to low- or zero-symmetry complexes over wide ranges of magnetic field strength and of temperature are not only ideally suited to mononuclear transition metal (and lanthanide) systems. They also provide the starting points for studies of magnetic exchange phenomena in dinuclear and small polynuclear compounds. The remainder of this article is devoted to an outline of current attempts to incorporate *chemically identifiable* features into the quantal treatment of the exchange phenomenon.

6.3 MAGNETIC EXCHANGE

A celebrated and important problem for solid state physics early this century was the nature of ferromagnetism. The molecular field theory of Weiss[33] sought to explain the phenomenon in terms of local molecular magnetic interactions but, for many, the problem then devolved into explaining the origin of the enormous magnitudes of the internal field required by the Weiss theory. Heisenberg[34] and Dirac[35] independently removed the problem and simultaneously demonstrated part of the power of quantum mechanics. The purely non-classical concepts of the indistinguishability of electrons and Pauli's exclusion principle lead inevitably, for example, to a prediction of spin-singlet and -triplet states in a hydrogen molecule. Hund's rules and the properties of the two-electron operator e^2/r_{ij} follow from these concepts. The particular feature of the Heisenberg and Dirac theories is their construction of an operator equivalent for the exchange process in terms of spin angular momentum operators. From this, they reproduced the more empirical results of the Weiss molecular-field model, but with the centrally important point that the fundamental process involved is electrostatic and not magnetic.

The exchange Hamiltonian for two electrons housed on (say) two centres with only one space orbital available for each electron, can be equivalenced by equation (6.47):

$$\mathscr{H} = \mathscr{J} \cdot \boldsymbol{s}_1 \cdot \boldsymbol{s}_2 \tag{6.47}$$

in which $\boldsymbol{s}_1$ and $\boldsymbol{s}_2$ are the one-electron spin operators for the two electrons and $\mathscr{J}$ is an exchange integral. The model at this stage does not require the electrons to be on different centres. Considerable confusion in inorganic chemistry has resulted from the belief that this spin-Hamiltonian is the true Hamiltonian, describing a magnetic ,'through-space' interaction. Part of the reason for this misunderstanding lies in the fact that equation (6.47) has the form of part of the classical dipole–dipole interaction expression and has been termed a 'dipole interaction Hamiltonian' in consequence. The name was meant to clarify but within chemistry it has confused.

Van Vleck extended equation (6.47) to cover the many-electron case[15]. In essence, for n electrons in n orthogonal space orbitals in two sets (two centres), we may write

$$\mathscr{H}' = J \cdot \boldsymbol{S}_1 \cdot \boldsymbol{S}_2 \tag{6.48}$$

in which the spin operators are now total operators for each centre. We note that the form of equation (6.48), involving the scalar product of two angular momentum operators, inevitably implies a Landé interval rule. The interaction between two iron(III) ions with $S=5/2$, for example, gives total spin angular momenta of $S_{tot}=5, 4, 3, 2, 1, 0$ and the energy diagram in Figure 6.14. Given the applicability of the Hamiltonian (6.48), which in turn depends only on the lack of orbital degeneracy implied by n electrons in n orbitals, the splitting pattern in Figure 6.14 must follow. This conclusion does not depend on any particular exchange pathway, i.e. region of space where the electrons can interact via e^2/r_{ij}, nor on the degree of covalency present. This latter point has been demonstrated by Löwdin[36], as we shall describe.

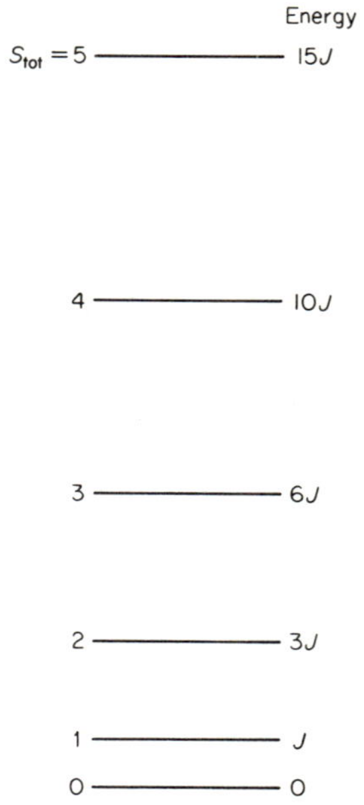

Figure 6.14 Landé interval rule for dipolar coupling of two spin-$\frac{5}{2}$ states.

Central to the Heisenberg–Dirac–Van Vleck (HDVV) approach is the question of the sign of the exchange. In the earliest forms of Heisenberg and Dirac, orthogonal orbitals led to $\mathcal{J}$ being a true exchange integral (6.49)

$$J_{ij}=\left\langle \varphi_i^{(1)}\varphi_j^{(2)}\left|\frac{e^2}{r_{12}}\right|\varphi_j^{(1)}\varphi_i^{(2)}\right\rangle \tag{6.49}$$

as in the usual Condon–Shortley–Slater theory of electron interactions in free ions[37]. However, the inevitable consequence of such integrals being positive is the emergence of Hund's rules and the relative stabilisation of levels with maximum spin multiplicity. In the context of magnetic exchange phenomena, the prediction is ferromagnetism. Antiferromagnetism is the more usual observation for insulators, however; further, the ground state of the hydrogen molecule, to which the model was first applied, is a singlet not a triplet. The deficiency in the approach, of course, is the neglect of overlap between orbitals on different centres and the whole phenomenon of chemical bonding. Under more realistic conditions, the electrons can be regarded[37] as moving in an attractive potential $-V(r)$, causing bonding, and we write

$$J_{ij}=\langle\psi_i^{(1)}\psi_j^{(2)}\left|\frac{e^2}{r_{12}}\right|\varphi_j^{(1)}\varphi_i^{(2)}\rangle-2\langle\varphi_i^{(1)}|\,\varphi_j^{(1)}\rangle\langle\varphi_j^{(2)}|\,V|\,\varphi_i^{(2)}\rangle$$
$$=J_{ij}^0-2S_{ij}V_{ij} \qquad (6.50)$$

Löwdin[36] has shown that under conditions of non-orthogonality between interacting orbitals, the HDVV approach is still appropriate, provided (*a*) that no orbital degeneracy is present and (*b*) that the 'exchange integral' is regarded as in equation (6.50) when it may take on positive or negative values, leading to ferro- and antiferro-magnetism, respectively. The second term in equation (6.50) is considered by Anderson[38] to predominate in most cases, as evidenced, for example, by the common occurrence of bonding electron pairs in molecules.

It is not the intention to review the whole field of magnetic exchange here. Suffice it to say that the foregoing provides the rationale behind the use of the spin–Hamiltonian (6.48) in inorganic chemists' treatment of antiferromagnetic exchange in transition metal dinuclear and other small clusters. In all cases, however, only systems with no ground state orbital degeneracy can be so treated. This apart, the nature of the spin-Hamiltonian as an operator equivalent renders the exchange *mechanism* irrelevent: it may be direct, involving overlap of metal orbitals, or indirect through intervening bridging groups as in superexchange theory. In common with our earlier remarks about crystal-field potential coefficients as parameters, however, the great deficiency of this approach is the lack of direct correlation to be had between J values and chemical bonding or structure. We confine our discussion therefore, to a review of a few studies in which chemical pathways are brought to the fore. We begin with a dinuclear, orbitally non-degenerate, complex of 'octahedral' chromium(III).

6.3.1 μ-Oxo-bis[pentamminechromium(III)]

The basic rhodo salt has been studied many times but only recently have reliable magnetic data in the range 80–300 K been obtained for a pure sample[39]. Glerup has published[40] an interpretation of the spectral and magnetic properties. At the phenomenological level of the HDVV spin Hamiltonian, Pedersen[39] obtained a good fit to experiment for $J=450\pm2\,\mathrm{cm}^{-1}$ for each chromium ion in a state with $S=3/2$. The absorption spectrum of the

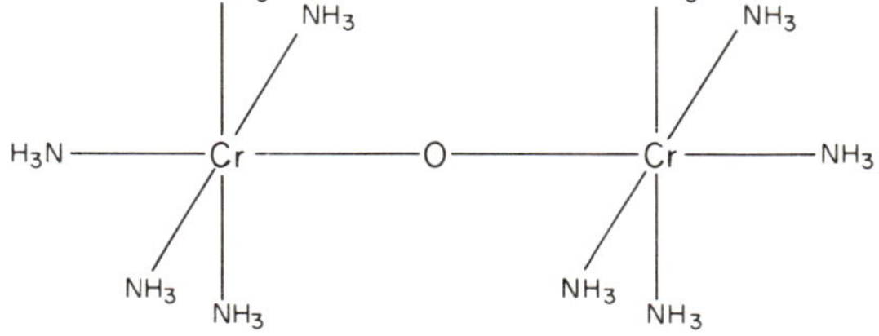

Figure 6.15 The μ-oxo-bis[pentamminechromium-(III)] molecule.

complex is relatively rich with broad bands similar to the d–d spin-allowed transitions in mononuclear chromium(III) systems, sharp 'spin-forbidden' bands allowed by exchange coupling, and also an interesting charge-transfer band not seen in monomers. Glerup was concerned to rationalise all of these features in terms of a bonding model for the dinuclear structure shown in Figure 6.15.

The three d electrons in octahedral chromium(III) are housed in the t_{2g} orbital triplet and Glerup's model explicitly ignores contributions from occupancy of the e_g set. A splitting of the t_{2g} orbitals is taken as resulting from the different π-bonding abilities of NH_3 and the shared O atom, within the spirit of the angular overlap model. A partial molecular orbital diagram for the whole complex may be drawn as in Figure 6.16, in which we use the notation

$$\zeta \sim xy, \; \xi \sim yz, \; \eta \sim xz \qquad (6.51)$$

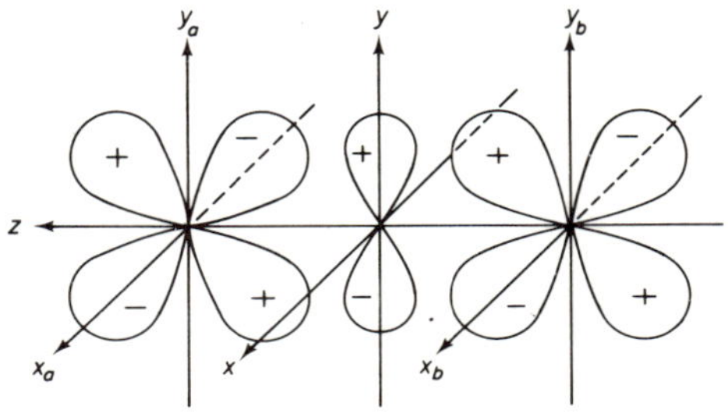

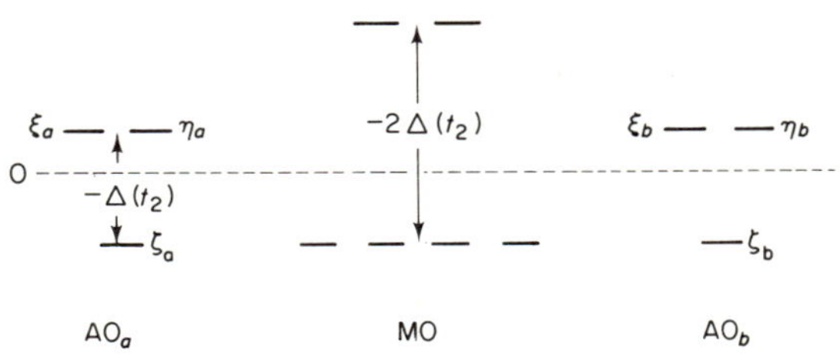

Figure 6.16 Molecular orbital diagram for the Cr—O—Cr structure.

Overlap between the chromium t_{2g} functions is achieved via the oxygen p_x and p_y orbitals. With the common z axis we see that the combination $\xi_b - \xi_a$ interacts with the p_y orbital on oxygen while the positive linear combination cannot; similar remarks apply to the combinations of η_b and η_a. In the angular overlap model all non-interacting orbitals are degenerate and there results the molecular orbital energies shown in Figure 6.16. In effect, the ligand field perturbation of the oxygen atom is half 'undone' by the formation of the bridging π-molecular orbitals. Finally, the baricentre rule gives the energy of the antibonding combinations as $-\frac{4}{3}\Delta(t_2)$ and we may construct the matrix

$$\begin{array}{c|cc|} \frac{1}{\sqrt{2}}(\xi_b-\xi_a) & -\frac{4}{3}\Delta(t_2) & 0 \\ \frac{1}{\sqrt{2}}(\xi_b+\xi_a) & 0 & \frac{2}{3}\Delta(t_2) \end{array} \quad (6.52)$$

or, in a more useful, non-diagonal form,

$$\begin{array}{c|cc|} \xi_a & -\frac{1}{3}\Delta(t_2) & \Delta(t_2) \\ \xi_b & \Delta(t_2) & -\frac{1}{3}\Delta(t_2) \end{array} \quad (6.53)$$

From this we observe that the interaction between the chromium ions may be expressed in this model by the one-electron matrix elements

$$\langle \xi_a | V | \xi_b \rangle = \langle \eta_a | V | \eta_b \rangle = \Delta(t_2) \quad (6.54)$$

where V is the ligand field operator. Direct metal–metal overlap has been ignored, but not metal–bridge. The application of equation (6.54) to the exchange problem now follows.

The Hamiltonian for interaction of the two chromium ions may be written:

$$\mathscr{H}' = V_a + \sum_{i<j}^{a} \frac{e^2}{r_{ij}} + V_b + \sum_{i<j}^{b} \frac{e^2}{r_{ij}} + V_{ab} \quad (6.55)$$

where V_a and V_b are the ligand field operators for centres a and b and V_{ab} is the one-electron interaction operator whose consequences were derived above [equation (6.54)]. In accordance with equation (6.50), in which one assumes dominance by the overlap bonding term in all cases where it is not identically zero by symmetry, Glerup neglects in equation (6.55) the two-electron, two-centre operator. He proceedes by forming product functions for the dimer for ground and relevant excited states. The ground term $^4A_{2g}$ from t_{2g}^3 in O_h symmetry transforms as $^4B_1$ in C_{4V} symmetry. The ground–ground direct product in the dinuclear D_{4h} symmetry gives

$$^4B_1 \times {}^4B_1 \rightarrow {}^1A_{1g} + {}^3A_{2u} + {}^5A_{1g} + {}^7A_{2u} \quad (6.56)$$

Consider now excited configurations resulting from the transfer of an electron from one chromium ion to the other. Glerup forms the direct product between ^3E terms arising from t_2^2 and t_2^4 configurations on the two centres:

$$[t_2^2\ {}^3E({}^3T_1)] \times [t_2^4\ {}^3E({}^3T_1)] \rightarrow {}^1A_{1g} + {}^3A_{2u} + {}^5A_{1g} + \text{others} \quad (6.57)$$

No spin septet can arise here. The terms in (6.56) and (6.57) can interact via the Hamiltonian (6.55), the degeneracies of the two sets only being raised by the two-centre operator V_{ab}. The one-centre energies of the two sets of terms result from the usual inter-electron repulsion and ligand-field effects. One function belonging to the $^5A_{1g}$ ground term is

$$\frac{1}{\sqrt{6}} \left| \xi_a\eta_a\zeta_a(\bar{\xi}_b\eta_b\zeta_b + \xi_b\bar{\eta}_b\zeta_b + \xi_b\eta_b\bar{\zeta}_b) - (\bar{\xi}_a\eta_a\zeta_a + \xi_a\bar{\eta}_a\zeta_a + \xi_a\eta_a\bar{\zeta}_a)\xi_b\eta_b\zeta_b \right|$$

and one belonging to $^5A_{1g}$ of the charge-transfer level is

$$\tfrac{1}{2}|\zeta_a\eta_a\xi_b\bar{\xi}_b\eta_b\zeta_b - \xi_a\zeta_a\xi_b\eta_b\bar{\eta}_b\zeta_b + \xi_a\bar{\xi}_a\eta_a\zeta_a\zeta_b\eta_b - \xi_a\eta_a\bar{\eta}_a\zeta_a\xi_b\zeta_b|$$

The matrix element between these functions under the one-electron operator V_{ab} may then be reduced to one-electron matrix elements of the type $\langle\eta_a|V_{ab}|\zeta_b\rangle$, etc. But we know these are all zero except for those in (6.54), according to the MO treatment above. In this way, the off-diagonal matrix elements between the charge-transfer states in (6.57) and the ground states in (6.56) may be reduced to

$$\begin{aligned}\langle{}^1A_{1g}\ \mathrm{gd}|V_{ab}|{}^1A_{1g}\ \mathrm{c.t.}\rangle &= -\frac{4}{\sqrt{3}}\Delta(t_2)\\ \langle{}^3A_{2u}\ \mathrm{gd}|V_{ab}|{}^3A_{2u}\ \mathrm{c.t.}\rangle &= -\frac{2\sqrt{10}}{3}\Delta(t_2)\\ \langle{}^5A_{1g}\ \mathrm{gd}|V_{ab}|{}^5A_{1g}\ \mathrm{c.t.}\rangle &= -\frac{4}{\sqrt{6}}\Delta(t_2)\end{aligned} \qquad (6.58)$$

Second-order perturbation theory then gives the energy of the A_{1g} level as depressed by $\frac{16}{3}[\Delta(t_2)]^2/E_{\mathrm{c.t.}}$, the ${}^3A_{2u}$ by $\frac{40}{9}[\Delta(t_2)]^2/E_{\mathrm{c.t.}}$ and that of the ${}^5A_{1g}$ by $\frac{8}{3}[\Delta(t_2)]^2/E_{\mathrm{c.t.}}$, as in Figure 6.17. The ratios of the splittings between the final singlet, triplet, quintet and septet states follow the Landé interval rule.

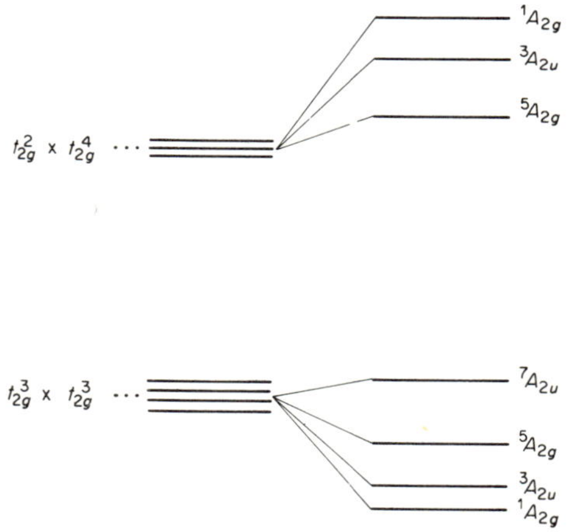

Figure 6.17 Landé interval rule resulting from configuration interaction.

Thus there results a relationship between J, deduced from the HDVV model, the crystal-field matrix element $\Delta(t_2)$, and the energy of the charge-transfer bond $E_{\mathrm{c.t.}}$. A likely estimate for $\Delta(t_2)$ is *ca.* 4000 cm^{-1}, from analogous mononuclear species. With $J=450$ cm^{-1}, $E_{\mathrm{c.t.}}$ is thus calculated as *ca.* 32 000 cm^{-1}, which is unusually low for this sort of charge-transfer band. However, there is an intense band in the spectrum at 35 940 cm^{-1} which is absent in mononuclear chromium(III) complexes. Accordingly, Glerup concludes that the origin of the magnetic exchange process is the admixture into the ground state of excited states from this charge-transfer level at 35 000 cm^{-1}. His paper goes on to deal with the spectrum, for which a good fit is obtained with his model. The point for our present purpose, however, is to note his interpretation of the antiferromagnetism in terms of the one-electron operator V_{ab}, in common with the general remarks of Anderson concerning the dominance of the bonding term over the two-electron terms in (6.50). This is an example of what we shall introduce below as kinetic exchange. We consider now the case of orbital degeneracy in a study by Lines *et al.*[41], during which a mechanistic approach to exchange is more fully developed.

6.3.2 Exchange in systems with orbital angular momentum

The most general exchange spin Hamiltonian which may be written for an orbital singlet ground state is

$$\mathscr{H} = -2J_{ij} \cdot \boldsymbol{S}_i \cdot \boldsymbol{S}_j + D_{ij} \cdot \boldsymbol{S}_i \times \boldsymbol{S}_j + \boldsymbol{S}_i \cdot \Gamma_{ij} \cdot \boldsymbol{S}_j \tag{6.59}$$

in which J_{ij} is the isotropic exchange integral of (6.48), D_{ij} is an antisymmetric vector coupling constant and Γ_{ij} is a symmetric anisotropic tensor coupling constant. The anisotropic and antisymmetric exchange terms arise from the effects of spin–orbit coupling, a perturbation treatment giving their orders of magnitude as

$$D \approx (\Delta g/g)J \quad \Gamma \approx (\Delta g/g)^2 J \tag{6.60}$$

where $\Delta g = |g-2|$. For ions with orbitally degenerate ground terms, however, the antisymmetric (and, in general, anisotropic) terms in (6.59) occur not merely as perturbations and there are no limits on their magnitudes

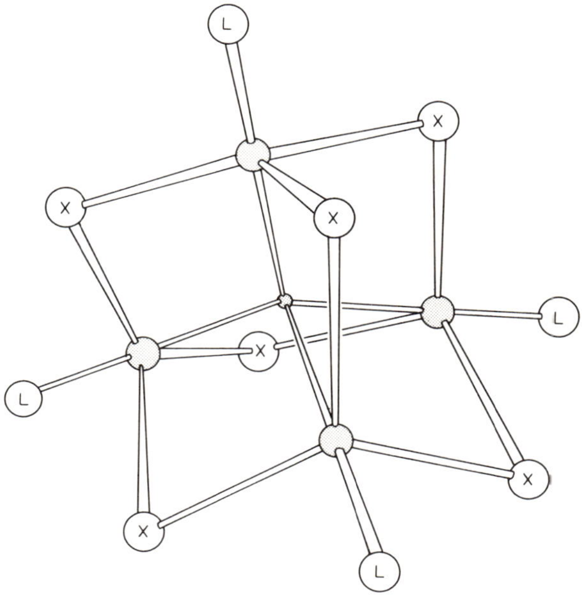

Figure 6.18 Molecular structure of the tetranuclear copper cluster.

relative to the isotropic term. Lines *et al.*[41] demonstrated this explicitly for some copper(II) cluster complexes.

The study concerns the tetranuclear clusters shown in Figure 6.18, where X=Cl, L=Ph_3PO; or X=Br, L=Ph_3PO or py. The mean moments $\bar{\mu}_{eff}$ of all these systems exhibit maxima in the 40–60 K temperature range and cannot be fitted quantitatively by the simple HDVV theory. Reasonable arguments are presented by Lines *et al.* to support an orbitally degenerate ground state for the individual copper(II) ions, in terms of the consequent inapplicability of the HDVV approach.

With neglect of spin–orbit coupling, a point-charge calculation for the essentially trigonal-bipyramidal copper(II) ions, suggests the possible ground state wavefunctions

$$\varphi(\pm 2)=(1+\gamma^2)^{-\frac{1}{2}}[|\pm 2>\pm\gamma|\mp 1>]$$
$$=2^{-\frac{1}{2}}(1+\gamma^2)^{-\frac{1}{2}}[d_{x^2-y^2}+\gamma d_{xz}\pm i(d_{xy}-\gamma d_{yz})] \quad (6.61)$$

where γ is a crystal-field mixing coefficient (note: only one required in this approximation): the axis definitions are as in Figure 6.19.

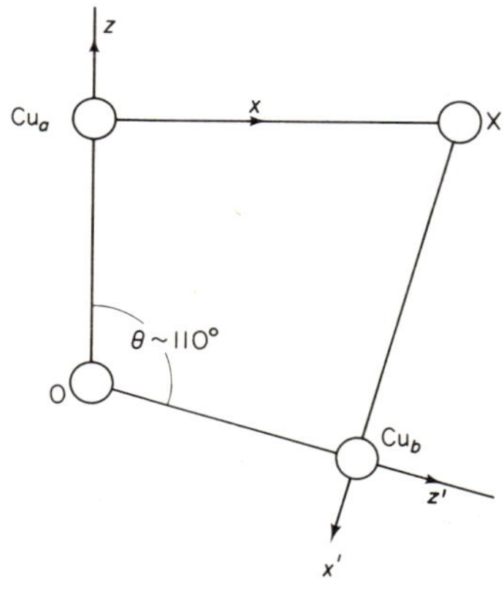

Figure 6.19 Coordinate frames for two copper atoms in the cluster.

The exchange problem begins with the construction of product wavefunctions for a copper pair (extension to the tetramer is dealt with later in the original paper, but need not concern us here). For example, consider two holes (1 and 2), one in a $\varphi_a(+2)\alpha$ or $\varphi_a(-2)\beta$ spin–orbital on copper *a* and the other in the equivalent orbitals of copper *b*; α and β refer to spin $\pm\frac{1}{2}$. Four Slater determinant wavefunctions may now be formed as in (6.62), in which suffixes 1 and 2 label the holes and primes denote spin functions on centre *b*:

$$\Phi_1=\frac{1}{\sqrt{2}}\Big\{[\varphi_a(+2)\alpha]_1[\varphi_b(+2)\alpha']_2-[\varphi_b(+2)\alpha']_1[\varphi_a(+2)\alpha]_2\Big\}$$

$$\Phi_2=\frac{1}{\sqrt{2}}\Big\{[\varphi_a(+2)\alpha]_1[\varphi_b(-2)\beta']_2-[\varphi_b(-2)\beta']_1[\varphi_a(+2)\alpha]_2\Big\} \qquad (6.62)$$

$$\Phi_3=\frac{1}{\sqrt{2}}\Big\{[\varphi_a(-2)\beta]_1[\varphi_b(+2)\alpha']_2-[\varphi_b(+2)\alpha']_1[\varphi_a(-2)\beta]_2\Big\}$$

$$\Phi_4=\frac{1}{\sqrt{2}}\Big\{[\varphi_a(-2)\beta]_1[\varphi_b(-2)\beta']_2-[\varphi_b(-2)\beta']_1[\varphi_a(-2)\beta]_2\Big\}$$

The exchange problem now devolves into the formation and diagonalisation of the matrix of Φ under an appropriate exchange Hamiltonian $\mathscr{H}_{\text{ex}}$. It is in the evaluation of the matrix elements $\langle\Phi|\mathscr{H}_{\text{ex}}|\Phi\rangle$ that we introduce Anderson's concepts of kinetic and potential exchange.

The ideas are contained largely in equation (6.50) and our earlier remarks concerning that 'exchange integral'. In essence, we consider that the second term in (6.50) expresses actual or 'incipient' bond formation between the interacting centres, whether directly or via bridging ligands as in super-exchange, and gives rise to an increase in the kinetic energy of antiparallel electrons. As such, this is the opposite of the true exchange effect. It is referred to as *kinetic exchange*[38]. The most simple prerequisite for it to occur is the presence of a pathway of orbital overlap between the interacting

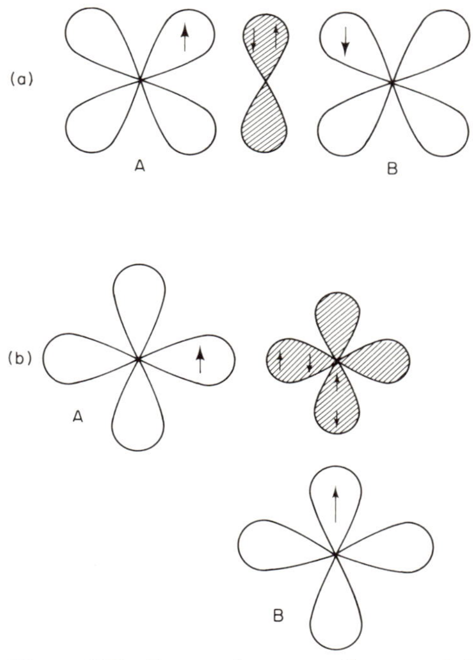

Figure 6.20 Superexchange pathways: (a) 'kinetic exchange' and (b) orthogonality leading to 'potential exchange'.

centres. From order-of-magnitude considerations, when the pathway overlap integrals are significant, the second (kinetic) term in (6.50) is considered to predominate, and its effect is to produce antiferromagnetism. True exchange, that is 'true' in the sense of the two-electron operator e^2/r_{ij}, is represented by the first term in (6.50), corresponding to a gain in potential energy of the system. As with Hund's rules, this *potential* exchange gives rise to ferromagnetism. A prerequisite for this interaction is one or more 'points of orthogonality' in the exchange pathway.

Consider the two superexchange pathways in Figure 6.20. One may view the kinetic, antiferromagnetic exchange situation in Figure 6.20a in broadly two ways. A MO picture involves a degree of delocalisation over the whole molecule, leading directly to the pairing of the metal d electrons. Alternatively, overlap of up-spin on A with the filled bridging π-orbital removes some down-spin from the bridging atom; the remaining excess up-spin on the bridge then overlaps with B only if that spin is 'down'. In either view, the metal–bridge overlap integral must be significant. In the case of potential, ferromagnetic exchange (Figure 6.20b), similar views may be taken so far as the individual metal–bridge overlaps are concerned. However, the two p-orbitals of the bridge (or the two molecular orbitals, if one prefers that description) are mutually orthogonal. In these circumstances, Hund's rules lead to the spin-parallel, ferromagnetic result. Here too, significant metal–bridge overlap integrals are required in order to bring electron spin density from the metal atoms into close proximity at the bridging atom where the exchange interaction e^2/r_{ij} may then be effective. A theoretical treatment of the possible relationships between the magnitudes of kinetic and potential exchange on the one hand, and transfer integrals, configuration interaction and Hartree–Fock theory on the other, has been given by Anderson[38].

Returning to the evaluation of $\langle\Phi|\mathscr{H}_{ex}|\Phi\rangle$, the first stage is their expansion into component parts like $\langle\varphi_a(+2)_1\varphi_b(+2)_2|\mathscr{H}_{ex}|\varphi_a(-2)_2\varphi_b(+2)_1\rangle$ multiplied by appropriate spin-overlap integrals. Lines *et al.*[41] neglect all such elements which involve only small or zero overlap integrals as evidenced by chemical intuition or symmetry. Explicitly, the overlap integrals $\langle x^2-y^2|x'y'\rangle$, $\langle x^2-y^2|y'z'\rangle$, $\langle xz|x'y'\rangle$ and $\langle xz|y'z'\rangle$, where the prime again indicates centre b, are taken as near zero. The remaining non-zero matrix elements then reduce to three different ones, parameterised as

$$\langle[x^2-y^2, xz]_1[x'^2-y'^2, x'z']_2|\mathscr{H}_{ex}|[x^2-y^2, xz]_2[x'^2-y'^2, x'z']_1\rangle \equiv 2J_1$$

$$\langle[xy, yz]_1[x'y', y'z']_2|\mathscr{H}_{ex}|[xy, yz]_2[x'y', y'z']_1\rangle \equiv 2J_2$$

$$\langle[x^2-y^2, xz]_1[x'y', y'z']_2|\mathscr{H}_{ex}|[x'^2-y'^2, x'z']_1[xy, yz]_2\rangle \equiv 2J_3$$

where $[xy, yz]=(1+\gamma^2)^{-\frac{1}{2}}(d_{xy}-\gamma d_{yz})$

$$[x^2-y^2, xz]=(1+\gamma^2)^{-\frac{1}{2}}(d_{x^2-y^2}+\gamma d_{xz}) \qquad (6.63)$$

The important point here is that the explicit form of $\mathscr{H}_{ex}$ is not known or assumed; rather, exchange pathways are parameterised empirically by J_1, J_2 and J_3. In the earlier distortion problem we favoured parameterisation of the Hamiltonian rather than of matrix elements in the interest of 'chemical transparency'. Similar reasons oblige us to adopt the reverse position in the exchange problem.

The major overlap in J_1 involves $\langle x^2-y^2|x'^2-y'^2\rangle$. The copper ions transfer spin via σ-bonds to the mutually orthogonal p_x and p_z on the chlorine (or bromine) atom and the pathway is thus one of potential exchange leading to ferromagnetism. The kinetic term is small because of the near orthogonality of x^2-y^2 and $x'^2-y'^2$ with the Cu—Cl—Cu angle being nearly 90°. Antiferromagnetic coupling is involved in the J_2 pathway, however, involving π-bonding via the chlorine ion ($\langle xy|x'y'\rangle$) and via the oxygen ligand ($\langle yz|y'z'\rangle$). Arguments were presented to suggest that J_3 is relatively small and, as Lines *et al.*[41] neglect J_3 in their initial treatment, we shall not discuss it further here.

With these assumptions concerning pathways, the Φ matrix is

$$
\begin{array}{ccccc}
 & |\Phi_1\rangle & |\Phi_2\rangle & |\Phi_3\rangle & |\Phi_4\rangle \\
\langle\Phi_1| & \frac{1}{4}(s^2-c^2)J^+ & \frac{1}{2}scJ^- & -\frac{1}{2}scJ^- & \frac{1}{2}s^2J^+ \\
\langle\Phi_2| & \frac{1}{2}scJ^- & \frac{1}{4}(c^2-s^2)J^+ & -\frac{1}{2}c^2J^+ & \frac{1}{2}scJ^- \\
\langle\Phi_3| & -\frac{1}{2}scJ^- & -\frac{1}{2}c^2J^+ & \frac{1}{4}(c^2-s^2)J^+ & -\frac{1}{2}scJ^- \\
\langle\Phi_4| & \frac{1}{2}s^2J^+ & \frac{1}{2}scJ^- & -\frac{1}{2}scJ^- & \frac{1}{4}(s^2-c^2)J^+
\end{array}
\tag{6.64}
$$

where

$$J^+=J_1+J_2 \qquad J^-=J_1-J_2 \tag{6.65}$$

$$c=\cos(\theta/2) \qquad s=\sin(\theta/2)$$

and θ in Figure 6.19 is ca. 110°. The trigonometric functions arise from the spin overlap integrals above. Using the approximate relationship

$$c^2-s^2=\cos\theta=-\tfrac{1}{3},\ 2sc=\sin\theta=2\sqrt{2}/3 \tag{6.66}$$

Lines *et al.*[41] were able to show that the matrix (6.64) can be regenerated in the 4×4 product spin-space, $\boldsymbol{s}_1\times\boldsymbol{s}_2$, for spin bases $s_1=s_2=\frac{1}{2}$, by the spin-Hamiltonian

$$
\begin{aligned}
\mathscr{H}_{12}=&-(\tfrac{1}{9})(J^++8J^-)(s_1{}^zs_2{}^z+s_1{}^xs_2{}^x)-J^+s_1{}^ys_2{}^y\\
&+(2\sqrt{2}/9)(J^+-J^-)(s_1{}^zs_2{}^x-s_1{}^xs_2{}^z)
\end{aligned}
\tag{6.67}
$$

This exchange Hamiltonian has a symmetric part of generally non-HDVV form but, most significantly, contains an antisymmetric term in the form $D\cdot\boldsymbol{s}_1\times\boldsymbol{s}_2$, which is not necessarily small.

We do not review this paper[41] further, much more being required to derive expressions for the whole tetranuclear structure and eventual successful comparison with experiment. The important points for our present discussion are to note the way in which orbitally degenerate systems may be tackled theoretically and the way in which a phenomenological spin-Hamiltonian follows from this treatment. In general, of course, it is not necessary to construct an operator equivalent like (6.67), the matrix (6.64) often being sufficient to permit calculation of magnetic properties. Finally, we note the chemical appeal of the model of Lines *et al.* in that chemically identifiable features of the system are incorporated into the exchange integrals J_1, J_2 and J_3.

6.3.3 Dinuclear octahedral cobalt(II) complexes

We end our discussion of exchange phenomena with an introduction to some current work in our own laboratory. Magnetic anisotropy has long been of interest and it seemed desirable to measure and interpret such properties in exchange-coupled systems, particularly those in which anisotropy in susceptibilities might arise from the exchange process as well as from distortion. Concern for the chemical content of information forthcoming from such studies made us reject the HDVV approach as too phenomenological even in those cases where it might be applicable. All this suggested the investigation of the difficult problem of exchange in octahedral cobalt(II) dinuclear complexes. There arises immediately the choice of system for study: for all our efforts are based on the hope that variation in J parameters can have recognisable effects on susceptibilities in the presence of similar variations in the ligand-field parameters describing the local distortions. The fewer ligand-field parameters the better, although one can rarely expect to find such ideal systems in this respect as that studied by Lines *et al.*[41]. As no mononuclear fragment within a dinuclear complex can be centrosymmetric, we note immediately the removal of one common symmetry element in distortion problems. The best one can aim for is axial symmetry. Using the ligand 2,5-di-(2-pyridyl)-3,4-diazahexa-2,4-diene (pmk), we have prepared the molecular ion $[Co_2(pmk)_3]^{4+}$, a crystal structure analysis of which confirms our expectation of near-axial symmetry[42]. The coordination is shown schematically in Figure 6.21 and the threefold helical nature of the complex is illustrated in Figure 6.22. Our quantum model for distortion and exchange in this molecule has been developed along the following lines.

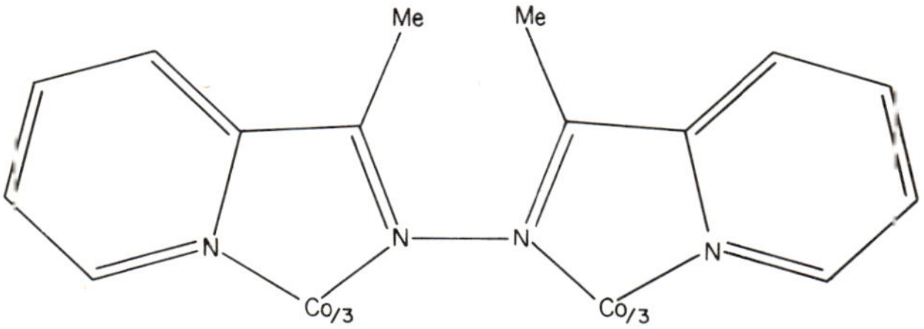

Figure 6.21 Coordination in $Co_2(pmk)_3^{4+}$.

The distortion problem is treated first. The ligand fields around each cobalt ions are taken to be equivalent and are treated within the angular overlap framework discussed earlier. The complete free-ion spin-quartet ($^4F+^4P$) terms are used as basis, but transformed into the equivalent real orbital basis for the ultimate convenience of treating real exchange pathways. Spin–orbit coupling is included simuleaneously with the interelectron repulsion and distortion Hamiltonians, as it must be for any realistic treatment of magnetic susceptibilities and anisotropies. The resulting matrix (6.68) is complex, as are the 40 eigenfunctions which result from the subsequent diagonalisation. Each eigenfunction ψ contains the 40 component functions χ (3-hole Slater determinants) of the original basis.

$$\mathscr{H}' = \sum_{i<j} \frac{e^2}{r_{ij}} + V_{\text{AOM}} + \sum_i \zeta l_i \cdot s_i \qquad (6.68)$$

The exchange problem requires the formation of product functions between the two centres. This implies a 1600 square matrix for the complete set of ψ from the distortion problem. Our first approximation, therefore, is to consider only the lowest few functions on each centre after distortion: the distortion

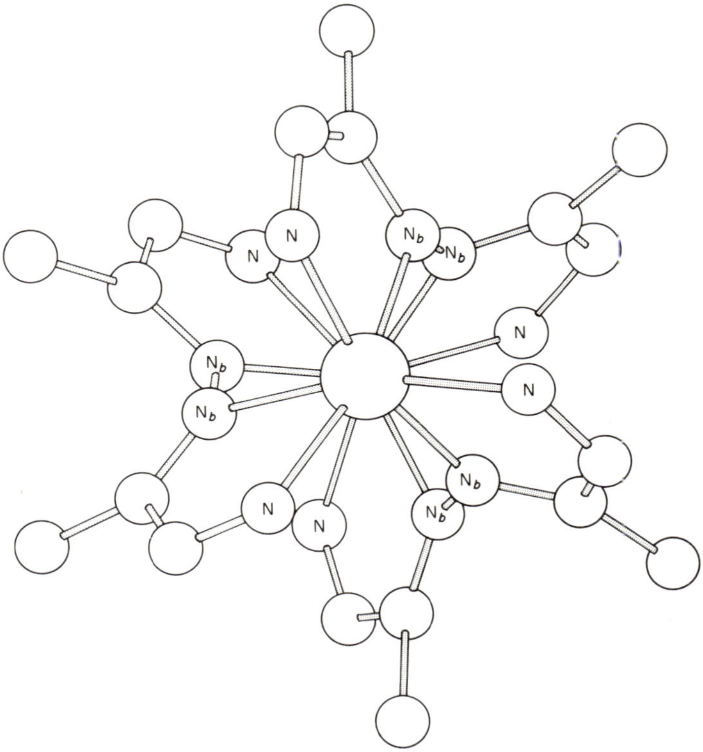

Figure 6.22 View down the approximate threefold axis (Co—Co vector) in $Co_2(pmk)_3^{4+}$.

calculation is completed first in order to recognise these lower energy states If we take the lowest six functions on each centre, for example, the exchange problem devolves into a 36×36 matrix. Each of the 36 product wavefunctions may be written as $\psi_i^A \psi_i^B$ for functions ψ on centres A and B. They are to be diagonalised under the operator:

$$\mathscr{H}'' = \sum_{i<j} \frac{e^2}{r_{ij}} + V_{\text{AOM}} + \sum_i \zeta l_i \cdot s_i + \mathscr{H}_{\text{ex}}$$

$$= \mathscr{H}' + \mathscr{H}_{\text{ex}} \qquad (6.69)$$

Matrix elements under $\mathscr{H}'$ are simple combinations of the eigenvalues resulting from the distortion calculation and so present no particular difficulties. Those under $\mathscr{H}_{\text{ex}}$, whose form we do not specify, are of the form

$$\langle \psi_I{}^{A}\psi_J{}^{\cdot}\mathscr{H}^{B}|_{ex}|\psi_K{}^{B}\psi_L{}^{A}\rangle$$

$$=\langle \sum_i^{40} c_{Ii}\chi_i{}^{A} \sum_j^{40} c_{Jj}\chi_j{}^{B}|\mathscr{H}_{ex}|\sum_k^{40} c_{Kk}\chi_k{}^{B} \sum_l^{40} c_{Ll}\chi_l{}^{A}\rangle \qquad (6.70)$$

Thus each of the 1296 ($=36\times 36$) matrix elements requires the evaluation, it seems, of 40^4 (some 2.5 million) matrix elements of the type

$$\langle \chi_i{}^{A}\chi_j{}^{B}|\mathscr{H}_{ex}|\chi_k{}^{B}\chi_l{}^{A}\rangle \qquad (6.71)$$

where the x are 3-hole Slater determinants, or anyway that number of products of the coefficient c in (6.70). While this might seem an ideal point to abandon the project, the general nature of the exchange Hamiltonian and the geometry of the molecule provide selection rules which cause most of these elements to vanish.

The exchange operator may have one- and two-electron parts. Accordingly, the products $\chi_i{}^{A}\chi_j{}^{B}$ and $\chi_k{}^{B}\chi_l{}^{A}$ in (6.71) must differ in no more than two orbitals. Further, as we are concerned with exchange integrals between the two centres rather than only on one [which latter have been included in (6.68)] the maximum number of orbital differences between any two χ is one. This furnishes a most powerful selection rule. A further 'effective selection rule' is introduced from assumptions concerning the nature of the exchange process. First, however, we note the factorisation of the exchange matrix elements (6.71) into a space-only part (exchange truly being a space-only effect) and spin-overlap integrals,

$$\langle a^{A}(1)b^{B}(1)|\mathscr{H}_{ex}|c^{B}(1)d^{A}(2)\rangle\langle s_a{}^{A}|s_c{}^{B}\rangle\langle s_b{}^{A}|s_d{}^{B}\rangle \qquad (6.72)$$

where a, b, c, d are now orbitals representing the possibly differing functions in χ after the operation of the first selection rule.

Each bridging moiety in this complex involves the structure shown in Figure 6.23. The essentially single-bond character of the N—N link is

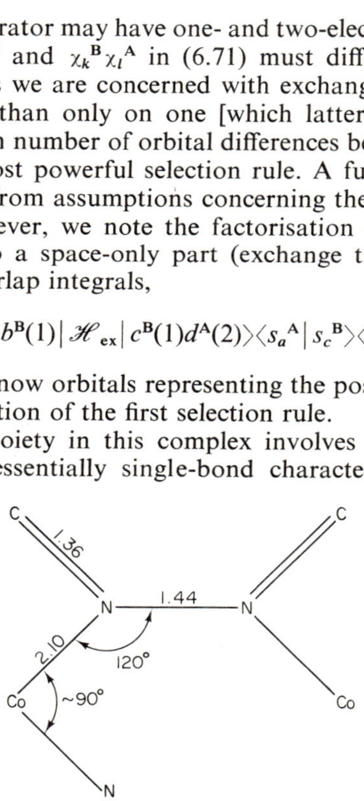

Figure 6.23 Geometry of the bridging moiety in $Co_2(pmk)_3{}^{4+}$

evidenced by the data in the figure, the short C=N bonds, and the dihedra. twist of the Co—N bonds by some 44°; the Co—Co distance is ca. 3.8 Å1 Accordingly, we assume the only significant exchange pathway in this molecule is via the σ-bonding framework in Figure 6.24.

Forming a trigonally quantized basis of real orbitals on each cobalt atom, the orbitals concerned with exchange may be labelled e_1 and e_2 from the e_g doublet of octahedral symmetry. There follows neglect of all integrals (6.72) which involve orbitals other than e_1 or e_2 on both centres A and B. This constitutes our second selection rule.

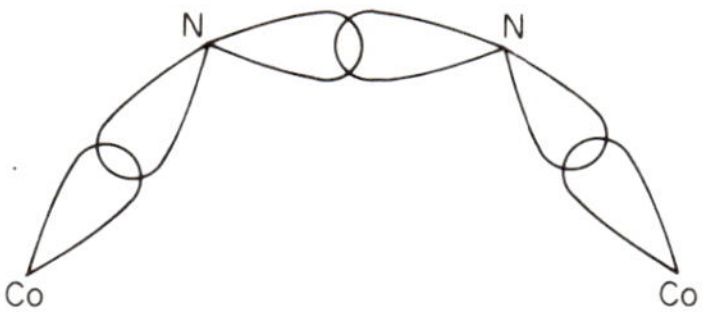

Figure 6.24 σ-bonding, kinetic exchange pathway in $Co_2(pmk)_3^{4+}$.

Systematic application of the two selection rules reduces the possible 2.5 million matrix elements to 7200, each of which involves up to 16 different integrals, corresponding to the 2^4 from $\langle e_i e_j | \mathscr{H}_{ex} | e_k e_l \rangle$ for $i, j, k, l = 1$ or 2. The hermitian properties of $\mathscr{H}_{ex}$ and the ultimate indistinguishability of electrons reduce these 16 to 7. Consider two of these by way of example:

$$\langle e_1(1)e_2(2) | \mathscr{H}_{ex} | e_1(2)e_2(1) \rangle$$
$$\langle e_1(1)e_2(2) | \mathscr{H}_{ex} | e_1(2)e_1(1) \rangle \qquad (6.73)$$

Our final assumption is to relate these, and indeed the other five, as follows All seven integrals have the —N—N— bridge unit as a common feature: they differ only in the extent to which the metal e orbitals overlap with the bridge. We may regard the overlap integrals M(e)—N(bridge) as furnishing the means of transferring metal electron spin density into the bridge region where the exchange coupling may be supposed to take place. This view is equivalent to factorising the integrals (6.73) as

$$\langle e_i(1)e_j(2) | \mathscr{H}_{ex} | e_k(2)e_l(1) \rangle = \mathscr{J} \cdot S_{e_i-N} \cdot S_{e_j-N} \cdot S_{e_k-N} \cdot S_{e_l-N} \qquad (6.74)$$

The relative magnitudes of the overlap integrals are simply the angular overlap integrals as furnished by the D matrix in the distortion angular overlap model. In this way, the 7200 matrix elements (6.70) are all reduced to calculable combinations of a single 'bridge exchange integral', $\mathscr{J}$.

Clearly, the eventual, efficient computation of the magnetic properties of a formally orbitally degenerate system like this is a long and difficult task. But the model introduced here at least has the virtue of providing a technique for a calculation, with a uniform degree of rigour, of the consequences of clearly identifiable chemical assumptions. In comparison with Levy's rigorous approach to exchange[43], in which he demonstrates the symmetry requirement of hundreds of exchange parameters, the present reduction to a single $\mathscr{J}$ parameter seems worthwhile.

As in our discussion of the distortion problem, models for magnetic properties of exchange-coupled systems have now developed beyond the purely phenomenological approach of simple operator equivalents and, though still empirical, strongly encourage the hope of obtaining chemical insight into bonding and structure, which for a time had faded.

Acknowledgements

The author wishes to thank Drs. M. G. Clark and R. G. Woolley and Prof. A. D. Buckingham for several important discussions and advice in connection with the proof of the generalised Van Vleck equation. Particular thanks go to two of my present collaborators, Dr. R. F. McMeeking, who programmed our distortion model, and Dr. P. D. W. Boyd for his work on the cobalt exchange problem.

References

1. Pauling, L. (1931). *J. Amer. Chem. Soc.*, **53,** 367, 3225; (1932). *ibid.*, **54,** 988
2. E.g., Nyholm, R.S. (1956). *Report to the Xth Solvay Council, Brussels*
3. Bleaney, B. and Stevens, K. W. H. (1953). *Reports Progr. Phys.*, **16,** 108
4. Stevens, K. W. H. (1953). *Proc. Roy. Soc.*, **A219,** 542
5. Figgis, B. N. (1961). *Trans. Faraday Soc.*, **57,** 204
6. Figgis, B. N., Lewis, J., Mabbs, F. E. and Webb, G. A. (1967). *J. Chem. Soc. A*, 1411
7. Figgis, B. N., Lewis, J., Mabbs, F. E. and Webb, G. A. (1967). *J. Chem. Soc. A*, 442
8. Figgis, B. N., Gerloch, M., Lewis, J., Mabbs, F. E. and Webb, G. A. (1968). *J. Chem. Soc. A*, 2086
9. Orgel, L. E. (1960). *Chemistry of Transition Metal Ions* (London: Methuen)
10. Bentley, R. B., Mabbs, F. E., Smail, W. R., Gerloch, M. and Lewis, J. (1970). *J. Chem. Soc. A*, 3003
11. Owsten, P. G. and Rowe, J. M. (1963). *J. Chem. Soc.*, 3411
12. Ferguson, J. (1964). *J. Chem. Phys.*, **40,** 3406
13. Figgis, B. N., Gerloch, M., Lewis, J. and Slade, R. C. (1968). *J. Chem. Soc. A*, 2028
14. Gerloch, M. (1968). *J. Chem. Soc. A*, 2023
15. Van Vleck, J. H. (1932). *Theory of Electric and Magnetic Susceptibilities* (Oxford: University Press)
16. Figgis, B. N., Gerloch, M. and Mason, R. (1969). *Proc. Roy. Soc.*, **A309,** 91
17. Gerloch, M., Lewis, J., Phillips, G. G. and Quested, P. N. (1970). *J. Chem. Soc. A*, 1941
18. Gerloch, M. and Miller, J. R. (1968). *Progr. Inorg. Chem.*, **10,** 1
19. Nyholm, R. S. (1967). *IUPAC Xth International Conference on Co-ordination Chemistry, Japan* (London: Butterworths)
20. Gladney, H. M. and Swalen, J. D. (1965). *J. Chem. Phys.*, **42,** 1999
21. Gerloch, M. and Slade, R. C. (1973). *Ligand Field Parameters* (Cambridge: University Press)
22. Gerloch, M. and Quested, P. N. (1971). *J. Chem. Soc. A*, 3729
23. Gerloch, M. and Mackey, D. J. (1970). *J. Chem. Soc. A*, 3030, 3040; (1971). *ibid.*, 2605, 2516, 3372; (1972). *ibid.*, 37, 42, 410, 415, 1555
24. Schäffer, C. E. and Jørgensen, C. K. (1965). *Mol. Phys.*, **9,** 401
25. Schäffer, C. E. (1968). *Structure and Bonding*, **5,** 68
26. Schäffer, C. E. (1967). *Proc. Roy. Soc.*, **A297,** 96
27. Schäffer, C. E. (1970). *Pure Appl. Chem.*, **24,** 361
28. Harnung, S. and Schäffer, C. E. (1974). *Structure and Bonding*, in the press
29. Racah, G. (1942). *Phys. Rev.*, **62,** 438
30. Judd, B. R. (1962). *Operator Techniques in Atomic Spectroscopy* (New York: McGraw-Hill)
31. Wybourne, B. G. (1965). *Spectroscopic Properties of Rare Earths* (New York: Interscience)
32. Gerloch, M. and Quested, P. N. (1971). *J. Chem. Soc. A*, 2308
33. Weiss, P. (1907). *J. Phys.*, **6,** 667
34. Heisenberg, W. (1926). *Z. Physik*, **38,** 411; (1928). *ibid.*, **49,** 619

35. Dirac, P. A. M. (1929). *Proc. Roy. Soc.*, **A123**, 714
36. Löwdin, P. O. (1962). *Rev. Mod. Phys.*, **34**, 80
37. Condon, E. U. and Shortley, G. H. (1935). *Theory of Atomic Spectra*, (Cambridge: University Press)
38. Anderson, P. W. (1963). *Solid State Phys.*, **14**, 99
39. Pedesen, E. (1972). *Acta Chem. Scand.*, **26**, 333
40. Glerup, J. (1972). *Acta Chem. Scand.*, **26**, 3775
41. Lines, M. E., Ginsberg, A. P., Martin, R. L. and Sherwood, R. C. (1972). *J. Chem. Phys.*, **57**, 1
42. Boyd, P. D. W., Gerloch, M. and Sheldrick, G. M. (1974). *J. Chem. Soc. Dalton Trans.*, 1097
43. Levy, P. M. (1968). *Phys. Rev. Lett.*, **20**, 1366

7
Hyperfine Interactions and Molecular Structure

M. G. CLARK
University Chemical Laboratory, Cambridge*

7.1 INTRODUCTION 240
7.1.1 *Purpose and scope* 240
7.1.2 *The hyperfine interactions* 240
7.1.3 *Experimental manifestations* 241

7.2 THEORY AND PARAMETERS 243
7.2.1 *The electric multipole expansion* 243
7.2.2 *Electric hyperfine interactions* 245
7.2.3 *Nuclear quadrupole coupling* 246
7.2.3.1 *The hamiltonian* 246
7.2.3.2 *Symmetrised parameters* 247
7.2.3.3 *The Sternheimer effect* 248
7.2.4 *E0 nuclear volume effects* 250
7.2.4.1 *The interaction* 250
7.2.4.2 *Isomer and isotope shifts* 252
7.2.5 *Magnetic hyperfine interactions* 253
7.2.5.1 *Magnetic multipoles* 253
7.2.5.2 *Magnetic dipole coupling* 254
7.2.5.3 *The contact hyperfine field* 255
7.2.5.4 *Hyperfine anomaly* 256
7.2.6 *Relativistic hyperfine operators* 256
7.2.7 *Atomic and nuclear parameters* 257
7.2.7.1 *The coupling constant dilemma* 257
7.2.7.2 *E2 moments of deformed nuclei* 258
7.2.7.3 $\Delta\langle r^k\rangle_n$ *and nuclear decay rates* 260

7.3 QUADRUPOLE COUPLING AND MOLECULAR STRUCTURE 261
7.3.1 *Parameters and data* 261
7.3.1.1 *Symmetry restrictions* 261

* Present address: Royal Radar Establishment, Malvern, Worcestershire

7.3.1.2 *Experimental constraints* 262
7.3.1.3 *Precision and reproducibility* 262
7.3.2 *Strategies and concepts* 263
7.3.2.1 *Molecular symmetry* 263
7.3.2.2 *Symmetry-breaking perturbations* 264
7.3.2.3 *Contributions to the EFG in covalent and ionic systems* 266

7.4 QUADRUPOLE COUPLING IN COVALENT MOLECULES 267
7.4.1 *Valence and lattice contributions* 267
7.4.1.1 *Effective population* 267
7.4.1.2 *Valence shell contributions* 269
7.4.1.3 *Overlap of filled orbitals* 270
7.4.1.4 *The 'lattice' contribution* 270
7.4.2 *The additive model* 271
7.4.2.1 *Characteristics of additivity* 271
7.4.2.2 *Significance of* $\eta \approx 1$ 274
7.4.2.3 *A molecular orbital approach* 276
7.4.2.4 *Partial field gradient parameters* 277
7.4.2.5 *The regression method applied to five-coordinate* Sn^{IV} 281
7.4.3 *Non-additive systems* 284
7.4.3.1 *General remarks* 284
7.4.3.2 *Trigonal-bipyramidal carbonyl complexes of* ($3d^8$) Fe^0 and Co^I 284
7.4.3.3 ($5d^2$) W^{IV} *and* ($5d^1$) W^V *octacyanotungstate anions* 287
7.4.4 *Distortions from ideal geometry* 290

ACKNOWLEDGEMENTS 292

7.1 INTRODUCTION

7.1.1 Purpose and scope

This review is concerned with the application of hyperfine studies to problems in electronic structure and bonding. As comprehensive coverage of this field would far exceed the limitations of a single review, we have elected to concentrate on the electric hyperfine interactions, and in particular on the quadrupole interaction. However, we have tried to give a unified account of the theoretical background to hyperfine studies, believing that such an approach is desirable.

7.1.2 The hyperfine interactions

The atomic nucleus[1,2] is a compact body with a characteristic radius of ca. $1.2A^{\frac{1}{3}} \times 10^{-13}$ cm, where A is the total number of nucleons (protons +

neutrons). Thus the natural unit of nuclear length is 10^{-13} cm, named the fermi, or in SI units the femtometre (both conveniently abbreviated to fm). The distribution of atomic electrons is much more diffuse, with a characteristic length of 10^{-8} cm, the familiar ångström unit.

Protons and electrons carry charges of $+e$ and $-e$, respectively. It is the electrostatic interaction of these charges which gives rise to the electric hyperfine interactions. The motions of the nucleons and the electrons set up current distributions, and the magnetic interaction between these currents gives the magnetic hyperfine interactions[3]. All observed hyperfine interactions, including those which arise because the electronic distribution penetrates the nucleus, can be understood in terms of these electromagnetic effects.

Typical hyperfine coupling energies are very much smaller than the electron–electron and nucleon–nucleon coupling energies (Table 7.1). Thus

Table 7.1 Comparison of typical hyperfine coupling energies with other electronic and nuclear energies

Interaction or motion	*Typical energy*/eV
Nucleon–nucleon	10^{6}
Electron–electron	1
Molecular vibration	10^{-1}
Molecular rotation	10^{-4}
Hyperfine	10^{-8}

the nuclear angular momentum is conserved to a very good approximation, and the nuclear states may be characterised by the angular momentum quantum numbers (I, M_I) and the inversion parity $\pi = \pm$.

To put this another way: the nucleus is only weakly coupled to its environment, with the result that the nuclear states are conveniently classified under the spherical point-symmetry group $O(3)$. It follows that the electric and magnetic couplings with the electrons may be expanded in terms of eigenfunctions of spherical symmetry to give the so-called multipole expansions[4]. By this method the details of nuclear structure are encapsulated in sets of magnetic and electric moments, of which the most familiar are the magnetic dipole and electric quadrupole moments. These quantities reflect the non-spherical nature of nuclei with non-zero spin I: existence of a dipole moment requires $I \geqslant \frac{1}{2}$, while a quadrupole requires $I \geqslant 1$.

7.1.3 Experimental manifestations

Hyperfine couplings may be detected by their effect on the energy levels of either the nucleus or the electrons. Although techniques such as low-temperature specific heat studies are sometimes useful[5], most widely-used experimental methods are spectroscopic in character. Indeed, the existence of nuclear magnetic moments was first proposed in 1924 by Pauli[6] to explain the already-known fact that the 'fine-structure' spin–orbit splitting of atomic spectral lines could sometimes be resolved into 'hyperfine structure'. About a decade later the nuclear quadrupole interaction was also recognised as a result of detailed analysis of atomic spectra[7]. The atomic and molecular beam techniques first developed during this period provide a method for precision studies of hyperfine coupling[8,9].

Following war-time developments in microwave technology, further spectroscopic probes of the electronic system were provided by electron paramagnetic resonance (e.p.r.)[10-12] and microwave spectroscopy[13], while in the nuclear magnetic resonance (n.m.r.)[11,14,15] experiment magnetic-dipole transitions between the Zeeman-split sublevels of a ground-state nucleus were observed directly. Nuclear resonance of quadrupolar nuclei in zero magnetic field[15,16], so-called nuclear quadrupole resonance (n.q.r.) or pure quadrupole resonance, is essentially a variant of the n.m.r. experiment despite its different name[15,17,18]. More recent developments have emphasised the advantages of combining different experiments (e.g. ENDOR[12,19]) and the application of sophisticated methods such as optical pumping[20].

All the above techniques are essentially restricted to the ground nuclear state. However, vector coupling of the (spin and orbital) angular momenta of Z protons and N neutrons, each with intrinsic spin $\frac{1}{2}$, according to the shell model[1,2], leads to systematic restrictions on the total spins of the nuclides (Table 7.2). From Table 7.2 it is seen that in their ground states many nuclides have spin zero, and thus will not show magnetic dipole or electric quadrupole interactions. Study of excited nuclear states substantially increases the range of nuclides available.

Table 7.2 Systematics of nuclear spin*

Z	N	*Number of stable nuclides*[2]	*Nuclear spin* *ground state*	*general state*
Even	even	163	zero	integer
Even	odd	54	$\frac{1}{2}$ to $\frac{9}{2}$	half-integer
Odd	even	49	$\frac{1}{2}$ to $\frac{9}{2}$	half-integer
Odd	odd	4	1 to 7	integer

* Z = no. of protons; N = no. of neutrons

Widespread interest in the application of nuclear spectroscopic techniques to chemistry and chemical physics seems to have followed the discovery and rapid development of the Mössbauer effect[21,22]. The stimulus to electric hyperfine studies has been particularly marked. Whereas in a tabulation[23] of n.q.r. frequencies obtained up to the end of 1966 most data referred to the halogen nuclei $^{35,37}Cl$, $^{79,81}Br$ and ^{127}I, Mössbauer spectroscopy had by 1965 already yielded considerable information[24] on quadrupole interactions of metal nuclei such as ^{57}Fe and ^{119}Sn. This increased interest in metal nuclei has been reinforced by developments in n.q.r. techniques[25] and wider application of other nuclear techniques, such as perturbed angular correlations[26, 27] and conversion electron spectroscopy[27, 28].

The proceedings of the 1966 NATO Advanced Study Institute provide a particularly useful survey of the whole range of experimental methods used in hyperfine studies[29], while other conference surveys have concentrated on hyperfine interactions in excited nuclei[30].

7.2 THEORY AND PARAMETERS

7.2.1 The electric multipole expansion

The energy of electrostatic interaction between a nuclear charge density $\rho_n(\boldsymbol{r})$ and an electronic charge density $\rho_e(\boldsymbol{r})$ is given by

$$W_E = \int \rho_n(\boldsymbol{r}) V_e(\boldsymbol{r})\, d^3\boldsymbol{r} = \int \rho_e(\boldsymbol{r}) V_n(\boldsymbol{r})\, d^3\boldsymbol{r} \tag{7.1}$$

where $V_n(\boldsymbol{r})$ and $V_e(\boldsymbol{r})$ are the potentials due to the nuclear and electronic charge distributions. Multipole expansion of this interaction is most elegantly achieved by use of the generalised three-dimensional δ functions described by Brink and Satchler[4]. The function $\delta^p_{LM}(\boldsymbol{r})$ picks out the coefficient F^p_{LM} in the series expansion of a function $f(\boldsymbol{r})$

$$f(\boldsymbol{r}) = \sum_{p=0}^{\infty} \sum_{L=0}^{\infty} \sum_{M=-L}^{L} F^p_{LM} r^{L+2p} C_{LM}(\theta,\phi) \tag{7.2}$$

where $C_{LM}(\theta,\phi)$ are the modified spherical harmonics[4], by the equation

$$F^{p*}_{LM} = \int \delta^p_{LM}(\boldsymbol{r}) f(\boldsymbol{r})^*\, d^3\boldsymbol{r} \tag{7.3}$$

The requirement for equation (7.2) is that $f(\boldsymbol{r})$ be regular at the origin, i.e. all partial derivatives non-singular*.

The $\delta^p_{LM}(\boldsymbol{r})$ are $(L + 2p)$th derivatives of $\delta(\boldsymbol{r})$, and, like $\delta(\boldsymbol{r})$ itself, are singular at the origin and zero everywhere else. Explicit formulae for the $\delta^p_{LM}(\boldsymbol{r})$ will not be required.

Since (Section 7.1.2) the nucleus is a compact body, W_E may be calculated by expanding $V_e(\boldsymbol{r})$ about an origin located at the nuclear centre of inversion,

$$V_e(\boldsymbol{r}) = \sum_{pLM} Q^p_{LM}(e) r^{L+2p} C^*_{LM}(\theta,\phi) \tag{7.4}$$

where the electronic multipole fields $Q^p_{LM}(e)$ are given, from (7.3), by

$$Q^p_{LM}(e) = \int \delta^p_{LM}(\boldsymbol{r}) V_e(\boldsymbol{r})\, d^3\boldsymbol{r} \tag{7.5}$$

* Mathematically, this is fairly restrictive. For example, e^{-r} cannot be expanded according to equation (7.2); the expansion

$$e^{-r} = 1 - r + \tfrac{1}{2}r^2 - \ldots$$

is a Taylor series in one dimension.

Inserting (7.4) into (7.1) gives

$$W_{\mathrm{E}} = \sum_{pLM} Q^{p}_{LM}(\mathrm{e}) Q^{p*}_{LM}(\mathrm{n}) \tag{7.6}$$

where

$$Q^{p}_{LM}(\mathrm{n}) = \int \rho_{\mathrm{n}}(\boldsymbol{r}) r^{L+2p} C_{LM}(\theta,\phi)\, \mathrm{d}^3\boldsymbol{r} \tag{7.7}$$

are the electric multipole moments of the nucleus.

From (7.5) and (7.6)

$$W_{\mathrm{E}} = \int \left[\sum_{pLM} Q^{p*}_{LM}(\mathrm{n}) \delta^{p}_{LM}(\boldsymbol{r}) \right] V_{\mathrm{e}}(\boldsymbol{r})\, \mathrm{d}^3\boldsymbol{r} \tag{7.8}$$

Thus, by comparing (7.8) and (7.1) it is seen that for the purpose of calculating W_{E} the true nuclear charge distribution $\rho_{\mathrm{n}}(\boldsymbol{r})$ has been replaced by the sum

$$\sum_{pLM} Q^{p*}_{LM}(\mathrm{n}) \delta^{p}_{LM}(\boldsymbol{r})$$

which is zero everywhere except at the origin. This is the essence of the multipole expansion.

In (7.4) the terms with $p = 0$ satisfy $\nabla^2 V = 0$, whereas those with $p > 0$ do not. Thus in (7.6) the terms with $p = 0$ correspond to interaction of the nuclear charge density with external electronic charge. These interactions would occur even for a point nucleus, whereas the terms with $p > 0$ correspond to interaction between overlapping nuclear and electronic charge densities, and thus allow for finite nuclear size. The latter interactions will be called nuclear volume effects. For given L the interaction with $p = 0$ will be denoted EL, and the volume effect $\left(\sum_{p>0}\right)$ will be denoted EL^{vol}.

Expressed in terms of $\rho_{\mathrm{e}}(\boldsymbol{r})$ the multipole fields $Q^{p}_{LM}(\mathrm{e})$ are given by[4]

$$Q_{LM}(\mathrm{e}) = \int \rho_{\mathrm{e}}(\boldsymbol{r}) r^{-(L+1)} C_{LM}(\theta,\phi)\, \mathrm{d}^3\boldsymbol{r} \qquad (p = 0) \tag{7.9}$$

$$Q^{p}_{LM}(\mathrm{e}) = -[2\pi/p(2p + 2L + 1)] \int \rho_{\mathrm{e}}(\boldsymbol{r}) \delta^{p-1}_{LM}(\boldsymbol{r})\, \mathrm{d}^3\boldsymbol{r} \quad (p > 0) \tag{7.10}$$

The quantum-mechanical multipole operators may now be written down immediately from (7.7), (7.9) and (7.10):

$$\hat{Q}^{p}_{LM}(\mathrm{n}) = e \sum_{\text{protons}} r_i^{L+2p} C_{LM}(\theta_i,\phi_i) \tag{7.11}$$

$$\hat{Q}_{LM}(\mathrm{e}) = \sum e_i r_i^{-(L+1)} C_{LM}(\theta_i,\phi_i) \qquad (p = 0) \tag{7.12}$$

$$\hat{Q}^{p}_{LM}(\mathrm{e}) = -[2\pi/p(2p + 2L + 1)] \sum \delta^{p-1}_{LM}(\boldsymbol{r}_i) \qquad (p > 0) \tag{7.13}$$

where the sum in (7.12) and (7.13) is over all charged particles (electrons and nuclei) except the interacting nucleus. The hamiltonian is then

$$\mathscr{H}_{\mathrm{E}} = \sum_{pLM} \hat{Q}^{p*}_{LM}(\mathrm{n}) \hat{Q}^{p}_{LM}(\mathrm{e}) \tag{7.14}$$

7.2.2 Electric hyperfine interactions

Hyperfine interaction energies are typically much smaller than the nuclear and electronic energy-level spacings (Table 7.1), and thus (subject to corrections discussed in Sections 7.2.3.3 and 7.2.5.3) may be calculated by first-order perturbation theory*. The wave functions may be written as products $|nIM\pi\rangle|e\varepsilon\rangle$, where $|nIM\pi\rangle$ $(M = -L, -L+1, \ldots, L)$ are the wave functions of a nuclear level n with spin I, parity π, and $|e\varepsilon\rangle$ $(\varepsilon = \varepsilon', \varepsilon'', \ldots)$ are the wave functions of an electronic level e. Then, from (7.14),

$$\langle nIM'\pi|\langle e\varepsilon'|\mathscr{H}_{\mathrm{E}}|e\varepsilon''\rangle|nIM''\pi\rangle = \sum_{pLM} \langle nIM'\pi|\hat{Q}^{p*}_{LM}(\mathrm{n})|nIM''\pi\rangle\langle e\varepsilon'|\hat{Q}^{p}_{LM}(\mathrm{e})|e\varepsilon''\rangle \tag{7.15}$$

where, by the Wigner–Eckart theorem[4]

$$\langle nIM'\pi|\hat{Q}^{p*}_{LM}(\mathrm{n})|nIM''\pi\rangle = (-1)^{I-M'}\begin{pmatrix} I & L & I \\ -M' & M & M'' \end{pmatrix}(2I+1)^{\frac{1}{2}}\langle nI\pi||\hat{Q}^{p*}_{L}(\mathrm{n})||nI\pi\rangle \tag{7.16}$$

Since, by (7.11), $\hat{Q}^{p}_{LM}(\mathrm{n})$ has parity $(-1)^L$ the reduced matrix element $\langle nI\pi||\hat{Q}^{p*}_{L}(\mathrm{n})||nI\pi\rangle$ is zero for L odd. Further, the $3-j$ symbol vanishes unless both $L \leqslant 2I$ and $M' = M + M''$. Thus a nucleus with spin I has only the $E0$ (monopole), $E2$ (quadrupole), ..., EL (2^L-pole), ... interactions, up to $L = 2I$ (I = integer) or $L = 2I - 1$ (I = integer + $\frac{1}{2}$).

It follows from the assumption that (I,M) are good quantum numbers in the unperturbed nucleus that a given EL moment can be characterised by a single number proportional to the reduced matrix element. For example, the spectroscopic quadrupole moment Q is defined by

$$\tfrac{1}{2}eQ = \langle nII\pi|\hat{Q}_{20}(\mathrm{n})|nII\pi\rangle \tag{7.17}$$

Thus the assumption, sometimes made when introducing this quantity, that the nucleus is 'axially symmetric' is redundant.

The $E0$ external ($p = 0$) interaction is the coulomb interaction of a point nucleus with the electrons, and is properly part of the zero-order hamiltonian. From (7.5) and (7.7) we see that, relative to this interaction, the magnitude of $Q^{p*}_{LM}(\mathrm{n})Q^{p}_{LM}(\mathrm{e})$ varies as $(r_{\mathrm{n}}/r_{\mathrm{e}})^{L+2p}$, showing that the importance of successive interactions decreases rapidly with increasing L. Thus only the $E2$ interaction is widely studied, although evidence for nuclear distortions corresponding to $E4$ and possibly $E6$ moments has been obtained in various nuclear scattering experiments[31,32], and more recently from studies of muonic x-rays[33]. Several authors have contemplated the possibility of observing $E4$ coupling spectroscopically[34].

The $E0^{\mathrm{vol}}$ interaction is well-established experimentally (Section 7.2.4). A volume effect with $L > 0$ gives level splittings of the same form as the corresponding EL interaction, but may, in principle, be detected through a hyperfine anomaly, i.e. apparent dependence of the nuclear moment on

* Space does not permit systematic discussion of effects due to off-diagonal matrix elements of the electric and magnetic hyperfine operators. The reader should not infer from this that they are unimportant. For example, the magnetic dipole operator in second order gives the so-called pseudoquadrupole interaction.

chemical environment, studied by measuring the ratio of the coupling energies of a pair of isotopes in various environments. In practice even the $E2$ anomaly does not seem to have been definitely observed in normal atoms, although such effects can be observed in muonic atoms, where the greater mass of the muon ($m_\mu \approx 207m_e$) leads to a much smaller Bohr radius and considerable enhancement of all electric hyperfine interactions[35].

7.2.3 Nuclear quadrupole coupling

7.2.3.1 The hamiltonian

Since the nuclear operators $\hat{Q}_{2M}(\mathrm{n})$ are only required to act within a manifold of fixed I they can be replaced by equivalent operators[4]

$$\tilde{Q}_{20}(\mathrm{n}) = [eQ/2I(2I-1)][3\hat{I}_z^2 - I(I+1)] \tag{7.18a}$$

$$\tilde{Q}_{2\pm1}(\mathrm{n}) = \mp(3/2)^{\frac{1}{2}}[eQ/2I(2I-1)][\hat{I}_z\hat{I}_\pm + \hat{I}_\pm\hat{I}_z] \tag{7.18b}$$

$$\tilde{Q}_{2\pm2}(\mathrm{n}) = (3/2)^{\frac{1}{2}}[eQ/2I(2I-1)](\hat{I}_\pm)^2 \tag{7.18c}$$

where Q is defined in (7.17) and $\hat{I}_\pm = \hat{I}_x \pm \mathrm{i}\hat{I}_y$.

The treatment of the electronic operators $\hat{Q}_{2M}(\mathrm{e})$ depends on details of the system concerned. If, as occurs for free atoms, ions and molecules in the gas phase, and for the f electrons of rare-earth ions even in condensed phases, the electronic system has a well-defined total angular momentum J, then, provided the splittings between J levels are much greater than the $E2$ interaction energy, the electronic operators may be replaced by equivalents exactly analogous to (7.18):

$$\tilde{Q}_{20}(\mathrm{e}) = [eq_J/2J(2J-1)][3\hat{J}_z^2 - J(J+1)] \tag{7.19a}$$

$$\tilde{Q}_{2\pm1}(\mathrm{e}) = \mp(3/2)^{\frac{1}{2}}[eq_J/2J(2J-1)][\hat{J}_z\hat{J}_\pm + \hat{J}_\pm\hat{J}_z] \tag{7.19b}$$

$$\tilde{Q}_{2\pm2}(\mathrm{e}) = (3/2)^{\frac{1}{2}}[eq_J/2J(2J-1)](\hat{J}_\pm)^2 \tag{7.19c}$$

where q_J is defined by [*cf.* (7.17)]

$$\tfrac{1}{2}eq_J = \langle JJ|\hat{Q}_{20}(\mathrm{e})|JJ\rangle \tag{7.20}$$

(Notice that the triangle rule for coupling angular momenta requires that $q_{\frac{1}{2}} = 0$).

The hamiltonian then becomes[8, 13]

$$\mathscr{H}_{E2} = [e^2q_JQ/2I(2I-1)J(2J-1)][3(\hat{\boldsymbol{I}}\cdot\hat{\boldsymbol{J}})^2 + \tfrac{3}{2}(\hat{\boldsymbol{I}}\cdot\hat{\boldsymbol{J}}) - I(I+1)J(J+1)] \tag{7.21}$$

which (in the absence of external fields) has eigenvalues

$$W_{E2}(F,I,J) = [e^2q_JQ/2I(2I-1)J(2J-1)][\tfrac{3}{4}K(K+1) - I(I+1)J(J+1)] \tag{7.22}$$

where $\boldsymbol{F} = \boldsymbol{I} + \boldsymbol{J}$ is the total angular momentum ($F = I+J,\ I+J-1,\ \ldots,\ |I-J|$) and

$$K = F(F+1) - I(I+1) - J(J+1) \tag{7.23}$$

Experiments on a system described by (7.21) yield values for the quadrupole coupling constant e^2q_JQ.

In molecule-fixed axes the $\hat{Q}_{2M}(\mathrm{e})$ are conveniently related to cartesian tensor operators $\hat{V}_{\alpha\beta}$

$$\hat{Q}_{20}(\mathrm{e}) = \tfrac{1}{2}\hat{V}_{zz} \tag{7.24a}$$

$$\hat{Q}_{2\pm1}(\mathrm{e}) = \mp(1/\sqrt{6})(\hat{V}_{xz} \pm \mathrm{i}\hat{V}_{yz}) \tag{7.24b}$$

$$\hat{Q}_{2\pm2}(\mathrm{e}) = (1/2\sqrt{6})(\hat{V}_{xx} \pm 2\mathrm{i}\hat{V}_{xy} - \hat{V}_{yy}) \tag{7.24c}$$

where

$$\hat{V}_{\alpha\beta} = \sum_i e_i r_i^{-5}(3x_{\alpha i}x_{\beta i} - r_i^2\delta_{\alpha\beta}) \tag{7.25}$$

the sum being over all particles (electrons and other nuclei) with charges e_i and coordinates $\boldsymbol{r}_i$. The $\hat{V}_{\alpha\beta}$ have two properties of note: they are symmetric ($\hat{V}_{\alpha\beta} = \hat{V}_{\beta\alpha}$) and traceless ($\hat{V}_{xx} + \hat{V}_{yy} + \hat{V}_{zz} = 0$). The expectation values $\langle\hat{V}_{\alpha\beta}\rangle \equiv V_{\alpha\beta}$ form a symmetric traceless second-rank cartesian tensor. It, or more strictly its negative, is called the electric field gradient (EFG) tensor. Allowing for the symmetry and zero trace, it has five independent components. The relationship between the quantity q_J and the components of the EFG tensor in molecule-fixed axes is well-documented[8, 13].

In condensed phases[17, 18], where any orbital angular momentum is usually at least partly quenched, it is convenient to write $\mathscr{H}_{E2}$ directly in terms of the $V_{\alpha\beta}$. In principal axes (x,y,z)

$$\mathscr{H}_{E2} = [eQ/2I(2I-1)][V_{xx}\hat{I}_x^2 + V_{yy}\hat{I}_y^2 + V_{zz}\hat{I}_z^2] \tag{7.26}$$

There are $3! = 6$ ways of assigning the labels x, y and z, so it is conventional[18] to label the principal axes so that

$$|V_{zz}| \geqslant |V_{yy}| \geqslant |V_{xx}| \tag{7.27}$$

and to define an axial component

$$eq \equiv V_{zz} \tag{7.28}$$

and an asymmetry parameter

$$\eta \equiv (V_{xx} - V_{yy})/V_{zz} \tag{7.29}$$

It follows[18] that $0 \leqslant \eta \leqslant 1$ and

$$\mathscr{H}_{E2} = [e^2qQ/4I(2I-1)][3\hat{I}_z^2 - I(I+1) + \tfrac{1}{2}\eta(\hat{I}_+^2 + \hat{I}_-^2)] \tag{7.30}$$

The coupling constant, e^2qQ, and η are measures of any asymmetry in the environment of the probe nucleus; η vanishes for axial (three-fold or greater) symmetry and both vanish for cubic or higher symmetries.

7.2.3.2 Symmetrised parameters

The characteristic equation of $V_{\alpha\beta}$

$$\det\{V_{\alpha\beta} - x\delta_{\alpha\beta}\} = 0$$

can be written

$$x^3 + xS_{02} - S_{03} = 0 \tag{7.31}$$

where the coefficient of x^2 vanishes because $V_{\alpha\beta}$ is traceless. Now the quantities S_{02} and S_{03} uniquely determine the principal values of $V_{\alpha\beta}$, and hence (q,η), and vice versa. Thus the 'symmetrised parameters'[36] (S_{02},S_{03}) provide an alternative parameterisation of the EFG to (q,η). Since S_{02},S_{03} are defined independently of choice of axes, they offer particular advantages in situations where enforcement of the ordering convention (7.27) is inconvenient. The notation S_{02},S_{03} is chosen for convenience in more general cases such as combined magnetic dipole and electric quadrupole interactions[36].

In terms of q and η the symmetrised parameters are

$$S_{02} = -\tfrac{3}{4}e^2q^2(1 + \tfrac{1}{3}\eta^2) \tag{7.32a}$$

$$S_{03} = \tfrac{1}{4}e^3q^3(1 - \eta^2) \tag{7.32b}$$

The energy-moment method[37], in which hyperfine parameters are calculated directly from line positions in n.m.r./n.q.r.[37] or Mössbauer[38] spectra, amounts to experimental determination of the symmetrised parameters[36]. Error in these experimental values of S_{02} and S_{03} can cause difficulties when transforming to (q,η) if as a result the condition for (7.31) to have three real roots is not satisfied*. This actually occurred unrecognised in an example used as an illustration[38] when the method was applied to Mössbauer spectroscopy.

The energy levels E_i of a nucleus with spin I in the presence of an EFG are the roots of the characteristic equation

$$\begin{aligned} 0 &= \det\{\langle IM|\mathscr{H}_{E2} - E\mathscr{I}|IM'\rangle\} \\ &= E^{2I+1} + c_2E^{2I-1} + \ldots + c_tE^{2I+1-t} + \ldots + c_{2I+1} \end{aligned}$$

where $\mathscr{I}$ is the unit operator and $c_1 = 0$ for $\mathscr{H}_{E2}$. It has been shown[36] that the c_t are polynomials in S_{02} and S_{03}, and that the form of these polynomials can often be deduced from dimensional analysis.

7.2.3.3 The Sternheimer effect

Filled deep-lying core atomic orbitals (AOs) are relatively unperturbed by chemical bonding, and, since they have spherical symmetry in the central field approximation, might be expected to make negligible contributions to the EFG at the nucleus. In fact this is not so; the core orbitals are distorted by the $E2$ field and make significant contributions to the EFG[39].

The main physical features may be seen by a simple perturbation argument which ignores exchange between core electrons and the polarising electronic distribution. If core electrons located at points $\boldsymbol{r}_c$ are polarised by a charge distribution $\rho(\boldsymbol{R})$, then the core wave function Φ is given, in atomic units, by

$$|\Phi\rangle = |\Phi_0\rangle - \sum_{N\neq 0} |\Phi_N\rangle\langle\Phi_N|\sum_c \int |\boldsymbol{r}_c - \boldsymbol{R}|^{-1}\rho(\boldsymbol{R})\,\mathrm{d}^3\boldsymbol{R}|\Phi_0\rangle\Delta_{N0}^{-1} + O(\Delta^{-2}) \tag{7.33}$$

* This is most likely to occur when η is small or zero, in which case the one real root obtained may still be an acceptable approximation to V_{zz}.

where $\Delta_{N0} = E_N - E_0$. The $E2$ multipole field due to the core electrons is then

$$Q_{2M}^{\text{core}}(\text{e}) = \langle \Phi | \hat{Q}_{2M}(\text{e}) | \Phi \rangle$$

$$= \langle \Phi_0 | \hat{Q}_{2M}(\text{e}) | \Phi_0 \rangle - 2\text{Re} \sum_{N \neq 0} [\langle \Phi_0 | \hat{Q}_{2M}(\text{e}) | \Phi_N \rangle \times$$

$$\langle \Phi_N | \sum_c \int |\boldsymbol{r}_c - \boldsymbol{R}|^{-1} \rho(\boldsymbol{R}) \, \text{d}^3\boldsymbol{R} | \Phi_0 \rangle \Delta_{N0}^{-1}] + O(\Delta^{-2}) \quad (7.34)$$

where the first term in (7.34) will in fact vanish as mentioned above, and Re = 'real part of'.

If $|\boldsymbol{r}_c - \boldsymbol{R}|^{-1}$ is expanded in terms of the modified spherical harmonics[4]

$$|\boldsymbol{r}_c - \boldsymbol{R}|^{-1} = \sum_k \sum_m (r_<^k / r_>^{k+1}) C_{km}^*(\boldsymbol{r}_<) C_{km}(\boldsymbol{r}_>) \quad (7.35)$$

the $k = 2$ terms are proportional to the EFG due to $\rho(\boldsymbol{R})$, and dominate the sum in (7.34). Taking only these terms and neglecting non-linear [$O(\Delta^{-2})$] terms in (7.34), we have for $R > r$ (external polarising charges)

$$Q_{2M}^{\text{core}}(\text{e}) = -\left[2\text{Re} \sum_{N \neq 0} \sum_m \langle \Phi_0 | \hat{Q}_{2M}(\text{e}) | \Phi_N \rangle \langle \Phi_N | \sum_c r_c^2 C_{2m}^*(\boldsymbol{r}_c) | \Phi_0 \rangle \Delta_{N0}^{-1} \right] Q_{2m}^{\text{ext}}(\text{e}) \quad (7.36)$$

where

$$Q_{2m}^{\text{ext}}(\text{e}) = \int R^{-3} C_{2m}(\boldsymbol{R}) \rho(\boldsymbol{R}) \, \text{d}^3\boldsymbol{R}$$

Provided $|\Phi_0\rangle$ and $|\Phi_N\rangle$ are eigenstates of angular momentum (in practice they are Slater determinants of AOs) the only non-zero term in (7.36) has $m = M$, and for spherical symmetry the part enclosed in square brackets becomes a scalar independent of M, denoted γ_{ext}. Thus the total EFG at the nucleus is given by

$$Q_{2M}(\text{e}) = (1 - \gamma_{\text{ext}}) Q_{2M}^{\text{ext}}(\text{e})$$

where

$$\gamma_{\text{ext}} = 2 \,\text{Re} \sum_{N \neq 0} \langle \Phi_0 | \hat{Q}_{2M}(\text{e}) | \Phi_N \rangle \langle \Phi_N | \sum_c r_c^2 C_{2M}^*(\boldsymbol{r}_c) | \Phi_0 \rangle \Delta_{N0}^{-1} \quad (7.37)$$

Similarly, if $R < r$ (internal polarising charges) then

$$Q_{2M}(\text{e}) = (1 - \gamma_{\text{int}}) Q_{2M}^{\text{int}}(\text{e})$$

with

$$\gamma_{\text{int}} = 2 \langle R^2 \rangle \langle R^{-3} \rangle^{-1} \,\text{Re} \sum_{N \neq 0} \langle \Phi_0 | \hat{Q}_{2M}(\text{e}) | \Phi_N \rangle \langle \Phi_N | \sum_c r_c^{-3} C_{2M}^*(\boldsymbol{r}_c) | \Phi_0 \rangle \Delta_{N0}^{-1} \quad (7.38)$$

where the forms of $\rho(\boldsymbol{R})$ considered in practice are such that

$$\frac{\langle R^2 \rangle}{\langle R^{-3} \rangle} = \frac{\int R^2 C_{2M}(\boldsymbol{R}) \rho(\boldsymbol{R}) \, \text{d}^3\boldsymbol{R}}{\int R^{-3} C_{2M}(\boldsymbol{R}) \rho(\boldsymbol{R}) \, \text{d}^3\boldsymbol{R}}$$

is independent of M. By inspection of (7.37) and (7.38) it is seen that $|\gamma_{\text{int}}| < |\gamma_{\text{ext}}|$.

If the unperturbed core states $|\Phi_0\rangle$, etc. are Slater determinants of AOs ϕ_{nlms} then, since only one-electron operators appear in (7.37) and (7.38), the $|\Phi_N\rangle$ each differ from $|\Phi_0\rangle$ by 'excitation' of a single AO. These excitations can be 'radial' (l unchanged) or 'angular' (l changed).

In practice it is convenient to define two Sternheimer factors on a different basis to the $R \lessgtr r$ condition used above. A factor denoted γ is used when the perturbing charge lies on atoms other than the one considered. Usually one ignores overlap of this charge with the core orbitals; γ is then denoted γ_∞ and equals γ_{ext} above. A second factor R_{nl} is used when the polarising EFG is due to asymmetric occupation of an (nl) valence shell of the atom in question. In general this involves both the $R < r$ and $R > r$ domains.

The contributions due to radial excitations are commonly calculated by unrestricted Hartree–Fock methods[40], or direct solution of the inhomogeneous Schrödinger equation for the perturbed wave functions[39, 41], and the usually less-important angular contributions obtained from a Thomas–Fermi model[39]. Other methods have also been employed[42].

Armstrong[43] has collected a number of calculated values for R in various atoms and ions. Feiock and Johnson[44] have calculated γ_∞ for closed-shell atoms and ions in the range $2 \leqslant Z \leqslant 92$ by use of relativistic Hartree–Fock–Slater wave functions. Sternheimer[45] has reviewed experimental evidence demonstrating the agreement between calculated R values and experimental results on hyperfine structure. Several recent calculations[46] have included many-body effects. These calculations gave a value for γ_∞ in Fe^{3+} rather close to previous estimates, but produced conflicting results for R in the $(2p^1)^2P$ state of Li.

Even when an atom is bound in a molecule it is customary to use an appropriate scalar atomic/ionic Sternheimer factor, although the factor could in principle acquire fourth-rank tensor character. This might be compared with, for example, the use of atomic/ionic Racah parameters or spin–orbit coupling constants in a molecular situation.

7.2.4 *E*0 nuclear volume effects

7.2.4.1 The interaction

From (7.6), (7.7) and (7.10) the $E0^{\text{vol}}$ interaction is given by

$$W_{E0}^{\text{vol}} = \sum_{p=1}^{\infty} Q_{00}^{p*}(\text{n}) Q_{00}^{p}(\text{e}) \tag{7.39}$$

where

$$Q_{00}^{p}(\text{n}) = \int \rho_{\text{n}}(\boldsymbol{r}) r^{2p}\, \text{d}^3\boldsymbol{r} = eZ\langle r^{2p}\rangle_{\text{n}} \tag{7.40}$$

and

$$Q_{00}^{p}(\text{e}) = -[2\pi/p(2p+1)] \int \rho_{\text{e}}(\boldsymbol{r}) \delta_{00}^{p-1}(\boldsymbol{r})\, \text{d}^3\boldsymbol{r} \tag{7.41}$$

If $\rho_e(\boldsymbol{r})$ is expanded about the origin by (7.2), then, neglecting terms with $L > 0$ since they do not contribute to monopole interactions, we may write

$$\rho_e(\boldsymbol{r}) = \rho_e(0)[1 - a_2r^2 + a_4r^4 - \ldots] + \text{neglected terms} \tag{7.42}$$

so that, by (7.3)

$$\int \rho_e(\boldsymbol{r})\delta_{00}^{p-1}(\boldsymbol{r})\, \mathrm{d}^3\boldsymbol{r} = (-1)^{p-1}a_{2p-2}\rho_e(0) \tag{7.43}$$

where $a_0 \equiv 1$. Writing $\rho_e(0) = -e|\Psi_e(0)|^2$, (7.39) becomes

$$\begin{aligned} W_{E0}^{\mathrm{vol}} &= 2\pi e^2 Z|\Psi_e(0)|^2 \sum_{p=1}^{\infty} [(-1)^{p-1}/p(2p+1)]\langle r^{2p}\rangle_n a_{2p-2} \\ &= \tfrac{2}{3}\pi e^2 Z|\Psi_e(0)|^2[\langle r^2\rangle_n - \tfrac{3}{10}\langle r^4\rangle_n a_2 + \tfrac{1}{7}\langle r^6\rangle_n a_4 - \ldots] \end{aligned} \tag{7.44}$$

The sum enclosed in square brackets in (7.44) is often approximated by a single term[35,47], essentially by the following argument. Since $W_E = \int\rho_n(\boldsymbol{r})V_e(\boldsymbol{r})\,\mathrm{d}^3\boldsymbol{r}$, it is seen from (7.44) that the potential $V_e(r)$ corresponding to the $L = 0$ part of $\rho_e(\boldsymbol{r})$ is

$$\begin{aligned} V_e(r) &= \tfrac{2}{3}\pi e|\Psi_e(0)|^2[r^2 - \tfrac{3}{10}a_2r^4 + \tfrac{1}{7}a_4r^6 - \ldots] \\ &= \tfrac{2}{3}\pi e|\Psi_e(0)|^2 f_1(r) \qquad \text{(say)} \end{aligned}$$

Hence we seek $f_2(r) \equiv br^k$ such that

$$\int_0^\infty \rho_n(r)f_1(r)r^2\,\mathrm{d}r \approx \int_0^\infty \rho_n(r)f_2(r)r^2\,\mathrm{d}r \tag{7.45}$$

where $\rho_n(r)$ is the $L = 0$ part of $\rho_n(\boldsymbol{r})$.

Integration by parts for $f(r)$ equal to either $f_1(r)$ or $f_2(r)$ gives

$$\int_0^\infty \rho_n(r)f(r)r^2\,\mathrm{d}r = \int_0^\infty g(r)\left[\int_0^r f(r')r'^2\,\mathrm{d}r'\right]\mathrm{d}r \tag{7.46}$$

where $g(r) = -\mathrm{d}\rho_n(r)/\mathrm{d}r$. Since (Figure 7.1) $g(r)$ is a sharply peaked symmetrical function, it is sufficient to approximate the integral enclosed by square brackets in (7.46) by a Taylor series accurate in the region where $g(r)$ has significant amplitude.

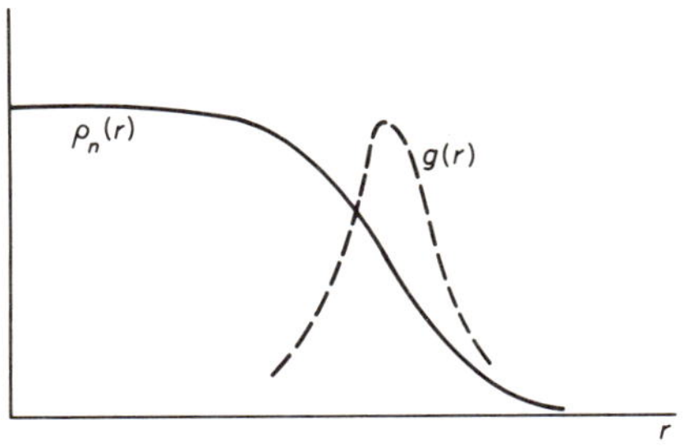

Figure 7.1 The functions $\rho_n(r)$ and $g(r)$

Wu and Wilets[35] suggest expanding about the point c_0 [$= eZ\langle r^{-2}\rangle_n / 4\pi\rho_n(0)$] such that

$$\int_0^\infty (r - c_0)g(r)\,dr = 0 \qquad (7.47)$$

Then, by (7.46),

$$\int_0^\infty \rho_n(r)f(r)r^2\,dr = \int_0^\infty g(r)\left[\int_0^{c_0} f(r')r'^2\,dr' + (r - c_0)f(c_0)c_0^2 + \tfrac{1}{2}(r - c_0)^2\{(d/dr)[f(r)r^2]\}_{r=c_0} + \ldots\right] dr \qquad (7.48)$$

where the second term in the expansion will vanish by virtue of (7.47).

Inserting (7.48) into both sides of (7.45) and equating terms one obtains, neglecting $O(a_4r^4)$,

$$k = 2 - (11a_2c_0^2/21) \qquad (7.49)$$

for $|a_2c_0^2| \ll 1$, and, to the same order,

$$b = 1 + a_2c_0^2[(110 \ln c_0 - 67)/210] \qquad (7.50)$$

Equation (7.44) is then written

$$W_{E0}^{\text{vol}} = \tfrac{2}{3}\pi e^2 Z|\Psi_e(0)|^2 b\langle r^k\rangle_n \qquad (7.51)$$

In the non-relativistic approximation, $a_2c_0^2$ is completely negligible, so that $k = 2$ and $b = 1$. This is not so when relativistic wave functions are considered, although present evidence is that k is unlikely to be reduced below about 1.7 even in the most extreme cases[48,195]. In principle, k and b might depend on the valence state of the atom through the parameter a_2. For a uniformly-charged spherical nucleus, Dunlop[195] obtains $k = 1 + (1 - a^2 Z^2)^{\frac{1}{2}}$, where a is the line-structure constant.

7.2.4.2 *Isomer and isotope shifts*

The $E0^{\text{vol}}$ interaction is usually observed spectroscopically by examining either the chemical shift of a Mössbauer transition between 'isomeric' nuclear states (isomer shift[48-50]) or the isotope dependence of electronic transitions (isotope shift[51]). In either case the shift in energy of the transition may be written, from (7.51), as

$$\delta_{\text{IS}} = \tfrac{2}{3}\pi e^2 Zb[\Delta|\Psi_e(0)|^2][\Delta\langle r^k\rangle_n] \qquad (7.52)$$

where $\Delta\langle r^k\rangle_n$ is the change in $\langle r^k\rangle_n$ between the nuclear states or isotopes involved, and $\Delta|\Psi_e(0)|^2$ is the change in electronic charge density at the nucleus. Since use of (7.52) as it stands requires relativistic atomic wavefunctions (which have only recently become available), correction factors have been devised[50,51] which can be tabulated once for all and used with non-relativistic estimates of $|\Psi_e(0)|^2$. Because of this, and use of the 'equivalent uniform-nucleus radius' $(R_u)^m = [(m + 3)/3]\langle r^m\rangle$, formulae actually used in the interpretation of experiments may appear different from (7.52).

In both isomer and isotope shift experiments the observed energy shift includes additional effects. The observed centre shift of a Mössbauer spectrum is the sum of the isomer shift given by (7.52) and a thermal red shift which can be regarded as arising from the second-order Doppler shift due to thermal vibration of the Mössbauer atom[52], or (equivalently[53]) as being due to the change in mass of the Mössbauer nucleus on emission or absorption[54]. In the case of the isotope shift, (7.52) gives only the so-called 'field shift', and other 'mass shifts' also contribute to the total isotope shift[51]. The 'normal mass shift' arises because the electronic mass in the kinetic energy operator should be replaced by the reduced mass, and the 'specific mass shift' from terms in the hamiltonian of the form $(2M)^{-1}\sum \hat{\boldsymbol{p}}_i\cdot\hat{\boldsymbol{p}}_j$, where $\hat{\boldsymbol{p}}_i, \hat{\boldsymbol{p}}_j$ are linear momentum operators for electrons i and j. Although the normal mass shift is readily calculated, the specific mass shift is more troublesome, particularly for $A < 60$ where it tends to dominate the field shift. Stacey[51] and Wu and Wilets[35] have discussed the problem of allowing for the specific mass shift.

The isomer shift in particular has attracted a good deal of attention, since it can be conveniently measured by Mössbauer spectroscopy, and is a valuable empirical parameter[21,22,55]. More detailed interpretation of isomer shifts tends to rely rather heavily on wavefunction calculations[196] (but see Section 7.2.7.3) and has been extensively reviewed, notably by Shirley[50] and Kalvius[48].

7.2.5 Magnetic hyperfine interactions

7.2.5.1 Magnetic multipoles

The magnetic hyperfine interaction energy W_{M} is given by

$$W_{\mathrm{M}} = -\int \boldsymbol{J}_{\mathrm{e}}(\boldsymbol{r})\cdot\boldsymbol{A}_{\mathrm{n}}(\boldsymbol{r})\,\mathrm{d}^3\boldsymbol{r} \tag{7.53}$$

where $\boldsymbol{J}_{\mathrm{e}}(\boldsymbol{r})$ and $\boldsymbol{A}_{\mathrm{n}}(\boldsymbol{r})$ are the total (orbital and spin) electronic current density and nuclear vector potential. As in the electric case the interaction can be expanded by use of generalised δ functions[4]. The resulting hamiltonian has the same general form as (7.14), namely

$$\mathscr{H}_{\mathrm{M}} = \sum_{L=0}^{\infty}\sum_{M=-L}^{L} \hat{M}^*_{LM}(\mathrm{n})\hat{M}_{LM}(\mathrm{e}) + \text{nuclear volume terms} \tag{7.54}$$

where the nuclear magnetic multipole moments $[M_{LM}(\mathrm{n})]_{M'M''}$ are given by [*cf.* (7.16)]

$$\begin{aligned}[M_{LM}(\mathrm{n})]_{M'M''} &= \langle nIM'\pi|\hat{M}_{LM}(\mathrm{n})|nIM''\pi\rangle \\ &= (-1)^{I-M'}\begin{pmatrix} I & L & I \\ -M' & M & M'' \end{pmatrix}(2I+1)^{\frac{1}{2}}\langle nI\pi||\hat{M}_L(\mathrm{n})||nI\pi\rangle \end{aligned} \tag{7.55}$$

Since $\hat{M}_{LM}(\mathrm{n})$ has parity $(-1)^{L+1}$ and the $3-j$ symbol must satisfy the triangle rule, a nucleus with spin I has only $M1$ (dipole), $M3$ (octupole), . . . , moments up to $L = 2I$ (I = integer + $\frac{1}{2}$) or $L = 2I - 1$ (I = integer).

These are exactly the moments which were forbidden in the electric case, and vice versa.

The ML moments are each characterised by a single parameter which is usually defined by taking $M = 0$, $M' = M'' = I$ in (7.55). In the $M1$ case

$$\langle nII\pi|\hat{M}_{10}(\mathrm{n})|nII\pi\rangle = \gamma_{\mathrm{n}}\hbar I = g_{\mathrm{n}}\beta_{\mathrm{n}}I \tag{7.56}$$

where γ_{n} is the magnetogyric ratio and g_{n} the g-factor of nucleus n, and β_{n} ($= e\hbar/2m_{\mathrm{p}}c$, where m_{p} is the protonic mass) is the nuclear magneton.

With increasing L the ML interactions decrease rapidly in magnitude, as in the electric case. Thus while it is difficult to imagine modern chemistry without the $M1$ interaction, magnetic octupole effects are accessible only to molecular-beam methods[8], and higher L interactions do not seem to have been observed.

7.2.5.2 *Magnetic dipole coupling*

The $M1$ coupling hamiltonian is

$$\mathscr{H}_{M1} = -\hat{\boldsymbol{\mu}}_{\mathrm{n}}\cdot\hat{\boldsymbol{B}} \tag{7.57}$$

where $\hat{\boldsymbol{\mu}}_{\mathrm{n}} = g_{\mathrm{n}}\beta_{\mathrm{n}}\hat{\boldsymbol{I}}$, and the hyperfine field operator can be written as a sum of orbital $\hat{\boldsymbol{B}}_l$, spin–dipole coupling $\hat{\boldsymbol{B}}_{\mathrm{sd}}$ and Fermi contact $\hat{\boldsymbol{B}}_{\mathrm{c}}$ contributions[12]

$$\hat{\boldsymbol{B}}_l = -2\beta \sum_{\text{electrons}} r_i^{-3}\hat{\boldsymbol{l}}_i \tag{7.58a}$$

$$\hat{\boldsymbol{B}}_{\mathrm{sd}} = -2\beta \sum_{\text{electrons}} [3r_i^{-5}(\hat{\boldsymbol{s}}_i\cdot\boldsymbol{r}_i)\boldsymbol{r}_i - r_i^{-3}\hat{\boldsymbol{s}}_i] \tag{7.58b}$$

$$\hat{\boldsymbol{B}}_{\mathrm{c}} = -(16\pi\beta/3) \sum_{\text{electrons}} \delta(\boldsymbol{r}_i)\hat{\boldsymbol{s}}_i \tag{7.58c}$$

This decomposition into orbital, 'classical' tensor coupling, and contact terms holds for all ML interactions[4], and is central to current interpretations of $M1$ coupling constants[40, 56-58].

The contact interaction reflects the density of unpaired electronic spin at the nucleus, while the orbital contribution arises from imperfect quenching of the orbital angular momentum, and the spin–dipole term from the dipole–dipole coupling of the nuclear spin with the electronic spin distribution. Although the contact term involves a δ function it is not a nuclear volume effect, and would occur even for a point nucleus.

Further treatment of the electronic operators in (7.58) depends on whether the electronic system has a well-defined total angular momentum J. If so, we write

$$\mathscr{H}_{M1} = \beta_{\mathrm{n}}g_{\mathrm{n}}\beta g_J(\hat{\boldsymbol{I}}\cdot\hat{\boldsymbol{J}}) \tag{7.59}$$

and

$$W_{M1}(F,I,J) = \tfrac{1}{2}\beta_{\mathrm{n}}g_{\mathrm{n}}\beta g_J[F(F+1) - I(I+1) - J(J+1)]$$

[*cf.* (7.21)–(7.23)]. This situation and its generalisations have been well documented[8, 13].

In condensed phases the electronic orbital angular momentum $\boldsymbol{L}$ is usually at least partly quenched. If $\boldsymbol{S}$ is also quenched, i.e. the M_S degeneracy is lifted either by magnetic ordering or application of an external field, then $\hat{\boldsymbol{B}}$ may be replaced by its expectation value to give

$$\mathscr{H}_{M1} = -g_n\beta_n\hat{\boldsymbol{I}}\cdot\boldsymbol{B} \tag{7.60}$$

Many paramagnetic systems can be described by the spin hamiltonian

$$\mathscr{H}_{M1} = \hat{\boldsymbol{I}}\cdot\boldsymbol{A}\cdot\hat{\boldsymbol{S}}' \tag{7.61}$$

where the fictitious-spin operator $\hat{S}'$ acts within a manifold of $(2S' + 1)$ electronic states. (This, and more complicated effective hamiltonians for $M1$ coupling, are described in standard texts[12,59]). In paramagnetic systems a hyperfine magnetic field in the strict sense can only be defined if an external polarising magnetic field is applied[60]. Nevertheless, if $\boldsymbol{A}$ is axially-symmetric with $|A_{\parallel}| \gg |A_{\perp}| \approx 0$, the quantity $-A_{\parallel}M_{S'}/g_n\beta_n$ is often termed the hyperfine field associated with the electronic state $|S'M_{S'}\rangle$, and if $\boldsymbol{A}$ is isotropic the quantity $AS'/g_n\beta_n$ is likewise the magnitude of a 'hyperfine field'.

7.2.5.3 The contact hyperfine field

The orbital and spin–dipole hyperfine fields due to unpaired valence electrons are modified by the core electrons[40] in an effect analogous to the Sternheimer effect (Section 7.2.3.3). Breakdown of the central-field approximation[43] also leads to a substantial contact interaction even in cases such as the $(3d^5)^6S$ ions Mn^{2+} and Fe^{3+}, in which the unpaired electrons apparently have zero density at the nucleus. This 'core polarisation' effect can be thought of in two ways[12]: either the ground (restricted Hartree–Fock) ^6S state of Mn^{2+} or Fe^{3+} is mixed by configuration interaction with other ^6S states which do have net unpaired spin density at the nucleus, or we may assign different radial wave functions to s↑ and s↓ electrons, reflecting their different exchange interactions with 3d↑. Each of these approaches is an approximation to the same 'perfect' calculation. The second approach has usually been adopted, and core polarisation contributions to the contact interaction have been calculated by unrestricted Hartree–Fock theory[40]. Recently the problem has begun to receive attention from many-body theorists[43,61].

The contact interaction is usually the dominant contribution to the observed hyperfine field. Further, the orbital and spin–dipole contributions can normally be calculated by making use of other experimental data. In many important cases, $\boldsymbol{B}_l$ can be calculated from the deviation of the g tensor from its spin-only value and $\boldsymbol{B}_{sd}$ from the observed EFG[40,57], although this approach can be very misleading if used when inappropriate[62]. Experimental[58] and theoretical[40] estimates of the contact field in $3d^n$ ions show that $B_c/2S$ is approximately constant at about -100 kG, increasing somewhat with increasing Z. (Similar results have been found for $3d^n4s^2$ atoms[9].) More significant variations in $B_c/2S$ arise from covalency in the metal–ligand bonds, increasing covalency leading to a decrease in $|B_c/2S|$. Thus studies of the $M1$ hyperfine field provide a convenient measure of covalency in 3d transition-metal complexes[40,57,58]. More recent work indicates that similar considerations may apply to 4d and 5d complexes[63].

7.2.5.4 Hyperfine anomaly

For completeness we mention briefly that the effect of nuclear size in the $M1$ interaction is well-known in the hyperfine anomaly[12,43,64], i.e. the ratio of the $M1$ coupling constants for two isotopes in the same environment is not equal to the ratio of their nuclear g factors. Some care must be exercised experimentally, since the same observation may also arise from higher-order effects or an isotope effect associated with the coupling of different zero-point motions to the electronic wave function[12]. It has been suggested that the hyperfine anomaly may be used to estimate the orbital component of the hyperfine field[64,65].

7.2.6 Relativistic hyperfine operators

The hamiltonian for a hyperfine interaction of rank K is a sum of one-electron operators:

$$\mathscr{H}_{TK} = \sum \hat{T}_K(\mathrm{n})\cdot\hat{T}_K(\mathrm{e}) \tag{7.62}$$

where $T = Q$ for the electric interaction and M for the magnetic. Now the electronic operator $\hat{T}_K(\mathrm{e})$ has the form

$$\hat{T}_K(\mathrm{e}) = \sum_{k_s,k_l} A_K^{(k_s k_l)}\hat{U}_K^{(k_s k_l)} \tag{7.63}$$

where $\hat{U}_K{}^{(k_s k_l)}$ is a tensor operator of rank k_s in spin space and k_l in orbital space. Since the electron has spin $\frac{1}{2}$ the only possible values of k_s are 0 and 1. Also, the triangle rule restricts K to integers in the range $k_s + k_l \geqslant K \geqslant |k_s - k_l|$, and if $\hat{U}_K{}^{(k_s k_l)}$ is to be hermitian, $(k_s + k_l + K)$ must be an even integer. These three conditions[66] imply that the only general solutions $(k_s,k_l)K$ are $(0,K)K$, $(1,K-1)K$ and $(1,K+1)K$. All three solutions are allowed for electrons with orbital angular momentum l if $2l \geqslant K + 1$. For $2l = K$ we have $(0,K)K$ and $(1,K-1)K$, while if $2l = K - 1$ only $(1,K-1)K$ is possible.

In the $M1$ interaction the solutions (0,1)1, (1,2)1 and (1,0)1 have the same forms as the orbital, spin–dipole and contact interactions, respectively. For the $E2$ case the three solutions are (0,2)2, (1,3)2 and (1,1)2, of which only (0,2)2 is allowed in the non-relativistic theory because $\hat{Q}_2(\mathrm{e})$ operates on spatial coordinates only. In relativistic theory spin and orbital motions are treated in the same basis, and all three solutions are allowed.

Sandars and Beck[66] showed that relativistic effects could be incorporated into the theory of hyperfine structure in the LS coupling scheme by defining effective operators $\hat{T}_K^{\mathrm{eff}}(\mathrm{e})$ such that

$$\langle\phi_{nljm}|\hat{T}_{KM}^{\mathrm{eff}}(\mathrm{e})|\phi_{nlj'm'}\rangle = \langle\phi_{nljm}^R|\hat{T}_{KM}(\mathrm{e})|\phi_{nlj'm'}^R\rangle \tag{7.64}$$

where ϕ_{nljm} and ϕ_{nljm}^R are LS coupled non-relativistic and relativistic orbitals, respectively. The coefficients $A_K^{(k_s k_l)}$ are then proportional to appropriate radial expectation values. Thus the only effect of relativity in the case of the $M1$ interaction is to alter the radial parameters: the forms of the three effective operators remain unchanged. In the $E2$ interaction the non-relativistic $\langle r^{-3} \rangle$ must be replaced by the appropriate relativistic parameter and two more, strictly relativistic, interactions appear.

Details of this elegant theory are described in the literature[9, 43, 66, 67], and the necessary radial parameters have been published for the lanthanides and actinides[67, 68]. The theory has proved valuable in detailed studies of hyperfine structure in heavier atoms[9] and ions[67].

7.2.7 Atomic and nuclear parameters

7.2.7.1 The coupling constant dilemma

Hyperfine coupling constants are products of nuclear moments and electronic hyperfine 'fields'. Quantitative interpretation of the hyperfine fields in terms of molecular structure is only possible if the other factors making up the coupling constant are already known.

In the case of magnetic dipole coupling it is possible to generate external magnetic fields of the same order of magnitude as internal $M1$ hyperfine fields. Further, the screening of the nuclear Zeeman interaction in diamagnetic molecules is well understood[8], and, being small, may be calculated *ab initio* in atoms and small molecules with acceptable accuracy. Thus determination of the sign and an estimate of the magnitude of the nuclear dipole moment does not usually present insurmountable problems for any nuclear state likely to be used in chemical hyperfine studies. The problems associated with establishing very accurate values for nuclear magnetic moments have recently been discussed[69] for the Mössbauer isotopes ^{57}Fe and ^{119}Sn.

The electric hyperfine interactions present quite a different problem. Laboratory EFGs are far too small to produce perceptible $E2$ coupling, while it is even more difficult to imagine how one might apply a known external charge density to modify the $E0^{vol}$ effect! Further, even if one could calculate the hyperfine 'field' due to the valence electrons this would be modified at the nucleus by polarisation of the core electrons. For example, each contribution to a measured $E2$ coupling constant is normally thought of as a product of three factors: a nuclear moment, a Sternheimer factor, and an EFG due to valence electrons or charges carried by other atoms in the molecule. Nuclear models lead to reliable independent estimates of the first factor only in the case of heavier 'deformed' nuclei (Section 7.2.7.2). The Sternheimer factor (Section 7.2.3.3) is at present only calculated by use of free-ion/atom wave functions. In the molecular case *ab initio* calculation of the EFG is only feasible in lighter molecules containing low-Z nuclei for which independent determination of nuclear $E2$ moments tends to be difficult. Further, it seems that covalency and overlap make important contributions to the EFG even in crystals of S-state ions[70], where calculation of the EFG by electrostatic lattice summation[71] might seem attractive.

Despite all these difficulties many determinations of $E2$ moments have been based on spectroscopic studies, with results tabulated in the literature[72]. There will, of course, be interdependence between the reported nuclear moments and the values taken for Sternheimer factors and other atomic parameters.

Several strategies attempt to bypass the difficulties mentioned above. The most familiar applies to the EFG set up by an incomplete l^n atomic shell: the coupling constant is expressed in terms of the 'coupling constant due to one l electron'

$$(e^2qQ)_l = -e^2Q(1 - R)\langle r^{-3}\rangle[2l(l + 1)/(2l - 1)(2l + 3)] \quad (7.65)$$

which is either determined experimentally or calculated for the free atom or ion. This approach fits well with the concept of effective population (Section 7.4.1.1) and has been particularly exploited in simple valence-bond interpretations of halogen and other non-metal coupling constants[13, 73, 74]. Recently, the troublesome question of $(e^2qQ)_p$ for ^{14}N (which does not have an atomic state suitable for experimental studies) has been resolved by use of *ab initio* calculations[75].

Another useful device is to compare isoelectronic isostructural compounds of different quadrupolar nuclei. The slope of a linear regression between the $E2$ coupling constants observed in the isoelectronic series is assumed to equal the ratio of the $(e^2qQ)_l$ parameters. If the atomic and nuclear parameters are known for one of the isotopes involved they can be estimated for the other. This method has been applied, for example, to $^{57}Fe/^{59}Co$ complexes[76], $^{57}Fe^{II}/^{99}Ru^{II}$ low-spin complexes[77], and $^{119}Sn^{IV}/^{121}Sb^{V}$ compounds[77].

Robinson[78] has derived a sum rule which relates the EFG at the nucleus to electric dipole matrix elements, thereby avoiding the use of Sternheimer factors. Twofold differentiation of the Schrödinger equation gives relationships between vibrational force constants and quadrupole coupling constants[79]. Sen and co-workers[80] have claimed that splitting of $p_{\frac{3}{2}}$ atomic core levels by an EFG could be observable in photoelectron spectra, and in anticipation of this have calculated the necessary Sternheimer factors for a number of rare-earth and other heavy ions[81].

7.2.7.2 *E2 moments of deformed nuclei*

In the nuclear shell model[1, 2] the individual nucleons move in an average potential due to their interactions. For realistic potentials, and with due allowance for spin–orbit coupling, sets of energy levels are obtained; those for the protons are at slightly higher energy than the neutron levels because of coulomb interactions, but are otherwise similar. When protons and neutrons are fed into their energy levels according to an aufbau procedure, closed shells occur at the 'magic numbers' 2, 8, 20, 28, 50, 82, 126, ... These closed shells are stable configurations with zero angular momentum. Thus a 'doubly magic' nucleus (N = magic, Z = magic) is an ideal 'spherical nucleus', i.e. the self-consistent potential is spherically symmetrical. However, addition of extra particles outside the closed shells perturbs the spherical potential. (This effect is more important here than in atomic shell-structure, since the atomic potential is dominated by the coulomb potential of the nuclear charge.) In this way the nucleus acquires collective motions which can be visualised as rotations and vibrations of the nuclear potential[2, 82]. Figure 7.2 shows the regions in which such 'deformed nuclei' can occur.

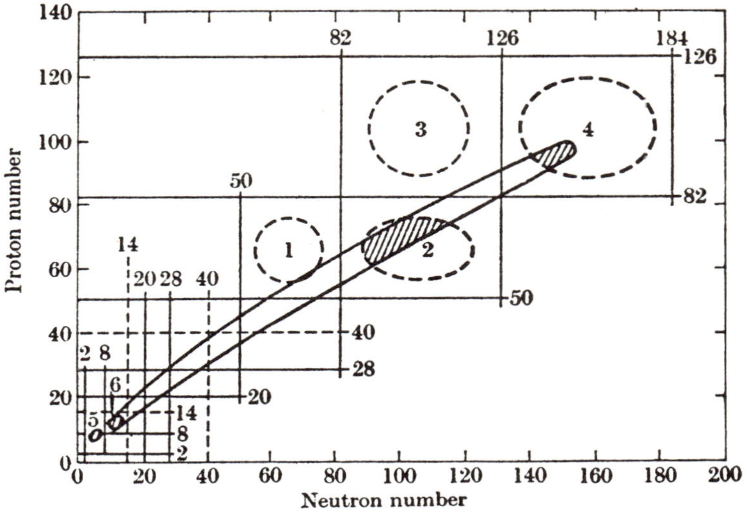

Figure 7.2 Nuclear periodic table showing closed shells (solid lines) and subshells (dashed lines). Banana-shaped curve approximately encloses nuclei which have been studied experimentally. Possible deformed-nucleus regions are numbered 1 to 6; regions where deformed nuclei have been observed experimentally are shown by hatching. (From Marshalek *et al.*[197], by courtesy of the American Institute of Physics.)

The most important regions are $150 \leqslant A \leqslant 195$ and $A > 220$. For these mass numbers reliable information on $E2$ moments can be obtained independently of spectroscopic coupling constants*.

It is often adequate for this purpose to consider the nucleus as a symmetric rotator[82]. The nuclear states fall into bands which can then be characterised by the quantum number K, with

$$I = K, K + 1, K + 2, \ldots$$

if $K \neq 0$, and if $K = 0$

$$I = \begin{cases} 0, 2, 4, \ldots \text{ for parity } + \\ 1, 3, 5, \ldots \text{ for parity } - \end{cases}$$

The shape of the proton distribution in the nucleus is characterised by the intrinsic quadrupole moment Q_0, defined in the internal coordinate system of the nucleus. If $Q_0 = 0$ the nucleus is spherical, if $Q_0 > 0$ it is prolate (cigar-shaped), and if $Q_0 < 0$ it is oblate (disc-shaped). The spectroscopic quadrupole moment Q [equation (7.17)], defined in laboratory axes, is given by [82]

$$Q = Q_0[3K^2 - I(I + 1)]/[(I + 1)(2I + 3)] \qquad (7.66)$$

* The origin-dependence of a point-of-view is well illustrated in Alder and Steffen's important review[82]: they regard spectroscopic determinations as 'model-independent' and nuclear ones as model-dependent; since our models are of *electronic* structure we would reverse these labels!

In deformed nuclei it is misleading to associate Q with nuclear shape. When $K = 0$ or $\frac{1}{2}$, $(Q/Q_0) \leqslant 0$ for all allowed I, while if $K \geqslant 1$, (Q/Q_0) will change from positive to negative with increasing I.

In this 'collective model' of the nucleus both diagonal and off-diagonal matrix elements of $\hat{Q}_{2M}(\mathrm{n})$ are related to Q_0. The off-diagonal elements determine transition probabilities between nuclear states. Thus a number of experimental methods yield information on Q_0, with results tabulated in the literature[83].

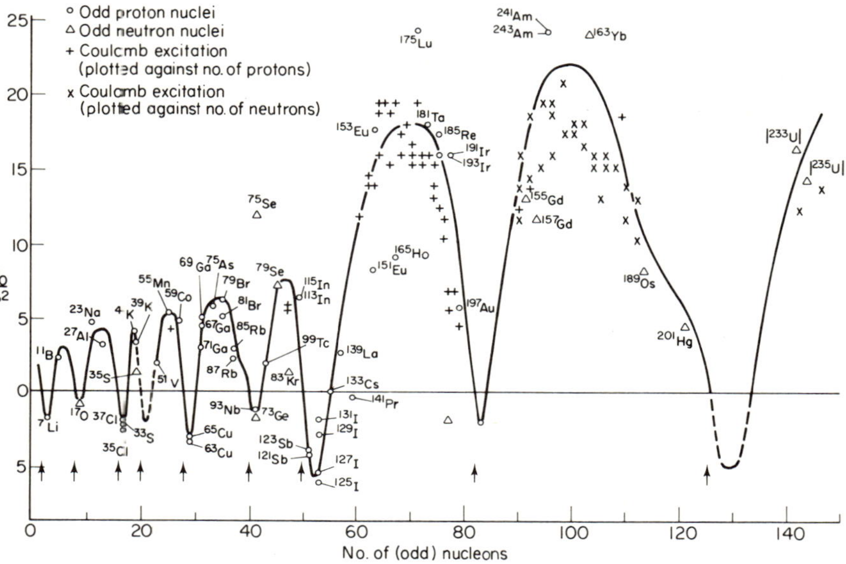

Figure 7.3 Plot of Q_0/R^2 ($R = 1.2 \times 10^{-13}$ $A^{\frac{1}{3}}$ cm) against Z for Z-odd/N-even nuclei, N for Z-even/N-odd nuclei, and both Z and N for even/even nuclei. Arrows mark closed shells and subshells. (From Townes[84], by courtesy of Springer-Verlag)

The nuclear shell structure leads to an oscillatory variation of Q_0 with increasing N or Z, but other factors (vibrational motions, coulomb interactions) impose a bias towards positive Q_0. The resulting systematic variations in Q_0 were noticed at an early stage in the study of deformed nuclei (Figure 7.3)[84]. Even the ground states of light nuclei, with Q_0 defined by putting $K = I$ in (7.66), seem to fit the variation shown in Figure 7.3, although the physical significance of Q_0 is less clear for these nuclei. These systematics allow the sign of Q_0 to be predicted in cases where it has not been determined experimentally.

7.2.7.3 $\Delta\langle r^k \rangle_{\mathrm{n}}$ *and nuclear decay rates*

The influence of chemical environment on nuclear decay rates is now well established[27, 28]. In particular, for most examples of electron capture or $M1$

internal conversion the decay rate is proportional to the electron density at the nucleus[28]. Thus if the coefficient of internal conversion (ratio of internally-converted to radiative decays[2]) for an (nl) electron is a_{nl} then

$$a_{nl}/a_{n'l'} = |\psi_{nl}(0)|^2/|\psi_{n'l'}(0)|^2 \tag{7.67}$$

This relationship was exploited, for example, by Bocquet *et al.*[85] in their determination of $\Delta\langle r^k\rangle_n$ for ^{119}Sn. If the difference in $a_{nl}/a_{n'l'}$ is determined by conversion electron spectroscopy for two chemical species with valence subshells nl and fixed core subshell $n'l'$, then

$$\sum_{nl} \Delta(a_{nl}/a_{n'l'}) = \delta_{IS}/[C\Delta\langle r^k\rangle_n|\psi_{n'l'}(0)|^2] \tag{7.68}$$

where δ_{IS} is the Mössbauer isomer shift given by (7.52) with $\frac{2}{3}\pi e^2 Zb = C$, and it is assumed that $|\psi_{n'l'}(0)|^2$ is not changed by changes in the valence electron densities. The quantity $C\Delta\langle r^k\rangle_n$ can be estimated from measured $\sum_{nl}\Delta(a_{nl}/a_{n'l'})$ and δ_{IS}, and calculated $|\psi_{n'l'}(0)|^2$.

The doubtful assumption that core-electron densities at the nucleus are invariant can be avoided if the (experimentally more difficult) measurement of relative change in total decay rate $\Delta\lambda/\lambda$ is undertaken. Then

$$\begin{aligned}\Delta\lambda/\lambda &= [a/(1+a)][\Delta \sum_i |\psi_i(0)|^2/\sum_i |\psi_i(0)|^2] \\ &= [a/(1+a)][\delta_{IS}/C\Delta\langle r^k\rangle_n \sum_i |\psi_i(0)|^2]\end{aligned} \tag{7.69}$$

where the sum is now over all subshells i. This equation requires only that $\Delta \sum_i |\psi_i(0)|^2$ should be a small fraction of $\sum_i |\psi_i(0)|^2$, which is certainly true since $\Delta\lambda/\lambda$ is typically of order 10^{-4}. If $\sum_i |\psi_i(0)|^2$ is calculated, $C\Delta\langle r^k\rangle_n$ can be estimated from measured $\Delta\lambda/\lambda$, δ_{IS} and total conversion coefficient a. Measurements of this kind have been performed for several nuclides[86], including the important Mössbauer isotopes ^{57}Fe and ^{119}Sn. A survey of experimental work in this area has been given by Emery[28].

Values of $\Delta\langle r^k\rangle_n$, derived by a variety of methods, have been tabulated in the literature[87].

7.3 QUADRUPOLE COUPLING AND MOLECULAR STRUCTURE

7.3.1 Parameters and data

7.3.1.1 Symmetry restrictions

The EFG tensor (Section 7.2.3.1) is characterised by five independent quantities, conveniently taken to be the three principal directions, the coupling constant e^2qQ and the asymmetry parameter η. The symmetrised parameters S_{02} and S_{03} (Section 7.2.3.2) are sometimes useful alternatives to e^2qQ and η.

If the symmetry of the nuclear site is orthorhombic or greater, the principal directions are determined by symmetry. If not, they can often be determined in the solid state by experiments on single crystals[17, 18, 88]. Gas phase studies lead to the diagonal components of the EFG in the principal axes of inertia[8, 13], but in low symmetry these axes may not coincide with the field gradient principal axes. Much of the available data on larger molecules has been obtained by n.m.r./n.q.r. or Mössbauer studies of polycrystalline samples, so where the symmetry is lower than orthorhombic the principal directions are often not known. Also, the presence of more than one molecule per unit cell may complicate the interpretation of single crystal studies in low-symmetry crystals.

When the site symmetry has an axis of order three or greater, η must vanish and the principal axes are directed along and perpendicular to the symmetry axis. If the symmetry is cubic or higher the EFG vanishes. The $E2$ interaction is also forbidden if $I < 1$ (Section 7.2.2.)

7.3.1.2 *Experimental constraints*

For $I = 1$ and $I \geqslant 2$ both S_{02} and S_{03} appear[36] in the characteristic equation of $\mathscr{H}_{E2}$, indicating that, in principle, both e^2qQ and η can be obtained from experiments on powdered samples without applying an external magnetic field. This is possible in practice in Mössbauer spectroscopy, but in n.m.r./n.q.r. spectroscopy only $|e^2qQ|$ and η can be determined (whether or not an external magnetic field is applied) unless one attains temperatures so low that kT is very much less than the splitting of the nuclear sublevels[15].

For $I = \frac{3}{2}$ only the single quantity S_{02} $[= -\frac{3}{4}e^2q^2(1 + \frac{1}{3}\eta^2)]$ can be determined by zero-magnetic-field experiments on powdered samples. Mössbauer effect experiments on oriented polycrystalline samples[89, 90], or Mössbauer Zeeman studies of powdered samples[91], are sufficient to determine the sign of the coupling constant, the latter method also yielding an estimate of η. Recently, the relative signs of two different Mössbauer nuclei in the same molecule have been determined by an oriented sample technique[92]. In n.m.r./n.q.r. spectroscopy Zeeman studies to determine $|e^2qQ|$ and η require the use of single crystals because of the greater sensitivity to inhomogeneous line-broadening characteristic of this technique[17]. Since $0 \leqslant \eta \leqslant 1$, $(1 + \frac{1}{3}\eta^2)^{\frac{1}{2}}$ is never more than 15.5% greater than unity, and possible corrections to $|e^2qQ|$ for non-zero η are sometimes ignored in studies of $I = \frac{3}{2}$ nuclei. This approximation is particularly tempting in the case of the halogen nuclei $^{35,37}Cl$ and $^{79,81}Br$ if it can be argued that the bond to monovalent halogen might be close to cylindrical symmetry. However, the interpretation of small variations in $|e^2qQ|$ with chemical environment could then be ambiguous.

7.3.1.3 *Precision and reproducibility*

A survey of independently duplicated measurements in the literature reveals that routine data on quadrupole coupling constants in complex molecules are rarely reproducible to better than the third significant figure, yet in the case of n.q.r. spectroscopy resonance frequencies may be measured and reported with a precision of five or more significant figures. This is clearly liable to produce an illusion of accuracy, and the temptation to 'interpret' differences of only a few percent in $E2$ coupling constants is best resisted unless the reproducibility of the data is above question. Precision and reproducibility are more nearly comparable in Mössbauer spectroscopy because of the lower precision achieved.

One useful indicator of reproducibility is the testing of internal and external consistency[90]. If observations $j = 1, 2, \ldots n$ yield values x_j for a given parameter and s_j for the associated standard deviation, then the expectation value $\bar{x}$ may be estimated by

$$\bar{x} = \left[\sum_{j=1}^{n} x_j/s_j^2\right] \bigg/ \left[\sum_{j=1}^{n} 1/s_j^2\right] \qquad (7.70)$$

Two independent expressions for the variance of $\bar{x}$ may be evaluated:

$$a_i^2 = \left[\sum_{j=1}^{n} 1/s_j^2\right]^{-1} \qquad (7.71)$$

and

$$a_e^2 = \left[\sum_{j=1}^{n} (x_j - \bar{x})^2/s_j^2\right] / \left[(n-1) \sum_{j=1}^{n} 1/s_j^2\right] \qquad (7.72)$$

Whereas a_i depends only on the standard deviations assigned to individual observations, a_e also depends on the scatter of the observations. Snedecor's F test can be used to check whether a_i^2 and a_e^2 can be taken as estimates of the same variance.

7.3.2 Strategies and concepts

7.3.2.1 Molecular symmetry

If a central atom M is bound to a number of ligands (which may be atoms or groups), the geometrical disposition of these bonds usually approximates to an idealised symmetry described by a coordination polyhedron, with the atom M located at the centre of the polyhedron and the bonds directed towards its vertices. We call the point symmetry group of this polyhedron the 'holosymmetry' of the atomic site[93]. If the ligands are all identical the idealised symmetry at M is the holosymmetry; if not, it is a subgroup of the holosymmetric group. Because of electronic or geometrical distortions arising from such factors as the detailed structure of polyatomic ligands or crystal packing forces, the true symmetry at the central atom may be less than the idealised symmetry.

For example, the cation in *trans*-$[Co(en)_2Cl_2]NO_3$ (en = ethylenediamine) has holosymmetry O_h and idealised local symmetry at Co of D_{2h}, or D_{4h} if the joining of the N atoms into bidentate chelates is irrelevant to the problem under consideration. Crystallographic studies[94] show that the actual symmetry of the cation is close to C_{2h} and that the site-symmetry of the Co position in the crystal structure is C_i. Clearly it is not easy to steer a safe course avoiding both laxity and pendantry when defining these various facets of molecular symmetry.

7.3.2.2 *Symmetry-breaking perturbations*

The magnitude and sign of q, and the value of η, each differ in their sensitivity to perturbations which lower the molecular symmetry[95], with, in general, $|q| < sgn\{q\} < \eta$, in order of increasing sensitivity.

These generalisations are forcibly illustrated by calculation of the EFG due to one particle (one or a pair of electrons or positive holes) in a d orbital in the absence of spin–orbit coupling. The most general d wave function is a linear combination involving all five of the conventional d orbitals: $d_{x^2-y^2}$, d_{z^2}, d_{xy}, d_{xz} and d_{yz}. However, by choice of principal axes this linear combination may be reduced to the form

$$\psi = d_{z^2} \cos a + d_{x^2-y^2} \sin a \tag{7.73}$$

The principal directions and the value of the parameter a vary with changes in the environment. In principal axes the EFG at the nucleus due to a particle of charge Z in the orbital ψ is[95]

$$V = \tfrac{4}{7}Z\langle r^{-3}\rangle_d(1-R)\begin{bmatrix} \cos(2a+\tfrac{2}{3}\pi) & 0 & 0 \\ 0 & \cos(2a-\tfrac{2}{3}\pi) & 0 \\ 0 & 0 & \cos 2a \end{bmatrix} \tag{7.74}$$

Consider a situation with axial idealised symmetry ($a = 0$); deviations from this ideal are reflected in non-zero a. From (7.74) we obtain (assuming $a > 0$ for convenience)

$$q(a) = q(0)\cos 2a = q(0)[1 - 2a^2 + \tfrac{2}{3}a^4 + O(a^6)] \tag{7.75}$$

and

$$\eta(a) = \sqrt{3}\tan 2a = 2\sqrt{3}a + (8\sqrt{3}/3)a^3 + O(a^5) \tag{7.76}$$

provided $a \leqslant \pi/12$. Clearly $\eta(a)$ varies much more rapidly than $q(a)$; for example, if $a = 0.1$ then $q(0.1) = 0.98q(0)$ but $\eta(0.1) = 0.351$.

As a increases to exceed $\pi/12$ the sign of the coupling constant changes and the principal axes must be reassigned if they are to satisfy the ordering convention [equation (7.27)]. However, since (without restriction on a) S_{02} is given by

$$S_{02} = -(12/49)[Z\langle r^{-3}\rangle_d(1-R)]^2 \tag{7.77}$$

which is independent of a, it is clear that

$$0.87|q(0)| \leq |q(a)| \leq |q(0)| \tag{7.78}$$

for all a.

Particular aspects of these general results have been developed in greater detail by several authors[96,97]; one example is given in Section 7.4.3.2.

A different physical situation which exemplifies the same general principle is provided by a crystal of molecular ions in which the species containing the quadrupolar nucleus has a high idealised symmetry. The true symmetry is usually lower because of the field due to the other ions in the crystal.

Consider, for example, the case of octahedral complex ions $[MA_5B]$ or $[\textit{trans}\text{-}MA_4B_2]$ subject to a small trigonal perturbation. The tetragonal idealised symmetry gives an axially-symmetric EFG

$$V = eq\begin{bmatrix} -\frac{1}{2} & 0 & 0 \\ 0 & -\frac{1}{2} & 0 \\ 0 & 0 & 1 \end{bmatrix} \tag{7.79}$$

and the perturbation may be written

$$\delta V^{\text{trig}} = eq\begin{bmatrix} 0 & \varepsilon & \varepsilon \\ \varepsilon & 0 & \varepsilon \\ \varepsilon & \varepsilon & 0 \end{bmatrix} \tag{7.80}$$

On solving the characteristic equation of $V + \delta V$, and expanding the roots as power series in ε, it is found that[95]

$$q(\varepsilon) = q(0)[1 + \tfrac{4}{3}\varepsilon^2 + \tfrac{8}{9}\varepsilon^3 + O(\varepsilon^4)] \tag{7.81}$$

$$\eta(\varepsilon) = 2\varepsilon - \tfrac{4}{3}\varepsilon^2 - \tfrac{32}{9}\varepsilon^3 + O(\varepsilon^4) \tag{7.82}$$

For $\varepsilon = 0.1$, equations (7.81) and (7.82) give $q(0.1) = 1.014q(0)$ and $\eta(0.1) = 0.183$, showing that a modest trigonal perturbation causes a substantial change in η, while leaving q virtually unchanged. The argument is readily generalised to the most general form for a non-tetragonal perturbation[95].

This effect is exemplified by the ^{59}Co n.q.r. data listed in Table 7.3, where large variations in η are observed although, to judge from the published crystal structure[101] of compound (ii), all the cations approximate to tetragonal symmetry. Notice particularly that in the two compounds with the

Table 7.3 Quadrupole coupling data for some octahedral complexes of low-spin $^{59}Co^{III}$

	*Compound**	$\lvert e^2qQ/h\rvert$/MHz	η	*Ref.*
(i)	$[Co(NH_3)_5Cl]SO_4{\cdot}H_2SO_4$	33.43	0.042	98
		34.39	0.082	
(ii)	$[Co(NH_3)_5Cl]Cl_2$	31.74	0.251	98
(iii)	$[\textit{trans}\text{-}Co(NH_3)_4Cl_2]Cl$	59.23	0.136	99
(iv)	$[Co(NH_3)_5CN]Cl_2$	26.57	0.185	100

* All data for crystalline state at room temperature; compound (i) has two inequivalent Co sites

same cation [(i) and (ii)] the η values differ by 0.2 whereas $|e^2qQ|$ varies by only a few percent. The perturbation undoubtedly arises from the 'lattice' contribution due to the other ions in the crystal. This is known[102] to vary with anion in $[\textit{trans}\text{-}Co(en)_2Cl_2]^+ X^-$ ($X = Cl{\cdot}HCl{\cdot}2H_2O$, Cl).

Arguments similar to the above have been used in studies[103] of the metal $E2$ couplings in $^{59}Co_2(CO)_8$ and $^{187}Re_2(CO)_{10}$.

The concept of symmetry-breaking perturbations has important implications for the interpretation of $E2$ coupling data in complex molecules. Clearly,

$|e^2qQ|$ is mainly telling us about the grosser features of molecular structure and bonding, and models for its interpretation need not take into account finer points which could be thought of as symmetry-breaking perturbations. An example of this is the additive model reviewed in Section 7.4.2; the η values in Table 7.3 do not preclude application of the additive model to octahedral complexes of low-spin Co^{III}, even though it predicts $\eta = 0$ for all the compounds in Table 7.3.

7.3.2.3 Contributions to the EFG in covalent and ionic systems

Since the EFG operator $\hat{V}_{\alpha\beta}$ [equation (7.25)] varies as r^{-3}, any asymmetry in the valence shell of the interacting atom is likely to dominate the EFG at its nucleus. Thus it is convenient to divide the EFG into so-called 'valence' and 'lattice' contributions, where the valence contribution arises from valence-shell asymmetry and the lattice contribution is due to charge asymmetry in more distant parts of the molecule or crystal. However, this decomposition is not easily put on a firm theoretical basis. Possible operational definitions are that the valence contribution is that part of the EFG, including consequent Sternheimer effects, which can conveniently be represented as arising from unequal 'effective population' (Section 7.4.1.1) of the probe-atom's valence AOs, and the lattice contribution is the remainder of the EFG.

In diamagnetic covalent molecules (Section 7.4) any EFG usually arises mainly from asymmetry in the valence shell of the probe atom[73, 74]. If this asymmetry can be related to other features of the electronic or geometrical structure of the molecule, then studies of the $E2$ coupling can yield important information on molecular structure.

The conceptual starting point for treatment of paramagnetic metal complexes is ligand-field theory[59]. In the ionic limit the valence EFG in, for example, distorted octahedral complexes of high-spin ($3d^6$) Fe^{2+}, configuration $t_{2g}{}^4e_g{}^2$, arises solely from inequalities in the thermal population of the crystal-field-split t_{2g} orbitals[104]. If these crystal-field splittings are very much less than kT the EFG tends to zero. For temperatures such that they are comparable with kT, the changes in thermal populations of the t_{2g} orbitals lead to a temperature dependence of the EFG which is characteristically much greater than that observed in diamagnetic molecules. If kT is very much less than the splitting of the t_{2g} orbitals the sixth d electron is 'frozen' into the lowest t_{2g} orbital, and the valence EFG is equal to the EFG set up by this electron. Covalency modifies this ionic picture both by changing the t_{2g} wave functions and by breaking the assumed (in the ionic limit) cubic symmetry of the $t_{2g}{}^3$ and $e_g{}^2$ subshells. The first of these effects can be taken into account by the 'orbital reduction factor'[59, 105] of ligand field theory.

According to simple crystal-field theory the lattice contribution $Q_{2M}(e)^{lat}$ should be proportional[104] to the crystal-field parameter B_{2M}. Used empirically this relationship appeared valid even in the extreme case of high-spin Fe^{2+} in square-planar coordination[90], where the lattice and valence contributions almost exactly cancel, with the lattice contribution actually having slightly larger magnitude[90, 106]. Perhaps more surprising is the agreement between the B_{20} parameters observed for Cr^{3+} impurity in $LiNbO_3$ and $LiTaO_3$

and the estimates of B_{20} calculated from the ^{93}Nb and ^{181}Ta quadrupole coupling constants in these materials[107]. These observations suggest that overlap and covalency might lead to proportional changes in the EFG and the empirical B_{2M} parameters.

A crystal of S-state ions (e.g. $^{27}Al_2O_3$) might appear conceptually simpler than either of the cases discussed above, since an S-state ion should have no valence EFG and the lattice EFG could be computed by an electrostatic 'lattice sum'[71] over all other ions in the crystal. However, recent work[70] has shown that overlap and covalency cause significant deviations from this picture.

7.4 QUADRUPOLE COUPLING IN COVALENT MOLECULES

7.4.1 Valence and lattice contributions

7.4.1.1 Effective population

In their classic paper, Townes and Dailey[73] emphasised that the EFG at the nucleus tends to be dominated by any asymmetry in the lowest non-s valence atomic orbitals (AOs), and that this effect is conveniently represented by unequal effective population of these orbitals.

Consider a system in which the electronic wave function is adequately approximated by a single Slater determinant

$$|\Psi\rangle = |\psi_1\psi_2\psi_3 \ldots\rangle \tag{7.83}$$

in which the ψ_i are orthonormal valence molecular spin orbitals, and the core AOs have not been explicitly included. In principal axes only two real, irreducible [under $O(3)$], EFG operators can have non-zero matrix elements, namely

$$\hat{V}_0 = \tfrac{1}{2}\sqrt{3}\hat{V}_{zz} \tag{7.84}$$

$$\hat{V}_{2c} = \tfrac{1}{2}(\hat{V}_{xx} - \hat{V}_{yy}) \tag{7.85}$$

These operators will be collectively denoted by $\hat{V}_\gamma$ ($\gamma = 0$, 2c). Recalling (Section 7.2.3.1) that $\hat{V}_\gamma$ is a sum of one-electron operators, the EFG at the nucleus of an atom M is given by

$$\langle\Psi|\hat{V}_\gamma^M|\Psi\rangle = (1 - R)\sum_i \langle\psi_i|\hat{v}_\gamma^M|\psi_i\rangle \tag{7.86}$$

where $\hat{v}_\gamma^M$ is the corresponding one-electron operator centred on atom M, and the polarisation of core AOs, allowed for by the Sternheimer factor R, is assumed to be linear, i.e.

$$(1 - R)[\langle\chi_1|\hat{v}_\gamma|\chi_1\rangle + \langle\chi_2|\hat{v}_\gamma|\chi_2\rangle] = (1 - R_1)\langle\chi_1|\hat{v}_\gamma|\chi_1\rangle + (1 - R_2)\langle\chi_2|\hat{v}_\gamma|\chi_2\rangle \tag{7.87}$$

(The dependence of R on the polarising-charge distribution will not always be shown explicitly.)

Let ψ_i be written as a linear combination

$$|\psi_i\rangle = \sum_N \sum_\nu |\phi_\nu^N\rangle\langle\phi_\nu^N|\psi_i\rangle \tag{7.88}$$

where the ϕ_ν^N are atomic spin orbitals with $\nu = \alpha, \beta, \ldots$ labelling orbitals centred on atoms $N = $ A, B, ... Then

$$\langle\Psi|\hat{V}_\gamma^M|\Psi\rangle = (1 - R)\sum_{N,\nu}\sum_{N',\nu'} \langle\phi_\nu^N|\hat{v}_\gamma^M|\phi_{\nu'}^{N'}\rangle N_{\nu\nu'}^{NN'} \tag{7.89}$$

where the effective populations $N_{\nu\nu'}^{NN'}$ are given by

$$N_{\nu\nu'}^{NN'} = \sum_i \langle\psi_i|\phi_\nu^N\rangle\langle\phi_{\nu'}^{N'}|\psi_i\rangle \tag{7.90}$$

Provided the one-centre integrals ($N = N' = M$) in (7.89) do not all vanish or cancel to zero, the three-centre integrals ($N \neq N' \neq M$) are negligible. Two-centre integrals of the type $N = N' \neq M$ can be considered part of the lattice contribution (Section 7.4.1.4). One-centre integrals are readily evaluated, and normally provide most of the valence contribution. However, the 'overlap' integrals $M = N \neq N'$ and $M = N' \neq N$ cause difficulties in a semi-empirical approach because they can be thought of as contributing both the remainder of the valence EFG and a part of the lattice EFG. The problem of dividing these 'overlap' contributions between the valence and lattice EFGs is, in essence, the troublesome problem of partitioning molecular charge densities[108].

Cotton and Harris[109] suggest use of the Mulliken approximation[110]

$$\phi_\mu^{M*}\phi_\nu^N = \tfrac{1}{2}\langle\phi_\mu^M|\phi_\nu^N\rangle[\phi_\mu^{M*}\phi_\mu^M + \phi_\nu^{N*}\phi_\nu^N] \tag{7.91}$$

Inserted into (7.89) this gives for the valence part of $\langle\Psi|\hat{V}_\gamma^M|\Psi\rangle$

$$[\langle\Psi|\hat{V}_\gamma^M|\Psi\rangle]_{\text{val}} = (1 - R)\sum_\mu\sum_{\mu'} \langle\phi_\mu^M|\hat{v}_\gamma^M|\phi_{\mu'}^M\rangle N_{\mu\mu'}^{\text{eff}} \tag{7.92}$$

where

$$N_{\mu\mu'}^{\text{eff}} = \sum_i [\langle\psi_i|\phi_\mu^M\rangle\langle\phi_{\mu'}^M|\psi_i\rangle + \delta_{\mu\mu'} \operatorname{Re} \sum_{N\neq M}\sum_\nu \langle\psi_i|\phi_\mu^M\rangle\langle\phi_\mu^M|\phi_\nu^N\rangle\langle\phi_\nu^N|\psi_i\rangle] \tag{7.93}$$

(Re = 'real part of'). The effective population $n_{\mu\mu' i}^{\text{eff}}$ associated with a particular MO ψ_i is

$$n_{\mu\mu' i}^{\text{eff}} = \langle\psi_i|[|\phi_\mu^M\rangle\langle\phi_{\mu'}^M| + \delta_{\mu\mu'} \operatorname{Re} \sum_{N\neq M}\sum_\nu |\phi_\mu^M\rangle\langle\phi_\mu^M|\phi_\nu^N\rangle\langle\phi_\nu^N|]|\psi_i\rangle \tag{7.94}$$

The Mulliken approximation has been closely examined in connection with its use in approximate MO theory[111], where, although popular, it is known to suffer drawbacks, the most important being lack of invariance under change of basis functions. Thus the 'success' or 'failure' of the Cotton–Harris approximation will depend on the choice of basis functions ϕ_ν^N. Not surprisingly, it has been found unsatisfactory in several cases, and alternative procedures have been suggested[112].

These difficulties are likely to be greatest with lighter atoms, where the dominance of one-centre contributions is less marked, and particularly with ^2H since the ϕ_μ^M in (7.89) are then 1s↑ and 1s↓, for which all one-centre EFG integrals are zero. Fortunately, for low-Z atoms 'exact' calculation of the EFG from *ab initio* wave functions is a practical alternative. For heavier atoms, including most metal compounds, the concept of effective population is valuable as a meeting point of experimental data and theoretical interpretation.

7.4.1.2 Valence shell contributions

In many semi-empirical interpretations of $E2$ coupling data it is sufficient to confine the sum in (7.92) to the lowest valence l-shell with $l \neq 0$. The axial component q_{val} is then for p-block elements

$$q_{\text{p}} = -\tfrac{4}{5}(1 - R_{\text{p}})\langle r^{-3}\rangle_{\text{p}}[N_z - \tfrac{1}{2}(N_x + N_y)] \qquad (7.95)$$

and for d-block elements

$$q_{\text{d}} = -\tfrac{4}{7}(1 - R_{\text{d}})\langle r^{-3}\rangle_{\text{d}}[N_{z^2} + \tfrac{1}{2}(N_{xz} + N_{yz}) - (N_{x^2-y^2} + N_{xy})] \qquad (7.96)$$

However, towards the end of a given row of transition elements one might anticipate that contributions from the nd shell (which is near to becoming a filled core shell) might fail to dominate the contribution from the $(n + 1)$p shell. Evidence for this is obtained by comparing the ^{197}Au Mössbauer quadrupole splittings observed[113-116] in two-coordinate linear halide complexes of Au^{I} ($5d^{10}$) with those in four-coordinate square-planar halide complexes of Au^{III} ($5d^8$). In the Au^{I} complexes the EFG arises mainly from the large imbalance in 6p-shell populations ($N_z \gg N_x = N_y$) caused by ligand→Au σ-bonding, and leading to quadrupole splittings of 4.1–4.6 mm s^{-1}*. However, in the Au^{III} species, ligand→Au σ-bonding leads to $N_x = N_y \gg N_z$, i.e. large positive $q_{6\text{p}}$, but this partly cancels the negative $q_{5\text{d}}$ arising from the aspherical $5d^8$ configuration ($N_{z^2} \approx N_{xz} = N_{yz} \approx N_{xy} \gg N_{x^2-y^2}$, assuming the ligands are on the x and y axes), leading to much smaller splittings in the range 0.75–1.7 mm s^{-1}.

In general the ratio $\langle r^{-3}\rangle_{n\text{d}}:\langle r^{-3}\rangle_{(n+1)\text{p}}$ is unreliable as a guide to the likely significance of $(n + 1)$p contributions to the EFG since, particularly in complexes with O_h or T_d holosymmetry, it is possible that the imbalance in effective populations of the $(n + 1)$p orbitals may be relatively small. For example, even in the asymmetrical tetragonal-pyramidal environment of five-coordinate Fe^{III} in bis(N,N-diethyldithiocarbamato)iron(III) chloride the calculated[117] 4p contribution was only a few percent of the total EFG although the calculated ratio $\langle r^{-3}\rangle_{3\text{d}}:\langle r^{-3}\rangle_{4\text{p}}$ was 2.76:1 for the wave functions used.

* For the 77.34 keV transition of ^{197}Au, 1 mm s^{-1} = 62.38 MHz. Numbers quoted are values of $|\tfrac{1}{2}ce^2qQ(1 + \tfrac{1}{3}\eta^2)^{\frac{1}{2}} E_\gamma^{-1}|$ at 4.2 K.

7.4.1.3 Overlap of filled orbitals

In this section we consider the interaction between filled non-orthogonal orbitals. One relevant example is the interaction of the filled ligand σ-orbitals with the doubly-occupied $5d_{z^2}$ orbital of W^{IV} in square-antiprismatic $W(CN)_8{}^{4-}$ (Section 7.4.3.3); another is lone-pair–lone-pair interactions.

The (normalised) overlapping orbitals χ_i can be formed into orthonormal molecular orbitals ψ_k in two steps: first they are orthogonalised[118, 119] to give orthonormal orbitals ϕ_j, and then the ϕ_j may undergo covalent interaction to yield the ψ_k. Since the orthogonalisation transformation is not unique neither is the decomposition into two steps, but the transformation of orthonormal ϕ_j into orthonormal ψ_k must be unitary, and therefore leaves unchanged the Slater-determinantal wave function $|\Psi\rangle$, and hence the EFG components $\langle\Psi|\hat{V}_\gamma|\Psi\rangle$. Thus covalent interaction between filled orbitals cannot change the EFG. This is correctly reflected in the effective populations $N_{\nu\nu'}^{NN'}$ [equation (7.90)] and $N_{\mu\mu'}^{\text{eff}}$ [equation (7.93)], since both are invariant under unitary transformation of the filled ψ_i.

However, the orthogonalisation transformation is not unitary, nor even unimodular. The contribution to the EFG from the interacting orbitals is given in terms of the χ_i by[120] (neglecting Sternheimer factors for clarity)

$$\sum_k \langle\psi_k|\hat{v}_\gamma|\psi_k\rangle = (\det \boldsymbol{S})^{-1} \sum_i \sum_j \langle\chi_i|\hat{v}_\gamma|\chi_j\rangle \operatorname{cof} S_{ij} \qquad (7.97)$$

where cof S_{ij} is the cofactor of S_{ij} in the overlap matrix

$$S_{ij} = \langle\chi_i|\chi_j\rangle \qquad (7.98)$$

and det $\boldsymbol{S}$, the determinant of $\boldsymbol{S}$, is assumed to be non-zero, i.e. the χ_i are linearly independent[119]. In general

$$(\det \boldsymbol{S})^{-1} \sum_i \sum_j \langle\chi_i|\hat{v}_\gamma|\chi_j\rangle \operatorname{cof} S_{ij} \neq \sum_r \langle\chi_r|\hat{v}_\gamma|\chi_r\rangle \qquad (7.99)$$

indicating that the overlap interaction between filled orbitals should be taken into account when calculating the EFG. However, in the Mulliken approximation[110]

$$\chi_i^*\chi_j = \tfrac{1}{2}S_{ij}(\chi_i^*\chi_i + \chi_j^*\chi_j) \qquad (7.100)$$

equation (7.99) becomes an equality, and the EFG is now just that given by superposition of the unorthogonalised filled orbitals.

The above results may be of limited application to real systems if mixing with empty excited orbitals is also important.

7.4.1.4 The 'lattice' contribution

It is often assumed that in a neutral covalent molecule the electrons are distributed between atoms so as to leave each atom fairly close to neutrality. In this case the lattice contribution at a central atom surrounded by ligands should be small. Even for the nearest-neighbour atoms the EFG due to the positive atomic cores tends to cancel with the two-centre integrals in (7.89).

Indeed, this cancellation is exploited in attempts to simplify semi-empirical calculations of EFGs by judicious sharing of the 'overlap' two-centre integrals between lattice and valence contributions[109, 112].

In a molecular ion it is impossible for all atoms to be neutral, and in lower-than-cubic symmetry a lattice EFG is inevitable. There is little empirical evidence on the magnitude of intramolecular contributions, although it is worth noting that the ^{119}Sn Mössbauer quadrupole splitting of 4.12 mm s^{-1} observed[121] for *trans*-$Me_2SnF_4^{2-}$ in $K_2[Me_2SnF_4]$ differs by only 6% from the splitting of 4.38 mm s^{-1} observed[122] in Me_2SnF_2, which contains the same [*trans*-Me_2SnF_4] unit because of F bridges[123]. The extramolecular contribution from other ions in a molecular crystal seems to be of similar magnitude, since coupling constants typically vary with counter-ion by about 5%, or at most 10%. Thus in molecular crystals of either neutral or charged units the lattice contribution can often be regarded as a symmetry-breaking perturbation (Section 7.3.2.2), affecting $sgn\{e^2qQ\}$ and (particularly) η rather more than $|e^2qQ|$ itself[95, 103].

The lattice contribution is a significant parameter in the n.q.r. spectroscopy of nuclei such as the halogens ^{35,37}Cl and ^{79,81}Br which tend to be located on the outside of the molecular unit. This vulnerability to intermolecular perturbation, combined with the sensitivity of n.q.r. makes them excellent empirical probes of the 'crystal-field effects' (to be distinguished from the ligand-field sense) recently reviewed by Weiss[124].

Most diamagnetic molecules do not have thermally-accessible excited electronic states. The temperature-dependence of the $E2$ coupling constant in molecular crystals is then due mainly to changes in the excitation of thermal vibrations, and is usually interpreted in terms of the Beyer theory, which emphasises the role of librational modes[17, 124]. Schempp and Silva[125] have recently reported n.q.r. evidence for a T^4 frequency dependence at low temperatures, indicating the influence of acoustic modes.

7.4.2 The additive model

7.4.2.1 Characteristics of additivity

Consider a probe atom M bound to ligands L = A, B, C, ... The EFG V at the nucleus of M is said to be additive if it can be written as a sum of contributions, one for each ligand:

$$V = V_A + V_B + V_C + \ldots \qquad (7.101)$$

For (7.101) to be useful there must be some degree of transferability, i.e. the same V_L (allowing for changes in orientation) must be associated with the M—L bond in at least two different molecules. The possibilities for transferability range from this minimal case to the maximal situation[126] in which the parameters of a partial field gradient (PFG) V_L are the same for all M—L bonds in molecules with a given holosymmetry.

It is convenient (although not strictly necessary[93, 127]) to assume that the PFGs associated with each M—L bond are axially symmetric. Local cartesian axes are defined with z directed along the M—L bond, and the PFG tensor in these axes is written

$$eq_L \begin{bmatrix} -\frac{1}{2} & 0 & 0 \\ 0 & -\frac{1}{2} & 0 \\ 0 & 0 & 1 \end{bmatrix}$$

The total EFG tensor in molecular axes is then calculated by writing down the PFG tensors, transforming each from its local axes to molecular axes, and summing the resulting tensors[93].

The EFG tensors obtained in this way for oct.-MA_5B and *trans*-oct.-MA_4B_2 are both axially symmetric with V_{zz} equal to $e(q_B - q_A)$ and $2e(q_B - q_A)$, respectively. In each case the EFG has one independent parameter, as required by the C_{4v} and D_{4h} idealised molecular symmetries. Thus, the 1:2 ratio of the coupling constants depends only on the transferability of the PFG parameter $(q_B - q_A)$.

The EFG calculated for *cis*-oct.-MA_4B_2 in the additive model is axially symmetric with $V_{zz} = -e(q_B - q_A)$, and thus also has only one independent parameter, although the C_{2v} molecular symmetry allows two. This arises because each individual PFG tensor $\boldsymbol{V}_L$ is centrosymmetric, so that the situations represented formally by M—L and $\frac{1}{2}$L—M—$\frac{1}{2}$L are indistinguishable in the additive approximation. Thus we define[93] an 'augmented' (hypothetical) system in which this formal process is applied to each ligand. This is shown for *cis*-oct.-MA_4B_2 in Figure 7.4, where it is seen that the augmented molecule has D_{4h} symmetry, thus restricting the number of independent parameters in the EFG to one.

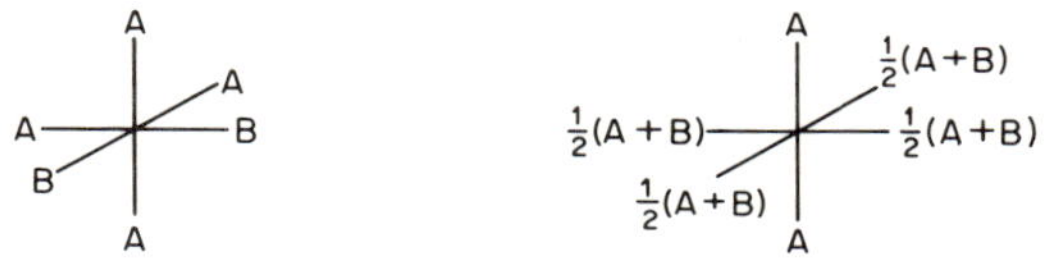

Figure 7.4 *cis*-oct.-MA_4B_2 and its augmented complex

In *mer*-oct.-MA_3B_3 the molecular symmetry is C_{2v} and the augmented symmetry D_{2h}. Both these groups allow two independent EFG components, yet the EFG calculated in the additive model again depends only on $(q_B - q_A)$:

$$e(q_B - q_A) \begin{bmatrix} \frac{3}{2} & 0 & 0 \\ 0 & -\frac{3}{2} & 0 \\ 0 & 0 & 0 \end{bmatrix}$$

This illustrates another aspect of the symmetry consequences of additivity: here non-classical 'colour symmetry' is leading to a reduction in the number of independent components[93, 128, 129].

Consider an operator θ which takes A to B and B to A. Under this operator the symmetry group of *mer*-oct.-MA_3B_3 is the type III dichromic point group[130]

$$D_{2d}(C_{2v}) = C_{2v} + \theta(D_{2d} - C_{2v})$$

i.e. operations which belong to both D_{2d} and C_{2v} are symmetry elements of the molecule, while operations which belong to D_{2d} but not C_{2v} become symmetry elements if multiplied by θ.

The effect of θ on the EFG is defined as follows[93, 129]. If we write

$$V = \sum V_{\mathrm{L}}(l) \tag{7.102}$$

where $V_{\mathrm{L}}(l)$ is the PFG tensor due to a ligand L situated on bond axis l, then for *mer*-oct.-MA_3B_3 V can be written

$$V(\mathrm{MA_3B_3}) = V_+(\mathrm{MA_3B_3}) + V_-(\mathrm{MA_3B_3}) \tag{7.103}$$

where

$$V_+ = \sum_l \tfrac{1}{2}[V_{\mathrm{A}}(l) + V_{\mathrm{B}}(l)] \tag{7.104}$$

and

$$V_- = \sum_{l_{\mathrm{A}}} \tfrac{1}{2}[V_{\mathrm{A}}(l) - V_{\mathrm{B}}(l)] + \sum_{l_{\mathrm{B}}} \tfrac{1}{2}[V_{\mathrm{B}}(l) - V_{\mathrm{A}}(l)] \tag{7.105}$$

These equations may be illustrated with formal 'molecules' in the same way as for the augmented symmetry; this is done in Figure 7.5. The quantities V_+ and V_- correspond to the two possible eigenvalues ($+1$ and -1) of θ, i.e. V_+ is invariant under θ and V_- changes sign. Note that θ is a unitary operator, and leads to unitary dichromic groups, in contrast to the magnetic groups formed using the antiunitary time-reversal operator[130].

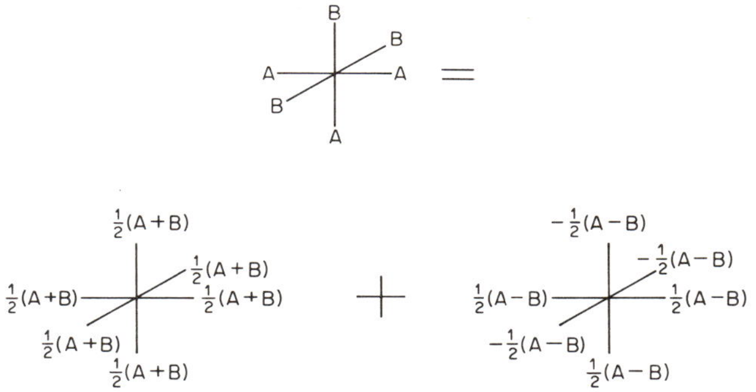

Figure 7.5 Illustration of equation (7.103) for *mer*-oct.-MA_3B_3

As is evident from Figure 7.5, V_+ has the full O_h holosymmetry of the molecule, implying that V_+ vanishes. The independent non-zero components of V_- must belong to the irreducible representation of D_{2d} in which elements common to D_{2d} and C_{2v} have character 1, and those belonging only to D_{2d} have character -1. Only V_{2c} [equation (7.85)] has this property, with the result that the EFG obtained has only one independent parameter and $\eta = 1$.

The additive model has been extensively applied to the interpretation of quadrupole coupling data, particularly in the Mössbauer spectroscopy of $^{119}Sn^{IV}$ and low-spin $^{57}Fe^{II}$ compounds[22, 55, 131, 132]. In experimental applications the PFG is often visualised in terms of devices such as notional point charges attached to the ligand positions or effective populations in L→M σ-bonds (the donated-charge model).

Additive models have also been used in the interpretation of other molecular properties such as polarisabilities[133] or ligand-field parameters[134]. Not surprisingly, concepts similar to those reviewed here have arisen independently in the literature of these other applications, often in superficially different forms and usually with different names.

7.4.2.2 *Significance of* $\eta \approx 1$

Various symmetry restrictions on the components of the EFG tensor were noted in Section 7.3.1.1. In principal axes these restrictions may be formulated in terms of the two irreducible components V_0 and V_{2c} [equations (7.84) and (7.85)], and the unit representation Γ_1 of the molecular symmetry group:

neither V_0 nor V_{2c} belongs to Γ_1 : $q = 0$ and $\eta = 0$
only V_0 belongs to Γ_1 : $q \neq 0$ and $\eta = 0$
both V_0 and V_{2c} belong to Γ_1 : $q \neq 0$ and $\eta \neq 0$

None of these possibilities specifically implies $\eta = 1$, and it is of interest to ask if any group-theoretical significance can be attached to unit α-symmetry.

The symmetry requirement for $\eta = 1$ is that, in appropriate axes, only V_{2c} belongs to Γ_1. This does not occur for any of the classical point-symmetry groups, implying that some kind of cryptosymmetry[130] must be involved. We shall consider only the case of a single antisymmetry element ϕ of order 2 ($\phi^2 = 1$), e.g. the operator θ introduced in Section 7.4.2.1, although it is likely that more complicated cryptosymmetries could yield similar results. The effect of ϕ is to reverse the sign of all components of the EFG; examples of physical situations leading to this cryptosymmetry are given below.

It is convenient to use irreducible EFG components defined in arbitrary axes (x,y,z) by[93] (7.84), and (7.85), $V_{2s} = V_{xy}$, $V_{1c} = V_{xz}$, and $V_{1s} = V_{yz}$. Consider the effect of a symmetry operator $\phi R(a)$ where $R(a)$ is a rotation through an angle a about z [or equally, since V is centrosymmetric, $\phi IR(a)$ where I is the inversion]; on equating the EFG tensor before and after applying the symmetry operator we obtain two non-trivial solutions[135]:

(A) $a = (n + \frac{1}{2})\pi$, $V_0 = V_{1c} = V_{1s} = 0$, V_{2c} and V_{2s} arbitrary
(B) $a = (2n + 1)\pi$, $V_0 = V_{2c} = V_{2s} = 0$, V_{1c} and V_{1s} arbitrary

By use of formulae[93] for the symmetrised parameters S_{02} and S_{03} in terms of the irreducible components we obtain, for solution (A),

$$S_{02} = -[(V_{2c})^2 + (V_{2s})^2] \neq 0, \; S_{03} = 0 \tag{7.106a}$$

and for solution (B)

$$S_{02} = -[(V_{1c})^2 + (V_{1s})^2] \neq 0, \; S_{03} = 0 \tag{7.106b}$$

Thus in both cases we have [see equation (7.32)] a non-zero EFG with $\eta = 1$.

Hence the presence of a cryptosymmetry axis ϕC_4, $\phi IC_4 = \phi S_4$, ϕC_2, or $\phi IC_2 = \phi\sigma$, is a condition for $\eta = 1$. Armed with this information the dichromic colour groups leading to $\eta = 1$ are readily enumerated, e.g. by use of the tables given in Birss[136].

Supplementary conditions are required to ensure that all components of the total EFG change sign under ϕ. In Section 7.4.2.1 we saw that the total EFG V can be written as a sum of a part V_+ which is invariant under ϕ, and a part V_- which changes sign. This resolution is always formally possible, but only gives new information if the EFG is additive[93, 129]. Thus the required supplementary conditions are that the EFG be additive, and that the holosymmetry (more strictly the 'augmented holosymmetry'[93]) be cubic or higher so that V_+ vanishes.

The example of $D_{2d}(C_{2v})$ symmetry in *mer*-oct.-MA_3B_3 was discussed in Section 7.4.2.1. The tetrahedral MA_2B_2 system also has $D_{2d}(C_{2v})$ symmetry under the operator θ, and the recent literature contains many instances of large η values associated with this symmetry.

A good example is provided by the bridged dimer (1) where M is a Group

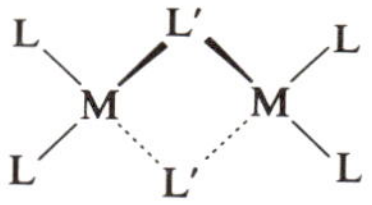

(1)

III element. Here each M site has effective $D_{2d}(C_{2v})$ symmetry leading to $\eta = 1$ for ideal tetrahedral coordination. Table 7.4 gives data for various examples with L = L′ (bridging and terminal L will still be electronically different). The η values are all less than unity, reflecting deviations from the ideal. Changes in L and L′ can cause these deviations to increase or decrease. The ^{27}Al compounds[138] [L = Et, L′ = EtO], [L = Me, L′ = Me_2N] and [L = Me, L′ = PhC≡C] have $\eta = 1.00$ (at 77 K), 0.957 (at 77 K) and 1.00 (at room temp.), respectively, whereas the halogen-bridged $(R_2AlX)_2$ species [L = alkyl, L′ = Cl, Br, I] have $\eta \approx 0.5$.

Other examples are ^{57}Fe Mössbauer Zeeman studies[142] at 4.2 K of ($3d^{10}$) $Fe(CO)_2(NO)_2$ ($\eta \approx 0.85$) and $Fe(Ph_3P)_2(NO)_2$ ($\eta \approx 0.76$), a similar ^{119}Sn Mössbauer Zeeman study[143] of $[\pi\text{-}C_5H_5Fe(CO)_2]_2SnCl_2$ ($\eta = 0.65$) and ^{182}W Mössbauer data at 4.2 K on several ($5d^0$) W^{VI} species[144], notably

Table 7.4 Asymmetry parameter of the EFG at M in various Group III L-bridged dimers M_2L_6

M	M_2L_6	η	*Temp.*/K	*Method*	*Ref.*
B	B_2H_6	0.844		calc.	137
^{27}Al	Al_2Me_6	0.78	77	n.q.r.	138
	Al_2Et_6	0.87	196	n.q.r.	138
	$Al_2Bu^i_6$	0.87	77	n.q.r.	138
	Al_2Ph_6	0.79	room	n.q.r.	138
	Al_2Br_6	0.73	77	n.q.r.	139
^{71}Ga	Ga_2Cl_6	0.87	room	n.q.r./Zeeman	140
^{115}In	In_2I_6	0.69	room	n.q.r.	141

$(NH_4)_2WS_4$ ($\eta = 0.76$), which is known[145] to have WS_4^{2-} tetrahedral units with two of the four sulphurs linked to bridging NH_4^+ ions to form —S—NH_4—S— chains.

Electric field gradients tabulated from the gaussian-basis-set SCF calculations of Snyder and Basch[137] show that the two lone pairs in an isolated H_2O molecule can be thought of as 'dummy ligands' since the EFG at O has $\eta = 0.766$. Gas-phase spectroscopic studies[198] of $HD^{17}O$ and $H_2{}^{33}S$ give $\eta = 0.748$ and $\eta = 0.6$, respectively.

Similar wave functions[137] also give $\eta = 0.685$ for the EFG at C in CH_2F_2. Incidentally, the relative values calculated for $|q|(1 + \frac{1}{3}\eta^2)^{\frac{1}{2}}$ at C in CH_3F, CH_2F_2 and CHF_3 are 0.588:0.676:0.524, respectively. When compared with the ratios 1:1.15:1 predicted by the additive model this gives some measure of the effects due to deviations from tetrahedral bond angles and non-additivity.

In general, three points must be borne in mind when interpreting large η values. First, the caveats concerning symmetry-breaking perturbations (Section 7.3.2.2) apply *a fortiori* when $\eta = 1$, since this situation is particularly vulnerable to any distortion from ideal geometry. Second, close to the probe nucleus the electronic charge distribution may not reflect the (straight-line) internuclear axes; i.e. the 'bonds' may be bent. Third, the presence of identifiable cryptosymmetry is only one possible cause of $\eta \sim 1$; another is distortion, while in some cases large η values seem to arise almost capriciously, with no single factor dominant. Further discussion relevant to the above-mentioned three points will be found in Section 7.4.4. Recent work by LaRossa and Brown is of interest both in this context and on the general question of transferability.

7.4.2.3 A molecular orbital approach

Consider[126] a closed-shell molecule consisting of n ligands L = A, B, . . . bound to a central atom M by n σ-bonds. Let the n orthonormal valence MOs $\psi_1, \psi_2, \ldots, \psi_n$ contain $2n$ electrons, and let the total wave function Ψ be approximated by a single Slater determinant

$$|\Psi\rangle = |\psi_{1\uparrow}\psi_{1\downarrow}\psi_{2\uparrow}\psi_{2\downarrow}\cdots\psi_{n\uparrow}\psi_{n\downarrow}\rangle \tag{7.107}$$

where closed inner-shell orbitals have not been explicitly included. In general each ψ_i is delocalised and may have non-zero amplitude at many points in the molecule.

However, the MOs ψ_i may be transformed into a set of n orthonormal localised orbitals (LOs)[146, 147] ϕ_L, chosen so that each ϕ_L is, as far as possible, localised in the region of a particular M—L bond[148]. The ϕ_L do not have well-defined orbital energies, and for this reason could not be used in a discussion of, say, ionisation processes. However, the components $V_{\alpha\beta}$ of the valence EFG at M are unchanged by the transformation to LOs, and may be written as a sum of one-electron matrix elements:

$$V_{\alpha\beta} = 2\sum_{L}(1 - R_L)\langle\phi_L|\hat{v}^M_{\alpha\beta}|\phi_L\rangle \tag{7.108}$$

where the Sternheimer factor R_L allows for polarisation of the core AOs of M by the EFG due to the charge distribution $\phi_L^*\phi_L$.

The quantity[126]

$$v_{\alpha\beta}(\mathrm{L}) = 2(1 - R_{\mathrm{L}})\langle\phi_{\mathrm{L}}|\hat{v}^{\mathrm{M}}_{\alpha\beta}|\phi_{\mathrm{L}}\rangle \qquad (7.109)$$

is the PFG due to L. If ϕ_{L} is well-localised into the region close to the M—L bond axis it seems reasonable to regard $v_{\alpha\beta}(\mathrm{L})$ as a property of the particular M—L bond. Of course, the other bonds will perturb the M—L bond to some extent. Thus the values of the PFG due to an M—L bond in a series of compounds with the same coordination polyhedron will form a distribution which may be characterised by the usual statistical parameters of mean, standard deviation, etc. The additive model is valid provided the PFG standard deviations are much smaller in magnitude than the EFG components.

Clearly, LOs provide the natural framework for discussion of additive EFGs. Further, in all cases where the additive model has been successfully applied to a significant number of compounds the σ MOs have been localisable, suggesting that the existence of a suitable localisation transformation is a necessary condition for additivity[126].

However, this condition is not satisfied for M→L π-bonding in low-spin Fe^{II} ($3d^6$) octahedrally coordinated by ligands with empty π orbitals. Nevertheless, recent experimental results[149-151] indicate that the additive model may be viable even for carbonyl complexes of Fe^{II}, although Bancroft and Libbey[150] discuss the possibility that the PFG due to carbonyl ligands varies with stereochemistry, in agreement with the localisation conjecture. Of course, it may be that π bonding has a relatively small effect on the EFG in these compounds; this would still be consistent with π bonding having a significant effect on the energy levels of the complex.

7.4.2.4 Partial field gradient parameters

The MO approach described in Section 7.4.2.3 has been used to understand the dependence of PFGs on coordination geometry[126, 152].

If local principal axes (x,y,z) are taken with z parallel (or most nearly parallel) to the M—L axis, then the PFG [equation (7.109)] is characterised by

$$2e[\mathrm{L}] = v_{zz}(\mathrm{L}) \qquad (7.110)$$

and

$$\eta_{\mathrm{L}} = [v_{xx}(\mathrm{L}) - v_{yy}(\mathrm{L})]/v_{zz}(\mathrm{L}) \qquad (7.111)$$

The LO ϕ_{L} can be written

$$\phi_{\mathrm{L}} = c_1 h_{\mathrm{L}} + c_2 \chi_{\mathrm{L}} \qquad (7.112)$$

where h_{L} is a normalised linear combination of AOs based on M, and χ_{L} a similar combination of ligand-based orbitals.

An additive scheme valid for all molecules belonging to a given holosymmetry can be set up on the basis of two assumptions[126]. First, it is assumed that the h_{L} are unchanged by substitution and thus have the form appropriate to the holosymmetric molecule, i.e. the hybrid orbitals of simple valence theory. Second, an effective population σ_{L} is defined by

$$2(1 - R_{\mathrm{L}})\langle\phi_{\mathrm{L}}|\hat{v}^{\mathrm{M}}_{\alpha\beta}|\phi_{\mathrm{L}}\rangle = (1 - R_h)\langle h_{\mathrm{L}}|\hat{v}^{\mathrm{M}}_{\alpha\beta}|h_{\mathrm{L}}\rangle\sigma_{\mathrm{L}} \qquad (7.113)$$

Table 7.5 Partial field gradient parameters for p and d valence electrons in important structural types

*Bond**	*Hybrid*	p *electrons†*	d *electrons†*	
tet	$h_z^{tet} = \frac{1}{2}s + \sqrt{\frac{3}{2}}p_z$	$[L]^{tet} = -\frac{3}{10}\langle r^{-3}\rangle_p \sigma_L^{tet}$	—	
tba	$h_z^{tba} = (\sqrt{\frac{1}{2}}\cos\theta)s + \sqrt{\frac{1}{2}}p_z + (\sqrt{\frac{1}{2}}\sin\theta)d_{z^2}$	$[L]^{tba} = -\frac{1}{5}\langle r^{-3}\rangle_p \sigma_L^{tba}$	$[L]^{tba} = -\frac{1}{7}\sin^2\theta\langle r^{-3}\rangle_d \sigma_L^{tba}$	
tbe	$h_z^{tbe} = (\sqrt{\frac{1}{3}}\sin\theta)s + \sqrt{\frac{2}{3}}p_z + (\frac{1}{2}\sqrt{\frac{1}{3}}\cos\theta)d_{z^2} - (\frac{1}{2}\cos\theta)d_{x^2-y^2}$	$[L]^{tbe} = -\frac{4}{15}\langle r^{-3}\rangle_p \sigma_L^{tbe}$	$[L]^{tbe} = +\frac{1}{21}\cos^2\theta\langle r^{-3}\rangle_d \sigma_L^{tbe}$	$\eta_L^{tbe} = 3$
oct	$h_z^{oct} = \sqrt{\frac{1}{6}}s + \sqrt{\frac{1}{2}}p_z + \sqrt{\frac{1}{3}}d_{z^2}$	$[L]^{oct} = -\frac{1}{5}\langle r^{-3}\rangle_p \sigma_L^{oct}$	$[L]^{oct} = -\frac{2}{21}\langle r^{-3}\rangle_d \sigma_L^{oct}$	

* tet = tetrahedral, tba = trigonal-bipyramidal-apical, tbe = trigonal-bipyramidal-equatorial, oct = octahedral

† General expressions for [L] and η_L are given in Ref. 126; $\eta_L = 0$ if not given explicitly; Sternheimer factors are implicit in radial averages, i.e. $\langle r^{-3}\rangle_l = (1 - R_l)\int_0^\infty u_l(r) r^{-3} u_l(r) r^2\, dr$

As in Sections 7.4.1.1 and 7.4.1.4 it can be argued that neglected two-centre integrals will tend to cancel any lattice contribution. In practice such cancellation occurs automatically, since σ_L is treated empirically.

General expressions have been calculated for the PFG parameters of tetrahedral (tet), trigonal-bipyramidal-apical (tba) and -equatorial (tbe), and octahedral (oct) bonds[126]. Some results derived from these calculations are given in Table 7.5. The hybrids in column 2 of Table 7.5 are referred to local axes with z along the M—L bond, thus giving h^{tbe} an unfamiliar form. The PFG parameters due to p and d imbalance are listed separately in columns 3 and 4; p-electron and d-electron PFGs must be combined by tensor addition. The s–d cross-terms in the general expressions[126] are omitted since $\langle r^{-3}\rangle_{sd}$ is usually negligible, and vanishes for (ns)–(nd) hydrogenic radial wave functions[153].

The PFGs have axial symmetry about the M—L bond ($\eta_L = 0$), except for d electrons in a tbe bond where $\eta_L = 3$, showing that the local axes chosen do not obey the ordering convention (7.27). This PFG tensor has axial symmetry about an axis parallel to the three-fold symmetry axis of the trigonal bipyramid.

The hybrids $\{h^{tet}\}$ and $\{h^{oct}\}$ span single equivalent sets under T_d and O_h, respectively. However, $\{h^{tba}\}$ and $\{h^{tbe}\}$ form distinct equivalent sets under D_{3h}, and a parameter θ ($0 \leqslant \theta \leqslant 2\pi$) is introduced to describe the distribution of s and d_{z^2} between apical and equatorial bonds[154].

It is evident from Table 7.5 that different PFG parameters must be assigned to tet, tba, tbe and oct bonds. This has been confirmed in detail for p electrons by studies of $^{119}Sn^{IV}$ Mössbauer quadrupole splittings[126,155,164]. The only d electron systems studied in detail have been octahedral low-spin $3d^6$, notably $^{57}Fe^{II}$ complexes[150,156]. In trigonal-bipyramidal coordination the partly-filled d shell gives a dominant non-additive contribution to the EFG (*cf.* Section 7.4.3.2), but d^0 and (rather unlikely) low-spin d^4 tetrahedral systems could in theory have additive EFGs based on d^3s hybrids. (In the tetrahedral $3d^{10}$ $^{57}Fe^{-II}$ complexes[142] the additive EFG must be based on sp^3 hybrids.)

Since the EFGs in tet.-MA_4 and oct.-MA_6 vanish by symmetry, only ($[L]^{tet} - [X]^{tet}$) and ($[L]^{oct} - [X]^{oct}$), with X a fixed 'reference ligand', are observable experimentally. Further, if M is a p-block element, the EFG in trig.-bipy.-MA_5 is axially symmetric with

$$q = 4[A]^{tba} - 3[A]^{tbe} = \tfrac{4}{5}\langle r^{-3}\rangle_p(\sigma_A^{tbe} - \sigma_A^{tba}) \qquad (7.114)$$

by column 3 of Table 7.6. Thus either ($[L]^{tba} - [X]^{tba}$) and ($[L]^{tbe} - \tfrac{4}{3}[X]^{tba}$) or ($[L]^{tba} - \tfrac{3}{4}[X]^{tbe}$) and ($[L]^{tbe} - [X]^{tbe}$) are the only independent experimental observables[126]. Analogous results will apply for other coordination polyhedra.

It is usual to measure PFGs in units of quadrupole splitting. Thus in ^{57}Fe or ^{119}Sn Mössbauer spectroscopy ($I = \tfrac{3}{2}$) the measured parameters are $\tfrac{1}{2}e^2|Q|([L]^{tet} - [X]^{tet})$, etc., multiplied by c/E_γ to put them into Doppler-shift units. These numerical parameters are often called 'partial quadrupole splittings' (PQSs). It is usually convenient to take X = halogen.

The relative nature of PQS values is often ignored in the literature, and [L] is written when ([L] — [X]) is meant. This does little harm in tetrahedral or octahedral systems; at most it obscures the reference ligand. However, when dealing with coordination polyhedra for which the holosymmetric EFG does not vanish by symmetry, or when comparing PFGs for different coordination polyhedra, it is essential to write out explicitly the actual experimental observables.

Evaluation of PQS parameters by systematic analysis of experimental data has been undertaken for $^{119}Sn^{IV}$ organometallics[126, 164] and low-spin $^{57}Fe^{II}$ complexes[150, 156]. In Sn^{IV} the relative parameters of key frequently-occurring halogen, alkyl and phenyl ligands were determined with particular care. The PQS values listed in Ref. 126 are unbiassed in the sense that they gave predicted quadrupole splittings which did not systematically under- or over-estimate the observed splittings.

For predictions based on these values a crude tolerance limit of 0.4 mm s^{-1} was estimated on the grounds that (a) it corresponded to roughly three times the population standard deviations estimated during the determination of key-ligand parameters, and (b) it was about 10% of the largest observed splittings, thus allowing an adequate margin for factors neglected in the additive model. This tolerance limit is the limit of acceptable disagreement between prediction and observation for a single datum. If n molecules known or assumed to have the same structure are tested simultaneously, the limit should be divided by $\sqrt{n}$. Further discussion of tolerance limits in applications of the additive model will be found in Section 7.4.2.5.

From the PQS values given in Ref. 126, and taking as an average of recent estimates[77, 157]*,

$$\Delta_p = -2ce^2Q(1 - R)\langle r^{-3}\rangle_p/5E_\gamma = +3.73 \text{ mm s}^{-1} \tag{7.115}$$

we calculate (R = alkyl, X = Cl, Br, F)

$$\sigma_R^{tet} - \sigma_X^{tet} = 0.98$$
$$\sigma_R^{oct} - \sigma_X^{oct} = 1.10$$

Likewise the splitting of about 0.7 mm s^{-1} observed[121] in $SnCl_5^-$ gives

$$|\sigma_{Cl}^{tbe} - \sigma_{Cl}^{tba}| = 0.19$$

These values are in agreement with intuitive expectations.

Bancroft and coworkers[150, 156] have obtained PQS values for octahedral complexes of low-spin Fe^{II}. The quantity tabulated was actually

$$\tfrac{1}{2}ce^2|Q|E_\gamma^{-1}([L] - [Cl]) - 0.30 \text{ mm s}^{-1}$$

because of an early decision[158] to assign Cl^- the parameter -0.30 mm s^{-1}. Values were carefully tested by comparing predicted and observed quadrupole splittings, but statistical tests for bias were not applied. A tolerance limit of 0.2 mm s^{-1} was suggested. The parameters were interpreted as measures of the difference between σ-donor and π-acceptor character of the ligands.

Mössbauer Zeeman studies of low-spin Fe^{II} complexes[156, 159] provided important confirmation that the quadrupole splittings in *trans*-MA_4B_2 and

* For the ^{119}Sn 23.88 keV Mössbauer resonance, Q is negative and 1 mm s^{-1} = 19.26 MHz.

cis-MA_4B_2 had opposite signs. More recently[160] this has also been confirmed for Sn^{IV} systems. However, several other *cis*-octahedral Sn^{IV} systems have opposite signs to those predicted, although the magnitudes of the splittings are about correct[121, 161]. This arises because the sign is more sensitive than the magnitude to distortions from ideal geometry[121].

7.4.2.5 The regression method applied to five-coordinate Sn^{IV}

Methods of data analysis used for octahedral and tetrahedral Sn^{IV} systems cannot easily be applied to trigonal-bipyramidal molecules because of the necessity to assign different partial field gradient parameters to apical and equatorial bonds. A recently reported regression method[155] overcomes this difficulty, and also offers a basis for establishing tolerance limits which provide a quantitative guide to the probable correctness of predictions based on the additive model.

Let the ^{119}Sn Mössbauer quadrupole splittings $[\frac{1}{2}ce^2qQ(1 + \frac{1}{3}\eta^2)^{\frac{1}{2}}E_\gamma^{-1}]$ in *trans*-oct.-R_2SnL_4, *equatorial*-R_3SnL_2 (I), *cis*-R_3SnL_2 (II) and *mer*-R_3SnL_2 (III) be denoted Δ_0, Δ_I, Δ_{II} and Δ_{III}, respectively. Then, from Table 7.5

$$\Delta_I = -[(\sigma_R^{tba} - \sigma_L^{tba})/(\sigma_R^{oct} - \sigma_L^{oct})]\Delta_0 + \Delta_p(\sigma_R^{tba} - \sigma_R^{tbe}) \quad (7.116)$$

where Δ_p is given by (7.115).

If R is fixed and L varied then, provided $(\sigma_R^{tba} - \sigma_L^{tba})/(\sigma_R^{oct} - \sigma_L^{oct})$ is roughly constant, there should be a linear regression between Δ_I and Δ_0. This is confirmed for the seven (R,L) combinations shown by filled circles in Figure 7.6, since with L monodentate x-ray studies show that structure (I) is adopted[162]. Remembering that, since $\sigma_R > \sigma_L$, Δ_0 is positive and Δ_I is negative, the data fit a linear regression

$$\Delta_I = -0.932\Delta_0 + 0.526 \text{ mm s}^{-1} \quad (7.117)$$

with correlation coefficient $r = -0.977$. This is a very good correlation; one factor which may have helped is a deliberate attempt to minimise counter-ion effects (Section 7.4.1.4) by using the Ph_4B^- anion in all cationic complexes. However, there is clearly ample reserve for situations where this tactic cannot be used.

It is not surprising that the ratio $(\sigma_R^{tba} - \sigma_L^{tba})/(\sigma_R^{oct} - \sigma_L^{oct})$ should be approximately constant, since h_z^{oct} and h_z^{tba} (Table 7.5) are similar in form and actually become equal if the parameter θ in h_z^{tba} takes the value 54°44′.

The intercept at $\Delta_0 = 0$ is the quadrupole splitting of the hypothetical species R_5Sn^-. However, the extrapolation back to $\Delta_0 = 0$ is a long one, with the result that the intercept is much more sensitive than the slope to small changes in the data. Thus if the two points with R = Ph are removed, the regression becomes

$$\Delta_I = -0.904\Delta_0 + 0.387 \text{ mm s}^{-1} \quad (7.118)$$

with correlation coefficient $r = -0.972$; a change of only 3% in the slope but about 30% in the intercept. (In fact the PFGs due to Ph and alkyl are distinguishable[126] and ought to be treated as such when sufficient data are available).

Clearly, Δ_I and Δ_0 have associated with them 'errors' arising from experiment and deficiencies in the additive model. These errors are assumed to be the same for both variables, and the regression lines considered are orthogonal mean-square regression lines[163]. Regression procedures based on the assumption that all error is concentrated in one variable only are clearly inappropriate.

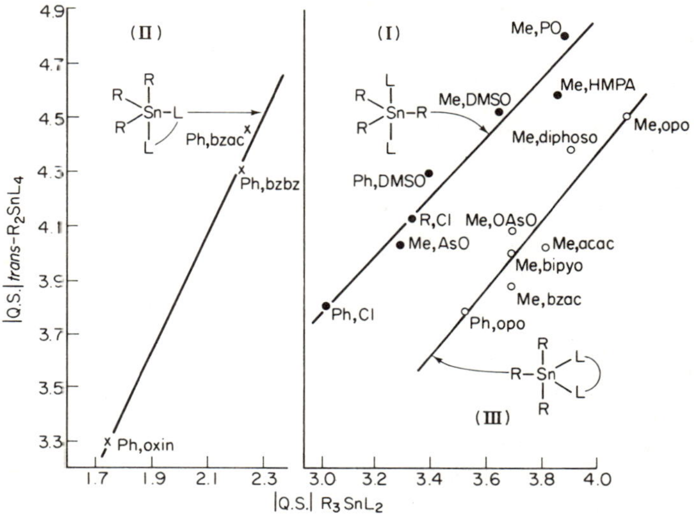

Figure 7.6 Magnitude of ^{119}Sn quadrupole splitting in *trans*-R_2SnL_4 (or twice *cis*-R_2SnL_4) plotted against magnitude of splitting in R_3SnL_2. Lines are least-squares fits to data. Units are mm s^{-1}. Ligand abbreviations: Me = CH_3, Ph = C_6H_5, R = alkyl, acac = acetylacetone anion, AsO = triphenylarsine oxide, bipyo = 2,2'-bipyridine *N*,*N'*-dioxide, bzac = benzoylacetone anion, bzbz = dibenzoylmethane anion, diphoso = 1,2-bis(diphenylphosphine oxide)ethane, DMSO = dimethyl sulphoxide, HMPA = hexamethylphosphoramide, OAsO = 1,2-bis(diphenylarsine oxide)-methane, opo = 1,2-bis(diphenylphosphine oxide)methane, oxin = 8-oxyquinoline, PO = triphenylphosphine oxide

Since the EFGs in (II) and (III) will be asymmetric, the regression of Δ_{II} or Δ_{III} against Δ_0 must be non-linear in general. Further, the signs of Δ_{II} and Δ_{III} cannot be firmly predicted when η is large, since then only a small change in the relative magnitudes of the principal components of the EFG may cause the principal axes to be permuted by the ordering convention (7.27)*. Thus it is best to consider only the magnitudes $|\Delta_{II}|$ and $|\Delta_{III}|$.

The available data lie in the region where $|\sigma_R^{tbe} - \sigma_R^{tba}|$ and $|\sigma_L^{tbe} - \sigma_L^{tba}|$

* In the additive model the principal *directions* are fixed for both (II) and (III), since the latter has C_{2v} symmetry, while although the former has only C_s symmetry the augmented complex belongs to D_{2h}.

are both much less than $|\sigma_R - \sigma_L|$. With these conditions the regressions can be approximated by the lines[164]

$$|\Delta_{II}| = (\sqrt{13}/6)[(\sigma_R^{tba} - \sigma_L^{tba})/(\sigma_R^{oct} - \sigma_L^{oct})]|\Delta_0| + c_{II} \qquad (7.119)$$

$$|\Delta_{III}| = (\sqrt{7}/3)[(\sigma_R^{tba} - \sigma_L^{tba})/(\sigma_R^{oct} - \sigma_L^{oct})]|\Delta_0| + c_{III} \qquad (7.120)$$

where c_{II} and c_{III} are functions of $\Delta_p(\sigma_R^{tba} - \sigma_R^{tbe})$ and $\Delta_p(\sigma_L^{tba} - \sigma_L^{tbe})$, with $c_{II} \neq c_{III} \neq |\Delta_p(\sigma_R^{tba} - \sigma_R^{tbe})|$, i.e. (7.119) and (7.120) have different intercepts at $\Delta_0 = 0$, and neither intercept is equal in magnitude to the quadrupole splitting in R_5Sn^-. The dependence of c_{II} and c_{III} on L is too weak to cause noticeable non-linearity in an R fixed, L variable, regression over the region where (7.119) and (7.120) are valid.

Regression lines for compounds assigned[155] to structure (II) (crosses in Figure 7.6) and structure (III) (open circles) are

$$|\Delta_{II}| = 0.448|\Delta_0| + 0.283 \text{ mm s}^{-1} \qquad (7.121)$$

$$|\Delta_{III}| = 0.704|\Delta_0| + 0.890 \text{ mm s}^{-1} \qquad (7.122)$$

with correlation coefficients $r = 0.992$ and $r = 0.932$, respectively. From (7.117), (7.121) and (7.122) the magnitudes of the observed slopes are in the ratio 1:0.481:0.756, in reasonable agreement with the theoretical values 1:0.601:0.882 given by (7.116), (7.119) and (7.120).

Quadrupole splittings in R_3SnL_2 span almost the entire range of splittings observed in tet.-R_3SnL and oct.-R_2SnL_4, emphasising that a particular range of quadrupole splittings cannot be uniquely associated with a particular stereochemistry, and that reliable structural information can only be deduced by use of either unbiassed PQS values (Section 7.4.2.4) or a regression method of the type outlined in this section.

Estimated population standard deviations measuring the scatter of data points about their regression line are 0.054, 0.039 and 0.062 mm s^{-1} for (7.117), (7.121), and (7.122), respectively. These standard deviations reflect both experimental errors in the data base and (assumed pseudo-random) errors due to deficiencies in the model.

The appropriate tolerance limits are lines parallel to the regression line lying on each side of it at distances of[165]

$$\pm\, s(1 + n^{-1})^{\frac{1}{2}} t(P, n-2)$$

where there are n data points, the estimated standard deviation is s, and $t(P,n-2)$ is the t-distribution P-percentage point on $(n-2)$ degrees of freedom. The probability that an additional observation properly belonging to the regression will lie outside the tolerance limits is less than $P/100$. In practice it is convenient to use limits of $\pm 3s$. These correspond to a P of between 5% and 2% for typical n, or over 2 units of support in likelihood terms[166].

In Figure 7.7 the data points (filled circles), regression line, and $\pm 3s$ tolerance limits for structure (I) are shown together with the data points (open circles) for structure (III). All the latter points lie outside the limits for (I). Further, if we assume in advance that they all belong to the same isomer of R_3SnL_2, our belief that this isomer is not (I) is stronger still, since the tolerance limits should then be divided by $\sqrt{N}$, where N (= 7 in this case)

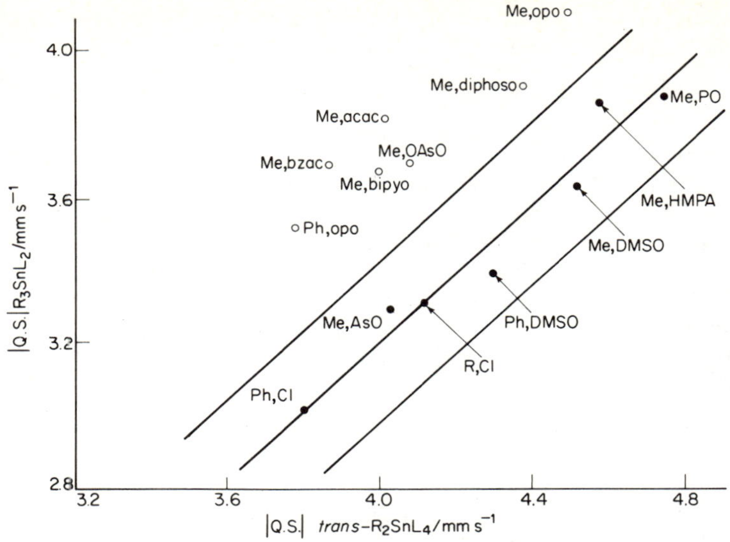

Figure 7.7 Regression line for data on structure (I) (filled circles) replotted with tolerance limits described in text. Open circles show data on structure (III)

is the number of points being jointly tested*. This illustrates how the tolerance limits can be adjusted to take into account other evidence or preconceptions.

Basano and Gamba[167] have, in a different context, advanced a statistical view of physical relationships which seems very apposite to any consideration of the additive model.

7.4.3 Non-additive systems

7.4.3.1 General remarks

In principle all systems are non-additive, but in practice some are intrinsically so, just as others are quite close to the additive idealisation. Among the former are open-shell molecules, typically transition-metal complexes, where the incomplete valence-shell gives rise to an EFG. Many paramagnetic species are best treated by ligand-field theory, but diamagnetic systems also occur and these lend themselves to simple MO arguments. Two examples of the latter are reviewed in Sections 7.4.3.2 and 7.4.3.3.

7.4.3.2 Trigonal-bipyramidal carbonyl complexes of ($3d^8$) Fe^0 *and* Co^I

Metal quadrupole coupling constants in pentacoordinate $3d^8$ carbonyl complexes have been studied by ^{57}Fe Mössbauer spectroscopy[89, 96, 168-170]

* The reviewer is indebted to A. J. Stone for a proof of this point based on the method of likelihood.

and ^{59}Co n.q.r.[171-173]. A schematic MO diagram for the holosymmetric system is shown in Figure 7.8. The numerical prefixes used to distinguish different orbitals belonging to the same representation are chosen to agree with Hillier's MO calculation[174] for $Fe(CO)_5$.

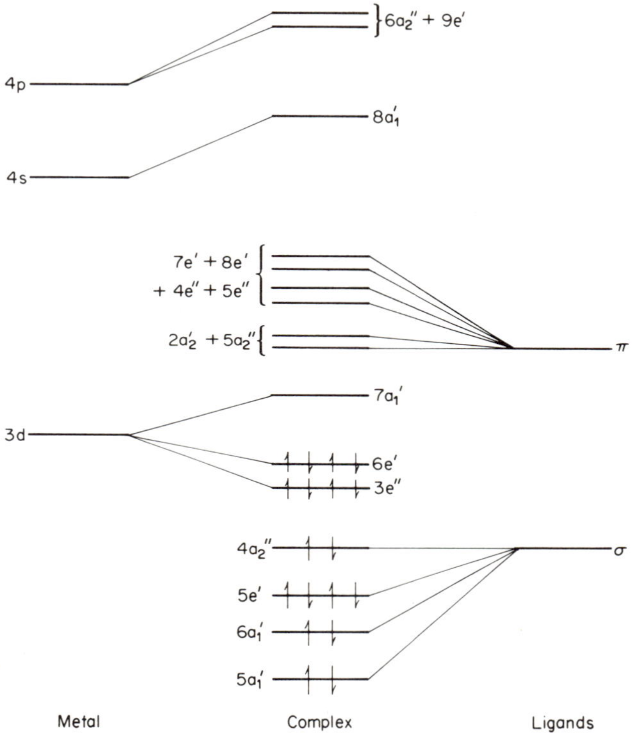

Figure 7.8 Schematic diagram of valence MO levels in a trigonal-bipyramidal $3d^8$ ML_5 complex such as $Fe(CO)_5$ [Symmetry species of the metal d orbitals are: $d_{z^2} = a_1'$, $(d_{x^2-y^2}, d_{xy}) = e'$, $(d_{xz}, d_{yz}) = e''$]

Where the three-fold symmetry is preserved the EFG, which must be axially symmetric, may be written[96] [*cf.* equation (7.96)]

$$V_{zz} = \tfrac{4}{7}e(1 - R_d)\langle r^{-3}\rangle_d(\pi - \sigma) \qquad (7.123)$$

with $\pi = 2N_{e'} - N_{e''}$ and $\sigma = N_{a_1'}$. In terms of the metal d AO effective populations, $N_{e''} = N_{xz} = N_{yz}$, $N_{e'} = N_{x^2-y^2} = N_{xy}$ and $N_{a_1'} = N_{z^2}$. The effective populations $N_{e'}$ and $N_{e''}$ are determined mainly by the extent to which the filled 3e″ and 6e′ orbitals deviate from atomic d character as a result of metal→ligand π-bonding. (Since 5e′ and 6e′ are both full, mixing of these orbitals does not change the EFG.) The effective population $N_{a_1'}$ measures the extent to which the (nominally empty) $3d_{z^2}$ metal orbital participates in the $5a_1'$ and $6a_1'$ σ MOs.

Table 7.6 Quadrupole coupling data for some pentacoordinate carbonyl complexes of Fe^0 and Co^I

Three-fold symmetry retained			*Three-fold symmetry broken*		
*Compound**	*Coupling constant†*	*η‡*	*Compound**	*Coupling constant†*	*η*
$Fe(CO)_5$	+2.60	0.0			
$(Ph_3P)Fe(CO)_4$	+2.49	0.0			
$(Ph_3P)_2Fe(CO)_3$	+2.74	0.0			
$(OC)_4Fe(diphos)Fe(CO)_4$	+2.46	0.0	$(diphos)Fe(CO)_3$	−2.12	0.8
$(Me_2N)_3PFe(CO)_4$	2.22				
$\{(Me_2N)_3P\}_2Fe(CO)_3$	2.30				
$Ph_3SnCo(CO)_4$	104.74	0.03			
$Ph_3SnCo(CO)_3P(OMe)_3$	104.19	0.04	$Ph_3SnCo(CO)_2\{P(OMe)_3\}_2$	104.20	0.37
$Ph_3SnCo(CO)_3P(OPh)_3$	110.44	0.27	$Ph_3SnCo(CO)_2\{P(OPh)_3\}_2$	95.40	0.55
$Ph_3GeCo(CO)_4$	109.98	0.05			
$Ph_3GeCo(CO)_3P(OPh)_3$	113.02	0.25			

* diphos = 1,2-bis(diphenylphosphino)ethane

† Data selected from Refs. 89, 96 and 168–173. Coupling constant is $ce^2qQ(1 + \frac{1}{3}\eta^2)^{\frac{1}{2}}/2E_\gamma$ in mm s^{-1} for ^{57}Fe or e^2qQ/h in MHz for ^{59}Co. Signs not given explicitly have not been experimentally determined. Measurements are in solid state at liquid nitrogen or room temperatures; temperature dependence in this range is 2% or less. For the ^{57}Fe 14.39 keV Mössbauer resonance 1 mm s^{-1} = 11.61 MHz.

‡ These values of η reflect deviations from perfect threefold symmetry for *apical–apical* isomer in solid state

It is anticipated[96] than $2 > \pi > \sigma > 0$, since in the absence of covalent bonding $\pi = 2$ and $\sigma = 0$, while $\pi < \sigma$ corresponds to a situation in which it would be difficult to justify the usual d^8 nominal configuration for the metal atom. Note that σ increases with increasing σ-donor character of the ligands, whereas π decreases with increasing π-acceptor character.

The quadrupole splitting in $Fe(CO)_5$ is[89,168] 2.60 mm s^{-1}, with V_{zz} positive, confirming that $\pi > \sigma$. Taking[72] $Q(^{57}Fe^m) = +0.2 \times 10^{-24}$ cm^2 and for $\langle r^{-3} \rangle_d$ the calculated[40] atomic value of 2.6×10^{25} cm^{-3},

$$(1 - R_d)(\pi - \sigma) = 0.58$$

compared with the value

$$(\pi - \sigma) = 1.54 - 1.23 = 0.31$$

obtained from the Mulliken orbital populations calculated by Schreiner and Brown[172,175]. The numerical uncertainties are such that the agreement is as good as could be expected.

Replacement of one or both apical carbonyls by other ligands will cause changes in $(\pi - \sigma)$ which may be rationalised in terms of the σ and π bonding qualities of the ligands[170,172,176]. As shown on the left-hand side of Table 7.6, replacement of CO by a tertiary phosphine causes only a small change in $(\pi - \sigma)$, and can thus be regarded as a perturbation of the electronic structure of the central metal atom. Thus when a tertiary phosphine is substituted for an *equatorial* CO we have exactly the kind of symmetry-breaking perturbation discussed in Section 7.3.2.2.

The analysis in Section 7.3.2.2 is readily extended to allow for complications of the real situation[96]. As in the idealised case, both η and $sgn\{e^2qQ\}$ are very much more sensitive to breaking of the threefold symmetry by substitution than is $|e^2qQ|$. The experimental results on the right-hand side of Table 7.6 confirm these predictions.

Unfortunately, the above predictions apply with equal force to all isomers in which the threefold symmetry is broken. Thus although η and $sgn\{e^2qQ\}$ are sensitive to lack of axial symmetry in these systems, they cannot easily be used to answer more detailed stereochemical questions. For example, the structure of $(diphos)Fe(CO)_3$ [Table 7.6] could not be firmly established by Mössbauer studies, although arguments based on $|e^2qQ|$ favoured[96] the isomer with *apical–equatorial* P atoms (rather than *equatorial–equatorial*), in agreement with the crystal structure reported[177] for $(diars)Fe(CO)_3$ [diars = 1,2-bis(dimethylarsino)benzene]. Other authors[178] also favour the *apical–equatorial* structure for tricarbonyl-1,2-bis(dimethylphosphinoethane)-iron(0). In solution these systems appear[169,178] to be non-rigid on an n.m.r. timescale, even when cooled to −80°C. Recent studies have also considered the $^{55}Mn^{-I}$ analogues[199].

7.4.3.3 ($5d^2$) W^{IV} *and* ($5d^1$) W^{V} *octacyanotungstate anions*

The octacyanotungstate anions $W(CN)_8^{n-}$ ($n = 3, 4$) offer another interesting example in which the EFG at the central metal atom arises mainly from an incomplete d shell[179]. Both W^{IV} ($5d^2$) and W^{V} ($5d^1$) species are known[180], and x-ray studies have revealed examples of both D_{2d} dodecahedral[181,182] and

D_{4d} square-antiprismatic[183, 184] coordination. Since the 100.1 keV Mössbauer resonance of ^{182}W is a $2^+ \leftrightarrow 0^+$ transition, $|e^2qQ|$, $sgn\{e^2qQ\}$ and η can all be determined from the observed spectra without recourse to special methods such as Zeeman experiments. Further, ^{182}W is a deformed nucleus (Section 7.2.7.2) with intrinsic quadrupole moment[83] $Q_0 \approx 6.4 \times 10^{-24}$ cm^2, corresponding to a spectroscopic Q of -1.8×10^{-24} cm^2 for the 2^+ 100.1 keV state. The sign is assumed from known systematics of deformed nuclei[82, 84] (Figure 7.3), and confirmed by theoretical calculation[185].

Figure 7.9 shows schematic MO diagrams for both dodecahedral and square-antiprismatic geometries. The ligand-based π orbitals have been

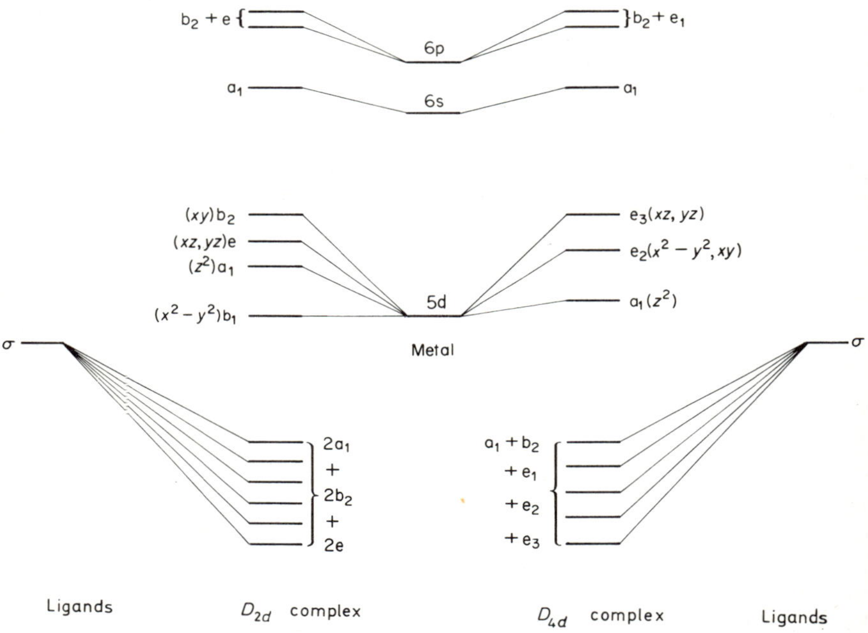

Figure 7.9 Schematic diagram of valence MO levels in dodecahedral and square-antiprismatic octacyanotungstate anions

omitted for clarity; they span $2A_1 + 2A_2 + 2B_1 + 2B_2 + 4E$ under D_{2d} and $A_1 + A_2 + B_1 + B_2 + 2E_1 + 2E_2 + 2E_3$ under D_{4d}. Although it is generally agreed that the lowest 5d-based MO is $(d_{x^2-y^2})b_1$ in dodecahedral coordination and $(d_{z^2})a_1$ in square-antiprismatic coordination, the order given in Figure 7.9 for the higher 5d-based orbitals is not the only one suggested in the literature[186, 187]. This does not affect interpretation of the EFG at W since these higher d orbitals are empty and thermally inaccessible in both W^{IV} and W^{V} anions. Spin–orbit coupling and, in the d^2 case, interelectronic repulsion, produce relatively small modifications to the simple orbital picture.

The EFG at W has been interpreted in terms of effective population of the metal 5d AOs[179]. The lowest 5d-based MO contains one electron in $W(CN)_8^{3-}$ and two in $W(CN)_8^{4-}$. These electrons give a positive contribution to eq in dodecahedral symmetry and a negative contribution in square-antiprismatic. The symmetry species given in Figure 7.9 show that in the dodecahedron the $(d_{x^2-y^2})b_1$ orbital is non-bonding if π bonding is neglected, whereas the $(d_{z^2})a_1$ orbital in the antiprism may also involve both the ligand σ orbitals (but see Section 7.4.1.3 in the d^2 case) and the metal 6s AO, implying greater deviation from pure 5d AO character in the latter case. Hence, unless the extent of π bonding is very different in the two geometries, the magnitude of the EFG set up by the 'd' electrons will be smaller in square-antiprismatic coordination.

The four nominally-empty higher 5d AOs can participate in the filled ligand-based σ MOs. In both geometries effective population of these 5d AOs gives a contribution to the EFG which opposes the contribution from the occupied 5d-based MO. This 'σ-orbital' contribution increases with increasing covalency.

This analysis of the EFG at W can be roughly quantified by assigning an effective population n_d to the lowest 5d AO and an average effective population $\bar{n}_\sigma$ to each of the other four d orbitals. Then, from (7.96)

$$q_d = \pm\tfrac{4}{7}(1 - R_d)\langle r^{-3}\rangle_d(n_d - \bar{n}_\sigma) \qquad (7.124)$$

where the + sign applies for the dodecahedron and − for the antiprism, and n_d will be smaller in the antiprism. Both D_{2d} and D_{4d} symmetries require $\eta = 0$.

Quadrupole coupling data for a number of compounds containing the octacyanotungstate anion are given in Table 7.7. Remembering that Q is predicted to be negative, it is seen that the signs of ce^2qQ/E_γ observed for

Table 7.7 ^{182}**W Mössbauer quadrupole coupling parameters of tungsten octacyanide anions**[179]

Compound	ce^2qQ/E_γ/mm s^{-1} *	*Structure*†
$K_4W(CN)_8 \cdot 2H_2O$	−16.02	dod[181, 182]
$K_4W(CN)_8$ frozen soln.‡	−16.48	dod[179, 189]
$Li_4W(CN)_8 \cdot nH_2O$	−15.90	dod[179]
$H_4W(CN)_8 \cdot 6H_2O$	+12.59	sqa[184]
$H_4W(CN)_8$ frozen soln.‡	+14.79	sqa[179]
$Cd_2W(CN)_8 \cdot 8H_2O$	+12.80 (η = 0.82)	sqa[179] §
$K_3W(CN)_8 \cdot H_2O$	~ 0	—
$Na_3W(CN)_8 \cdot 4H_2O$	~ 0	sqa[183]
$Ag_3W(CN)_8$	~ 0	—
$\{Co(NH_3)_6\}W(CN)_8$	~ 0	—

* All measurements at liquid-helium temperature. Estimated errors ±0.5 mm s^{-1} for frozen-solution spectra and ±0.2 mm s^{-1} for other resolved splittings; $\eta \sim 0$ when not given explicitly. Upper limit for unresolved quadrupole coupling in single-line spectra is $|ce^2qQ/E_\gamma|$ = 2 mm s^{-1}. For the ^{182}W 100.1 keV Mössbauer resonance 1 mm s^{-1} = 80.74 MHz

† dod = dodecahedral, sqa = square-antiprismatic

‡ 0.5 M aqueous solution + small amount of glycerol to promote glassification

§ X-Ray studies[188] have revealed square-antiprismatic geometry for $Mo(CN)_8^{4-}$ in $Cd_2[Mo(CN)_8] \cdot 2N_2H_4 \cdot 4H_2O$

known dodecahedral[181,182] $K_4W(CN)_8{\cdot}2H_2O$ and known square-antiprismatic[184] $H_4W(CN)_8{\cdot}6H_2O$ are in agreement with (7.124) if $n_d > \bar{n}_\sigma$. Alternatively, since $n_d > \bar{n}_\sigma$ would certainly be anticipated for the d^2 species, these data can be regarded as confirming the nuclear theory prediction of $sgn\{Q\}$. Notice also that $|ce^2qQ/E_\gamma|$ is slightly less in $H_4W(CN)_8{\cdot}6H_2O$, as predicted.

Clearly ^{182}W Mössbauer spectroscopy provides a convenient method for distinguishing the dodecahedral and square-antiprismatic configurations of $W(CN)_8^{4-}$. In Table 7.7 the structural assignments for $Li_4W(CN)_8{\cdot}nH_2O$ and $Cd_2W(CN)_8{\cdot}8H_2O$ are made on this basis. (The large η observed in the latter compound is discussed in Section 7.4.4). The frozen-solution data in Table 7.7 suggest that in both $K_4W(CN)_8$ and $H_4W(CN)_8$ the solid-state geometry is retained in solution, in agreement with more recent studies[189] of the vibrational spectra of $K_4M(CN)_8$ (M = W, Mo), but contrary to earlier suggestions[190].

None of the $W(CN)_8^{3-}$ anions in Table 7.7 showed resolvable quadrupole splitting, suggesting that in the d^1 anions n_d and $\bar{n}_\sigma$ [equation (7.124)] approximately cancel. This is consistent with the high degree of covalency expected in cyanide complexes of third-row transition elements.

Consider, as a crude first approximation, a 'rigid-orbital model' in which the MOs are unchanged on removing a d electron from $W(CN)_8^{4-}$ to obtain $W(CN)_8^{3-}$. Then $\bar{n}_\sigma$ will be the same in both oxidation states, and n_d may be written 2κ in the former and κ in the latter, where κ is an orbital-reduction factor[59,105] for the lowest d orbital. The absence of quadrupole splitting in $W(CN)_8^{3-}$ indicates that $\kappa \approx \bar{n}_\sigma$. Thus applying (7.124) to dodecahedral $W(CN)_8^{4-}$

$$\kappa|\tfrac{4}{7}ce^2(1 - R_d)\langle r^{-3}\rangle_d QE_\gamma^{-1}| \approx 16 \text{ mm s}^{-1} \qquad (7.125)$$

With[45,191] $(1 - R_d) = 1.4$ and $\langle r^{-3}\rangle_d = 11$ a.u. (an average of calculated values[192,193] for free W^{4+} and W^{5+}), (7.125) gives $\kappa = 0.35$, while for $\langle r^{-3}\rangle_d = 8$ a.u. (an average of calculated values[193,194] for free neutral W), $\kappa = 0.48$.

7.4.4 Distortions from ideal geometry

In the above sections the relationship between quadrupole coupling and electronic and geometrical structure has been discussed in terms of idealised molecular symmetries so that symmetry-based arguments could be used with greatest effect. However, the internuclear axes in individual compounds may show distortions from the ideal geometry. At present the effect of such distortions on the EFG is not completely understood.

In specific molecules the EFG can be calculated from a molecular wave function if the nuclear positions are known. For example, the calculations of Snyder and Basch[137] mentioned in Section 7.4.2.2 are a useful guide to the effects in small molecules. However, such calculations cannot be a complete answer, since they have a strong element of hindsight. Also, for large molecules (and many of the molecules considered in this review are large in this context) the basis set used in any calculation will necessarily be incomplete, and will usually be chosen to describe the orbital energies and regions of interatomic overlap with good accuracy. This is not necessarily the best choice for calculation of a quantity such as the EFG which weights charge density close to a particular nucleus.

A further point emerges if the electronic structure is described by localised orbitals chosen so as to minimise the sum of exchange integrals[146, 148]. Then it can be argued that minimisation of electrostatic interactions between electrons in different bonds is most likely to determine the bond directions close to the nucleus where interelectron separations are least. This will tend to favour the idealised high-symmetry arrangement of bond directions. Such arguments coupled with empirical experience have led to repeated suggestions[74, 126, 138] that coupling constants in distorted molecules are often better understood in terms of the ideal geometry than on the basis of the observed nuclear positions. The implication is that bonds are bent even in nominally unstrained molecules.

Visualisation of the additive model (Section 7.4.2) as a 'point-charge model', in which the EFG arises from notional point charges at the ligand sites, has tempted several workers to calculate the effect of angular distortions by placing the charges at the observed nuclear positions. In this procedure the only effect of angular distortion is to change the angles between the PFG principal axes and the molecular axes. The parameter [L] [equation (7.110)] remains constant; but since the hybrids h_z (Table 7.5) certainly vary as their angular disposition is changed this can only be correct if exactly compensating changes occur in the σ_L [equation (7.113)]. Unfortunately it is difficult to construct a good experimental test, since in this model additional disposable parameters appear on distortion[126].

The effects of distortion in tetrahedral $^{119}SnA_2B_2$ and $^{119}SnA_3B$ systems have been calculated allowing for changes in h_z, but keeping the σ_L constant[126]. However, theory and experiment were in better agreement if corrections for distortion were ignored, supporting the bent-bond hypothesis mentioned above. Recently, further work on these systems has been reported[201].

All models which make use of crystallographic data suffer from the problem of bond bending. In addition, attempts to gain information on distortions by comparing observed and calculated η values are open to the criticism (Section 7.3.2.2) that η is very sensitive to small perturbations not allowed for in the model. In general, those qualitative conclusions which hold in all reasonable models are the most convincing.

Finally we mention one general result which is useful in certain cases. Suppose a particular distortion path is characterised by parameters $\{a_1, a_2, \ldots\} = \boldsymbol{\alpha}$ and that the EFG tensor $\boldsymbol{V}$ is a continuous function of $\boldsymbol{\alpha}$. This implies that the symmetrised parameters S_{02} and, in particular, S_{03} are continuous functions of $\boldsymbol{\alpha}$. Now if the distortion path joins two ideal geometries in which the EFG has opposite signs, then S_{03}, which has the same sign as eq, must vanish at least once on the path. At the zeros of S_{03}

$$q^3(1 - \eta^2) = 0$$

which has the solutions $q = 0$ (i.e. $\boldsymbol{V} = 0$) or $\eta = 1$. Thus if a distortion path leads to a change in the sign of the EFG, there is at least one point where either $\boldsymbol{V} = 0$ or $\eta = 1$.

The dodecahedral and square-antiprismatic $W(CN)_8^{4-}$ anions (Section 7.4.3.3) provide an application of this result[179], since the two ideal geometries give EFGs of opposite sign. Thus the large η observed in $Cd_2W(CN)_8{\cdot}8H_2O$ (Table 7.7) is indicative of a path passing through $\eta = 1$. This is supported by detailed arguments[179], which suggest that Mössbauer experiments on distorted $W(CN)_8^{4-}$ anions will unambiguously characterise them as dodecahedron-like ($eq > 0$) or square-antiprism-like ($eq < 0$).

ACKNOWLEDGEMENTS

The reviewer is indebted to Prof. A. D. Buckingham for his interest and encouragement; to Drs. G. M. Bancroft, A. G. Maddock, and A. J. Stone for many helpful discussions; and to many other colleagues, both in Cambridge and elsewhere, for information and enlightenment.

References

1. Elton, L. R. B. (1965). *Introductory Nuclear Theory*, 2nd ed. (London: Pitman)
2. Reid, J. M. (1972). *The Atomic Nucleus*, (Harmondsworth, Middlesex: Penguin)
3. Casimir, H. B. G. (1936). *On the Interaction between Atomic Nuclei and Electrons*, (Haarlem: Teyler's Tweede Genootschap); reprinted (1963). (San Francisco: W. H. Freeman)
4. Brink, D. M. and Satchler, G. R. (1968). *Angular Momentum*, 2nd ed. (Oxford: Clarendon Press)
5. Lounasmaa, O. V. (1967). *Hyperfine Interactions*, Chap. 10 (A. J. Freeman and R. B. Frankel, editors) (New York: Academic)
6. Pauli, W. (1924). *Naturwiss.*, **12,** 741
7. Schüler, H. and Schmidt, Th. (1935). *Z. Phys.*, **94,** 457. Casimir, H. B. G. (1935). *Physica*, **2,** 719
8. Ramsey, N. F. (1956). *Molecular Beams* (Oxford: Clarendon Press)
9. Childs, W. J. (1973). *Case Stud. Atom. Phys.*, **3,** 215; Zorn, J. C. and English, T. C. (1973). *Adv. Atom. Molec. Phys.*, **9,** 243
10. Zavoisky, E. (1945). *J. Phys. USSR*, **9,** 245
11. Carrington, A. and McLachlan, A. D. (1967). *Introduction to Magnetic Resonance* (New York: Harper and Row)
12. Abragam, A. and Bleaney, B. (1970). *Electron Paramagnetic Resonance of Transition Ions* (Oxford: Clarendon Press)
13. Gordy, W., Smith, W. V. and Trambarulo, R. F. (1953). *Microwave Spectroscopy* (New York: Wiley); Townes, C. H. and Schawlow, A. L. (1955). *Microwave Spectroscopy* (New York: McGraw-Hill)
14. Bloch, F., Hansen, W. W. and Packard, M. (1946). *Phys. Rev.*, **70,** 474. Purcell, E. M., Torrey, H. C. and Pound, R. V. (1946). *Phys. Rev.*, **69,** 37
15. Abragam, A. (1961). *The Principles of Nuclear Magnetism* (Oxford: Clarendon Press)
16. Dehmelt, H. G. and Krüger, H. (1950). *Naturwiss*, **37,** 111
17. Das, T. P. and Hahn, E. L. (1958). *Solid State Phys. Suppl.*, **1,** 1
18. Cohen, M. H. and Reif, F. (1957). *Solid State Phys.*, **5,** 321
19. Kwiram, A. L. (1971). *Ann. Rev. Phys. Chem.*, **22,** 133
20. Happer, W. (1972). *Rev. Mod. Phys.*, **44,** 169
21. Mössbauer, R. L. (1958). *Z. Phys.*, **151,** 124; (1958). *Naturwiss.*, **45,** 538; (1959). *Z. Naturforsch.*, **14a,** 211; Frauenfelder, H. (1963). *The Mössbauer Effect* (New York: Benjamin); Greenwood, N. N. and Gibb, T. C. (1971). *Mössbauer Spectroscopy* (London: Chapman and Hall)
22. Bancroft, G. M. (1973). *Mössbauer Spectroscopy* (London: McGraw-Hill)

23. Biryukov, I. P., Voronkov, M. G. and Safin, I. A. (1969). *Tables of Nuclear Quadrupole Resonance Frequencies* (Jerusalem: Israel Program for Scientific Translations)
24. Muir, A. H., Ando, K. J. and Coogan, H. M. (1966). *Mössbauer Effect Data Index 1958–1965* (New York: Wiley)
25. Chihara, H. and Nakamura, N. (1972). *MTP International Review of Science, Physical Chemistry Series One*, Vol. 4, Chap. 4 (London: Butterworths); (1974). *Advances in Nuclear Quadrupole Resonance* Vol. 1 (J. A. S. Smith, editor) (London: Heyden); Brown, T. L. (1974). *Accounts Chem. Res.*, **7,** 408
26. Haas, H. and Shirley, D. A. (1973). *J. Chem. Phys.*, **58,** 3339
27. Vargas, J. I. (1972). *MTP International Review of Science, Inorganic Chemistry, Series One*, Vol. 8, Chap. 2 (London: Butterworths)
28. Emery, G. T. (1972). *Ann. Rev. Nucl. Sci.*, **22,** 165
29. (1967). *Hyperfine Interactions* (A. J. Freeman and R. B. Frankel, editors) (New York: Academic Press)
30. (1968). *Hyperfine Structure and Nuclear Radiations* (E. Matthias and D. A. Shirley, editors) (Amsterdam: North-Holland); (1971). *Hyperfine Interactions in Excited Nuclei* (G. Goldring and R. Kalish, editors) (London: Gordon and Breach)
31. Hendrie, D. L., Glendenning, N. K., Harvey, B. G., Jarvis, O. N., Duhm, H. H., Saudinos, J. and Mahoney, J. (1968). *Phys. Lett. B*, **26,** 127; de Swiniarski, R., Glashausser, C., Hendrie, D. L., Sherman, J., Bacher, A. D. and McClatchie, E. A. (1969). *Phys. Rev. Lett.*, **23,** 317; Kubo, K., Broglia, R. A., Riedel, C. and Udagawa, T. (1970). *Phys. Lett. B*, **32,** 29; Aponick, A. A., Chesterfield, C. M., Bromley, D. A. and Glendenning, N. K. (1970). *Nucl. Phys. A*, **159,** 367
32. Moss, J. M., Terrien, Y. D., Lombard, R. M., Brassard, C., Lorseaux, J. M. and Resmini, F. (1971). *Phys. Rev. Lett.*, **26,** 1488; Hendrie, D. L. (1973). *ibid.*, **31,** 478; Bemis, C. E., McGowan, F. K., Ford, J. L. C., Milner, W. T., Stelson, P. H. and Robinson, R. L. (1973). *Phys. Rev. C*, **8,** 1466
33. Davidson, J. P., Close, D. A. and Malanify, J. J. (1974). *Phys. Rev. Lett.*, **32,** 337
34. Mahler, R. J. (1966). *Phys. Rev.*, **152,** 325; Mooberry, E. S., Spiess, H. W. and Sheline, R. K. (1972). *J. Chem. Phys.*, **57,** 813; Hammerle, R. H., van Ausdal, R. and Zorn, J. C. (1972). *ibid.*, **57,** 4068; Sternheimer, R. M., (1961). *Phys. Rev. Lett.*, **6,** 190; (1966). *Phys. Rev.*, **146,** 140
35. Wu, C. S. and Wilets, L. (1969). *Ann. Rev. Nucl. Sci.*, **19,** 527; Kim, Y. N. (1971). *Mesic Atoms and Nuclear Structure* (Amsterdam: North-Holland)
36. Clark, M. G. (1971). *J. Chem. Phys.*, **54,** 697
37. Brown, L. C. and Parker, P. M. (1955). *Phys. Rev.*, **100,** 1764
38. Williams, P. G. L. and Bancroft, G. M. (1969). *Chem. Phys. Lett.*, **3,** 110; (1970). *Molec. Phys.*, **19,** 717; (1971). *Mössbauer Effect Method.*, **7,** 39
39. Sternheimer, R. M. (1950). *Phys. Rev.*, **80,** 102; (1951). *ibid.*, **84,** 244; (1954). *ibid.*, **95,** 736
40. Freeman, A. J. and Watson, R. E. (1965). *Magnetism*, Vol. IIA, Chap. 4 (G. T. Rado and H. Suhl, editors) (New York: Academic); Watson, R. E., and Freeman, A. J. (1967). *Hyperfine Interactions*, Chap. 2 (A. J. Freeman and R. B. Frankel, editors) (New York: Academic)
41. Foley, H. M., Sternheimer, R. M. and Tycko, D. (1954). *Phys. Rev.*, **93,** 734; Sternheimer, R. M. and Foley, H. M. (1956). *ibid.*, **102,** 731; Sternheimer, R. M. (1967). *ibid.*, **164,** 10
42. Dalgarno, A. (1962). *Advan. Phys.*, **11,** 281
43. Armstrong, L. (1971). *Theory of the Hyperfine Structure of Free Atoms* (New York: Wiley)
44. Feiock, F. D. and Johnson, W. R. (1969). *Phys. Rev.*, **187,** 39
45. Sternheimer, R. M. (1972). *Phys. Rev. A*, **6,** 1702
46. Lyons, J. D., Pu, R. T. and Das, T. P. (1969). *Phys. Rev.*, **178,** 103; Nesbet, R. K. (1970). *Phys. Rev. A*, **2,** 661; Hameed, S. and Foley, H. M. (1972). *ibid.*, **6,** 1399; Ray, S. N., Lee, T. and Das, T. P. (1974). *ibid.*, **9,** 93
47. Ford, K. W. and Wills, J. G. (1969). *Phys. Rev.*, **185,** 1429
48. Kalvius, G. M. (1971). *Hyperfine Interactions in Excited Nuclei*, Vol. 2, 523 (G. Goldring and R. Kalish, editors) (Gordon and Breach: London)
49. Kistner, O. C. and Sunyar, A. W. (1960). *Phys. Rev. Lett.*, **4,** 412
50. Shirley, D. A. (1964). *Rev. Mod. Phys.*, **36,** 339; (1969). *Ann. Rev. Phys. Chem.*, **20,** 25

51. Stacey, D. N. (1966). *Rept. Progr. Phys.*, **29,** 171
52. Pound, R. V. and Rebka, G. A. (1960). *Phys. Rev. Lett.*, **4,** 274
53. Clark, M. G. and Stone, A. J. (1969). *Phys. Lett. A*, **30,** 144
54. Josephson, B. D. (1960). *Phys. Rev. Lett.*, **4,** 341
55. Bancroft, G. M. and Platt, R. H. (1972). *Advan. Inorg. Chem. Radiochem.*, **15,** 59. Parish, R. V. (1972). *Progr. Inorg. Chem.*, **15,** 101
56. Marshall, W. and Johnson, C. E. (1962). *J. Phys. Radium*, **23,** 733
57. Johnson, C. E. (1967). *Symp. Faraday Soc.*, **1,** 7
58. Geschwind, S. (1967). *Hyperfine Interactions*, Chap. 6 (A. J. Freeman and R. B. Frankel, editors) (New York: Academic)
59. Griffith, J. S. (1961). *The Theory of Transition-Metal Ions* (Cambridge: Cambridge University Press)
60. Wickman, H. H., Klein, M. P. and Shirley, D. A. (1966). *Phys. Rev.*, **152,** 345
61. Ray, S. N., Lee, T. and Das, T. P. (1973). *Phys. Rev. B*, **8,** 5291
62. Lang, G. and Dale, B. W. (1973). *J. Phys. C*, **6,** L80
63. McMillan, J. A. and Halpern, T. (1971). *J. Chem. Phys.*, **55,** 33; McMillan, J. A. and Munic, G. C. (1972). *ibid.*, **56,** 113
64. Perlow, G. J. (1971). *Hyperfine Interactions in Excited Nuclei*, Vol. 2, 651 (G. Goldring and R. Kalish, editors) (London: Gordon and Breach)
65. Perlow, G. J., Henning, W., Olson, D. and Goodman, G. L. (1969). *Phys. Rev. Lett.*, **23,** 680
66. Sandars, P. G. H. and Beck, J. (1965). *Proc. Roy. Soc. A*, **289,** 97; Sandars, P. G. H. and Dodsworth, B. (1967). *Hyperfine Interactions*, Chap. 4 (A. J. Freeman and R. B. Frankel, editors) (New York: Academic)
67. Dunlap, B. D. (1971). *Mössbauer Effect Method*, **7,** 123
68. Rosen, A. (1969). *J. Phys. B*, **2,** 1257; Lewis, W. B. (1970). *Proc. 16th Colloq. Ampere*, Bucharest; Lewis, W. B., Mann, J. B., Liberman, D. A. and Cromer, D. T. (1970). *J. Chem. Phys.*, **53,** 809; Mann, J. B. and Waber, J. T. (1973). *Atom. Data*, **5,** 201
69. Brooks, J. S., Williams, J. M. and Webster, P. J. (1973). *J. Phys. D*, **6,** 1403
70. Sharma, R. R. and Bung Ning Teng (1971). *Phys. Rev. Lett.*, **27,** 679. Sharma, R. R. (1970). *ibid.*, **25,** 1622; (1971). *ibid.*, **26,** 563; Sawatzky, G. A., van der Woude, F. and Hupkes, J. (1971). *J. Phys. (Paris)*, **32,** Cl-276; Betsuyaku, H. (1969). *J. Chem. Phys.*, **51,** 2546
71. Artman, J. O. (1971). *Mössbauer Effect Method.*, **7,** 187
72. Fuller, G. H. and Cohen, V. W. (1969). *Nucl. Data A*, **5,** 433; Shirley, V. S. (1971). *Hyperfine Interactions in Excited Nuclei*, Vol. 4, 1255 (G. Goldring and R. Kalish, editors) (London: Gordon and Breach)
73. Townes, C. H. and Dailey, B. P. (1949). *J. Chem. Phys.*, **17,** 782
74. Lucken, E. A. C. (1969). *Nuclear Quadrupole Coupling Constants* (London: Academic); Schempp, E. and Bray, P. J. (1970). *Physical Chemistry*, Vol. 4, Chap. 11 (H. Eyring, D. Henderson and W. Jost, editors) (New York: Academic); Lucken, E. A. C. (1972). *Fortsch. Chem. Forsch.*, **30,** 155
75. O'Konski, C. T. and Ha, T.-K. (1972). *J. Chem. Phys.*, **56,** 3169
76. Harris, C. B. (1968). *J. Chem. Phys.*, **49,** 1648; Bancroft, G. M. (1971). *Chem. Phys. Lett.*, **10,** 449
77. Bancroft, G. M., Butler, K. D. and Libbey, E. T. (1972). *J. Chem. Soc. Dalton Trans.*, 2643
78. Robinson, E. J. (1969). *Phys. Rev. Lett.*, **22,** 579
79. Anderson, A. B., Handy, N. C. and Parr, R. G. (1969). *J. Chem. Phys.*, **50,** 3634
80. Gupta, R. P. and Sen, S. K. (1972). *Phys. Rev. Lett.*, **28,** 1311; Sen, S. K. (1969). *Nucl. Instr. Method.*, **72,** 321; Sen, S. K., Salie, D. L. and Tomchuk, E. (1972). *Phys. Rev. Lett.*, **28,** 1295
81. Gupta, R. P., Rao, B. K. and Sen, S. K. (1971). *Phys. Rev. A*, **3,** 545; Gupta, R. P. and Sen, S. K. (1973). *ibid.*, **7,** 850
82. Alder, K. and Steffen, R. M. (1964). *Ann. Rev. Nucl. Sci.*, **14,** 403; Yoshida, S. and Zamick, L. (1972). *Ann. Rev. Nucl. Sci.*, **22,** 121
83. Löbner, K. E. G., Vetter, M. and Hönig, V. (1970). *Nucl. Data A*, **7,** 495; Christy, A. and Häusser, O. (1972). *Nucl. Data Tables*, **11,** 281
84. Townes, C. H., Foley, H. M. and Low, W. (1949). *Phys. Rev.*, **76,** 1415; Townes, C. H. (1958). *Handb. Phys.*, **38/1,** 377

85. Bocquet, J.-P., Chu, Y. Y., Kistner, O. C., Perlman, M. L. and Emery, G. T. (1966). *Phys. Rev. Lett.*, **17,** 809; Emery, G. T. and Perlman, M. L. (1970). *Phys. Rev. B*, **1,** 3885
86. Raff, U., Alder, K. and Baur, G. (1972). *Helv. Phys. Acta*, **45,** 427; Rüegsegger, R. and Kündig, W. (1972). *Phys. Lett. B*, **39,** 620; (1973). *Helv. Phys. Acta*, **46,** 165
87. Kienle, P., Kalvius, G. M. and Ruby, S. L. (1968). *Hyperfine Structure and Nuclear Radiations*, Appendix A (E. Matthias and D. A. Shirley, editors) (Amsterdam: North-Holland); Shenoy, G. K. and Kalvius, G. M. (1971). *Hyperfine Interactions in Excited Nuclei*, Vol. 4, 1201 (G. Goldring and R. Kalish, editors) (London: Gordon and Breach); Boehm, F. (1974). *Atom. Data Nucl. Data Tables*, **14,** 479
88. Zory, P. (1965). *Phys. Rev.*, **140,** A1401; Gibb, T. C. (1974). *J. Phys. C*, **7,** 1001
89. Kienle, P. (1963). *Phys. Verhandl.*, **3,** 33
90. Clark, M. G., Bancroft, G. M. and Stone, A. J. (1967). *J. Chem. Phys.*, **47,** 4250
91. Reiff, W. M. (1973). *Coord. Chem. Rev.*, **10,** 37
92. Libbey, E. T. and Bancroft, G. M. (1973). *J. Chem. Soc. Chem. Commun.*, 503
93. Clark, M. G. (1971). *Molec. Phys.*, **20,** 257
94. Ooi, S. and Kuroya, H. (1963). *Bull. Chem. Soc. Jap.*, **36,** 1083
95. Clark, M. G. (1972). *Chem. Phys. Lett.*, **13,** 316
96. Clark, M. G., Cullen, W. R., Garrod, R. E. B., Maddock, A. G. and Sams, J. R. (1973). *Inorg. Chem.*, **12,** 1045
97. Ganiel, U. (1969). *Chem. Phys. Lett.*, **4,** 87; Eicher, H. and Trautwein, A. (1969). *J. Chem. Phys.*, **50,** 2540; Cosgrove, J. G. and Collins, R. L. (1971). *ibid.*, **55,** 4238 [see also: Dale, B. W. (1972). *J. Chem. Phys.*, **56,** 4721; Cosgrove, J. G. and Collins, R. L. (1972). *ibid.*, **56,** 4721]
98. Watanabe, I., Tanaka, H. and Shimizu, T. (1970). *J. Chem. Phys.*, **52,** 4031
99. Watanabe, I. and Yamagata, Y. (1967). *J. Chem. Phys.*, **46,** 407
100. Spiess, H. W., Haas, H. and Hartmann, H. (1969). *J. Chem. Phys.*, **50,** 3057
101. Shigeta, Y., Komiyama, Y. and Kuroya, H. (1963). *Bull. Chem. Soc. Jap.*, **36,** 1159
102. Brill, T. B. and Hugus, Z. Z. (1970). *J. Phys. Chem.*, **74,** 3022
103. Mooberry, E. S., Spiess, H. W., Garrett, B. B. and Sheline, R. K. (1969). *J. Chem. Phys.*, **51,** 1970; Mooberry, E. S., Spiess, H. W. and Sheline, R. K. (1969). *ibid.*, **51,** 3932; Mooberry, E. S., Pupp, M., Slater, J. L. and Sheline, R. K. (1971). *ibid.*, **55,** 3655; Mooberry, E. S., Spiess, H. W. and Sheline, R. K. (1972). *ibid.*, **57,** 813
104. Ingalls, R. (1964). *Phys. Rev.*, **133,** A787
105. Gerloch, M. and Miller, J. R. (1968). *Progr. Inorg. Chem.*, **10,** 1
106. Johnson, C. E. (1971). *Proceedings of the Conference on the Application of the Mössbauer Effect (Tihany 1969)*, 663 (I. Dézsi, editor) (Budapest: Akadémiai Kiadó)
107. Clark, M. G., DiSalvo, F. J., Glass, A. M. and Peterson, G. E. (1973). *J. Chem. Phys.*, **59,** 6209
108. Bader, R. F. W. and Beddall, P. M. (1972). *J. Chem. Phys.*, **56,** 3320; Roby, K. R. (1974). *Molec. Phys.*, **27,** 81
109. Cotton, F. A. and Harris, C. B. (1966). *Proc. Nat. Acad. Sci. USA*, **56,** 12
110. Mulliken, R. S. (1949). *J. Chim. Phys.*, **46,** 497, 675
111. Nicholson, B. J. (1970). *Advan. Chem. Phys.*, **18,** 249
112. O'Konski, C. T. and Ha, T.-K. (1968). *J. Chem. Phys.*, **49,** 5354; Bonaccorsi, R., Scrocco, E. and Tomasi, J. (1969). *ibid.*, **50,** 2940; White, W. D. and Drago, R. S. (1970). *ibid.*, **52,** 4717
113. Williams, A. F. (1973). Unpublished dissertation
114. Bartunik, H. D., Potzel, W., Mössbauer, R. L. and Kaindl, G. (1970). *Z. Phys.*, **240,** 1
115. Charlton, J. S. and Nichols, D. I. (1970). *J. Chem. Soc. A*, 1484
116. Faltens, M. O. and Shirley, D. A. (1970). *J. Chem. Phys.*, **53,** 4249
117. de Vries, J. L. K. F., Keijzers, C. P. and de Boer, E. (1972). *Inorg. Chem.*, **11,** 1343
118. Messiah, A. (1961). *Quantum Mechanics*, Vol. 1, 173 (Amsterdam: North-Holland)
119. Löwdin, P.-O. (1956). *Advan. Phys.*, **5,** 1
120. Slater, J. C. (1963). *Quantum Theory of Molecules and Solids*, Vol. 1, Appendix 9 (New York: McGraw-Hill)
121. Parish, R. V. and Platt, R. H. (1970). *Inorg. Chim. Acta*, **4,** 65; Parish, R. V. and Johnson, C. E. (1971). *J. Chem. Soc. A* 1906
122. Davies, A. G., Smith, L. and Smith, P. J. (1970). *J. Organomet. Chem.*, **23,** 135; Herber, R. H. and Chandra, S. (1970). *J. Chem. Phys.*, **52,** 6045

123. Schlemper, E. O. and Hamilton, W. C. (1966). *Inorg. Chem.*, **5,** 995; Rush, J. J. and Hamilton, W. C. (1966). *ibid.*, **5,** 2238
124. Weiss, A. (1972). *Fortsch. Chem. Forsch.*, **30,** 1
125. Schempp, E. and Silva, P. R. P. (1973). *Phys. Rev. B*, **7,** 2983; *J. Chem. Phys.*, **58,** 5116
126. Clark, M. G., Maddock, A. G. and Platt, R. H. (1972). *J. Chem. Soc. Dalton Trans.* 281
127. Clark, M. G. (1969). *Discuss. Faraday Soc.*, **47,** 144
128. Heesch, H. (1930). *Z. Krist.*, **73,** 325
129. Clark, M. G. (1970). *Chem. Phys. Lett.*, **6,** 558
130. Cracknell, A. P. (1969). *Rept. Progr. Phys.*, **32,** 633; Niggli, A. (1964). *Advan. Struct. Res. Diffract. Method.*, **1,** 199
131. Bancroft, G. M. (1973). *Coord. Chem. Rev.*, **11,** 247
132. Sams, J. R. (1972). *MTP International Review of Science, Physical Chemistry Series One*, Vol. 4, Chap. 3 (London: Butterworths)
133. Smith, R. P. and Mortensen, E. M. (1960). *J. Chem. Phys.*, **32,** 502; Le Fèvre, R. J. W. (1965). *Advan. Phys. Org. Chem.*, **3,** 1
134. Jørgensen, C. K. (1971). *Modern Aspects of Ligand Field Theory* (Amsterdam: North-Holland); Newman, D. J. (1971). *Advan. Phys.*, **20,** 197
135. Clark, M. G. (1974). Unpublished results
136. Birss, R. R. (1964). *Symmetry and Magnetism* (Amsterdam: North-Holland)
137. Snyder, L. C. and Basch, H. (1972). *Molecular Wave Functions and Properties* (New York: Wiley)
138. Dewar, M. J. S., Patterson, D. B. and Simpson, W. I. (1973). *J. Chem. Soc. Dalton Trans.*, 2381; (1971). *J. Amer. Chem. Soc.*, **93,** 1030
139. Casabella, P. A., Bray, P. J. and Barnes, R. G. (1959). *J. Chem. Phys.*, **30,** 1393
140. Peterson, G. E. and Bridenbaugh, P. M. (1969). *J. Chem. Phys.*, **51,** 238
141. Barnes, R. G., Segel, S. L., Bray, P. J. and Casabella, P. A. (1957). *J. Chem. Phys.*, **26,** 1345
142. Mazak, R. A. and Collins, R. L. (1969). *J. Chem. Phys.*, **51,** 3220
143. Gibb, T. C., Greatrex, R. and Greenwood, N. N. (1972). *J. Chem. Soc. Dalton Trans.*, 238
144. Maddock, A. G., Platt, R. H., Williams, A. F. and Gancedo, R. (1974). *J. Chem. Soc. Dalton Trans.*, 1314
145. Sasvari, K. (1963). *Acta Crystallogr.*, **16,** 719
146. Lennard-Jones, J. (1949). *Proc. Roy. Soc. A*, **198,** 1, 14; Hall, G. G. and Lennard-Jones, J. (1950). *ibid.*, **202,** 155; Lennard-Jones, J. and Pople, J. A. (1950). *ibid.*, **202,** 166; Pople, J. A. (1950). *ibid.*, **202,** 323
147. Hall, G. G. (1950). *Proc. Roy. Soc. A*, **202,** 336
148. Edmiston, C. and Ruedenberg, K. (1963). *Rev. Mod. Phys.*, **34,** 457
149. Dominelli, N., Wood, E., Vasudev, P. and Jones, C. H. W. (1972). *Inorg. Nucl. Chem. Lett.*, **8,** 1077
150. Bancroft, G. M. and Libbey, E. T. (1973). *J. Chem. Soc. Dalton Trans.*, 2103
151. Vasudev, P. and Jones, C. H. W. (1973). *Can. J. Chem.*, **51,** 405
152. Maddock, A. G. and Platt, R. H. (1971). *J. Chem. Soc. A*, 1191
153. Pasternack, S. and Sternheimer, R. M. (1962). *J. Math. Phys.*, **3,** 1280
154. Cotton, F. A. (1963). *Chemical Applications of Group Theory*, 116 (London: Wiley)
155. Bancroft, G. M., Kumar Das, V. G., Sham, T. K. and Clark, M. G. (1974). *J. Chem. Soc. Chem. Commun.*, 236
156. Bancroft, G. M., Mays, M. J. and Prater, B. E. (1970). *J. Chem. Soc. A*, 956; Bancroft, G. M., Garrod, R. E. B. and Maddock, A. G. (1971). *ibid.*, 3165
157. Maddock, A. G. and Platt, R. H. (1971). *J. Chem. Phys.*, **55,** 1490; Micklitz, H. and Barrett, P. H. (1972). *Phys. Rev. B*, **5,** 1704
158. Bancroft, G. M., Mays, M. J. and Prater, B. E. (1969). *Discuss. Faraday Soc.*, **47,** 136
159. Bancroft, G. M., Garrod, R. E. B., Maddock, A. G., Mays, M. J. and Prater, B. E. (1970). *Chem. Commun.*, 200
160. Poller, R. C. and Ruddick, J. N. R. (1972). *J. Chem. Soc. Dalton Trans.*, 555
161. Goodman, B. A., Greenwood, N. N., Jaura, K. L. and Sharma, K. K. (1971). *J. Chem. Soc. A*, 1865
162. Ho, B. Y. K. and Zuckerman, J. J. (1973). *J. Organomet. Chem.*, **49,** 1

163. Cramér, H. (1946). *Mathematical Methods of Statistics*, Sec. 21.6 (Princeton: Princeton University Press)
164. Bancroft, G. M., Clark, M. G., Kumar Das, V. G. and Sham, T. K. (1975). *J. Chem. Soc. Dalton Trans.*, to be published
165. Arley, N. and Buch, K. R. (1950). *Probability and Statistics*, Sec. 11.8 (London: Chapman and Hall)
166. Edwards, A. W. F. (1972). *Likelihood*, Sec. 9.5 (Cambridge: Cambridge University Press)
167. Basano, L. and Gamba, A. (1971). *J. Statist. Phys.*, **3,** 237
168. Collins, R. L. and Pettit, R. (1963). *J. Amer. Chem. Soc.*, **85,** 2332; *J. Chem. Phys.*, **39,** 3433
169. Cullen, W. R., Harbourne, D. A., Liengme, B. V. and Sams, J. R. (1969). *Inorg. Chem.*, **8,** 1464
170. Carroll, W. E., Deeney, F. A., Delaney, J. A. and Lalor, F. J. (1973). *J. Chem. Soc. Dalton Trans.*, 718
171. Nesmeyanov, A. N., Semin, G. K., Bryukhova, E. V., Anisimov, K. N., Kolobova, N. E. and Khandozhko, V. N. (1969). *Bull. Acad. Sci. USSR, Div. Chem. Sci.*, 1792
172. Brown, T. L., Edwards, P. A., Harris, C. B. and Kirsch, J. L. (1969). *Inorg. Chem.*, **8,** 763
173. Boyd, T. E. and Brown, T. L. (1974). *Inorg. Chem.*, **13,** 422
174. Hillier, I. H. (1970). *J. Chem. Phys.*, **52,** 1948
175. Schreiner, A. F. and Brown, T. L. (1968). *J. Amer. Chem. Soc.*, **90,** 3366
176. Spencer, D. D., Kirsch, J. L. and Brown, T. L. (1970). *Inorg. Chem.*, **9,** 235; Graybeal, J. D., Ing, S. D. and Hsu, M. W. (1970). *ibid.*, **9,** 678
177. Brown, D. S. and Bushnell, G. W. (1967). *Acta Crystallogr.*, **22,** 296
178. Akhtar, M., Ellis, P. D., MacDiarmid, A. G. and Odom, J. D. (1972). *Inorg. Chem.*, **11,** 2917
179. Clark, M. G., Gancedo, J. R., Maddock, A. G. and Williams, A. F. (1975). *J. Chem. Soc. Dalton Trans.*, 120
180. Lippard, S. J. (1967). *Progr. Inorg. Chem.*, **8,** 109; Parish, R. V. (1966). *Advan. Inorg. Chem. Radiochem.*, **9,** 315
181. Hoard, J. L. and Nordsieck, H. H. (1939). *J. Amer. Chem. Soc.*, **61,** 2853; Hoard, J. L., Hamor, T. A. and Glick, M. D. (1968). *ibid.*, **90,** 3177
182. Baadsgaard, H. and Treadwell, W. D. (1955). *Helv. Chim. Acta*, **38,** 1669
183. Bok, L. D. C., Leipoldt, J. G. and Basson, S. S. (1970). *Acta Crystallogr. B*, **26,** 684
184. Basson, S. S., Bok, L. D. C. and Leipoldt, J. G. (1970). *Acta Crystallogr. B*, **26,** 1209
185. Kumar, K. and Baranger, M. (1968). *Nucl. Phys. A*, **122,** 273
186. Perumareddi, J. R., Liehr, A. D. and Adamson, A. W. (1963). *J. Amer. Chem. Soc.*, **85,** 249
187. Parish, R. V. and Perkins, P. G. (1967). *J. Chem. Soc. A*, 345
188. Chojnacki, J., Grochowski, J., Lebioda, L., Oleksyn, B. and Stadnicka, K. (1969). *Roczniki Chem.*, **43,** 273
189. Long, T. V. and Vernon, G. A. (1971). *J. Amer. Chem. Soc.*, **93,** 1919
190. Hartman, K. O. and Miller, F. A. (1968). *Spectrochim. Acta*, **24A,** 669; Parish, R. V., Simms, P. G., Wells, M. A., Woodward, L. A. (1968). *J. Chem. Soc. A*, 2882
191. Murakawa, K. (1962). *J. Phys. Soc. Jap.*, **17,** 891
192. Freeman, A. J., Mallow, J. V. and Bagus, P. S. (1970). *J. Appl. Phys.*, **41,** 1321
193. Stone, A. J. (1973). Unpublished results
194. Lu, C. C., Carlson, T. A., Malik, F. B., Tucker, T. C. and Nestor, C. W. (1971). *Atom. Data*, **3,** 1; Froese Fischer, C. (1972). *ibid.*, **4,** 301
195. Dunlap, B. D. (1972). *Phys. Rev. A*, **6,** 2057
196. Duff, K. J. (1974). *Phys. Rev. B*, **9,** 66
197. Marshalek, E., Person, L. W. and Sheline, R. K. (1963). *Rev. Mod. Phys.*, **35,** 108
198. Burrus, C. A. and Gordy, W. (1953). *Phys. Rev.*, **92,** 274; Verhoeven, J., Dymanus, A. and Bluyssen, H. (1969). *J. Chem. Phys.*, **50,** 3330
199. Pribula, C. D., Brown, T. L. and Münck, E. (1974). *J. Amer. Chem. Soc.*, **96,** 4149
200. Larossa, R. A. and Brown, T. L. (1974). *J. Amer. Chem. Soc.*, **96,** 2072
201. Dickinson, R. J., Parish, R. V., Rowbotham, P. J., Manning, A. R. and Hackett, P. (1975). *J. Chem. Soc. Datlon Trans.*, 424

8
Equilibrium Properties of Molecular Fluids

P. A. EGELSTAFF and C. G. GRAY
University of Guelph

and

K. E. GUBBINS
University of Florida

8.1	INTRODUCTION	300
8.2	INTERMOLECULAR FORCES	302
	8.2.1 *Models*	302
	8.2.2 *Spherical harmonic expansion for* $u(\mathbf{R}_{12}\omega_1\omega_2)$	303
8.3	EQUILIBRIUM STATISTICAL MECHANICS	304
	8.3.1 *Configurational distribution functions*	304
	8.3.2 *Bulk phase thermodynamic functions*	307
	8.3.3 *Interfacial thermodynamic properties*	308
	8.3.4 *The structure factor*	308
	8.3.5 *The mean squared torque*	309
	8.3.6 *Dielectric constant and Kerr constant*	310
	8.3.7 *Other properties*	310
8.4	LOW DENSITY GASES	311
8.5	INTEGRAL EQUATION METHODS	311
	8.5.1 *Percus–Yevick theory*	312
	8.5.2 *The mean spherical model*	314
8.6	SCALED PARTICLE THEORY	314
8.7	PERTURBATION THEORY	318
	8.7.1 *General expansion for the angular pair correlation function*	318

8.7.2 *Expansions of Pople and Gubbins and Gray* 319
8.7.3 *Effective isotropic potentials* 325
8.7.4 *Expansion of Sung and Chandler* 326
8.7.5 *Expansion of Bellemans* 328
8.7.6 *Non-spherical reference potentials* 329
8.7.7 *Generalised van der Waals model* 330

8.8 COMPUTER SIMULATION RESULTS 332

8.9 EXPERIMENTAL STUDIES 335
8.9.1 *Virial coefficients* 335
8.9.2 *Equation of state, configurational energy and entropy* 336
8.9.3 *Mean orientational coefficients* 337
8.9.4 *The liquid structure factor* 339

8.10 CONCLUSIONS 341

8.11 APPENDIX: THE INTERACTION COEFFICIENTS $\varepsilon(\Lambda;n_1n_2;R_{12})$ 342

ACKNOWLEDGEMENTS 343

8.1 INTRODUCTION

Most of the literature prior to 1970 on the statistical mechanics of fluids assumes that molecules are approximately spherical, with intermolecular forces acting between molecular centres. The experimentalist, on the other hand, must deal with the real world, in which most fluids are composed of non-spherical polyatomic molecules having non-central forces. This difference between the experimental and theoretical situations has led in the past to the loss of useful information contained in the experimental data. In this paper we review the present understanding of the relationship between theory and experiment for equilibrium polyatomic fluids. The literature on this subject does not appear to have been reviewed comprehensively in the recent past. Various individual topics have been covered in Refs. 1–5, and quantum effects are discussed by Friedman[7]. Much of the recent work on atomic fluids has been reviewed in a book edited by Singer[6].

In discussing the theory we assume that the system potential energy $\mathscr{U}(\boldsymbol{R}^N\omega^N)$ depends only on the positions of the centres of mass $\boldsymbol{R}^N \equiv \boldsymbol{R}_1 \ldots \boldsymbol{R}_N$ for the N molecules, and their molecular orientations $\omega^N \equiv \omega_1 \ldots \omega_N$. The vibrational coordinates of the molecules are assumed to be dynamically and statistically independent of the centre of mass and orientation coordinates. In addition, we assume that internal rotations are either absent, or are independent of the $\boldsymbol{R}^N$ and ω^N coordinates. Finally, we assume classical statistical mechanics. These assumptions should be quite realistic for many fluids composed of simple molecules, e.g. N_2, CO, CO_2, CH_4, etc. The assumption that vibrational states are independent of the rotational and translational states should not lead to serious error for molecules in which the separation between vibrational energy states greatly exceeds that between rotational states; this is usually the case for simple molecules. Furthermore, the effect of density on characteristic vibrational frequencies of molecules can be determined spectroscopically[8], and is usually small. Classical mechanics should apply for both translational and rotational motions for all molecular fluids except some of those containing hydrogen at low temperatures (particularly H_2 itself). The model used here may not give good results for more complicated molecules, however, because flexing vibrational modes and internal rotational states are expected to depend on molecular configurations.

In considering fluids in equilibrium, we can distinguish three principal cases: (a) isotropic, homogeneous fluids (e.g. liquid or compressed gas states of N_2, O_2, CH_4, etc., in the absence of an external field), (b) anisotropic, homogeneous fluids (e.g. polyatomic fluid in the presence of a uniform field, liquid crystals), and (c) inhomogeneous fluids (e.g. the interfacial region). These fluid states have been listed in order of increasing complexity; thus, more independent variables are involved in cases (b) and (c), and consequently the evaluation of the necessary distribution functions is more difficult.

In what follows we have not attempted to review the considerable literature that exists either on liquid crystals, or on hydrogen-bonded fluids. Of the many models proposed for the liquid crystal phase transition, only a few are based on the statistical mechanical methods used for simple fluids. Of particular note are methods based on the virial expansion, the integral equation techniques, and scaled particle theory; these applications are mentioned at appropriate places in the sections that follow. The statistical mechanical theories of water have been reviewed recently[4].

Polyatomic fluids can be studied by theoretical methods (Sections 8.2–8.7), by computer simulation of a model fluid (Section 8.8), or by direct experiment (Section 8.9). These methods complement each other, and progress in understanding such fluids should be facilitated by combining these three approaches. Section 8.2 briefly reviews models used for the intermolecular pair potential. Relationships for observable physical properties in terms of this potential and the appropriate distribution functions are given in Section 8.3. Various methods for evaluating the distribution functions (and hence observable properties through the equations of Section 8.3) are considered in Sections 8.4–8.7. The computer simulation studies considered in Section 8.8 play an intermediate role between theory and experiment. On the one hand, comparing theory and computer simulation results provides an unambiguous test of approximations in the theory; in such a test, the computer simulation results are to be regarded as 'experimental data'. The second use of the computer results is to provide information about the intermolecular potential, by comparison with experimental data for real fluids; in this case the computer simulation acts as the 'theory'. Finally, methods for learning about angular correlation effects between molecules by direct experiment are considered in Section 8.9.

8.2 INTERMOLECULAR FORCES

8.2.1 Models

There is a vast literature (for recent reviews, *cf.* Refs. 9, 10, 12–15, 43) on the determination of the intermolecular potential of simple (monatomic) molecules by *ab initio*, semi-empirical and empirical means. For diatomic and polyatomic molecules semi-empirical methods have been used extensively; there have been, especially at short range, rather few *ab initio* calculations (*cf.* the above references and Refs. 17–21, 25–27, 45, 151–153, 167, 168).

For liquids one needs to know two facts. (a) What is the pair potential? (b) How large are the three-body and higher-body potentials? For molecules the many-body potentials have been studied very little[21], and are usually not taken into account in liquid calculations, except indirectly by use of an 'effective pair potential'[22,23,33].

Information about the pair potential can be obtained from single-molecule experiments or from binary collision experiments (e.g. molecular beam scattering, Kerr effect, depolarised light scattering, field-gradient birefringence, transport coefficients, virial coefficients, pressure-broadening, pressure-induced absorption, dielectric constant). These experiments, when combined with a suitable theory, yield values of the parameters (e.g. Lennard-Jones constants, polarisabilities, dipole moments, quadrupole moments, octupole moments, etc) occurring in the expressions for the intermolecular potential. The shape of the 'hard core' of the molecule is usually determined from the molecular structure (which is often obtained experimentally by electron diffraction).

We consider here molecules which are permanently in their ground electronic and ground vibrational states. The intermolecular pair potential $u(\boldsymbol{R}_{12}\omega_1\omega_2)$ is thus assumed to depend only on the intermolecular separation $\boldsymbol{R}_{12}$ and on the molecular orientations ω_1 and ω_2. (Here $\omega \equiv \phi\theta\chi$ denotes for molecules of general shape the Euler angles, specifying the orientation of a molecule relative to an arbitrary space-fixed frame of reference; for linear molecules we have $\omega \equiv \theta\phi$, the polar angles of the symmetry axis).

At long range the interaction energy is classified as multipolar, dispersion or induction energy. The induction interaction energy is of order (α/σ^3) times the corresponding multipolar energy, where α is the polarisability and σ the Lennard-Jones diameter, and since this factor is typically of order 0.05, we neglect induction interaction here*. At short range the interaction energy is referred to as the overlap energy.

For some applications it is convenient to divide the pair potential $u(12)$ into isotropic and anisotropic parts,

$$u(\boldsymbol{R}_{12}\omega_1\omega_2) = u_0(R_{12}) + u_a(\boldsymbol{R}_{12}\omega_1\omega_2) \tag{8.1}$$

where the isotropic part u_0 is defined as the unweighted orientational average of the full potential,

* Induction forces may make an appreciable contribution to thermodynamic properties, however, as shown recently by McDonald[119].

$$u_0(R_{12}) = \langle u(\boldsymbol{R}_{12}\omega_1\omega_2)\rangle_{\omega_1\omega_2} = (1/\Omega^2)\int \mathrm{d}\omega_1\,\mathrm{d}\omega_2\, u(\boldsymbol{R}_{12}\omega_1\omega_2) \quad (8.2)$$

with $\Omega = 4\pi$ or $8\pi^2$ for linear or non-linear molecules respectively.

The calculation to date (both theoretical and machine) have involved reasonably simple models for the pair potential. For example, for N_2 one might choose u_0 to be a Lennard-Jones potential, and u_a as the sum of the longest-range multipolar term (quadrupole–quadrupole) plus the simplest short-range anisotropic overlap term. This neglects other terms at long and short range, and assumes that simple additivity holds at intermediate range.

We discuss briefly some other models of intermolecular pair potentials that have been employed. The 'atom–atom' model[34-38,40-42] assumes the intermolecular potential is a sum of interatomic potentials of the constituent atoms. The interatomic potential is usually taken as hard-sphere, Lennard-Jones or exponential form. It would seem intuitively that this model misses the multipole interaction at long range, and is too asymmetric at short range. Detailed criticism of this model is given in Refs. 18, 28 and 44. A refinement of the atom–atom model is to add a point dipole interaction to the sum of Lennard-Jones interactions[46]. For completeness we mention the spherical shell potential[4,47], the Kihara model with cores of various shapes[4,11], the Corner multi-force centres potential[48] (similar to the atom–atom potential above), the loaded and rough sphere models used in transport theory[49], and the Gaussian model[50]. Several highly simplified potential models have proved useful for testing statistical mechanical theories. These have included hard spheres with embedded point dipoles[51,52] or multipoles[53,54], and rigid bodies of various shapes[11]. Rigid shapes used have included cubes, tetrahedra, octahedra, rods, discs, cylinders, ellipsoids, spherocylinders, and dumbells. For water, both a point charge model[3] and a Gaussian model[55] have been used.

8.2.2 Spherical harmonic expansion for $u(\boldsymbol{R}_{12}\omega_1\omega_2)$[16,30-32]

For many calculations it is convenient to expand the intermolecular pair potential $u(\boldsymbol{R}_{12}\omega_1\omega_2)$ in terms of products of (generalised) spherical harmonics of the orientation of the molecules and of the intermolecular axis. This expansion is a sum over all values of $\Lambda \equiv l_1 l_2 l$ of terms of the form

$$u(\Lambda) = \sum \varepsilon(\Lambda; n_1 n_2; R_{12})\, C(\Lambda; m_1 m_2 m)\, D^{l_1}_{m_1 n_1}(\omega_1)^*\, D^{l_2}_{m_2 n_2}(\omega_2)^*\, Y_{lm}(\omega_{12})^* \quad (8.3)$$

where the summation is over the m and n values. In equation (8.3), $D^l_{mn}(\omega) \equiv D^l_{mn}(\phi\theta\chi)$ is a generalised spherical harmonic[24,31], $Y_{lm}(\omega) \equiv Y_{lm}(\theta\phi)$ is a spherical harmonic, $C(\Lambda; m_1 m_2 m) \equiv C(l_1 l_2 l; m_1 m_2 m)$ is a Clebsch–Gordan coefficient, and ε is an expansion coefficient. The $\Lambda = 000$ term is the isotropic potential $u_0(R_{12})$, and $\Lambda \neq 000$ terms comprise the anisotropic potential $u_a(\boldsymbol{R}_{12}\omega_1\omega_2)$. Properties of the ε coefficients, and explicit equations for these for some specific types of intermolecular forces, are given in the Appendix.

8.3 EQUILIBRIUM STATISTICAL MECHANICS

Using the model outlined in the introduction, the canonical partition function can be written

$$Q = \Lambda^{-3N} q_v^N q_r^N Z \tag{8.4}$$

where $\Lambda = h/(2\pi mkT)^{\frac{1}{2}}$ is the inverse of the translational kinetic energy partition function, N is the number of molecules, and q_v and q_r are the single molecule vibrational and rotational partition functions, respectively. The configurational partition function Z is given by

$$Z = \frac{1}{N!} \int \langle \exp [-\beta \mathscr{U}(R^N \omega^N)] \rangle_{\omega 1 \cdots \omega N} \, d\boldsymbol{R}^N \tag{8.5}$$

where $\beta = 1/kT$; $\boldsymbol{R}_i$ and ω_i are the centre of mass and orientational coordinates of molecule i. The angle brackets indicate an unweighted average over the orientations. For any function, a, of the configuration*

$$\langle a(R^N \omega^N) \rangle_{\omega 1 \cdots \omega N} = \frac{1}{\Omega^N} \int a(R^N \omega^N) \, d\omega^N \tag{8.6}$$

In what follows we use the canonical ensemble wherever possible, because of the notational simplicity. Some relations (e.g. the compressibility hierarchy and the perturbation theory expressions for the distribution functions) are more easily derived using the grand canonical ensemble; references to detailed derivations are given for such cases below.

Observed configurational properties can be expressed in terms of distribution functions, defined in the next sub-section. In deriving these relations we shall assume that, for homogeneous fluids in the absence of external fields, the potential energy is pairwise additive,

$$\mathscr{U}(\boldsymbol{R}^N \omega^N) = \sum_{i<j} u(\boldsymbol{R}_{ij} \omega_i \omega_j) \tag{8.7}$$

where $u(\boldsymbol{R}_{ij} \omega_i \omega_j)$ is the pair potential for the isolated ij molecular pair. Equation (8.7) neglects higher multibody potential terms. At least for atomic fluids, existing studies[3] suggest that three-body forces contribute only a few percent to the configurational internal energy at the triple point. For many molecular systems (N_2, CH_4, etc.) a similar situation should hold, but for large and flexible molecules this may not be very satisfactory. In the presence of an external field, a sum of single particle potentials $u(\boldsymbol{R}_i \omega_i)$ must be added to the right hand side of (8.7).

8.3.1 Configurational distribution functions

Orientation-dependent distribution functions are defined[2,4] in an analogous way to the usual centre of mass distribution functions for monatomic

* In equation (8.4), q_r^N is the rotational partition function for N molecules of ideal gas at temperature T, and contains a factor Ω^N. Some authors, e.g. Pople[56], define Z without the multiplying factor Ω^{-N} in front; their q_r^N is therefore just the kinetic energy part of the rotational partition function, and does not contain Ω^N as a factor.

molecules. The specific angular distribution function of order N is $\mathscr{P}(R^N\omega^N)$, and gives the probability density of observing the configuration $(\boldsymbol{R}^N\omega^N)$. It is given by:

$$\mathscr{P}(\boldsymbol{R}^N\omega^N) = \frac{\exp[-\beta\mathscr{U}(\boldsymbol{R}^N\omega^N)]}{\int \exp[-\beta\mathscr{U}(\boldsymbol{R}^N\omega^N)]\,\mathrm{d}\boldsymbol{R}^N\mathrm{d}\omega^N} \tag{8.8}$$

Properties are usually expressed in terms of lower order (two-, three- or four-molecule) angular distribution functions. The generic (or un-normalised probability) distribution function $f(\boldsymbol{R}_1\ldots\boldsymbol{R}_h\omega_1\ldots\omega_h)$ is defined so that $f(\boldsymbol{R}^h\omega^h)\,\mathrm{d}\boldsymbol{R}_1\ldots\mathrm{d}\boldsymbol{R}_h\mathrm{d}\omega_1\ldots\mathrm{d}\omega_h$ is proportional to the probability that a subset of h molecules has the configuration in which one molecule is in the element $(\mathrm{d}\boldsymbol{R}_1\mathrm{d}\omega_1)$ about $(\boldsymbol{R}_1\omega_1)$, another is in $(\mathrm{d}\boldsymbol{R}_2\mathrm{d}\omega_2)$ about $(\boldsymbol{R}_2\omega_2)$, and so on. It is given by

$$f(R^h\omega^h) = \frac{N!}{(N-h)!}\,\frac{\int \exp(-\beta\mathscr{U})\,\mathrm{d}\boldsymbol{R}_{h+1}\ldots\mathrm{d}\boldsymbol{R}_N\,\mathrm{d}\omega_{h+1}\ldots\mathrm{d}\omega_N}{\int \exp(-\beta\mathscr{U})\,\mathrm{d}\boldsymbol{R}_1\ldots\mathrm{d}\boldsymbol{R}_N\,\mathrm{d}\omega_1\ldots\mathrm{d}\omega_N} \tag{8.9}$$

If the molecules are uncorrelated, then $f(\boldsymbol{R}^h\omega^h)$ is a product of the one-particle distribution functions $f(\boldsymbol{R}_1\omega_1)f(\boldsymbol{R}_2\omega_2)\ldots f(\boldsymbol{R}_h\omega_h)$. It is therefore convenient to define an *angular correlation function*, $g(\boldsymbol{R}_1\ldots\boldsymbol{R}_h\omega_1\ldots\omega_h)$ by

$$f(\boldsymbol{R}_1\ldots\boldsymbol{R}_h\omega_1\ldots\omega_h) = f(\boldsymbol{R}_1\omega_1)f(\boldsymbol{R}_2\omega_2)\ldots f(\boldsymbol{R}_h\omega_h)g(\boldsymbol{R}_1\ldots\boldsymbol{R}_h\omega_1\ldots\omega_h) \tag{8.10}$$

In the special case of an *isotropic and homogeneous fluid* the probability density that a molecule has coordinates $(\boldsymbol{R}_1\omega_1)$ is independent of the values of both $\boldsymbol{R}_1$ and ω_1. Putting $h = 1$ in (8.9) and integrating over $\boldsymbol{R}_1\omega_1$ therefore gives

$$f(\boldsymbol{R}_1\omega_1) = \rho/\Omega \tag{8.11}$$

where $\rho = N/V$ is the bulk number density. Under these conditions, (8.10) becomes

$$f(\boldsymbol{R}^h\omega^h) = (\rho^h/\Omega^h)\,g(\boldsymbol{R}^h\omega^h) \tag{8.12}$$

Furthermore, the two-particle distribution or correlation function will depend only on the separation and orientation coordinates $\boldsymbol{R}$ and ω relative to $\boldsymbol{R}_1$ and ω_1, i.e. $f(\boldsymbol{R}_1\omega_1\boldsymbol{R}_2\omega_2)$ is invariant under rotations and translations of the pair of molecules, and can be denoted by $f(\boldsymbol{R}_{12}\omega_1\omega_2)$ or $f(\boldsymbol{R}_{12}\omega_{12})$.

In addition to the angular distribution and correlation functions defined above, it is sometimes convenient to use the homogeneous fluid 'total' angular correlation function, $h(\boldsymbol{R}_{12}\omega_1\omega_2)$, and the 'direct' angular correlation function, $c(\boldsymbol{R}_{12}\omega_1\omega_2)$. These are defined by an extension of the usual definition for monatomic fluids[2,4]:

$$h(\boldsymbol{R}_{12}\omega_1\omega_2) = g(\boldsymbol{R}_{12}\omega_1\omega_2) - 1 \tag{8.13}$$

$$h(\boldsymbol{R}_{12}\omega_1\omega_2) = c(\boldsymbol{R}_{12}\omega_1\omega_2) + \rho\int \langle c(\boldsymbol{R}_{13}\omega_1\omega_3)h(\boldsymbol{R}_{23}\omega_2\omega_3)\rangle_{\omega_3}\,\mathrm{d}\boldsymbol{R}_3 \tag{8.14}$$

The direct correlation function offers the advantage that it usually has essentially the same range as the intermolecular potential. Several theories based on approximations applied to the cluster expansion of $c(\boldsymbol{R}_{12}\omega_1\omega_2)$ are discussed in Section 8.5.

Distribution functions for molecular centres, irrespective of orientations, are obtained by integrating the corresponding angular distribution function over orientations. These functions are defined by:

$$f(\boldsymbol{R}^h) = \int f(\boldsymbol{R}^h\omega^h)\,\mathrm{d}\omega^h \tag{8.15}$$

$$g(R^h) = \langle g(\boldsymbol{R}^h\omega^h)\rangle_{\omega 1..\omega h} \tag{8.16}$$

The centres pair correlation function is of particular importance.

It is similarly possible to define distribution functions for molecular orientations, irrespective of the location of molecular centres[1]. These functions play a central role in the theory of anisotropic fluids and have been discussed by Buckingham[1]. An important function is the singlet probability density $\mathscr{P}(\omega)$; for a homogeneous fluid this is

$$\mathscr{P}(\omega_1) = (1/\rho) f(\boldsymbol{R}_1\omega_1) \tag{8.17}$$

[For an isotropic, homogeneous fluid, $\mathscr{P}(\omega_1) = \Omega^{-1}$ is a constant, from (8.11)]. Knowing $\mathscr{P}(\omega_1)$ one can obtain the quantities

$$F_l = \langle P_l(\cos\theta_1)\rangle = \int P_l(\cos\theta_1)\mathscr{P}(\omega_1)\,\mathrm{d}\omega_1 \tag{8.18}$$

F_1 is called the 'polarisation' and F_2 the 'alignment'[57,58]. Here P_l is the lth Legendre polynomial, $P_1(x) = x$, $P_2(x) = (1/2)(3x^2 - 1)$. The quantity F_1 is involved in the dielectric constant, and the Kerr constant involves a linear combination of F_1 and F_2 (see Section 8.3.6).

The angular pair correlation function $g(\boldsymbol{R}_{12}\omega_1\omega_2)$ plays a central role in the pairwise additivity theory of homogeneous fluids. Here $g(\boldsymbol{R}_{12}\omega_1\omega_2)$ is given by (8.9) and (8.12) as (for homogeneous, isotropic fluids):

$$g(\boldsymbol{R}_{12}\omega_1\omega_2) = \frac{N(N-1)\Omega^2}{\rho^2}\frac{\int \mathrm{e}^{-\beta\mathscr{U}}\,\mathrm{d}\boldsymbol{R}_3..\mathrm{d}\boldsymbol{R}_N\,\mathrm{d}\omega_3..\mathrm{d}\omega_N}{\int \mathrm{e}^{-\beta\mathscr{U}}\,\mathrm{d}\boldsymbol{R}_1..\mathrm{d}\boldsymbol{R}_N\,\mathrm{d}\omega_1..\mathrm{d}\omega_N}$$

$$= \frac{\Omega^2}{\rho^2}\sum_{i\neq j}\langle\delta(\boldsymbol{R}_i - \boldsymbol{R}_1)\delta(\omega_i - \omega_1)\delta(\boldsymbol{R}_j - \boldsymbol{R}_2)\delta(\omega_j - \omega_2)\rangle \tag{8.19}$$

and is normalised to unity (neglecting terms of order N^{-1}) at large R_{12} for all angles. The δ function form is usually used in the theory of radiation scattering[16].

Most of the properties considered below are related to averages over $g(\boldsymbol{R}_{12}\omega_1\omega_2)$. We remark, however, that certain thermodynamic functions (which can be related to fluctuations in the number density) can be related directly to the centres pair correlation function; the best known example is the isothermal compressibility (see Section 8.3.2).

8.3.2 Bulk phase thermodynamic functions

Thermodynamic functions for a homogeneous fluid are obtained from the partition function in the usual way. Thus the configurational Helmholtz free energy, A^c, is given by

$$A^c = -kT \ln Z \tag{8.20}$$

where Z is given by (8.5). The equation for configurational energy U^c, is

$$U^c = \langle \mathscr{U}^c \rangle = \tfrac{1}{2}\, \rho N \int \langle u(\boldsymbol{R}_{12}\omega_1\omega_1) g(\boldsymbol{R}_{12}\omega_1\omega_2)\rangle_{\omega_1\omega_2}\, \mathrm{d}\boldsymbol{R}_{12} \tag{8.21}$$

The pressure equation for an isotropic fluid is obtained from the relation $P = kT(\partial \ln Z/\partial V)$ as

$$P = \rho kT - \tfrac{1}{6}\rho^2 \int \langle \frac{\partial u(\boldsymbol{R}_{12}\omega_1\omega_2)}{\partial R_{12}} g(\boldsymbol{R}_{12}\omega_1\omega_2)\rangle_{\omega_1\omega_2} R_{12}\, \mathrm{d}\boldsymbol{R}_{12} \tag{8.22}$$

Equations (8.21) and (8.22) assume pairwise additivity, and reduce to the usual forms for monatomic fluids when angular correlations are absent. Other thermodynamic functions are obtained from the usual equations of classical thermodynamics, e.g.

$$S = (U - A)/T \text{ and } C_v = (\partial U/\partial T)_v$$

It is also possible to derive a series of equations called the compressibility hierarchy, for the density derivatives of the distribution functions. Although these may be derived using the canonical ensemble[59], the derivation is much simplified if the grand canonical ensemble is used. The angular distribution function of order h is given, in the grand canonical ensemble, by

$$f(\boldsymbol{R}^h\omega^h) = \frac{1}{\Xi} \sum_{N \geqslant h} \frac{q_v^N q_r^N \mathrm{e}^{N\beta\mu}}{\Lambda^{3N}(N-h)!\,\Omega^N} \int \mathrm{e}^{-\beta\mathscr{U}_N}\, \mathrm{d}\boldsymbol{R}^{N-h}\, \mathrm{d}\omega^{N-h} \tag{8.23}$$

where

$$\Xi(\beta V\mu) = \sum_{N \geqslant 0} \frac{q_v^N q_r^N \mathrm{e}^{N\beta\mu}}{\Lambda^{3N}\, N!\, \Omega^N} \int \mathrm{e}^{-\beta\mathscr{U}_N}\, \mathrm{d}\boldsymbol{R}^N\, \mathrm{d}\omega^N \tag{8.24}$$

Differentiating (8.23) with respect to μ and converting the $\partial/\partial\mu$ derivative to $\partial/\partial\rho$ in the usual way[4] gives, for $h = 1$ (homogeneous fluid)

$$kT\rho^2\chi \frac{\partial}{\partial\rho} f(\boldsymbol{R}_1\omega_1) = f(\boldsymbol{R}_1\omega_1) + \int [f(\boldsymbol{R}_{12}\omega_1\omega_2) - f(\boldsymbol{R}_1\omega_1)f(\boldsymbol{R}_2\omega_2)]\, \mathrm{d}\boldsymbol{R}_2\, \mathrm{d}\omega_2 \tag{8.25}$$

Here $\chi = \rho^{-1}(\partial\rho/\partial P)$ is the isothermal compressibility. Integrating both sides of (8.25) over ω_1 and using (8.15) gives the well-known *compressibility equation*,

$$kT\rho\chi = 1 + \rho \int [g(R) - 1]\, \mathrm{d}\boldsymbol{R} \tag{8.26}$$

which holds for either isotropic or anisotropic fluids, provided that they are homogeneous. For $h = 2$, differentiation of (8.23) with respect to μ gives, for the homogeneous fluid:

$$kT\chi \frac{\partial}{\partial\rho}[\rho^2 g(R_{12})] = 2g(R_{12}) + \rho\int [g(R_{12}R_{13}R_{23}) - g(R_{12})]\, \mathrm{d}\boldsymbol{R}_3 \tag{8.27}$$

Higher order members of the compressibility hierarchy are readily derived from (8.23); the equations are coupled, since each one relates a correlation function $g(\boldsymbol{R}^h)$ to the next higher function $g(\boldsymbol{R}^{h+1})$. These equations assume the fluid to be homogeneous, but do not rest on the assumption of pairwise additivity or on any assumption about the angle dependence of the potential. A simple alternative method of derivation of the compressibility hierarchy for the $g(\boldsymbol{R}^h)$ has been given by Egelstaff *et al.*[60]; this method relates the $g(\boldsymbol{R}^h)$ to the density correlation functions $\langle\rho(0)\rho(\boldsymbol{R})\rho(\boldsymbol{R}')\ldots\rangle$.

8.3.3 Interfacial thermodynamic properties

We consider a plane interface between two fluid phases (e.g. liquid/gas or isotropic liquid/liquid crystal phases) lying in the x–y plane. The transition layer between the two phases will be inhomogeneous and anisotropic (in the sense that in general the molecules will have a preferred orientation). The singlet and pair distribution functions of interest for such an interfacial region will be $f(z_1\omega_1)$ and $f(z_1\boldsymbol{R}_{12}\omega_1\omega_2)$, respectively.

The Kirkwood–Buff[61,62] derivation of the rigorous equation for interfacial tension γ in monatomic fluids has been extended to molecular fluids by Gray and Gubbins[63]. The result is

$$\gamma = \frac{\Omega^2}{4}\int_{-\infty}^{+\infty} \mathrm{d}z_1 \int \mathrm{d}\boldsymbol{R}_{12}\, \left\langle f(z_1\boldsymbol{R}_{12}\omega_1\omega_2)\left[R_{12}\frac{\partial u(12)}{\partial R_{12}} - 3z_{12}\frac{\partial u(12)}{\partial z_{12}}\right]\right\rangle_{\omega_1\omega_2} \tag{8.28}$$

where $u(12) \equiv u(\boldsymbol{R}_{12}\omega_1\omega_2)$. The simplest model for the pair distribution function in the transition region is that of Fowler[61,64]. For a gas/liquid interface this model assumes that the pair distribution function takes the bulk liquid phase value when both molecules are in the liquid, and vanishes when either molecule is in the gas. For molecular liquids the surface tension in the Fowler model is[63]

$$\gamma = \frac{\pi}{8}\rho^2 \int_0^{\infty} \mathrm{d}R_{12}\, R_{12}^4 \left\langle g(\boldsymbol{R}_{12}\omega_1\omega_2)\frac{\partial u(\boldsymbol{R}_{12}\omega_1\omega_2)}{\partial R_{12}}\right\rangle_{\omega_1\omega_2} \tag{8.29}$$

8.3.4 The structure factor

Structure factors determined by neutron or x-ray diffraction for molecular fluids show substantial differences from those for monatomic fluids[16]. These differences arise from intramolecular, as well as intermolecular effects. Steele and Pecora[67] gave a formal treatment of the structure factor for x-rays and neutrons. The theory of neutron diffraction from molecular fluids has been considered for some special cases by Sears[65] and by Egelstaff *et al.*[66] and

Narten[68] has discussed the situation for x-ray diffraction. A detailed treatment of the theory for neutron diffraction structure factors for molecular fluids has been recently given by Gubbins *et al.*[16].

In the static approximation, the differential cross-section for neutrons is the sum of incoherent and coherent contributions. The coherent part of the cross-section is proportional to the structure factor $S(q)$, which is given by[16]

$$S(q) = S_{\text{intra}}(q) + S_{\text{inter}}(q) \tag{8.30}$$

Here $S_{\text{intra}}(q)$ arises entirely from intramolecular effects and is given by

$$S_{\text{intra}}(q) = \left(\sum_{\alpha} \bar{b}_\alpha\right)^{-2} \sum_{\alpha\beta} \bar{b}_\alpha \bar{b}_\beta \frac{\sin (q\, r_{\alpha\beta})}{q\, r_{\alpha\beta}} = \langle |F(\boldsymbol{q}\, \omega_1)|\rangle^2_{\omega 1} \tag{8.31}$$

where $r_{\alpha\beta}$ is the internuclear $\alpha\beta$ distance for a given molecule, $\bar{b}_\alpha$ is the mean coherent scattering length for nucleus a, $\boldsymbol{q}$ is the scattering vector, and sums over a and β are over nuclei in a single molecule. The term $S_{\text{inter}}(q)$ in (8.3) contains all intermolecular correlation effects, and is given by[16]

$$S_{\text{inter}}(q) = \rho \int \exp (i\boldsymbol{q}\cdot\boldsymbol{R}_{12}) \langle [g(\boldsymbol{R}_{12}\omega_1\omega_2) - 1] F(\boldsymbol{q}\omega_1) F(\boldsymbol{q}\omega_2)^* \rangle_{\omega 1 \omega 2}\, \mathrm{d}\boldsymbol{R}_{12} \tag{8.32}$$

where

$$F(\boldsymbol{q}\, \omega_i) = \left(\sum_{\alpha} \bar{b}_\alpha\right)^{-1} \sum_{\alpha} \bar{b}_\alpha \exp (i\boldsymbol{q}\cdot\boldsymbol{r}_{ci\alpha}) \tag{8.33}$$

is proportional to the scattering amplitude for molecule i in the orientation ω_i; here $\boldsymbol{r}_{ci\alpha}$ is the vector from the centre of molecule i to nucleus a. If we assume that the electron density distribution is rigidly coupled to the nucleus, the x-ray scattering formulae are the same as (8.30)–(8.33) provided we interpret the indices a,β to denote electrons rather than nuclei. Equation (8.32) is not a simple Fourier transform, as for the monatomic case, so that $g(\boldsymbol{R}_{12}\omega_1\omega_2)$ cannot be obtained from the experimental $S_{\text{inter}}(q)$. Equation (8.32) reduces to the usual equation involving $g(R)$ when the molecules are monatomic. In the special case when the molecules rotate freely, i.e. $\mathscr{U}$ is independent of the molecular orientations, (8.30) and (8.32) simplify to

$$S(q) = S_{\text{intra}}(q) + \rho \langle F(\boldsymbol{q}\omega_i)\rangle^2_{\omega_i} \int \exp (i\boldsymbol{q}\cdot\boldsymbol{R})[g(R) - 1]\, \mathrm{d}\boldsymbol{R} \tag{8.34}$$

where $\langle F(\boldsymbol{q}\omega_i)\rangle_{\omega_i}$ depends on the molecular bond lengths and angles, and is given by

$$\langle F(\boldsymbol{q}\omega_i)\rangle_{\omega_i} = \sum_{\alpha} \bar{b}_\alpha \frac{\sin (q\, r_{ci\alpha})}{rq_{ci\alpha}} / \sum_{\alpha} \bar{b}_\alpha \tag{8.35}$$

8.3.5 The mean squared torque

From the hypervirial theorem[7], the mean squared torque on molecule 1 is

$$\langle \tau_1^2 \rangle = kT \langle \nabla^2_{\omega 1} \mathscr{U} \rangle \tag{8.36}$$

where ∇_{ω}^{2} is the angular Laplacian. With the assumption of pairwise additivity of intermolecular potentials, (8.36) becomes

$$\langle \tau_1^2 \rangle = kT\rho \int \langle \nabla_{\omega 1}^2\, u(\boldsymbol{R}_{12}\omega_1\omega_2) g(\boldsymbol{R}_{12}\omega_1\omega_2) \rangle_{\omega_1\omega_2}\, \mathrm{d}\boldsymbol{R}_{12} \tag{8.37}$$

If $u(\boldsymbol{R}_{12}\omega_1\omega_2)$ is expanded in products of generalised spherical harmonics, as in (8.3), it follows that[31]

$$\langle \tau_1^2 \rangle = -kT\rho \sum_{\Lambda} l_1(l_1+1) \int \langle u(\Lambda)\, g(\boldsymbol{R}_{12}\omega_1\omega_2) \rangle_{\omega_1\omega_2}\, \mathrm{d}\boldsymbol{R}_{12} \tag{8.38}$$

We note that the integral in (8.38) is the same one that appears in the equation for the configurational internal energy (8.21). Because of the factor $l_1(l_1 + 1)$ in (8.38), the mean squared torque is more sensitive to the higher harmonic components of the potential than is the configurational internal energy. The mean squared torque provides a direct and unambiguous measure of the strength of the anisotropic intermolecular forces, since it vanishes for atomic fluids.

8.3.6 Dielectric constant and Kerr constant

For linear molecules the dielectric and Kerr constants involve the polarisation F_1 and the alignment F_2, respectively [cf. (8.18)]; if molecular hyperpolarisability effects are included[58], the Kerr constant also involves F_1. The Kirkwood-type theory[1,57,58,70] for F_l, in which the fluctuations in the collision-induced dipole moments are neglected, relates it to the so-called angular correlation parameters G_l:

$$G_l = \rho \int \langle P_l(\cos \gamma_{12}) g(\boldsymbol{R}_{12}\omega_1\omega_2) \rangle_{\omega_1\omega_2}\, \mathrm{d}\boldsymbol{R}_{12} \tag{8.39}$$

Here γ_{12} is the angle between the axes of molecules 1 and 2. G_1 might be termed the 'mutual polarisation' and G_2 the 'mutual alignment'. The quantity G_2 occurs in theories of depolarised light scattering[1,71] and the quadratic electric-field effect[58,169,172] in n.m.r. Also it is necessary to exercise care in deriving the G values from experiment, as usually there are other contributions to the measured quantity.

8.3.7 Other properties

The above compilation is by no means exhaustive. As examples of other properties of interest that are related to $g(\boldsymbol{R}_{12}\omega_1\omega_2)$ we mention the mean squared force[76], the infrared and Raman vibrational shifts[8], and collision-induced absorption[77]. Several of these other properties have been considered by Buckingham[1].

The equations given above are readily extended to mixtures. Twu *et al.* have given equations for the mean squared torque[69] and for thermodynamic properties[166]. Certain mixture thermodynamic properties (which can be related to fluctuations in the number density) can be expressed directly in terms of the centres distribution functions[72-75]. Examples are isothermal compressibility, partial molar volume, and concentration derivatives of the chemical potential of a component.

8.4 LOW DENSITY GASES

The expressions given for observable properties in Section 8.3 can only be evaluated if the angular correlation functions $g(\mathbf{R}^h\omega^h)$ are known. In this and the following three sections we review methods for evaluating these functions, particularly the pair function $g(\mathbf{R}_{12}\omega_1\omega_2)$.

At low to moderate densities it is useful to expand the angular correlation functions as power series in the number density. Detailed derivations are given in many places[4,15].

Early work (up until 1967) on calculations of the virial coefficients for orientation-dependent potential models has been reviewed by Mason and Spurling[15]. Sweet and Steele[40,46] have calculated spherical harmonic contributions to the low density limit of the angular pair correlation function, $g^{(0)}(\mathbf{R}_{12}\omega_1\omega_2)$, for both a diatomic Lennard-Jones (12,6) and a Kihara core potential for the case of linear molecules; in addition, they made similar calculations for a potential consisting of the diatomic Lennard-Jones model plus a point dipole. Similar calculations have been made by Chen and Steele[78] for linear dumbell molecules composed of two rigidly connected hard spheres. Harmonic contributions to $g^{(0)}(\mathbf{R}_{12}\omega_1\omega_2)$ and $g^{(1)}(\mathbf{R}_{12}\omega_1\omega_2)$ were calculated, together with the Percus–Yevick approximation for $g^{(2)}(\mathbf{R}_{12}\omega_1\omega_2)$; these results were used to compute the second, third and fourth virial coefficients for this potential model. Third virial coefficients for hard spheres with embedded point dipoles have been recently calculated by Watts[52].

The virial expansion for the free energy and distribution functions of an anisotropic fluid has been considered by several authors[79-85].

8.5 INTEGRAL EQUATION METHODS

Included in the methods[2,4,6] discussed in this section are the Yvon–Born–Green (YBG), Percus–Yevick (PY), hypernetted chain (HNC) and mean spherical model (MSM) theories. These involve different approximation methods for the angular correlation functions; in each case this yields an integral equation, which must be solved (numerically in most cases) for the potential model of interest. The YBG theory is based on applying the superposition approximation for the triplet function $g(\mathbf{R}^3\omega^3)$ in the first member of the YBG hierarchy of equations for the functions $g(\mathbf{R}^h\omega^h)$. The other three methods involve approximations for the direct angular correlation function, $c(\mathbf{R}_{12}\omega_1\omega_2)$. The integral equation methods are useful for higher densities, where the density expansion of the previous section breaks down. Although these approximations did not originate from any common basis or set of ideas, their relationship may to some extent be understood[2,86,89] by examining the full graphical expansion of $g(\mathbf{R}_{12}\omega_1\omega_2)$. An exact treatment would include all graphs to all orders in the density. The integral equation methods omit certain classes of graphs that are difficult to evaluate, while including other classes of graphs to all orders; the theories differ in the classes of graphs omitted. The PY and MSM theories have both been applied to fluids with orientation dependent forces, but the YBG and HNC approaches seem to have received little or no attention.

8.5.1 Percus–Yevick theory

In the PY approximation[87], the direct angular pair correlation function is related to $g(\boldsymbol{R}_{12}\omega_1\omega_2)$ by

$$c(\boldsymbol{R}_{12}\omega_1\omega_2) = g(\boldsymbol{R}_{12}\omega_1\omega_2)\,\{1 - \exp[\beta u(\boldsymbol{R}_{12}\omega_1\omega_1)]\} \tag{8.40}$$

Substitution of this result into (8.14) yields the PY integral equation for $g(\boldsymbol{R}_{12}\omega_1\omega_2)$:

$$g(\boldsymbol{R}_{12}\omega_1\omega_2)\exp\,[\beta u(\boldsymbol{R}_{12}\omega_1\omega_2)] = 1 + \rho\int \mathrm{d}\boldsymbol{R}_3\,\langle g(\boldsymbol{R}_{13}\omega_1\omega_3)\{1 - \exp[\beta u(\boldsymbol{R}_{13}\omega_1\omega_3)]\}\{g(\boldsymbol{R}_{23}\omega_2\omega_3) - 1\}\rangle_{\omega_3} \tag{8.41}$$

It is seen from (8.40) that in the PY approximation $c(\boldsymbol{R}_{12}\omega_1\omega_2)$ has the same range as $u(\boldsymbol{R}_{12}\omega_1\omega_2)$; it can be shown[88] that (8.72) is the only consistent way of requiring that the range of c be the same as that of u.

Chen and Steele[90] have used the PY theory to calculate the pair correlation function and equation of state for a fluid composed of linear dumbell molecules. The angular pair correlation function was expanded in a series of spherical harmonics and (8.41) then becomes a set of coupled integral equations. This set of equations was truncated by neglecting higher harmonic terms, and solved numerically. The resulting plot of the compressibility factor $P/\rho kT$ against reduced density for rigid dumbells is similar to, but somewhat above, the corresponding plot for rigid spheres (Figure 8.1).

Using the same procedure as Chen and Steele, Morrison[91] has solved the PY equation for the first few spherical harmonic contributions to $g(\boldsymbol{R}_{12}\omega_1\omega_2)$ for homonuclear diatomic molecules, using an atom–atom Lennard-Jones potential; the density range $\rho^* = 0.1$–1.2 was covered at $T^* = 0.75$, 1.00 and 1.30. The first two coefficients in the spherical harmonic expansion were found to be sufficient for good results at these state conditions.

Ben-Naim[3,92] has proposed an approximate form of the PY equation for molecular fluids, and has applied it to the case of liquid water using a point-charge model for the potential. In order to simplify (8.41) he assumes that the function $y(\boldsymbol{R}_{12}\omega_1\omega_2)$, defined by

$$g(\boldsymbol{R}_{12}\omega_1\omega_2) = \exp\,[-\beta u(\boldsymbol{R}_{12}\omega_1\omega_2)]y(\boldsymbol{R}_{12}\omega_1\omega_2) \tag{8.42}$$

is independent of the orientations of molecules 1 and 2. Thus the essence of Ben-Naim's approximation is

$$y(\boldsymbol{R}_{12}\omega_1\omega_2) \approx y(R_{12}) \tag{8.43}$$

Substituting (8.43) into (8.41) yields a simplified PY equation. We note here that (8.43) is exact in (a) the low density limit and (b) cases in which the anisotropic forces are sufficiently weak [see equation (8.67)]. However, Monte Carlo calculations[93] indicate that (8.43) fails for large anisotropies in the potential (see Figure 8.9). Thus the simplified PY equation cannot be expected to give good results for water; for weak anisotropies, for which (8.43) may work, it is undoubtedly easier to use perturbation theory (Section 8.7).

Ben-Naim[55,94] has solved the full PY equation for a two-dimensional system of 'water-like' molecules; the potential model adopted consisted of a Lennard-Jones part plus a hydrogen-bonding part.

Recently, Chandler and Andersen[133,134] have proposed an integral equation for the pair correlation function in a fluid composed of molecules which

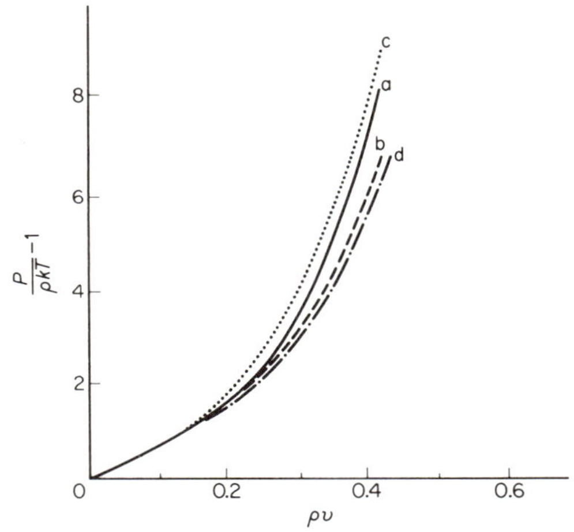

Figure 8.1 Compressibility factors for the RISM model for dumbells compared to the PY theory for hard spheres. (a) PY compressibility equation, (b) PY pressure equation, (c) RISM for $R/\sigma = 0.6$ compressibility equation, (d) RISM for $R/\sigma = 0.6$ pressure equation (the ordinate scale of figure 3 of Ref. 135 has been increased by a factor $\sqrt{2}$). v is the particle volume

consist of several rigidly connected hard spheres. The derivation of this result (called the RISM equation) is based on a functional expansion similar to that used in deriving PY theory. Lowden and Chandler[135] have discussed the numerical solution of the RISM equation and have presented some results. Figure 8.1 compares the compressibility factors for dumbells, with $R/\sigma = 0.6$, to the Percus–Yevick hard sphere data. It can be seen that the difference between the virial and compressibility equation results is somewhat greater for the RISM model, although of the same magnitude as for the PY case.

Workman and Fixman[100] have generalised the PY theory to the case of anisotropic fluids. In a similar study, Wulf[79] used the Kirkwood integral equations as a starting point to obtain the Maier–Saupe theory[99].

8.5.2 The mean spherical model

The MSM integral equation was proposed by Lebowitz and Percus[95], and is a generalisation to continuum fluids of the spherical model for Ising spin systems. The model applies only to fluids in which the molecules have rigid cores of diameter d,

$$\begin{aligned} u(\boldsymbol{R}_{12}\omega_1\omega_2) &= \infty & R_{12} < d \\ &= u_t(\boldsymbol{R}_{12}\omega_1\omega_2) & R_{12} > d \end{aligned} \tag{8.44}$$

where u_t is the 'tail' and is assumed to be finite. From (8.44) we must have

$$g(\boldsymbol{R}_{12}\omega_1\omega_2) = 0 \qquad R_{12} < d \tag{8.45}$$

The MSM approximation then states that

$$c(\boldsymbol{R}_{12}\omega_1\omega_2) = -\beta u_t(\boldsymbol{R}_{12}\omega_1\omega_2) \quad R_{12} > d \tag{8.46}$$

where c is the direct correlation function. The MSM integral equation is obtained by using (8.45) and (8.46) in the Ornstein–Zernike relation (8.14). For $u_t = 0$ the MSM reduces to the PY equation for hard spheres. For a multicomponent system in which $d = 0$ and u_t is the Coulomb potential, the MSM reduces to the Debye–Hückel theory of electrolytes.

Wertheim[51] has solved the MSM for a fluid of hard spheres with permanent electric dipole moments, and has applied the solution to the evaluation of the dielectric constant. The final result for $g(\boldsymbol{R}_{12}\omega_1\omega_2)$ consists of a spherically symmetric term and two other terms with different dependences on the orientations of the two dipole moments. The spherically symmetric part is the solution of the PY equation for hard spheres; the angle-dependent terms involve the solution of two integral equations. The extension of Wertheim's solution to the case of hard spheres with arbitrary multipole moments has been given by Blum[53,54]. For hard spheres with dipoles the MSM has recently been compared with both Monte Carlo calculations[96,97] and with perturbation theory[96-98]; some of these results are given in Table 8.2 (p. 323). The MSM is in poor agreement with the Monte Carlo results. Alderman and Deutch[51a] have extended Wertheim's solution to the case of mixtures.

8.6 SCALED PARTICLE THEORY

Scaled particle theory provides a method of calculating the free energy change on adding a rigid solute particle to a system which already contains N rigid particles. It was first developed by Reiss *et al.*[101] for hard spheres, and was extended to hard sphere mixtures by Lebowitz *et al.*[102]. For rigid spheres of diameter σ, one starts by adding an imaginary sphere of diameter $L' = -\sigma$, since this requires no work (the excluded volume of interaction between centres of the added sphere and of those already present is zero), and then calculates the work of increasing ('scaling') the size of this sphere to diameter $L' = +\sigma$. It is possible to calculate this work exactly in the region $-\sigma \leqslant L' \leqslant 0$; the additional work required to increase L' from 0 to $+\sigma$ is estimated by making a Taylor series expansion about $L' = 0$. Scaled particle theory is limited to calculating certain of the macroscopic properties (e.g. thermodynamic functions) and cannot yield the distribution functions themselves. It is also restricted to fluids of rigid particles, and cannot deal with attractive intermolecular potentials. Within these limitations, however, it is easy to use and yields an accurate description up to quite high densities; for rigid spheres the equation of state obtained is identical to that obtained by the PY theory used in the compressibility equation.

Gibbons[103,104] has extended the scaled particle theory to fluids of rigid convex bodies (*cf.* also Ref. 170). Consider a fluid containing N convex particles of characteristic length L in volume V at temperature T; the configurational integral $Z(NVT)$ is given by (8.5). If a convex particle of characteristic length L' is now added to the system, the configuration integral changes to $Z(NVT; L')$. Here L' might be the radius in the case of spheres, the length of the side in the case of cubes or tetrahedra, etc.; $L' = 0$ corresponds to a point particle. If $L' \leqslant 0$ the volumes from which the particle L' is excluded by the other particles do not overlap, and $Z(NVT; L')$ can be evaluated exactly as

$$Z(NVT; L') = (N+1)^{-1} (V - N\langle v_{LL'}\rangle_{\omega'})\, Z(NVT) \quad \text{for } L' \leqslant 0 \tag{8.47}$$

where $\langle v_{LL'}\rangle_{\omega'}$ is the volume from which the centre of the L' particle is excluded by an L particle, averaged over all possible orientations of particle L'. The geometry of convex bodies[11] gives this quantity as

$$\langle v_{LL'}\rangle_{\omega'} = \mathscr{V}_L + \mathscr{V}_{L'} + (1/4\pi)(M_L S_{L'} + M_{L'} S_L) \tag{8.48}$$

where $\mathscr{V}$, S and M are the volume, surface area and the mean curvature integrated over the surface, respectively ($M/4\pi$ is called the *mean radius* by some authors). Kihara[11,105] and Gibbons[103,104] give expressions for $\mathscr{V}$, S and M in terms of L for various convex shapes.

Equations (8.47) and (8.48) may be used to calculate the free energy $A(L')$ for adding a particle of size $L' = 0$; this value is then extrapolated to the value $A(L)$ for a particle of the full size L by the same procedure as for rigid spheres. This yields the chemical potential and hence, through the usual thermodynamic relations, the other thermodynamic functions. Gibbons[103,104] has derived these results for mixtures. If all particles in the mixture are of the same shape, the equation of state is found to be

$$\frac{\beta P}{\rho} = \frac{1}{1-Y} + \frac{B^2 C}{3\rho(1-Y)^3} + \frac{AB}{\rho(1-Y)^2} \tag{8.49}$$

where

$$\rho = \sum_{\alpha=a}^{r} \rho_\alpha$$

$$A = \frac{1}{4\pi} \sum_{\alpha=a}^{r} \rho_\alpha M_\alpha$$

$$B = \sum_{\alpha=a}^{r} \rho_\alpha S_\alpha$$

$$C = \sum_{\alpha = a}^{r} \rho_\alpha (M_\alpha/4\pi)^2$$

$$Y = \sum_{\alpha = a}^{r} \rho_\alpha \mathscr{V}_\alpha$$

Equation (8.49) reduces to the usual PY result when the hard sphere expressions are used for M, S and $\mathscr{V}$.

Gibbons[103] has used scaled particle theory to calculate virial coefficients up to the fifth for convex molecules of various shapes. Considerable variations in the magnitude of the virial coefficients occur with change in shape, the values always being larger than for rigid spheres of the same volume. The compressibility factor $\beta P/\rho$ has also been calculated for various shapes as a function of reduced density. The curves for non-spherical molecules always lie above the curve for rigid spheres; thus, for a given density the pressure is greater in the system of non-spherical molecules than for rigid spheres. From these calculations Gibbons has concluded that it is not possible to replace a non-spherical particle by an equivalent sphere which will reproduce both the equation of state and the virial coefficients of the non-spherical particle.

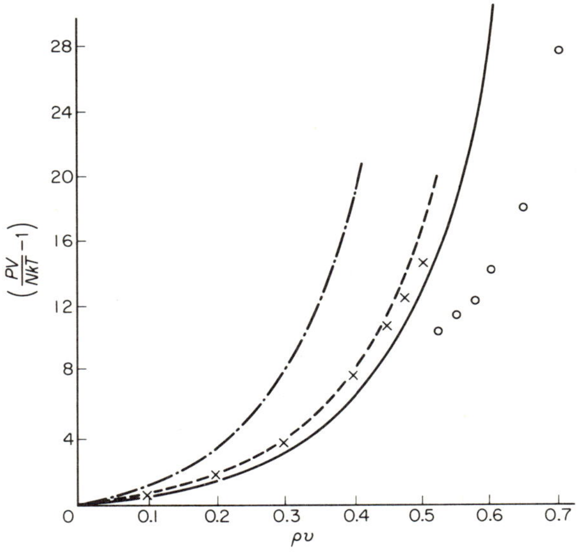

Figure 8.2 Equation of state for convex rigid particles; v is particle volume and $\rho = N/V$. Points are Monte Carlo data for prolate spherocylinders, $R^* = R/\sigma = 1$ (where R is length of cylinder, excluding semispherical caps, and σ is diameter of cylinder); × fluid phase; ○ solid phase. Lines are scaled particle theory calculations based on equation (8.49) for various shapes; ——, rigid spheres; ---, prolate spherocylinders, $R^* = 1$; —— - —— - ——, regular tetrahedra. (From Few and Rigby[106] and Gibbons[103], by courtesy of the American Institute of Physics and Taylor and Francis)

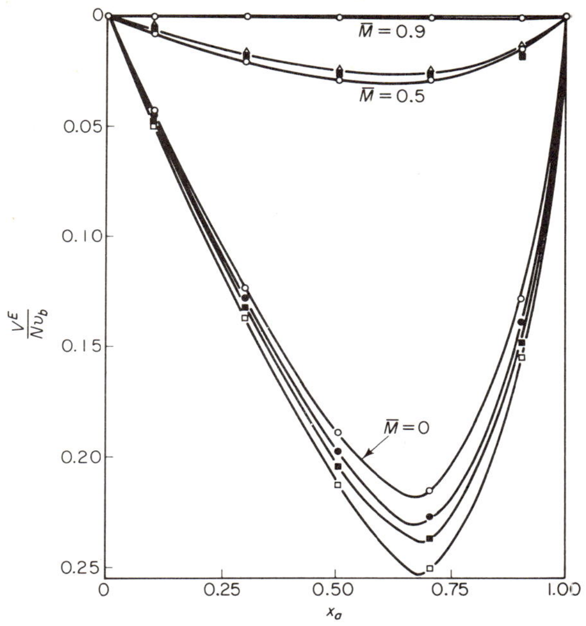

Figure 8.3 Excess volume for a binary mixture of convex particles of species a and b at $Pv_b/kT = 1.0$, from scaled particle theory. Here $\bar{M} = M_a/M_b$ is the ratio of mean radii. $\bar{M} = 1$ corresponds to an ideal solution ($V^E = 0$). Points are for various shapes: ○, spheres; ●, ellipsoids, eccentricity = 0.75; □, cubes; ■, cylinders, length = twice radius; △, tetrahedron (From Gibbons[104], by courtesy of Taylor and Francis)

Few and Rigby[106], and more recently Vieillard-Baron[107], have carried out Monte Carlo calculations of the equation of state for a system of rigid spherocylinders, and have used the results to test scaled particle theory (Figure 8.2). Excellent agreement is obtained throughout the fluid region. The Monte Carlo results also suggest a fluid–solid phase transition at a reduced density of about 0.5–0.55.

Equation (8.49) has also been applied by Gibbons[104] to the calculation of excess volumes (V^E) for binary mixtures at fixed temperature and pressure. Typical results are shown in Figure 8.3. The main conclusions from these calculations were that (a) V^E is always negative, (b) V^E depends strongly on particle size ratio and pressure, less strongly (but still significantly) on particle shape, and (c) there is no simple relationship between V^E and particle shape. The excess Gibbs free energy was also found to be negative in all cases studied. It follows that there can be no fluid–fluid phase separation for these mixtures, at least according to scaled particle theory.

Cotter and Martire[108,109], and also Lasher[110], have used scaled particle theory to study a model of the nematic–isotropic phase transition occurring in liquid crystals. To test for a liquid crystal phase transition they equated expressions for the chemical potentials of the anisotropic and isotropic phases, and also equated expressions for the pressures. Rigby[111] has proposed a modified van der Waals equation of state for non-spherical molecules, which is an extension of scaled particle theory. It is discussed in Section 8.7.7.

8.7 PERTURBATION THEORY

The results of Sections 8.4 and 8.5 can be regarded as perturbation treatments. There, the reference system is the perfect gas, and the full potential is taken as the perturbation. The PY and related theories involve a selective resummation of terms of all orders in the potential. These methods are also called 'cluster expansion', 'graphical expansion', and 'functional expansion' techniques, which distinguish them from the version of perturbation theory which we now describe.

In what is traditionally called perturbation theory, one takes as reference a system in which the molecules interact with some reference potential u_0 which is different from the potential u for the real fluid; the properties of the real fluid are then expanded about those of the reference system. Pople[56] was the first to apply this method to molecular fluids. He expanded the Helmholtz free energy, using as reference a system of molecules with an isotropic potential (see Section 8.7.2 below). A more general approach is to expand the distribution functions themselves, and this is discussed in the next section.

8.7.1 General expansion for the angular pair correlation function[112–114]

We consider a uniform fluid in which pairwise additivity holds, and in which the pair potential is $u(\boldsymbol{R}_{12}\omega_1\omega_2;\, a)$. Here a is some perturbation parameter (to be defined precisely later) such that

$$u(\boldsymbol{R}_{12}\omega_1\omega_2;\, a = 0) = u_0(\boldsymbol{R}_{12}\omega_1\omega_2) \tag{8.50}$$

$$u(\boldsymbol{R}_{12}\omega_1\omega_2;\, a = 1) = u(\boldsymbol{R}_{12}\omega_1\omega_2) \tag{8.51}$$

where u_0 and u are the pair potentials in the reference and real systems, respectively. The angular pair correlation function for this fluid is now expanded in a Taylor series about $a = 0$. On subsequently putting $a = 1$ to regain the real system, one obtains

$$g(\boldsymbol{R}_{12}\omega_1\omega_2) = g_0(\boldsymbol{R}_{12}\omega_1\omega_2) + g_1(\boldsymbol{R}_{12}\omega_1\omega_2) + g_2(\boldsymbol{R}_{12}\omega_1\omega_2) + \ldots \tag{8.52}$$

where $g_0(\boldsymbol{R}_{12}\omega_1\omega_2)$ is the correlation function for the reference system, $g_1(\boldsymbol{R}_{12}\omega_1\omega_2)$ is the first order perturbation, and so on. The expression for $g_1(\boldsymbol{R}_{12}\omega_1\omega_2)$ is[112-114]

$$g_1(12) = \left(\frac{\partial g(12)}{\partial a}\right)_{\alpha=0} = -\beta\left[\frac{\partial u(12)}{\partial a}\right]_{\alpha=0} g_0(12)$$

$$- \beta\rho\int\left\langle\left[\left(\frac{\partial u(13)}{\partial a}\right)_{\alpha=0} + \left(\frac{\partial u(23)}{\partial a}\right)_{\alpha=0}\right] g_0(123)\right\rangle_{\omega_3} \mathrm{d}\boldsymbol{R}_3$$

$$-\tfrac{1}{2}\beta\rho^2\int\left\langle\left(\frac{\partial u(34)}{\partial a}\right)_{\alpha=0}[g_0(1234)-g_0(12)\,g_0(34)]\right\rangle_{\omega_3\omega_4}\mathrm{d}\boldsymbol{R}_3\mathrm{d}\boldsymbol{R}_4$$

$$+\beta\frac{\partial}{\partial\rho}\left(\rho^2 g_0(12)\right)\left(\int\left\langle\left(\frac{\partial u(34)}{\partial a}\right)_{\alpha=0}g_0(34)\right\rangle_{\omega_3\omega_4}\mathrm{d}\boldsymbol{R}_{34}\right.$$

$$\left.+\tfrac{1}{2}\rho\int\left\langle\left(\frac{\partial u(34)}{\partial a}\right)_{\alpha=0}[g_0(345)-g_0(34)]\right\rangle_{\omega_3\omega_4\omega_5}\mathrm{d}\boldsymbol{R}_{34}\mathrm{d}\boldsymbol{R}_{45}\right) \qquad (8.53)$$

where $g_0(12)\equiv g_0(\boldsymbol{R}_{12}\omega_1\omega_2)$, etc.

An alternative perturbation expansion is obtained by expanding the function $y(\boldsymbol{R}_{12}\omega_1\omega_2)$ given by (8.42). In place of (8.52) one obtains

$$g(\boldsymbol{R}_{12}\omega_1\omega_2)=\exp\left[-\beta u(\boldsymbol{R}_{12}\omega_1\omega_2)\right]\left[y_0(\boldsymbol{R}_{12}\omega_1\omega_2)+y_1(\boldsymbol{R}_{12}\omega_1\omega_2)+\ldots\right] \qquad (8.54)$$

where

$$y_1(\boldsymbol{R}_{12}\omega_1\omega_2)=\exp\left[\beta u_0(\boldsymbol{R}_{12}\omega_1\omega_2)\right]\left\{\beta\left(\frac{\partial u(12)}{\partial a}\right)_{\alpha=0}g_0(\boldsymbol{R}_{12}\omega_1\omega_2)+g_1(\boldsymbol{R}_{12}\omega_1\omega_2)\right\} \qquad (8.55)$$

with g_1 given by (8.53).

The corresponding perturbation expansion for the Helmholtz free energy is obtained from (8.20) and (8.5) as

$$A^{\mathrm{c}}=A_0+A_1+A_2+\ldots. \qquad (8.56)$$

where A_0 is the value for the reference system, and

$$A_1=\tfrac{1}{2}\rho^2\int\left\langle\left(\frac{\partial u(12)}{\partial a}\right)_{\alpha=0}g_0(\boldsymbol{R}_{12}\omega_1\omega_2)\right\rangle_{\omega_1\omega_2}\mathrm{d}\boldsymbol{R}_1\,\mathrm{d}\boldsymbol{R}_2 \qquad (8.57)$$

$$A_2=\tfrac{1}{4}\rho^2\int\left\langle\left(\frac{\partial^2 u(12)}{\partial a^2}\right)_{\alpha=0}g_0(\boldsymbol{R}_{12}\omega_1\omega_2)\right\rangle_{\omega_1\omega_2}\mathrm{d}\boldsymbol{R}_1\,\mathrm{d}\boldsymbol{R}_2$$

$$+\tfrac{1}{4}\rho^2\int\left\langle\left(\frac{\partial u(12)}{\partial a}\right)_{\alpha=0}g_1(\boldsymbol{R}_{12}\omega_1\omega_2)\right\rangle_{\omega_1\omega_2}\mathrm{d}\boldsymbol{R}_1\,\mathrm{d}\boldsymbol{R}_2 \qquad (8.58)$$

The term of order k in (8.56) corresponds to the term of order $k-1$ in (8.52). Perturbation expansions for other macroscopic properties can be obtained from the appropriate expressions in Section 8.3, together with equation (8.53).

8.7.2 Expansions of Pople[56] and Gubbins and Gray[114]

These authors have taken the reference system to be one in which the molecules interact with an isotropic potential defined by

$$u_0(R_{12})=\langle u(\boldsymbol{R}_{12}\omega_1\omega_2)\rangle_{\omega_1\omega_2} \qquad (8.59)$$

They adopt the parameterisation

$$u(\boldsymbol{R}_{12}\omega_1\omega_2;\,a)=u_0(R_{12})+a\,u_{\mathrm{a}}(\boldsymbol{R}_{12}\omega_1\omega_2) \qquad (8.60)$$

where $u_a(\boldsymbol{R}_{12}\omega_1\omega_2) \equiv u(\boldsymbol{R}_{12}\omega_1\omega_2) - u_0(R_{12})$ is the anisotropic part of the potential. This choice of reference and parameterisation yields the important property

$$\left\langle \left(\frac{\partial u(\boldsymbol{R}_{12}\omega_1\omega_2;\, a)}{\partial a} \right)_{\alpha=0} \right\rangle_{\omega_1\omega_2} = \langle u_a(\boldsymbol{R}_{12}\omega_1\omega_2)\rangle_{\omega_1\omega_2} = 0 \tag{8.61}$$

so that (8.53), (8.57) and (8.58) simplify to[114]:

$$g_1(12) = -\beta u_a(12) g_0(12) - \beta\rho \int \langle u_a(13) + u_a(23)\rangle_{\omega_3} g_0(123)\, \mathrm{d}\boldsymbol{R}_3 \tag{8.62}$$

$$A_1 = 0 \tag{8.63}$$

$$A_2 = \tfrac{1}{4}\rho^2 \int \langle u_a(12) g_1(12)\rangle_{\omega_1\omega_2}\, \mathrm{d}\boldsymbol{R}_1\, \mathrm{d}\boldsymbol{R}_2 \tag{8.64}$$

Similarly

$$A_3 = \tfrac{1}{6}\rho^2 \int \langle u_a(12) g_2(12)\rangle_{\omega_1\omega_2}\, \mathrm{d}\boldsymbol{R}_1\, \mathrm{d}\boldsymbol{R}_2 \tag{8.65}$$

When (8.63) and (8.64) together with (8.62) are substituted into (8.56), Pople's[56] expansion for the free energy is obtained.

For those anisotropic intermolecular potentials which involve spherical harmonics of order $l \neq 0$ in the orientation ω_3 (e.g. all multipolar interactions for neutral molecules), then $\langle u_a(\boldsymbol{R}_{13}\omega_1\omega_3)\rangle_{\omega_3} = 0$ and the last term in (8.62) vanishes; thus the triplet function need not be known in these cases. Equation (8.52) becomes, to first order

$$g(\boldsymbol{R}_{12}\omega_1\omega_2) = g_0(R_{12})[1 - \beta u_a(\boldsymbol{R}_{12}\omega_1\omega_2)] \tag{8.66}$$

Similarly, (8.54) for this case gives $y = y_0$ to first order, i.e.

$$g(\boldsymbol{R}_{12}\omega_1\omega_2) = \exp\,[-\beta u_a(\boldsymbol{R}_{12}\omega_1\omega_2)] g_0(R_{12}) \tag{8.67}$$

For anisotropic potentials without $l = 0$ terms the second order contribution to $g(\boldsymbol{R}_{12}\omega_1\omega_2)$ has also been evaluated[114]. It is:

$$\begin{aligned} g_2(12) = \tfrac{1}{2}\beta^2 \Big\{ & u_a(12)^2\, g_0(12) \\ & + \rho \int \langle [u_a(13) + u_a(23)]^2\rangle_{\omega_3}\, g_0(123)\, \mathrm{d}\boldsymbol{R}_3 \\ & + \tfrac{1}{2}\rho^2 \int \langle u_a(34)^2\rangle_{\omega_3\omega_4} [g_0(1234) - g_0(12)\, g_0(34)]\, \mathrm{d}\boldsymbol{R}_3\, \mathrm{d}\boldsymbol{R}_4 \\ & - \tfrac{1}{2}\frac{\partial}{\partial\rho}[\rho^2 g_0(12)]\, [2 \int \langle u_a(34)^2\rangle_{\omega_3\omega_4}\, g_0(34)\, \mathrm{d}\boldsymbol{R}_{34} \\ & + \rho \int \langle u_a(34)^2\rangle_{\omega_3\omega_4} [g_0(345) - g_0(34)]\, \mathrm{d}\boldsymbol{R}_{34}\, \mathrm{d}\boldsymbol{R}_{45}] \Big\} \end{aligned} \tag{8.68}$$

When there are no $l = 0$ harmonic terms, the expressions for A_2 and A_3 also simplify. From (8.64) and (8.66):

$$A_2 = \tfrac{1}{4}\beta\rho^2 \int \langle u_a(12)^2\rangle_{\omega_1\omega_2}\, g_0(12)\, \mathrm{d}\boldsymbol{R}_1\, \mathrm{d}\boldsymbol{R}_2 \tag{8.69}$$

We notice that A_2 is inherently negative. In fact, one can show[154] that $A \leqslant A_0$ is a rigorous result, valid to all orders. The next higher perturbation term in (8.56) is obtained from (8.65) and (8.68) as

$$A_3 = \tfrac{1}{12}\beta^2\rho^2 \int \langle u_a(12)^3\rangle_{\omega_1\omega_2}\, g_0(12)\, d\boldsymbol{R}_1\, d\boldsymbol{R}_2$$
$$+ \tfrac{1}{6}\beta^2\rho^3 \int \langle u_a(12)u_a(13)u_a(23)\rangle_{\omega_1\omega_2\omega_3}\, g_0(123)\, d\boldsymbol{R}_1\, d\boldsymbol{R}_2\, d\boldsymbol{R}_3 \qquad (8.70)$$

For the special case when the molecules are linear and $u_a(12)$ is just the dipole–dipole term (or any term in the multipole series involving an odd multipole), the quantity $\langle u_a(12)^3\rangle_{\omega_1\omega_2}$ vanishes.

Equations (8.66) and (8.67) have been tested against Monte Carlo results for dipolar and quadrupolar interactions[93]. Better results are obtained using (8.66), which gives good agreement with the Monte Carlo data for values of μ^* $(\equiv\mu/(\varepsilon\sigma^3)^{\frac{1}{2}})$ and Q^* $(\equiv Q/(\varepsilon\sigma^5)^{\frac{1}{2}})$ up to about 0.4.

Perturbation expansions for any given property may be generated by substituting (8.52) in the appropriate expression of Section 8.3, with g_1 given by (8.62), and so on. For a potential of the form (8.3), explicit expressions have been derived for the first non-vanishing contribution to the various thermodynamic functions[116], for the structure factor[16] and for the mean squared torque[69]. Comparisons with experiment are also given in these references. Tests against Monte Carlo results have been made for some of these properties[69,115]. Figure 8.4 shows such a test of the second-order theory (i.e. to the first non-vanishing perturbation term) for the mean squared torque.

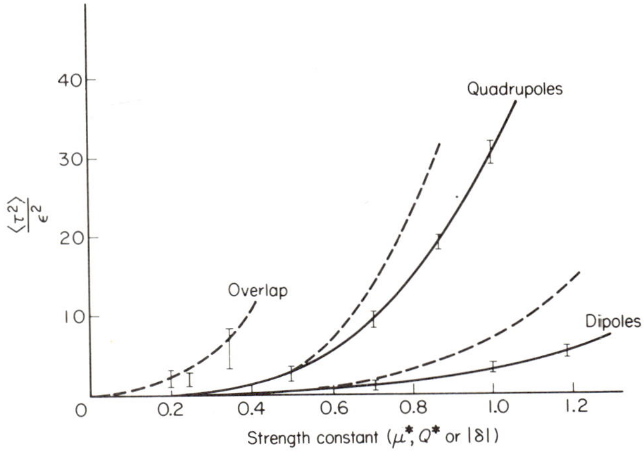

Figure 8.4 Mean squared torques for a liquid of linear molecules, $\rho^* = 0.800$, $T^* = 0.719$, u_0 = Lennard-Jones potential. Points and solid lines are Monte Carlo values, dashed lines are perturbation theory results (to first non-vanishing order) obtained by substituting (8.66) in (8.38). The anisotropic overlap potential used was equation (8.119) (From Twu *et al.*[69] by courtesy of Taylor and Francis.)

Several authors[96-98] have recently studied the second non-vanishing contribution to the free energy, A_3, for a fluid of dipolar molecules interacting with the potential $u = u(000) + u(112)$ [cf. (8.3)];

$$u(\boldsymbol{R}_{12}\omega_1\omega_2) = u_0(R_{12}) + (\mu^2/R_{12}^3)\,[s_1s_2c_{12} - 2c_1c_2] \tag{8.71}$$

where $s_i = \sin\theta_i$, $c_i = \cos\theta_i$, and $c_{12} = \cos\phi_{12}$. Here $\theta_i\phi_i$ refer to the intermolecular axis. With this potential, A_3 is given by the second term of (8.70), with u_a being the dipole–dipole interaction; performing the averaging over orientations in (8.70) yields

$$\langle u_a(12)u_a(13)u_a(23)\rangle_{\omega_1\omega_2\omega_3} = \frac{\mu^6}{9}\left[\frac{1 + 3\cos a_1 \cos a_2 \cos a_3}{(R_{12}R_{13}R_{23})^3}\right] \tag{8.72}$$

where a_1, a_2 and a_3 are the angles of the triangle formed by the centres of molecules 1, 2 and 3. McDonald[119] has carried out Monte Carlo calculations for a dense fluid in which the molecules obey (8.71), with u_0 being the Lennard-Jones (12,6) model, and has compared these results with the Pople expansion terminated at the A_3 term. Some of his results are given in Table 8.1. At this large reduced dipole moment ($\mu^* = 1$) the series terminated at the second order term is in serious error. When the third order term is included the agreement is improved. However, the results are still unsatisfactory in that the predicted density dependence is too weak for both U_a and P_a. It appears that inclusion of the A_3 term in the free energy series provides only a relatively small extension of the range of validity of the theory, and that the series convergence is slow for large values of the anisotropic strength parameter. Thus at $\rho^* = 0.85$, $T^* = 1.15$, $\mu^* = 1$ we have $U_3/U_2 \approx 0.6$. and $P_3/P_2 \approx 0.6$. McDonald[119] has also calculated the effect of polarisability on the thermodynamic properties, and finds that such contributions are appreciable, particularly for the free energy and the pressure (of the order 30% of the dipole–dipole contribution). A similar study has been carried out by Patey and Valleau[96,97], who studied the potential (8.71) but with u_0 being the hard sphere model. They compared Monte Carlo results for such a fluid with the Pople expansion terminated at the A_3 term, and found poor agreement for large dipole moments.

The terms A_2 and A_3 are negative and positive, respectively. In view of the slow convergence for highly anisotropic potentials, Stell *et al.*[118] have suggested the following simple Padé approximation for the free energy, based on the two terms that are known,

$$A = A_0 + A_2\left[\frac{1}{(1 - A_3/A_2)}\right] \tag{8.73}$$

Stell *et al.*[118] have used (8.73) to calculate excess free energies and the liquid–vapour coexistence curve for a Stockmayer fluid with $\mu^* = 1$. This Padé approximation has also been compared with the Monte Carlo results of McDonald[119], Patey and Valleau[96,97] and Verlet and Weis[120]. Some of their results are shown in Tables 8.1 and 8.2. Equation (8.73) is found to give excellent results for most properties for all of the state conditions studied. The greatest discrepancy seems to occur for the internal energy of the Stockmayer fluid (Table 8.1), where values of U_a obtained from (8.73) are too negative by about 17%; however, (8.73) is superior to the truncated

Table 8.1 Test of Pople expansion for the Stockmayer potential

ρ^*	$U_2/N\varepsilon$†	$(U_2 + U_3)/N\varepsilon$ ‡	$(U_a/N\varepsilon)_{Pade}$ §	$(U_a/N\varepsilon)_{mc}$ ‖	$\beta P_2/\rho$ †	$\beta(P_2 + P_3)/\rho$ ‡	$(\beta P_a/\rho)_{Pade}$ §	$(\beta P_a/\rho)_{mc}$ ‖
	$kT/\varepsilon = 1.15$, $\mu^* = \mu/(\varepsilon\sigma^3)^{\frac{1}{2}} = 1.0$							
0.60	−1.13	−0.68	−0.80	−0.66	−0.65	−0.32	−0.41	−0.41
0.70	−1.36	−0.70	−0.90	−0.80	−0.88	−0.38	−0.54	−0.53
0.80	−1.63	−0.70	−1.02	−0.86	−1.14	−0.43	−0.68	−0.59
0.85	−1.79	−0.69	−1.08		−1.29	−0.46	−0.76	

† Second order term obtained by differentiating equation (8.69)
‡ Third order term obtained by differentiating equation (8.70)
§ Padé approximation, from equation (8.73); $U_a = U - U_0$, where U is the configurational internal energy for the system and U_0 is the value for the Lennard-Jones reference
‖ Monte Carlo results; the estimated statistical errors are about 0.08 in $\beta P_a/\rho$ and 0.05 in $U_aN\varepsilon$.

Table 8.2 Contribution of dipole–dipole forces to thermodynamic properties for a fluid of hard spheres with embedded point dipoles at $\rho/\rho_{cp} = 0.59$, where ρ_{cp} is the close-packed density

T_μ^*†	μ/D at $T = 298.15$ K, $d = 3.0$ Å	$-U_a/NkT$ MSM‡	$-U_a/NkT$ Padé§	$-U_a/NkT$ MC	$C_{v,a}/Nk$ MSM	$C_{v,a}/Nk$ Padé	$C_{v,a}/Nk$ MC	$\beta P_a/\rho$ MSM	$\beta P_a/\rho$ Padé	$\beta P_a/\rho$ MC
17.781	0.25	0.0036	0.0060	0.0064 ± 0.0004	0.0035	0.0061	0.0057 ± 0.0003	0.0018	0.0045	−0.03
1.976	0.75	0.226	0.368	0.348 ± 0.006	0.168	0.308	0.238 ± 0.011	0.102	0.265	0.58
0.711	1.25	1.20	1.87	1.79 ± 0.01	0.631	1.160	0.935 ± 0.038	0.468	1.26	1.35
0.363	1.75	3.21	4.64	4.80 ± 0.04	1.288	2.075	2.068 ± 0.083	1.09	2.92	3.08

† $T_\mu^* \equiv d^3kT/\mu^2$, where d is hard sphere diameter
‡ Mean spherical model
§ From (8.73), by applying the usual thermodynamic relations

series in predicting the density dependence of U_a. For hard spheres with embedded point dipoles it is possible to compare (8.73) with the mean spherical model (MSM). As seen from Table 8.2, the MSM gives poor results even for small dipole moments. A detailed comparison of the MSM and (8.73) has recently been carried out for hard spheres with embedded point dipoles by Rushbrooke *et al.*[98]. Their investigation shows that (8.73) does have some theoretical justification. In particular, for a fluid obeying (8.71) it is known[121,122] that A varies as μ^4 for small μ, and as μ^2 in the large μ limit. It is clear from (8.73) that the Padé approximation satisfies both this large and small μ limit, and thus serves as an interpolation formula for intermediate μ values. The Monte Carlo results of Patey and Valleau[96,97] (Table 8.2) appear to confirm this interpretation, since agreement with the Padé results is best at large and small μ values, and poorest in between.

The term $g_1(12)$ of (8.62) vanishes on averaging over orientations because of (8.61). Thus the centres pair correlation function defined in (8.16) is

$$g(R_{12}) = g_0(R_{12}) + \langle g_2(12) \rangle_{\omega_1 \omega_2} + \cdots \tag{8.74}$$

Monte Carlo results[115,119] show that for dipolar and quadrupolar liquids

$$g(R_{12}) \approx g_0(R_{12}) \tag{8.75}$$

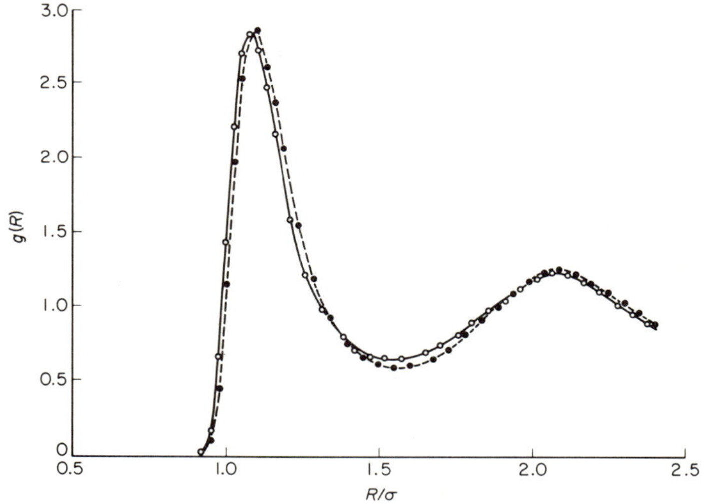

Figure 8.5 Effect of dipole–dipole forces on centre correlation function for linear molecules, $\rho^* = 0.800$, $T^* = 0.719$, u_0 is the Lennard-Jones potential. Solid line is for $\mu^{*2} = 1.4$, dashed lines are for isotropic Lennard-Jones molecules ($\mu^* = 0$) (From Wang *et al.*[115] by courtesy of North Holland Publishing Co.)

even for reduced moments μ^* and Q^* as large as unity (Figure 8.5). This suggests a useful correlation for liquid isothermal compressibilities[123], since from (8.26) and (8.75)

$$kT\rho\chi \approx kT\rho\chi_0 \tag{8.76}$$

8.7.3 Effective isotropic potentials

Any isotropic potential $u(R)$ may be written

$$u(R) = u_0(R) + u_p(R) \tag{8.77}$$

and the free energy and radial distribution function expanded in powers of βu_p to give[4,6]

$$A' = A_0 + A_1' + A_2' + \ldots \tag{8.78}$$

$$g' = g_0 + g_1' + g_2' + \ldots \tag{8.79}$$

(Here the primes serve to differentiate these perturbation terms from the corresponding ones in the Pople expansion). If we now choose u_p to be

$$u_p(R_{12}) = -\tfrac{1}{2}\beta\langle u_a(R_{12}\omega_1\omega_2)^2\rangle_{\omega_1\omega_2} \tag{8.80}$$

where u_a is an anisotropic potential containing no $l = 0$ terms and obeying (8.60) and (8.61), then it is readily shown that (a) A_1' in (8.78) is identical to the Pople A_2 as given by (8.69); (b) g_1' in (8.79) is identical to the Pople $\langle g_2\rangle_{\omega_1\omega_2}$ as given by angle averaging of (8.68). Thus, to first non-vanishing order in the perturbation, the thermodynamics and the centres pair correlation function for a system obeying (8.1) are equivalent (for $l \neq 0$ terms in u_a) to those for a system with the temperature dependent isotropic potential given by (8.77) and (8.80). Perturbation methods that are now well developed for fluids with isotropic potentials, e.g. the Verlet–Weis theory[124], may therefore be used to estimate properties of fluids of anisotropic molecules. Stell *et al.*[117,118] have used this approach to calculate properties of fluids in which u_a is the dipole–dipole, dipole–quadrupole and dipole–octupole contribution.

This approach is particularly useful when the R-dependence of $\langle u_a^2\rangle_{\omega_1\omega_2}$ is the same as that for u_0 (or part thereof); choosing the Lennard-Jones (LJ) potential for u_0, examples of such situations are when u_a is the dipolar interaction ($\langle u_a^2\rangle_{\omega_1\omega_2}$ varies as R^{-6}) or the anisotropic London dispersion interaction ($\langle u_a^2\rangle_{\omega_1\omega_2}$ varies as R^{-12}). Thus, taking u_a to be the dipolar interaction and u_0 the LJ model we have $u_p = -\beta\langle u_a^2\rangle_{\omega_1\omega_2}/2 = -\beta\mu^4/3R^6$, and (8.77) can be rearranged to the LJ form

$$u'(R) = 4\,\varepsilon'\,[(\sigma'/R)^{12} - (\sigma'/R)^6] \tag{8.81}$$

where the potential parameters ε' and σ' are now temperature dependent and are given by

$$\sigma' = \sigma(1 + \chi)^{-1/6}$$

$$\varepsilon' = \varepsilon(1 + \chi)^2$$

where

$$\chi = (\mu^{*4}/12T^*)$$

and $\mu^* = \mu/(\varepsilon\sigma^3)^{\frac{1}{2}}$ and $T^* = kT/\varepsilon$. Since (8.81) is conformal with the Lennard-Jones potential, one should be able to use the corresponding states principle to calculate the properties of weakly polar fluids from the known properties of Lennard-Jones fluids.

Corresponding states correlations based on (8.81), and generalisations of these ideas to more realistic anisotropic potentials, have been developed by

Rowlinson and Cook[2,125-127] and by Stell *et al.*[117,118]. Comparisons with experiment are given in the book by Rowlinson[2]. Poole and Aziz[128] have applied this method to velocity of sound data for molecular fluids.

8.7.4 Expansion of Sung and Chandler[129]

These authors consider a fluid having purely repulsive intermolecular pair potentials $u_{\text{rep}}(\boldsymbol{R}_{12}\omega_1\omega_2)$, and expand the function $y(\boldsymbol{R}_{12}\omega_1\omega_2)$ for this system about the value $y_d(R_{12})$ for a fluid of hard spheres of diameter d. This method is essentially a generalisation of the Weeks–Chandler–Andersen perturbation theory[130], which gives excellent results for monatomic liquids[124]. Sung and Chandler[129] base their perturbation theory on a functional Taylor series[131], which they call the 'blip function' expansion. However, an alternative derivation[113] of their result is to make the following choice for $u(\boldsymbol{R}_{12}\omega_1\omega_2; a)$ in the equations of Section 8.7.1:

$$\exp[-\beta u(\boldsymbol{R}_{12}\omega_1\omega_2;a)] = \exp[-\beta u_d(R_{12})] + a\{\exp[-\beta u_{\text{rep}}(\boldsymbol{R}_{12}\omega_1\omega_2)] - \exp[-\beta u_d(R_{12})]\} \tag{8.82}$$

Here $u_d(R_{12})$ is the pair potential for hard spheres of diameter d. If (8.82) is used in (8.57) we see that A_1 vanishes provided that d is chosen to satisfy

$$\int \langle \exp[-\beta u_{\text{rep}}(\boldsymbol{R}_{12}\omega_1\omega_2)] - \exp[-\beta u_d(R_{12})]\rangle_{\omega_1\omega_2}\, y_d(R_{12})\,\mathrm{d}\boldsymbol{R}_{12} = 0 \tag{8.83}$$

Equation (8.54) is used to obtain $g(\boldsymbol{R}_{12}\omega_1\omega_2)$. The expression for y_1 may be obtained from (8.55), (8.53) and (8.82). However, Sung and Chandler assume that the zeroth order term in (8.54) is sufficient; this approximation is found to give good results for monatomic liquids[113]. Thus, their equation is

$$g_{\text{rep}}(\boldsymbol{R}_{12}\omega_1\omega_2) = \exp[-\beta\, u_{\text{rep}}(\boldsymbol{R}_{12}\omega_1\omega_2)]y_d(R_{12}) \tag{8.84}$$

with d given by (8.83).

In order to apply this perturbation scheme to fluids with some general pair potential $u(\boldsymbol{R}_{12}\omega_1\omega_2)$, having both attractive and repulsive regions, Sung and Chandler put

$$g(\boldsymbol{R}_{12}\omega_1\omega_2) \approx g_{\text{rep}}(\boldsymbol{R}_{12}\omega_1\omega_2) \tag{8.85}$$

in analogy with the atomic case, where it is called the 'high temperature approximation'. In (8.85), g_{rep} is for a fluid with molecules interacting with the repulsive part of $u(\boldsymbol{R}_{12}\omega_1\omega_2)$, i.e. with

$$\begin{aligned} u_{\text{rep}}(\boldsymbol{R}_{12}\omega_1\omega_2) &= u(\boldsymbol{R}_{12}\omega_1\omega_2) + \varepsilon(\omega_1\omega_2) \qquad & R_{12} < R_{\text{min}}(\omega_1\omega_2) \\ &= 0 & R_{12} > R_{\text{min}}(\omega_1\omega_2) \end{aligned} \tag{8.86}$$

where $\varepsilon(\omega_1\omega_2)$ is the potential well depth, which occurs at the separation $R_{\text{min}}(\omega_1\omega_2)$ for the orientations ω_1 and ω_2. The equation analogous to (8.85) for monatomic liquids gives good results near the triple point[113,124,130]. Equations (8.85) and (8.84) give poor results for strong quadrupolar forces, as

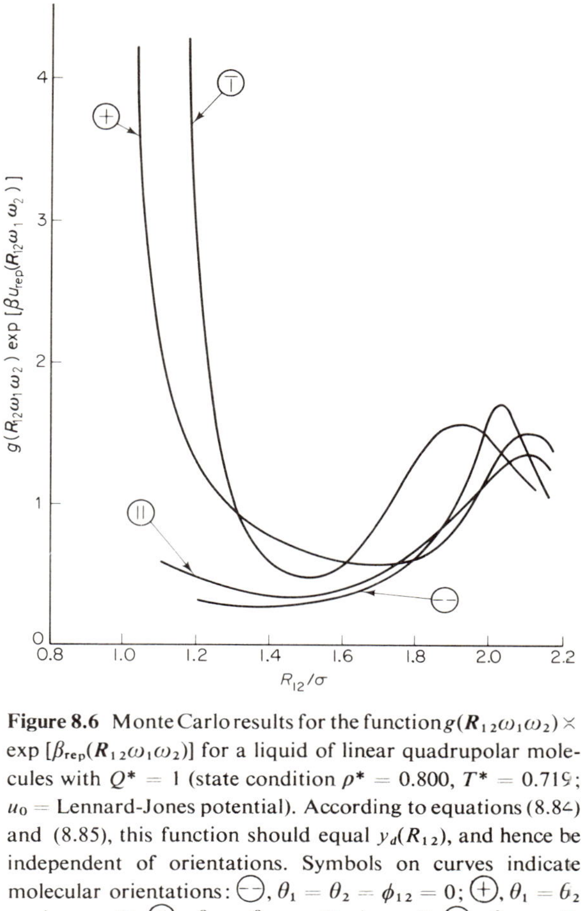

Figure 8.6 Monte Carlo results for the function $g(\boldsymbol{R}_{12}\omega_1\omega_2) \times \exp[\beta_{rep}(\boldsymbol{R}_{12}\omega_1\omega_2)]$ for a liquid of linear quadrupolar molecules with $Q^* = 1$ (state condition $\rho^* = 0.800$, $T^* = 0.719$; u_0 = Lennard-Jones potential). According to equations (8.84) and (8.85), this function should equal $y_d(R_{12})$, and hence be independent of orientations. Symbols on curves indicate molecular orientations: ⊖, $\theta_1 = \theta_2 = \phi_{12} = 0$; ⊕, $\theta_1 = \theta_2 = \phi_{12} = \pi/2$; (II), $\theta_1 = \theta_2 = \pi/2$, $\phi_{12} = 0$; (T), $\theta_1 = \pi/2$, $\theta_2 = \phi_{12} = 0$ (From Wang *et al.*[93], by courtesy of North Holland Publishing Co.)

shown in Figure 8.6. Recently, Steele and Sandler[132] have compared (8.84) with the solution of the PY equation for rigid dumbells, and found good agreement. Their results, in conjunction with those shown in Figure 8.6, suggest that the Sung–Chandler expansion works well for short-range repulsive anisotropic forces, but that (8.85) breaks down when long-range attractive anisotropic forces are present.

In addition to the expansion of Sung and Chandler, other perturbation methods based on cluster expansions have been considered by Chandler and Andersen[133-135]; these have been applied to fluids in which (a) the molecules consist of rigid spheres in which are embedded interaction sites, and (b) the molecules consist of several rigidly connected hard spheres, with perturbation interactions between sites on different molecules.

8.7.5 Expansion of Bellemans[136]

The methods considered above all involve an expansion in powers of the strength of the perturbing potential. Such an approach is unsuitable for a fluid of rigid non-spherical molecules (of the type considered in Section 8.6). Bellemans[136] has proposed that for such fluids the expansion be made in terms of the anisotropy of the rigid core; the reference fluid is one of hard spheres. His expansion may be derived from the equations of Section 8.7.1 by introducing a pair potential $u(12)$*:

$$\begin{aligned} u(\boldsymbol{R}_{12}\omega_1\omega_2; a) &= 0 \qquad R_{12} > d(\omega_1\omega_2; a) \\ &= \infty \qquad R_{12} < d(\omega_1\omega_2; a) \end{aligned} \tag{8.87}$$

where $d(\omega_1\omega_2;\ a)$ is the shortest distance of approach of the centres for two molecules with orientations ω_1 and ω_2; $a = 1$ corresponds to the full potential. The following parameterisation is now introduced

$$d(\omega_1\omega_2; a) = d_0 + a\,\gamma(\omega_1\omega_2)d_0 \tag{8.88}$$

where d_0 is a hard sphere (reference fluid) diameter defined by

$$d_0 \equiv \langle d(\omega_1\omega_2)\rangle_{\omega_1\omega_2} \tag{8.89}$$

and $\gamma(\omega_1\omega_2) = [d(\omega_1\omega_2) - d_0]/d_0$. From (8.89),

$$\langle \gamma(\omega_1\omega_2)\rangle_{\omega_1\omega_2} = 0 \tag{8.90}$$

Thus $u(\boldsymbol{R}_{12}\omega_1\omega_2;\ a = 0) = u_0(R_{12})$, the potential for hard spheres of diameter d_0. From (8.87)–(8.90) it is readily shown that

$$\left(\frac{\partial u(12)}{\partial a}\right)_{\alpha=0} = \frac{d_0}{\beta}\exp\left[\beta u_0(R_{12})\right]\delta(R_{12} - d_0)\,\gamma(\omega_1\omega_2) \tag{8.91}$$

$$\left\langle\left(\frac{\partial u(12)}{\partial a}\right)_{\alpha=0}\right\rangle_{\omega_1\omega_2} = 0 \tag{8.92}$$

$$\left(\frac{\partial^2 u(12)}{\partial a^2}\right)_{\alpha=0} = \beta\left(\frac{\partial u(12)}{\partial a}\right)^2_{\alpha=0} - \frac{d_0^2}{\beta}\exp\left[\beta u_0(12)\right]\delta'(R_{12} - d_0)\gamma(\omega_1\omega_2)^2 \tag{8.93}$$

where $\delta'(R_{12} - d_0)$ is the derivative of the Dirac delta function.

From (8.53) and (8.91) we obtain

$$\begin{aligned} g_1(\boldsymbol{R}_{12}\omega_1\omega_2) &= -d_0 g_0(R_{12})\delta(R_{12} - d_0)\,\gamma(\omega_1\omega_2) \\ &\quad - \rho d_0 \int [\delta(R_{13} - d_0)\,\langle\gamma(\omega_1\omega_3)\rangle_{\omega_3} \\ &\quad + \delta(R_{23} - \mathrm{d}_0)\langle\gamma(\omega_2\omega_3)\rangle_{\omega_3}]g_0(R_{12}R_{13}R_{23})\,\mathrm{d}\boldsymbol{R}_3 \end{aligned} \tag{8.94}$$

* We assume a coordinate system in which the polar axis is along R_{12}; thus $d(\omega_1\omega_2;\ a)$ is independent of the direction of $\boldsymbol{R}_{12}$.

However, (8.52) is unlikely to give a satisfactory description of $g(\boldsymbol{R}_{12}\omega_1\omega_2)$ because of the discontinuity at $d(\omega_1\omega_2)$. For spherically symmetric repulsive forces it is known that the expansion of $y(R)$ gives good results[113,124,137] and one might therefore expect (8.54) to give better results than (8.52) for rigid non-spherical molecules. From (8.54), (8.55) and (8.94), the $y(\boldsymbol{R}_{12}\omega_1\omega_2)$ expansion of (8.54) gives, to first order:

$$g(\boldsymbol{R}_{12}\omega_1\omega_2) = \exp\{-\beta[u(\boldsymbol{R}_{12}\omega_1\omega_2) - u_0(R_{12})]\}g_0(R_{12})$$
$$-\rho d_0\int[\delta(R_{13} - d_0)\,\langle\gamma(\omega_1\omega_3)\rangle_{\omega 3}$$
$$+\,\delta(R_{23} - d_0)\langle\gamma(\omega_2\omega_3)\rangle_{\omega 3}]g_0(R_{12}R_{13}R_{23})\,\mathrm{d}\boldsymbol{R}_3 \qquad (8.95)$$

Equations (8.94) and (8.95) have not yet been tested.

From (8.92) and (8.57) it is seen that $A_1 = 0$. The first non-vanishing perturbation term in the free energy expansion (8.56) is A_2; from (8.58), (8.91) and (8.93):

$$\frac{A_2}{NkT} = \pi\rho d_0^2\langle\gamma(\omega_1\omega_2)^2\rangle_{\omega_1\omega_2}\left\{\frac{\partial}{\partial R_{12}}\,[R_{12}^2 y_0(R_{12})]\right\}_{R_{12}=d_0}$$
$$-\,\tfrac{1}{2}\rho^2 d_0^2\int\langle\gamma(\omega_1\omega_2)\gamma(\omega_1\omega_3)\rangle_{\omega_1\omega_2\omega_3}\,\delta(R_{12} - d_0)\,\times$$
$$\delta(R_{13} - d_0)g_0(R_{12}R_{13}R_{23})\,\mathrm{d}\boldsymbol{R}_{12}\,\mathrm{d}\boldsymbol{R}_3 \qquad (8.96)$$

The only calculations using this theory seem to be those made by Bellemans[136] for prolate ellipsoids. He obtained $g_0(R_{12})$ and $g_0(R_{12}R_{13}R_{23})$ by Monte Carlo calculation for two densities in the dense fluid region and one in the solid region; the perturbation correction to A_0 was calculated from (8.96) as:

$$\begin{aligned}(A_2/NkT\varepsilon^2) &= 0.21 \pm 0.03 \text{ for } \rho^* = 0.7725 \text{ (fluid)}\\ &= 0.38 \pm 0.08 \quad\ \ \rho^* = 0.8369 \text{ (fluid)}\\ &= 4.56 \pm 0.09 \quad\ \ \rho^* = 1.0000 \text{ (solid)}\end{aligned}$$

Here $\varepsilon = (a - b)/b$, where a and b are the principal axes ($a > b$). Thus, for small ε, e.g. $\varepsilon = 0.1$, the perturbation term A_2/NkT is negligible (of the order 0.002) in the fluid region, so that $A^c = A_0$ is a good approximation. The perturbation term is larger for solids, however.

8.7.6 Non-spherical reference potentials

For strongly anisotropic pair potentials it seems unlikely that *any* choice of spherically symmetric reference potential will give good convergence. Thus for a liquid of quadrupolar molecules with $Q^* = 1$ the first peak in $g(\boldsymbol{R}_{12}\omega_1\omega_2)$ for the most probable orientation exceeds that for the Lennard-Jones fluid by more than a factor of 5 (see Figure 8.8). Mo and Gubbins[138] have therefore suggested using a reference fluid of rigid non-spherical molecules interacting with

$$\begin{aligned}u_0(\boldsymbol{R}_{12}\omega_1\omega_2) &= 0 \qquad R_{12} > d(\omega_1\omega_2)\\ &= \infty \qquad R_{12} < d(\omega_1\omega_2)\end{aligned} \qquad (8.97)$$

The potential for the real system, $u(\boldsymbol{R}_{12}\omega_1\omega_2)$, is divided into repulsive and attractive regions, and these are treated by separate expansions, as in the Weeks–Chandler–Anderson[130] theory. When first order terms are retained in each of these expansions, the resulting equation for the free energy is

$$A = A_0 + 2\pi\rho N\int_0^{\infty} \mathrm{d}R_{12}\, R_{12}^2 \langle \exp[-\beta u_{\mathrm{rep}}(\boldsymbol{R}_{12}\omega_1\omega_2)]y_0(\boldsymbol{R}_{12}\omega_1\omega_2) \times u_{\mathrm{p}}(\boldsymbol{R}_{12}\omega_1\omega_2)\rangle_{\omega_1\omega_2} \tag{8.98}$$

where $d(\omega_1\omega_2)$ in (8.97) has been chosen to satisfy[130]

$$\int_0^{\infty} \mathrm{d}\boldsymbol{R}_{12}R_{12}^2 \{\exp[-\beta u_{\mathrm{rep}}(\boldsymbol{R}_{12}\omega_1\omega_2)]-\exp[-\beta u_0(\boldsymbol{R}_{12}\omega_1\omega_2)]\}y_0 \times (\boldsymbol{R}_{12}\omega_1\omega_2) = 0 \tag{8.99}$$

and

$$\begin{aligned} u_{\mathrm{p}}(\boldsymbol{R}_{12}\omega_1\omega_2) &= -\,\varepsilon(\omega_1\omega_2) && R_{12} \leqslant R_{\mathrm{min}}(\omega_1\omega_2) \\ &= u(R_{12}\omega_1\omega_2) && R_{12} > R_{\mathrm{min}}(\omega_1\omega_2) \end{aligned} \tag{8.100}$$

while $u_{\mathrm{rep}}(\boldsymbol{R}_{12}\omega_1\omega_2)$ is given by (8.86). Here $\varepsilon(\omega_1\omega_2)$ and $R_{\mathrm{min}}(\omega_1\omega_2)$ are the magnitude and separation distance respectively of $u(R_{12}\omega_1\omega_2)$ at the minimum of the pair potential (for fixed $\omega_1\omega_2$). The values of A_0 and y_0 in (8.98) may be estimated from the Belleman's expansion, using (8.96) and (8.95), respectively; alternatively, scaled particle theory could be used for A_0.

This expansion apparently gives good results for both dipolar and quadrupolar liquids even for quite large values of reduced multipole moments. Some comparisons with Monte Carlo data are shown in Table 8.3.

Table 8.3 Contribution of anisotropic forces to configurational internal energy for a liquid with potential $u_0 + u_a$, where u_0 is the Lennard-Jones potential†

Type of anisotropy	*Anisotropic strength*	*Monte Carlo‡* $U_a/N\varepsilon$	*Equation* (8.98)
Dipole	$(\mu^*)^2 = 0.5$	-0.38 ± 0.13	-0.373
	$(\mu^*)^2 = 1.0$	-1.18 ± 0.12	-0.884
Quadrupole	$(Q^*)^2 = 0.25$	-0.32 ± 0.11	-0.310
	$(Q^*)^2 = 0.5$	-0.99 ± 0.12	-0.915
	$(Q^*)^2 = 0.75$	-1.97 ± 0.16	-1.652

† All values at $\rho\sigma^3 = 0.800$, $kT/\varepsilon = 0.719$, where ε and σ are Lennard-Jones parameters
‡ From Wang *et al.*[115]

Sandler[139] has also recently proposed the use of a reference system of non-spherical molecules, using a different perturbation scheme from the above. He has carried out some preliminary calculations for molecules interacting with a potential $u_0 + u_a$, where u_0 is the potential for rigid diatomic molecules composed of two overlapping hard spheres, and u_a is the dipole or quadrupole interaction. Properties of the reference system are estimated from the Sung–Chandler[129] theory.

8.7.7 Generalised van der Waals model

In the theory of atomic liquids the van der Waals model has been used successfully for many years to interpret thermodynamic properties, especially the equation of state. This model is readily derived from equations of the type used in previous sections by writing $g_0(12) \approx 1$, $g_0(123) \approx 1$, etc.—that is, we assume that terms involving the fluctuations of the g_0 functions may be neglected in comparison with those involving the mean value of the g_0 function. Thus the model assumes that the range of the perturbation potential is long compared with the hard core.

Consider a potential $u(\boldsymbol{R}_{12}\omega_1\omega_2;\, a)$ as in (8.50) and (8.51), with the parameterisation $u(12;a) = u_0(12) + au_p(12)$. We further choose the separation so that $u_p(12)$ vanishes for small separations, $R_{12} < \sigma(\omega_1\omega_2)$ (corresponding to separations for which $g_0 \approx 0$). When this potential is used in (8.53), (8.57) and (8.58), and all g_0 functions are put equal to unity, we have

$$A_1 = \tfrac{1}{2}\rho N \int \langle u_p(12)\rangle_{\omega_1\omega_2}\, \mathrm{d}\boldsymbol{R}_{12} \tag{8.101}$$

$$\begin{aligned} A_2 = & -\tfrac{1}{4}\beta\rho N \int \langle u_p(12)^2\rangle_{\omega_1\omega_2}\, \mathrm{d}\boldsymbol{R}_{12} \\ & -\tfrac{1}{2}\beta\rho^2 N \int \langle u_p(12)\, u_p(13)\rangle_{\omega_1\omega_2\omega_3}\, \mathrm{d}\boldsymbol{R}_{12}\, \mathrm{d}\boldsymbol{R}_3 \\ & + \tfrac{1}{2}\beta\rho^2 N \left[\int \langle u_p(12)\rangle_{\omega_1\omega_2}\, \mathrm{d}\boldsymbol{R}_{12}\right]^2 \end{aligned} \tag{8.102}$$

where $u_p(ij) = u_p(\boldsymbol{R}_{ij}\omega_i\omega_j)$. In the usual van der Waals treatment this series is terminated at A_1, because in the limit $u_p \ll kT$, the higher terms can be neglected and because, in practice, the error introduced by the approximation $g_0 \approx 1$ may be of similar size to the higher terms. If this were done here we would find the familiar result:

$$A = A_0 + Na_1\rho \tag{8.103}$$

where $a_1 = \frac{1}{2}\int \langle u_p(12)\rangle_{\omega_1\omega_2}\, \mathrm{d}\boldsymbol{R}_{12}$. This is a generalisation of the usual formula. For this case the anisotropy appears only in the term A_0 (i.e. hard core anisotropy), and if the anisotropic part of u_p is small this is a reasonable result.

However, if the anisotropic part of u_p is large, the leading anisotropic term in A_2 may be of the same size as $a_1\rho N$ and should be included. This term may be found by writing:

$$\langle u_p\rangle_{\omega_1\omega_2} = u_{ip};\ u_p - u_{ip} = u_{ap} \tag{8.104}$$

and taking the case $\langle u_{ap}^2\rangle_{\omega_1\omega_2} \gg u_{ip}^2$. From equation (8.102) we find:

$$\begin{aligned} A_2 = & -(\beta\rho N/4)\int \langle u_{ap}(12)^2\rangle_{\omega_1\omega_2}\, \mathrm{d}\boldsymbol{R}_{12} \\ & -(\beta\rho^2 N/2)\int \langle u_{ap}(12)\, u_{ap}(13)\rangle_{\omega_1\omega_2\omega_3}\, \mathrm{d}\boldsymbol{R}_{12}\, \mathrm{d}\boldsymbol{R}_{13} + 0\,(u_{ip}^2) \end{aligned} \tag{8.105}$$

The second term vanishes for spherical harmonics of order $l \neq 0$, and in the general case should be small compared to the first term. We therefore arrive at the corrected van der Waals expression:

$$A = A_0 + N(a_1 - \beta a_2)\rho \tag{8.106}$$

where $a_2 = \frac{1}{4}\int \langle u_{ap}(12)^2\rangle_{\omega_1\omega_2}\, d\boldsymbol{R}_{12}$.

From this result one finds:

$$P = P_0 + (a_1 - \beta a_2)\rho^2 \tag{8.107}$$

Several choices are possible for the reference system. If we make the Pople choice for u_0 [*cf.* (8.59)], then A_0 and P_0 are for a system of spherical molecules, and the constant a_1 vanishes, giving

$$P = P_0 - \beta a_2 \rho^2 \tag{8.108}$$

In such an equation P_0 could be obtained from corresponding states relationships for atomic fluids. Equation (8.108) might be a suitable approximation for fluids having a potential $u_0 + u_a$, where u_0 is the Lennard-Jones potential and u_a is a sum of multipole interactions.

Another approach is to choose a non-spherical reference potential, e.g. convex rigid bodies [*cf.* equation (8.87)]. If the A_2 term in the free energy expansion is neglected, (8.107) then becomes:

$$P = P_0 - a_1 \rho^2 \tag{8.109}$$

where P_0 is for the system of rigid convex bodies. This equation of state has been studied by Rigby[111], who uses scaled particle theory [equation (8.49)] for P_0. Similar ideas have been used by Kihara *et al.*[140] to study the solubility of gases in liquids.

While the usefulness of (8.103) in the study of atomic systems is well established, the usefulness of the corrected equation (8.106) has not been examined for general molecules.*

8.8 COMPUTER SIMULATION RESULTS

The most direct method for evaluating the theories of the previous sections is comparison with computer simulation results (by the Monte Carlo or molecular dynamics techniques) for a fluid of non-spherical molecules. Relatively few such studies have yet been reported for simple polyatomic fluids, although such studies are currently in progress by several groups.

Vieillard-Baron *et al.*[141,142] have studied a two-dimensional system of hard ellipses using the Monte Carlo method. In addition to the usual fluid–solid phase transition, they also found another first order transition from a disoriented phase to one in which the orientation of the ellipses showed long-range order. For small eccentricities this second phase transition occurred in the solid range. For larger eccentricities it occurred in the fluid range; the oriented fluid is thus analogous to a nematic liquid crystal. Few and Rigby[106] have studied the equation of state for a system of hard prolate spherocylinders. Their results were shown in Figure 8.2. Vieillard-Baron[143] has also studied this system recently.

Studies of fluids in which linear molecules interact with an atom–atom potential are in progress, and some preliminary results have been reported.

* Mo and Gubbins have recently applied this generalised van der Waals approach to CO_2, using a potential which accounts for the large quadrupole moment.

McDonald and Singer[144] have simulated liquid nitrogen by the Monte Carlo method using an atom–atom Lennard-Jones potential model. Barojas *et al.*[38] have made a detailed molecular dynamics study of a system of 500 homonuclear diatomic molecules interacting via an atom–atom Lennard-Jones potential. They calculate the pressure, configurational energy, atom–atom pair correlation functions and static structure factors (and also time-dependent properties) for nine different state conditions. With appropriate choices for the Lennard-Jones parameters, the results are found to be in good agreement with experimental data for nitrogen. O'Brien[145] has reported a molecular dynamics study for nitrogen dissolved in argon, using an atom–atom potential; he obtains the mean squared torque, in addition to angular time correlation functions.

Wang *et al.*[93,115] have made a Monte Carlo study of a liquid in which the pair potential is of the type (1) with u_0 being the Lennard-Jones potential. For u_a they considered point dipole, point quadrupole, and anisotropic overlap interactions; the model used for anisotropic overlap consisted of the first two terms in a spherical harmonic expansion [$\Lambda = 202 + 022$ in (8.3), with ε given by (8.119)]. They calculated the angular pair correlation function

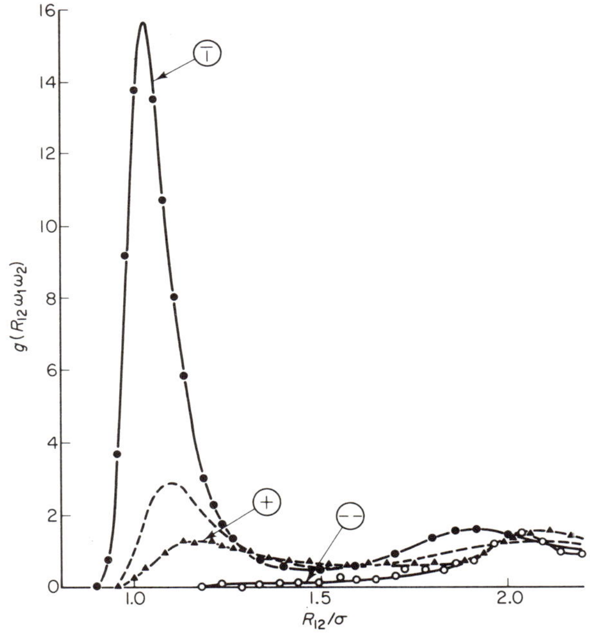

Figure 8.7 Monte Carlo results for the angular pair correlation function for linear quadrupolar molecules, $Q^* = 1$. Dashed line is $g_0(R_{12})$ for the Lennard-Jones fluid, while points and solid curves are results for the fluid of quadrupolar molecules. Intermolecular potential, state condition, and symbols are the same as in Figure 8.6 (From Wang *et al.*[93], by courtesy of North Holland Publishing Co.)

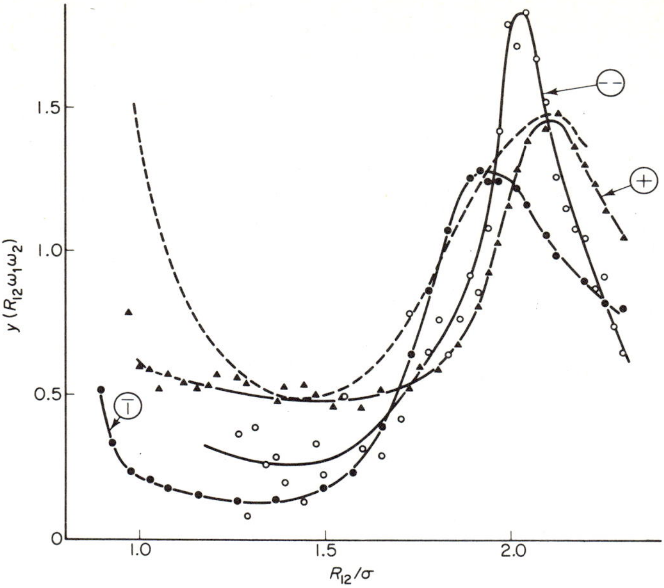

Figure 8.8 Monte Carlo results for $y(\boldsymbol{R}_{12}\omega_1\omega_2)$, for the same fluid as shown in Figure 8.6 (From Wang *et al.*[93], by courtesy of North Holland Publishing Co.)

$g(\boldsymbol{R}_{12}\omega_1\omega_2)$, the centres correlation function, and various equilibrium properties for several values of the strength of u_a in each case. Typical results are shown in Figures 8.7 and 8.8 (see also Figures 8.4 and 8.5 and Table 8.3). Recently McDonald[119], Patey and Valleau[96,97], and Verlet and Weis[120] have all carried out computer simulation studies for the thermodynamic properties of a fluid with potential $u_0 + u_a$, where u_a is the point dipole interaction; u_0 is either the Lennard-Jones[119,120] or hard sphere[96,97,120] interaction. Some of their results have been given in Tables 8.1 and 8.2. Harp and Berne[146] have used the molecular dynamics method to simulate CO and N_2 using a Stockmayer potential.

Rahman and Stillinger[147-149] have made a molecular dynamics simulation of liquid water, using an effective pair potential consisting of a Lennard-Jones part plus an orientation-dependent part based on a point charge model. They calculate both equilibrium and non-equilibrium properties at several temperatures. Their main conclusions are: (a) there is no evidence of the formation of clusters, 'icebergs', or of partition of the molecules into two or more classes; (b) the liquid structure at 307 K consists of a network; (c) as the temperature rises the order breaks down, and there is a rapid increase in freedom of molecular motions; (d) the average binding energy per H seems to be about 2.5 kcal mol^{-1}; (e) there is no need to introduce the idea of a hydrogen bond, as simple pair potential ideas work satisfactorily. Monte Carlo studies of some properties of water molecules have been reported in Refs. 150 and 171.

8.9 EXPERIMENTAL STUDIES

In this section we review briefly the comparison between theory and experiment for a number of equilibrium properties. For some (e.g. virial coefficients) comparisons via the intermolecular potential have been made for many years, while for others (e.g. the structure factor) such comparisons are comparatively recent. Many of the theories discussed in the previous sections are either of recent origin or have been developed fully only in recent years, and therefore have been compared to experimental data in a limited way so far. Thus although in some cases there is an extensive experimental literature (e.g. equation of state) there are often few theoretical data to compare. Examples have been chosen, occasionally from among several possible, to illustrate a variety of equilibrium properties.

8.9.1 Virial coefficients

We give two examples to illustrate the properties (molecular shape and multipolar interaction) of interest in this review. (a) Fluids of hard (non-spherical) objects. Chen and Steele[78] have used the Percus–Yevick approximation to calculate the virial coefficients for hard 'dumbells'. A 'dumbell' consists of two spheres of diameter σ and centres a distance R apart. Rigby[111] has compared scaled particle theory to calculate virial coefficients for hard spherocylinders (diameter σ and length R), with Monte Carlo calculations. Table 8.4 is a summary of these results. It can be seen that the third coefficient is given to better than 1% but the fourth coefficient is overestimated by ~3% even at the relatively small value of $R/\sigma = 0.4$. This discrepancy in the SPT theory increases at larger R/σ values being ~10% at 0.8, while the PY results of Chen and Steele fall below the Monte Carlo data at $R/\sigma = 0.6$. The latter may be due to difficulties in applying the PY theory.

Table 8.4 Virial coefficients for hard objects

*Coefficient**	R/σ	*Spheres*	P/σ	*Dumbells*[78] *PY*	*Dumbells*[78] *MC*	*Spherocylinders*[111] *SPT*	*Spherocylinders*[111] *MC*
B_2	0	1.000	0.4	1.0528	—	1.0375	—
B_3	0	0.625	0.4	0.679	0.684	0.663	0.665
B_4	0	0.287	0.4	0.323	0.318	0.318	0.301

	R/σ	*Dumbells*[78] *PY*	*Dumbells*[78] *MC*	R/σ	*Spherocylinders*[111] *SPT*	*Spherocylinders*[111] *MC*
B_2	0.6	1.1185	—	0.8	1.1091	—
B_3	0.6	0.750	0.757	0.8	0.738	0.740
B_4	0.6	0.359	0.359	0.8	0.361	0.336

* B is given exactly by both theories; non-dimensional units are used

(b) Real fluids.
The magnitude of multipole moments may be obtained from a study of the second virial coefficient as a function of temperature. In Table 8.5 we summarise the data of Spurling *et al.*[155] on methane and Datta and Singh[156] on carbon dioxide who compared calculated second virial coefficients with experimental values.

Table 8.5 Experimental and calculated second virial coefficients*

	Methane[155]			*Carbon Dioxide*[156]	
Temp./K	*Values of B_2/cm^3 mol^{-1}* Expt.	*Calc.*	*Temp./K*	*Values of B_2/cm^3 mol^{-1}* Expt.	*Calc.*
108.45	−365.0	−357.0	265.4	−156.6	−154.9
149.1	−188.0	−188.5	321.1	−102.2	−101.6
223.6	−82.6	−84.4	396.5	−59.3	−59.6
273.16	−53.4	−54.1	449.5	−38.5	−41.9
373.16	−21.3	−21.8	570.9	−13.8	−15.8
473.16	−4.0	−5.2			
573.16	+6.8	+4.9			

* $\Omega = \pm 3.7 \times 10^{-34}$ e.s.u. $Q = \pm 4.33 \times 10^{-26}$ e.s.u.

It can be seen that there is fair agreement. The uncertainty in the values of these moments arises from uncertainties in the relative sizes of the various terms in the potential: a fit similar to that in the table may be obtained with somewhat different parameters for the central and anisotropic parts of the potential. For this reason the error in such values of Q and Ω may be ~20%.

8.9.2 Equation of state, configurational energy and entropy

The bulk phase thermodynamic functions have been described in Section 8.3.2, and the perturbation theory expressions for them were given in Section 8.7. As an example of the application of these results we consider nitrogen along the vapour pressure curve and also at high pressure. Table 8.6 shows a comparison between theory and experiment[157,158] for a number of properties. This table is taken from the work of Ananth *et al.*[116], who also made similar comparisons for oxygen and methane. For nitrogen and oxygen anisotropic terms involving multipole, dispersion and overlap contributions were considered, while for methane only the octupole term was included. The overall agreement is very satisfactory, although a large contribution arises in these cases from the zero order (argon-like) term in the perturbation series; at the triple point the anisotropic contribution to the energy is ~7% and for the pressure it is ~50%. The largest disagreement in this table occurs near the critical point and this may be due to the non-analyticity of the critical point or to errors in the method of evaluating the theoretical expressions.

Table 8.6 Comparison of calculated thermodynamic functions of dense fluid N_2 with experiment[116]

Compressed liquid

T/K	$\tilde{\rho}$/mol l^{-1}	P/atm Expt.	P/atm Calc.	U^c/NkT Expt.	U^c/NkT Calc.	S^r/Nk* Expt.	S^r/Nk* Calc.
90	26.63	10	12.73	−6.146	−6.106	−3.005	−2.916
	26.89	30	32.48	−6.212	−6.164	−3.060	−2.966
	27.14	50	53.07	−6.274	−6.219	−3.109	−3.015
	27.69	100	104.2	−6.415	−6.341	−3.222	−3.124
	28.59	200	206.8	−6.632	−6.537	−3.398	−3.311
	29.96	400	414.5	−6.924	−6.818	−3.658	−3.611
	30.97(a)	600	613.4	−7.118	−7.007	−3.884	−3.843

Saturated liquid

T/K	$\tilde{\rho}$/mol l^{-1} Expt.	$\tilde{\rho}$/mol l^{-1} Calc.	P/atm Expt.	P/atm Calc.	U^c/NkT Expt.	U^c/NkT Calc.	S^r/Nk* Expt.	S^r/Nk* Calc.
120	18.80	19.24	24.84	23.03	—	−3.23	—	−1.68
110	22.18	22.29	14.52	13.39	−4.12	−4.09	−2.20	−2.09
100	24.57	24.57	7.72	7.13	−5.06	−5.01	−2.63	−2.48
90	26.65	26.50	3.53	3.32	−6.11	−6.08	−2.97	−2.89
80	28.39	28.23	1.35	1.26	−7.41	−7.39	−3.37	−3.35
70	29.97	29.82	0.38	0.35	−9.11	−9.13	−3.90	−3.94

* Superscript r means the residual property, defined as the value of the dense fluid at N, V and T minus the ideal gas value at N, V and T

8.9.3 Mean orientation coefficients (equations 8.26, 8.37 and 8.39)

A set of coefficients involving the weighting of $g(\boldsymbol{R}_{12}\omega_1\omega_2)$ in a number of different ways has been described under the physical names for such coefficients (e.g. compressibility, mutual polarisation, mutual alignment and mean square torque). Experimental values for them compared to perturbation theory predictions will be discussed in this section.

The isothermal compressibility for a number of liquids along their saturated vapour pressure curves is shown in Figure 8.9 (taken from Brelvi and O'Connell[159]). Gubbins and O'Connell[160] have pointed out that the similarity between the values for different liquids shown in this diagram arises because only the centres correlation function appears in equations (8.75) and (8.76). At a high density the size of a molecule is sufficient to determine $g(R)$ to a first approximation, and therefore if the value of the molecular diameter, σ, is determined from the molecular volume the compressibility should scale with $\rho\sigma^3$. This result confirms that the zero order term in the perturbation expansion of Section 8.7.2 is the only term of importance for this property.

In Section 8.3.6 it was pointed out that an experimental value of the mutual polarisation G_1 could be obtained from the static dielectric constant, while the mutual alignment G_2 can be obtained from measurements of optical birefringence. Molecules with a permanent dipole moment are frequently studied for G_1 while molecules with anisotropic polarisabilities

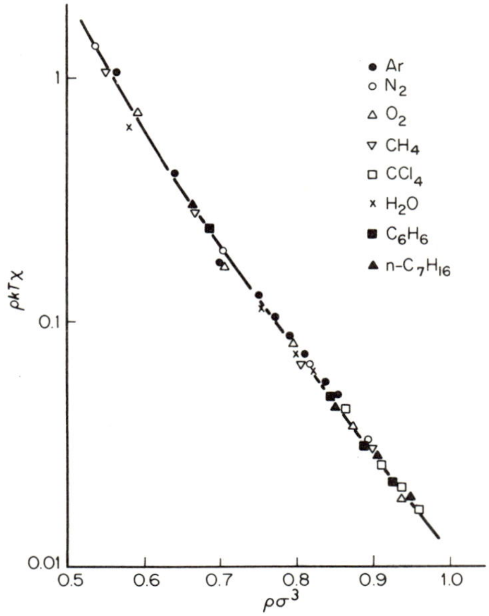

Figure 8.9 Isothermal compressibility of saturated liquids; the σ values were derived from characteristic molecular volumes (From Brelvi and O'Connell[159], by courtesy of the American Institute of Chemical Engineers)

Table 8.7 Experimental values of orientational correlation coefficients for various molecular liquids at 25 °C

Liquid	G_1*	G_2†
NH_3	+0.65	
NMe_3	+0.5	
MeCl	−0.1	
PhCl	−0.4	
C_6H_6		−0.45
CS_2		−0.3
n-Hexane		−0.2
Experimental errors ~20%		

* Ref. 1

† Ref. 161

Table 8.8 Values of mean squared torques for saturated liquids, in units of 10^4 cm^{-2}

Liquid	*Temp./*K	*I.r. value*	*Raman value*	*Calculated value*
Carbon monoxide	80	6.9*		13.7
Oxygen	88		8.0†	0.95
Methane	90	1.4†	2.2†	3.25
Methane	117	8.2‡		3.15

* Ref. 162
† Ref. 163
‡ Ref. 164

are frequently studied for G_2. This means that values of both coefficients are not generally available for the same molecular liquid. Table 8.7 gives data on these quantities for a number of molecular liquids, and through the use of empirical models (e.g. Ref. 161) estimates for G_1 and G_2 in broad agreement with experiment may be obtained. The zero and first order terms in the perturbation theory of Gubbins and Gray for $g(\boldsymbol{R}_{12}\omega_1\omega_2)$ with anisotropic multipolar forces vanish for both coefficients. It is necessary to use a second order theory to obtain a non-vanishing result and this does not seem to have been carried through as yet*. It would be useful also to study the effect of molecular shape (hard core anisotropy) on these coefficients.

While the former properties (χ, G_1,G_2) are strongly related to either the zero or second term in the perturbation series for $g(12)$, we now turn to a property—the mean square torque—strongly related to the first term in this series. We consider the expression (8.38) for $\langle\tau^2\rangle$, with $g_1(12)$ from (8.62) replacing $g(12)$ since the $g_0(12)$ term makes no contribution. Ananth *et al.*[116] have compared calculations based on this model with values of $\langle\tau^2\rangle$ deduced from infrared and Raman band moments. Table 8.8 is a summary of their results. Both the experimental and calculated entries in this table have large errors—the error in the calculation arising, for example, from uncertainties in the multipole moments. Agreement to about a factor of two may be considered satisfactory in these cases. Thus oxygen is the only case where the lack of agreement is serious, and this may be due either to neglect of the hexadecapole term in the calculation or because of experimental uncertainties.

8.9.4 The liquid structure factor

This property was discussed in Section 8.3.4, and several of the theories described in Sections 8.5 and 8.7 have been used to evaluate it for a simple diatomic molecule. Several authors have shown that the anisotropic terms make a small contribution to $S(Q)$, relative to isotropic terms, in the cases of nitrogen and chlorine. Ananth *et al.*[116] compared the Pople, Gubbins, Gray perturbation theory to experimental neutron diffraction data at the triple point; Lowden and Chandler[135] compared the RISM model results to the Monte Carlo data for a two-centre model of nitrogen by Barojas *et al.*[38]. Morrison[91] used the PY model to calculate the structure factor of chlorine,

* This has now been done[173].

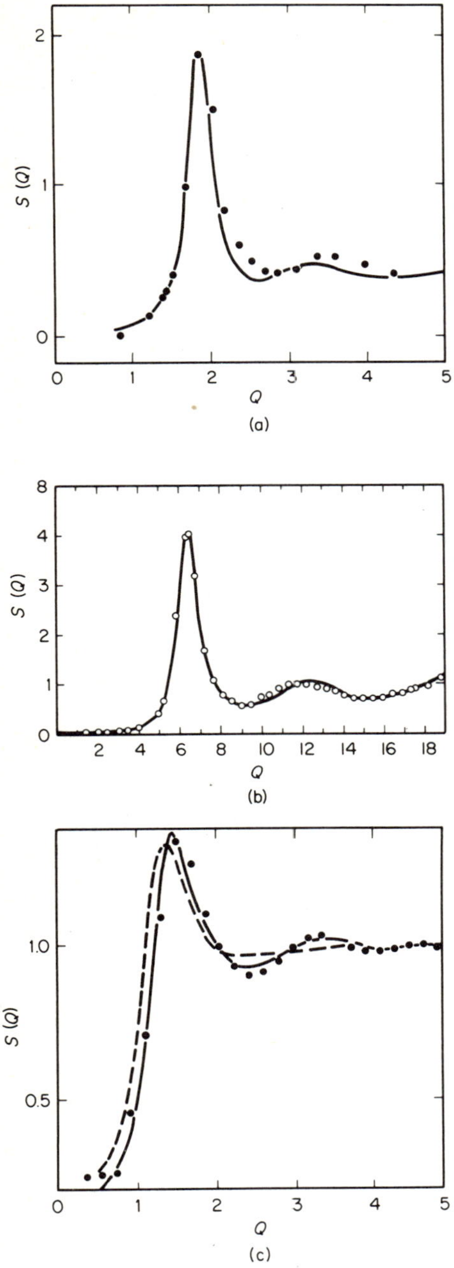

Figure 8.10 Several calculations for the structure factor of a simple molecular liquid (N_2 or Cl_2).
(a) The full line is the Pople, Gubbins, Gray theory (Ref. 116) and the points are the experimental values at the triple point of nitrogen. In this figure, $S(Q)$ per molecule is plotted compared with $S(Q)$ per atom in figures (b) and (c).
(b) The solid line is obtained from the RISM model, and the circles are the molecular dynamics results (Ref. 38) for a thermodynamic state which is near the triple point of liquid N_2, $\rho\sigma^3 = 0.696$ and $(\beta\varepsilon)^{-1} = 1.46$.
(c) The dashed line is the free rotation model, the points are the PY calculations of Morrison[91] and the full line is the Sung–Chandler theory (for $L = 0.50\sigma$) evaluated by Sandler *et al.*[37]. The state is for Cl_2 with $(\beta\varepsilon)^{-1} = 0.75$ and $\rho\sigma^3 = 0.3$

which is quite similar to that for nitrogen, and Sandler *et al.*[37] evaluated the Sung–Chandler theory for this case. These results are shown in Figures 8.10(a)–(c). It can be seen that all methods give a fair description of the data, and that the anisotropic terms are small. It would be useful to compare the predictions with one another in an attempt to identify significant differences. However, these differences appear to be of the same order as the errors in the calculations.

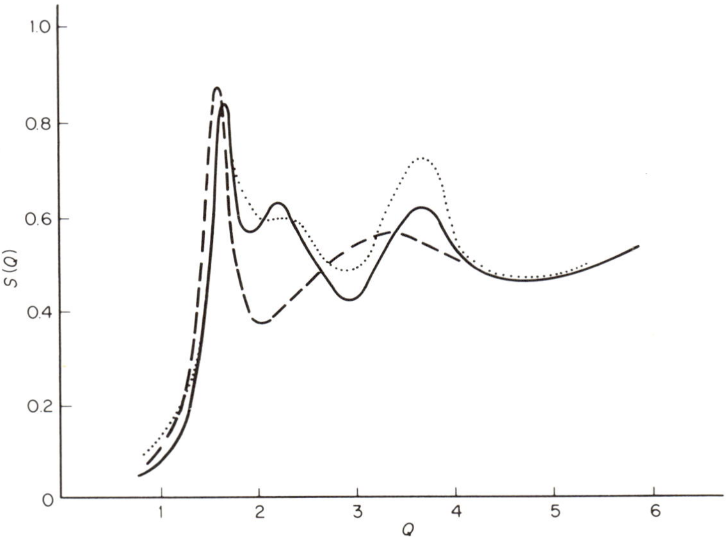

Figure 8.11 The structure factor for liquid bromine (Refs. 16 and 165): the dotted line is experimental, the full line is the perturbation theory and the dashed line is the free rotation model

A better case is that of bromine because it has a larger quadrupole moment, and the anisotropic term is greater. Gubbins *et al.*[16] have compared the predictions of the perturbation theory with the experimental data of Ascarelli and Cagliotti[165], as shown in Figure 8.11. A comparison of Figures 8.10 and 8.11 verifies the greater size of the anisotropic term for bromine, which is given by the difference between full line and the dashed line. Also it can be seen that the zero and first order terms of this theory are sufficient to give a reasonable approximation to the data, although some small differences remain.

8.10 CONCLUSIONS

The most successful theoretical approach to properties of molecular fluids at the present time is through the perturbation expansions discussed in Section 8.7. During the period when this review was written these methods were in a state of particularly rapid development. For highly anisotropic molecules the two most promising approaches so far are the Padé approximation to the Pople expansion[118], equation (8.73), and the perturbation expansion about a non-spherical reference potential[138]. These and similar methods will be the subject of considerable study in the near future. The scaled particle theory, although limited in application to thermodynamic properties for fluids of rigid bodies, is also useful since such fluids may be very suitable as reference substances. In view of the significant computational effort needed to evaluate terms in these perturbation expansions, approximate versions of these theories may also be useful. While the effective potential concept (Section 8.7.3) has been widely used, van der Waals models for molecular fluids (Section 8.7.7) have been little studied, and warrant more attention.

8.11 APPENDIX: THE INTERACTION COEFFICIENTS $\varepsilon(\Lambda;n_1n_2;R_{12})$

The interaction (8.3) is manifestly invariant under simultaneous rotation of ω_1, ω_2 and ω_{12}. Since u is real the expansion coefficient ε satisfies

$$\varepsilon(\Lambda;\underline{n}_1\underline{n}_2;R) = (-)^{l_1+l_2+l+n_1+n_2}\,\varepsilon(\Lambda;n_1n_2;R)^* \tag{8.110}$$

where $\underline{n} = -n$. For identical molecules, (8.3) is invariant under $\omega_1 \rightarrow \omega_2$, $\omega_2 \rightarrow \omega_1$, $\omega_{12} \rightarrow -\omega_{12}$, leading to

$$\varepsilon(l_1l_2l;n_1n_2;R) = (-)^{l_1+l_2}\,\varepsilon(l_2l_1l;n_2n_1;R) \tag{8.111}$$

The coefficients ε satisfy further conditions[30,31] for molecules of symmetrical shape. For example, for linear molecules $\varepsilon(\Lambda;n_1n_2;R) = 0$ unless $n_1 = n_2 = 0$. From this fact and the relation[24]

$$D^l_{m0}(\phi\theta\chi)^* = [4\pi/(2l+1)]^{\frac{1}{2}}\,Y_{lm}(\theta\phi)$$

(8.3) reduces to a sum over three ordinary spherical harmonics. For linear molecules one also finds from parity considerations that $\varepsilon(\Lambda;00;R) = 0$ unless $(l_1 + l_2 + l)$ is even. For symmetrical linear molecules (e.g. CO_2) $\varepsilon = 0$ unless l_1 and l_2 are even. Conditions on the ε coefficients for tetrahedral (e.g. CH_4) and octahedral (e.g. SF_6) molecules are given in Refs. 16, 30 and 31 [*cf.* also (8.113) and (8.114) below].

Listed below are some of the expansion coefficients for a pair of identical molecules of simple shape[16].

Multipolar, linear:

$$\varepsilon_{\mathrm{mult}}(112;00;R) = -2\,(6\pi/5)^{\frac{1}{2}}\,(\mu^2/R^3) \tag{8.112}$$

$$\varepsilon_{\mathrm{mult}}(224;00;R) = \tfrac{2}{3}\,(70\pi)^{\frac{1}{2}}\,(Q^2/R^5)$$

Multipolar, tetrahedral:

$$\varepsilon_{\mathrm{mult}}(336;22;R) = -\,\varepsilon_{\mathrm{mult}}(336;\underline{2}2;R) = (168/5)\,(33\pi/91)^{\frac{1}{2}}\,(\Omega^2/R^7) \tag{8.113}$$

Multipolar, octahedral:

$$\varepsilon_{\mathrm{mult}}(448;00;R) = (14/5)^{\frac{1}{2}}\,\varepsilon_{\mathrm{mult}}(448;40;R) = 6\,(1430\pi/17)^{\frac{1}{2}}\,(\Phi^2/R^9) \tag{8.114}$$

Dispersion, linear:

$$\varepsilon_{\mathrm{dis}}\,(202;00;R) = -8\,(\pi/5)^{\frac{1}{2}}\,\varepsilon\kappa(\sigma/R)^{6} \tag{8.115}$$

$$\varepsilon_{\mathrm{dis}}\,(224;00;R) = -48\,(2\pi/35)^{\frac{1}{2}}\,\varepsilon\kappa^{2}(\sigma/R)^{6} \tag{8.116}$$

Dispersion, tetrahedral

$$\varepsilon_{\mathrm{dis}}\,(303;20;R) = \mathrm{i}\,4(4\pi/7)^{\frac{1}{2}}\,\beta_{7}\,\varepsilon(\sigma/R)^{7} \tag{8.117}$$

Overlap linear:

$$\varepsilon_{\mathrm{over}}\,(101;00;R) = 4\,(4\pi/3)^{\frac{1}{2}}\,\delta_{1}\varepsilon(\sigma/R)^{12} \tag{8.118}$$

$$\varepsilon_{\mathrm{over}}\,(202;00;R) = 16\,(\pi/5)^{\frac{1}{2}}\,\delta_{2}\varepsilon(\sigma/R)^{12} \tag{8.119}$$

Overlap, tetrahedral:

$$\varepsilon_{\mathrm{over}}\,(303;20;R) = \mathrm{i}4(4\pi/7)^{\frac{1}{2}}\,\delta_{3}\varepsilon(\sigma/R)^{12} \tag{8.120}$$

Overlap, octahedral:

$$\varepsilon_{\mathrm{over}}(404;00;R) = 4(4\pi/9)^{\frac{1}{2}}\,(14/5)^{\frac{1}{2}}\,\delta_{4}\varepsilon(\sigma/R)^{12} \tag{8.121}$$

In the above relations, μ, Q, Ω and Φ are the dipole, quadrupole, octupole, and hexadecapole moments respectively as defined in Ref. 20, $\kappa = (a_{\parallel} - a_{\perp})/(a_{\parallel} + 2a_{\perp})$ is the anisotropy of the polarisability, and the δ_i are dimensionless overlap parameters. In (8.117), $\beta_7 = -8A/(30)^{\frac{1}{2}}a\sigma$ where $a = (a_{\parallel} + 2a_{\perp})/3$ is the mean polarisability, and A is the quadrupole polarisability[13,39]. The overlap potentials above are model potentials chosen for simplicity. As discussed in Ref. 16, there are restrictions on the magnitudes of the δ_i to ensure that the total overlap potential be positive.

The other non-vanishing coefficients for these cases can be obtained from those written down together with the symmetry properties (8.110) and (8.111).

Acknowledgements

It is a pleasure to thank I. R. McDonald, J. P. Valleau, G. Stell, J. Veillard-Baron, S. I. Sandler and L. Verlet for sending results prior to publication. A helpful discussion with George Stell concerning the Padé approximation is gratefully acknowledged, and discussions with R. L. Henderson and S. S. Wang have also been helpful.

References

1. Buckingham, A. D. (1967). *Discuss. Faraday Soc.*, **43,** 205
2. Rowlinson, J. S. (1969). *Liquids and Liquid Mixtures*, 2nd. Ed., Chap. 8 (London: Butterworths)
3. Ben-Naim, A. and Stillinger, F. H. (1972). *Water and Aqueous Solutions* (R. A. Horne, editor) 295 (Wiley: New York); Ben-Naim, A. (1972). *Water. A Comprehensive Treatise. Volume 1. The Physics and Physical Chemistry of Water*, (F. Franks, editor) 413 (New York: Plenum)
4. Reed, T. M. and Gubbins, K. E. (1973). *Applied Statistical Mechanics* (New York: McGraw–Hill)

5. Deutch, J. M. (1973). *Ann. Rev. Phys. Chem.*, **24,** 301
6. (1973). *Specialist Periodical Report in Statistical Mechanics*, Vol. 1 (Singer, K. editor) (London: Chemical Society)
7. Friedman, H. (1962). *Adv. Chem. Phys.*, **4,** 225
8. For a review, see e.g. Robin, M. B. (1968). *Simple Dense Fluids*, 215 (H. L. Frisch and Z. W. Salsburg, editors) (New York: Academic Press)
9. Certain, P. R. and Bruch, L. W. (1972). *MTP International Review of Science, Physical Chemistry, Series One*, Vol. 1, *Theoretical Chemistry* (W. Byers Brown, editor) (London: Butterworths)
10. Torrens, M. (1972). *Interatomic Potentials* (New York: Academic Press)
11. Kihara, T. (1970). *Physical Chemistry, An Advanced Treatise.* Volume V, *Valency*, 663 (H. Eyring, editor) (New York: Academic Press)
12. Margenau, H. and Kestner, N. R. (1969). *Theory of Intermolecular Forces* (Oxford: Pergamon Press)
13. (1967). *Intermolecular Forces* (J. O. Hirschfelder, editor) (New York: Wiley)
14. Buckingham, A. D. and Utting, B. D. (1970). *Ann. Rev. Phys. Chem.*, **21,** 287
15. Mason, E. A. and Spurling, T. H. (1969). *The Virial Equation of State* (Oxford: Pergamon Press)
16. Gubbins, K. E., Gray, C. G., Egelstaff, P. A. and Ananth, M. S. (1973). *Mol. Phys.*, **25,** 1353
17. Krauss, M. and Mies, F. H. (1965). *J. Chem. Phys.*, **42,** 2503
18. Roberts, C. S. (1963). *Phys. Rev.*, **131,** 203
19. Buckingham, A. D. (1970). *Physical Chemistry*, Vol. 4, Chap. 8 (D. Henderson, editor) (New York: Academic Press)
20. Stogryn, D. E. and Stogryn, A. D. (1966). *Mol. Phys.*, **11,** 371
21. Stogryn, D. E. (1971). *Molec. Phys.*, **22,** 81
22. Casanova, G., Dulla, R. J., Jonah, D. A., Rowlinson, J. S. and Saville, G. (1970). *Mol. Phys.*, **18,** 589
23. Dulla, R. J., Rowlinson, J. S. and Smith, W. R. (1971). *Mol. Phys.*, **21,** 299
24. Rose, M. E. (1957). *Elementary Theory of Angular Momentum* (New York: Wiley)
25. Gordon, M. D. and Secrest, D. (1970). *J. Chem. Phys.*, **52,** 120
26. Sanghaff, P., Gordon, R. G., and Karlus, M. (1971). *J. Chem. Phys.*, **54,** 2126
27. Kochanski, E. (1973). *J. Chem. Phys.*, **58,** 5823
28. Morris, J. M. (1973). *Aust. J. Chem.*, **26,** 649
29. Eisenberg, D. and Kauzmann, W. (1969). *The Structure and Properties of Water* (Oxford: University Press)
30. Gray, C. G. (1968). *Can. J. Phys.*, **46,** 135
31. Armstrong, R. L., Blumenfeld, S. M. and Gray, C. G. (1968). *Can. J. Phys.*, **46,** 1331
32. Gray, C. G. and Van Kranendonk. (1966). *Can. J. Phys.*, **44,** 2411
33. Stillinger, F. H. (1972). *J. Chem. Phys.*, **57,** 1750
34. de Boer, J. (1942). *Physica*, **9,** 363
35. Brout, R. (1954). *J. Chem. Phys.*, **22,** 934
36. Herzfeld, K. F. and Litovitz, T. A. (1959). *Absorption and Dispersion of Ultrasonic Waves*, 303 (New York: Academic Press)
37. Sandler, S. I., Das Gupta, A. and Steele, W. A. (1974). *J. Chem. Phys.*, **61,** 1326
38. Barojas, J., Levesque, D. and Quentrec, B. (1973). *Phys. Rev.*, **7,** 1092
39. Buckingham, A. D. (1968). *J. Chem. Phys.*, **48,** 3827
40. Sweet, J. R. and Steele, W. A. (1967). *J. Chem. Phys.*, **47,** 3029
41. Berne, B. J. and Forster, D. (1971). *Ann. Rev. Phys. Chem.*, **22,** 563
42. Berne, B. J., Pechulsas, P. and Harp, G. (1968). *J. Chem. Phys.*, **49,** 3125
43. Leanas, V. B. (1973). *Usp. Khim.*, **15,** 266
44. Talsazowagi, K. (1963). *Suyopl. Progr. Theoret. Phys.*, **25**
45. Lischa, H. (1973). *Chem. Phys. Lett.*, **20,** 448
46. Sweet, J. R. and Steele, W. A. (1968). *J. Chem. Phys.*, **50,** 668
47. De Rocco, A. G. and Hoover, W. G. (1962). *J. Chem. Phys.*, **36,** 916
48. Corner, J. (1948). *Proc. Roy. Soc.*, **A192,** 275
49. Chapman, S. and Cowling, T. G. (1970). *Mathematical Theory of Nonuniform Gases*, 3rd ed, 217 (Cambridge: University Press)
50. Berne, B. J. and Pechukas, P. (1972). *J. Chem. Phys.*, **56,** 4213
51. Wertheim, M. S. (1971). *J. Chem. Phys.*, **55,** 4291

51a. Adelman, S. A. and Deutch, J. M. (1973). *J. Chem. Phys.*, **59**, 3971
52. Watts, R. O. (1972). *Molec. Phys.*, **23**, 445
53. Blum, L. (1972). *J. Chem. Phys.*, **57**, 1862
54. Blum, L. (1973). *J. Chem. Phys.*, **58**, 3295
55. Ben-Naim, A. (1971). *J. Chem. Phys.*, **54**, 3682
56. Pople, G. A. (1954). *Proc. Roy. Soc.*, **A221**, 498
57. Buckingham, A. D. and Graham, C. (1971). *Molec. Phys.*, **22**, 335
58. Ramshaw, J. D., Schaefer, D. W., Waugh, J. S. and Deutch, J. M. (1971). *J. Chem. Phys.*, **54**, 1239
59. Lebowitz, J. L. and Percus, J. K. (1961). *Phys. Rev.*, **122**, 1675
60. Egelstaff, P. A., Gray, C. G. and Gubbins, K. E. (1971). *Phys. Lett.*, **37A**, 321
61. Kirkwood, J. G. and Buff, F. P. (1949). *J. Chem. Phys.*, **17**, 338
62. Buff, F. P. (1952). *Z. Elektrochem.*, **56**, 311
63. Gray, C. G. and Gubbins, K. E. (1975). *Molec. Phys.*, in press
64. Fowler, R. H. (1937). *Proc. Roy. Soc.*, **A159**, 229
65. Sears, V. F. (1966). *Can. J. Phys.*, **44**, 1279, 1299
66. Egelstaff, P. A., Page, D. I. and Powles, J. C. (1971). *Molec. Phys.*, **20**, 881
67. Steele, W. A. and Pecora, R. (1965). *J. Chem. Phys.*, **42**, 1863
68. Narten, A. H. (1972). *J. Chem. Phys.*, **56**, 5681
69. Twu, C. H., Gray, C. G. and Gubbins, K. E. (1975). *Molec. Phys.*, **29**, 713
70. Kirkwood, J. G. (1939). *J. Chem. Phys.*, **7**, 911
71. Kielich, S. (1967). *J. Chem. Phys.*, **46**, 4090
72. Kirkwood, J. G. and Buff, F. P. (1951). *J. Chem. Phys.*, **19**, 774
73. Buff, F. P. and Brout, R. (1955). *J. Chem. Phys.*, **23**, 458
74. Buff, F. P. (1955). *J. Chem. Phys.*, **23**, 419
75. O'Connell, J. P. (1971). *Molec. Phys.*, **20**, 27
76. Gray, C. G., Gubbins, K. E. and Wang, S. S. (1974). *Chem. Phys. Lett.*, **26**, 610
77. Gray, C. G. (1971). *J. Phys. B*, **4**, 1661
78. Chen, Y. D. and Steele, W. A. (1969). *J. Chem. Phys.*, **50**, 1428; **52**, 5284
79. Wulf, A. (1971). *J. Chem. Phys.*, **55**, 4512
80. Onsager, L. (1949). *Ann. N.Y. Acad. Sci.*, **51**, 627
81. Zwanzig, R. (1963). *J. Chem. Phys.*, **39**, 1714
82. Lakatos, K. (1970). *J. Stat. Phys.*, **2**, 121
83. Runnels, L. K. and Colvin, C. (1970). *J. Chem. Phys.*, **53**, 4219; Hubbard, J. B. and Runnels, L. K. (1972). *J. Chem. Phys.*, **56**, 536
84. Brenner, S. L., McQuarrie, D. A. and Olivares, D. (1973). *J. Chem. Phys.*, **59**, 2596
85. Wulf, A. and De Rocco, A. G. (1971). *J. Chem. Phys.*, **55**, 12
86. Stell, G. (1964). *The Equilibrium Theory of Classical Fluids* (H. L. Frisch and J. L. Lebowitz, editors) (New York: Benjamin)
87. Percus, J. K. and Yevick, G. J. (1958). *Phys. Rev.*, **110**, 1
88. Rowlinson, J. S. (1965). *Molec. Phys.*, **9**, 217; **10**, 533; (1967). *Discuss. Faraday Soc.*, **43**, 243
89. Stell, G. (1963). *Physica*, **29**, 517
90. Chen, Y. D. and Steele, W. A. (1971). *J. Chem. Phys.*, **54**, 703
91. Morrison, P. F. (1972). *Ph.D. Thesis*, California Institute of Technology
92. Ben-Naim, A. (1970). *J. Chem. Phys.*, **52**, 5531
93. Wang, S. S., Egelstaff, P. A., Gray, C. G. and Gubbins, K. E. (1974). *Chem. Phys. Lett.*, **24**, 453
94. Ben-Naim, A. (1972). *Molec. Phys.*, **24**, 705, 723
95. Lebowitz, J. L. and Percus, J. K. (1966). *Phys. Rev.*, **144**, 251
96. Patey, G. N. and Valleau, J. P. (1973). *Chem. Phys. Lett.*, **21**, 297
97. Patey, G. N. and Valleau, J. P. (1974). *J. Chem. Phys.*, in the press
98. Rushbrooke, G. S., Stell, G. and Hoye, J. S. (1973). *Molec. Phys.*, **26**, 1199
99. Maier, W. and Saupe, A. (1959). *Z. Naturforsch.*, **14a**, 882; (1960). *ibid.*, **15a**, 287
100. Workman, H. and Fixman, M. (1973). *J. Chem. Phys.*, **58**, 5024
101. Reiss, H., Frisch, H. L. and Lebowitz, J. L. (1959). *J. Chem. Phys.*, **31**, 369
102. Lebowitz, J. L., Helfand, E. and Praestgaard, E. (1965). *J. Chem. Phys.*, **43**, 774
103. Gibbons, R. M. (1969). *Molec. Phys.*, **17**, 81
104. Gibbons, R. M. (1970). *Molec. Phys.*, **18**, 809
105. Kihara, T. (1953). *J. Phys. Soc. Jap.*, **8**, 686

106. Few, G. A. and Rigby, M. (1973). *Chem. Phys. Lett.*, **20,** 433
107. Vieillard-Baron, J. (1974). *Molec. Phys.*, **28,** 809
108. Cotter, M. A. and Martire, D. E. (1970). *J. Chem. Phys.*, **52,** 1902; 1909
109. Cotter, M. A. and Martire, D. E. (1970). *J. Chem. Phys.*, **53,** 4500
110. Lasher, G. (1970). *J. Chem. Phys.*, **53,** 4141
111. Rigby, M. (1972). *J. Phys. Chem.*, **76,** 2014
112. Buff, F. P. and Schindler, F. M. (1958). *J. Chem. Phys.*, **29,** 1075
113. Gubbins, K. E., Smith, W. R., Tham, M. K. and Tiepel, E. W. (1971). *Molec. Phys.*, **22,** 1089
114. Gubbins, K. E. and Gray, C. G. (1972). *Molec. Phys.*, **23,** 187
115. Wang, S. S., Gray, C. G., Egelstaff, P. A. and Gubbins, K. E. (1973). *Chem. Phys. Lett.*, **21,** 123
116. Ananth, M. S., Gubbins, K. E. and Gray, C. G. (1974). *Molec. Phys.*, **28,** 1005
117. Stell, G., Rasaiah, J. C. and Narang, H. (1972). *Molec. Phys.*, **23,** 393
118. Stell, G., Rasaiah, J. C. and Narang, H. (1974). *Molec. Phys.*, **27,** 1393
119. McDonald, I. R. (1974). *J. Phys. C*, **7,** 1225
120. Verlet, L. and Weis, J. J. (1974). *Molec. Phys.*, **28,** 665
121. Onsager, L. (1939). *J. Phys. Chem.*, **43,** 189
122. Stell, G. (1974). Personal communication
123. Gubbins, K. E. and O'Connell, J. P. (1974). *J. Chem. Phys.*, **60,** 3449
124. Verlet, L. and Weis, J. J. (1972). *Phys. Rev. A*, **5,** 939
125. Cook, D. and Rowlinson, J. S. (1953). *Proc. Roy. Soc.*, **A219,** 405
126. Rowlinson, J. S. (1954). *Trans. Faraday Soc.*, **50,** 647
127. Rowlinson, J. S. (1955). *Trans. Faraday Soc.*, **51,** 1317
128. Poole, G. R. and Aziz, R. A. (1972). *Can. J. Phys.*, **50,** 721
129. Sung, S. and Chandler, D. (1972). *J. Chem. Phys.*, **56,** 4989
130. Weeks, J. D., Chandler, D. and Andersen, H. C. (1971). *J. Chem. Phys.*, **54,** 5237
131. Andersen, H. C., Weeks, J. D. and Chandler, D. (1971). *Phys. Rev. A*, **4,** 1597
132. Steele, W. A. and Sandler, S. I. (1974). *J. Chem. Phys.*, **61,** 1315
133. Chandler, D. and Andersen, H. C. (1972). *J. Chem. Phys.*, **57,** 1930
134. Chandler, D. (1973). *J. Chem. Phys.*, **59,** 2742
135. Lowden, L. J. and Chandler, D. (1973). *J. Chem. Phys.*, **59,** 6587
136. Bellemans, A. (1968). *Phys. Rev. Lett.*, **21,** 527
137. Wang, S. S., Egelstaff, P. A. and Gubbins, K. E. (1973). *Molec. Phys.*, **25,** 461
138. Mo, K. C. and Gubbins, K. E. (1974). *Chem. Phys. Lett.*, **27,** 144; (1975). *J. Chem. Phys.*, in press
139. Sandler, S. I. (1974). *Molec. Phys.*, **28,** 1207
140. Kihara, T. and Jhon, M. S. (1970). *Chem. Phys. Lett.*, **7,** 559; Kihara, T., Yamazaki, K., Jhon, M. S. and Kim, U. R. (1971). *ibid.*, **9,** 62; Kihara, T. and Yamazaki, K. (1971). *ibid.*, **11,** 62
141. Levesque, D., Schiff, D. and Vieillard-Baron, J. (1969). *J. Chem. Phys.*, **51,** 3625
142. Vieillard-Baron, J. (1972). *J. Chem. Phys.*, **56,** 4729
143. Vieillard-Baron, J. (1974). Preprint
144. McDonald, I. and Singer, K. (1973). *Chem. Brit.*, **9,** 54
145. O'Brien, E. F. (1973). *Molec. Phys.*, **26,** 453
146. Harp, G. D. and Berne, B. J. (1970). *Phys. Rev. A*, **2,** 975
147. Rahman, A. and Stillinger, F. H. (1971). *J. Chem. Phys.*, **55,** 3336
148. Stillinger, F. H. and Rahman, A. (1972). *J. Chem. Phys.*, **57,** 1281
149. Stillinger, F. H. and Rahman, A. (1974). *J. Chem. Phys.*, **60,** 1545
150. Barker, J. A. and Watts, R. O. (1973). *Molec. Phys.*, **26,** 789
151. Yarkony, D. R., O'Neill, S. V., Schaefer, H. F., III, Baskin, C. P. and Bender, C. F. (1974). *J. Chem. Phys.*, **60,** 855
152. Patch, R. W. (1973). *J. Chem. Phys.*, **59,** 6468
153. Tsapline, B. and Kutzelnigg, W. (1973). *Chem. Phys. Lett.*, **23,** 173
154. Gray, C. G. and Henderson, R. L. (1975). *Molec. Phys.*, in press
155. Spurling, T. H., de Rocco, A. G. and Storvick, T. S. (1968). *J. Chem. Phys.*, **48,** 1006
156. Datta, K. K. and Singh, Y. (1971). *J. Chem. Phys.*, **55,** 3541
157. Din, F. (1961). *Thermodynamic Function for Gases*, Vol. 1–3 (London: Butterworths)
158. Frisch, H. L. and Salsberg, Z. W. (1968). *Simple Dense Fluids* (New York: Academic Press)

159. Brelvi, S. W. and O'Connell, J. P. (1972). *Amer. Inst. Chem. Eng.*, **18,** 1239
160. Gubbins, K. E. and O'Connell, J. P. (1974). *J. Chem. Phys.*, **60,** 3449
161. Dezelić, G. (1970). *Pure Appl. Chem.*, **23,** 327
162. Gordon, R. G. (1963). *J. Chem. Phys.*, **39,** 2788; **41,** 1819
163. Blumenfeld, M. (1969). *Molecular Dynamics and Structure of Solids* (R. S. Carter and J. J. Rush, editors) 441 (N.B.S. Special Publication 301) (Washington: N.B.S.)
164. Cabana, A., Bardoux, R. and Chamberland, A. (1969). *Can. J. Chem.*, **47,** 2915
165. Ascarelli, P. and Cagliotti, G. (1966). *Nuovo Cim.*, **43,** 375
166. Twu, C. H., Gubbins, K. E. and Gray, C. G. (1975). *Mol. Phys.*, **29,** 713
167. Green, S. (1974). *J. Chem. Phys.*, **60,** 2654
168. McMahan, A. K., Beck, H. and Krumhansl, J. A. (1974). *Phys. Rev.*, **A9,** 1852
169. Hilbers, C. W. and Maclean, C. (1969). *Molec. Phys.*, **16,** 275
170. Boublick, T. (1974). *Molec. Phys.*, **27,** 1415
171. Sarkisov, G. N., Dashevsky, V. G. and Malenkov, G. G. (1974). *Molec. Phys.*, **27,** 1249
172. Hilbers, C. W. and MacLean, C. (1972). *NMR, Basic Principles and Progress*, **7,** 1
173. Gray, C. G. and Gubbins, K. E. (1975). *Molec. Phys.*, in press

9
Structure/Property Relationship in High Polymers

G. ALLEN and D. C. WATTS
University of Manchester

9.1 INTRODUCTION 350
9.1.1 *States of matter* 350
9.1.2 *Molecular weight distributions* 351
9.1.3 *Polymer chain configurations* 351

9.2 STATICS AND DYNAMICS OF INDIVIDUAL POLYMER CHAINS 352
9.2.1 *Chain statistics* 352
9.2.1.1 *Phantom chains* 353
9.2.1.2 *Real chains* 355
9.2.1.3 *Short-range interactions* 355
9.2.1.4 *Long-range interactions* 356
9.2.1.5 *Experimental studies of chain conformation* 357
9.2.1.6 *Entropy of a single chain* 358
9.2.2 *The dynamics of individual chains* 358
9.2.2.1 *Modes of motion of polymer chains* 359
9.2.2.2 *Relaxation of rod-like molecules* 359
9.2.2.3 *The normal modes of flexible chains* 361
9.2.2.4 *Segmental motions of flexible chains* 363

9.3 POLYMERS IN BULK 365
9.3.1 *Rubbers* 365
9.3.1.1 *Kinetic theory and thermodynamics of rubber elasticity* 365
9.3.1.2 *Dynamic properties* 369
9.3.2 *Glasses* 374
9.3.2.1 *The glass temperature* 374
9.3.2.2 *Effect of stress on* T_g 376
9.3.2.3 *Static theories of the glass transition* 376
9.3.2.4 *Transition temperatures and chemical structure* 378
9.3.2.5 *Dynamic properties* 378

9.3.3 *Crystalline polymers* 318
9.3.3.1 *Relation between chemical structure and crystal structure* 382
9.3.3.2 *Morphology* 384
9.3.3.3 *Dynamic properties* 386

9.4 CONCLUSION 390

9.1 INTRODUCTION

We shall attempt to review the physical properties of polymerised materials in terms of our understanding of the behaviour of individual macromolecules and of assemblies of macromolecules. The review will be concerned with aspects of static and dynamic properties. It will consider effects which are a consequence of the existence of macromolecules and in some instances phenomena which arise from localised motions either of very short segments of the backbone or of side groups.

A survey of the whole span of polymerised matter would lead, within the space available, to little more than a catalogue of classified materials. For this reason the review is restricted to a discussion of the properties of synthetic polymers and even so it is necessary to consider materials in which the parent polymer molecule is essentially linear, *viz.* —R—R—R—R—R—. This simplification enables us to concentrate on properties which are a direct consequence of the presence of long chain molecules. At the same time it avoids discussion of branched molecules. Statistical methods and characterisation techniques have not yet developed to a point at which properties of assemblies of branched macromolecules can be quantitatively distinguished from properties of assemblies of corresponding linear macromolecules.

9.1.1 States of matter

Linear polymers with molecular weights $>10^5$ exist in three states of matter, as rubbers (elastomers), glasses and partially crystalline materials.

(a) The rubber state is essentially the liquid state for high molecular weight polymers. It is characterised by long-range elasticity and rapid response to an external stress in a temperature interval for which the upper limit is determined by the chemical thermal stability of the molecules and the lower limit by glass formation or extensive crystallisation. To obtain materials with long-term dimensional stability it is necessary to cross-link the assembly of chains into a three-dimensional network, using approximately one cross-link per 100 monomer units.

(b) The glass state is amorphous and displays the properties of an amorphous solid including short-range elasticity. For linear organic polymers the upper limit is in the region of 250 °C and only just overlaps the lower melting temperatures of inorganic glasses. On the whole, organic polymer glasses are more tractable and more versatile than the inorganic counterparts, though they are inherently softer and have poorer solvent resistance.

(c) The crystalline state is usually only partly crystalline and thus has an amorphous rubbery or glassy component. Melting temperatures for organic polymer range from $-55\,^{\circ}\mathrm{C}$ to 300–400 °C. The properties associated with this state are determined very much by the extent of crystallisation and the degree of orientation achieved in the overall sample during the crystallisation process.

9.1.2 Molecular weight distributions

Polymerisation processes produce high molecular weight polymers in any one of these three states. However, it must be recognised at the outset that each polymer sample contains a statistical distribution of molecular weights arising from the chemistry of the process. A given property may be related to a particular moment of the distribution and this is often very important in comparing the properties of samples of different origins. For example, many static properties in bulk samples (e.g. coefficients of expansion or compressibility) and colligative properties in solution correlate with the number-average molecular weight ($\bar{M}_{\mathrm{n}} = \sum n_i M_i / \sum n_i$). Flow properties of melts and light scattering in solution correlate with the weight-average molecular weight ($\bar{M}_{\mathrm{w}} = \sum n_i M_i^2 / \sum n_i M_i$). We shall not discuss molecular weight distributions further but simply note that the appropriate average is essential to the understanding of a particular property. The methods of characterisation of molecular weight distributions would command a chapter in its own right.

9.1.3 Polymer chain configurations

The detailed molecular structure of polymer chains is a variable of general importance when considering properties of polymers. Simple monomers such as styrene, methyl methacrylate, ethylene oxide, butadiene, etc. produce chains which are essentially linear, but ethylene and vinyl acetate readily produce branched structures and it is essential to this review that properties of the more linear products are available. Some monomers can produce isomeric structures by linking up a proportion of units in a head-to-head or tail-to-tail configuration rather than the head-to-tail configuration which usually dominates the repeat structure.

Of much greater relevance to this review is the fact that repeat units possessed of assymetric centres in the backbone of the macromolecules give rise to iso-, syndio- or a-tactic stereochemical sequences (Figure 9.1) and of course distributions of these sequences along a chain. Thus, whereas different isomers can only exist through branching in the polyethylene chain, and linear structures such as poly(methylene oxide) and poly(vinyl fluoride) have the possibility of head-to-head and tail-to-tail linkages, polymers with asymmetric centres can in addition produce chains of different and varying

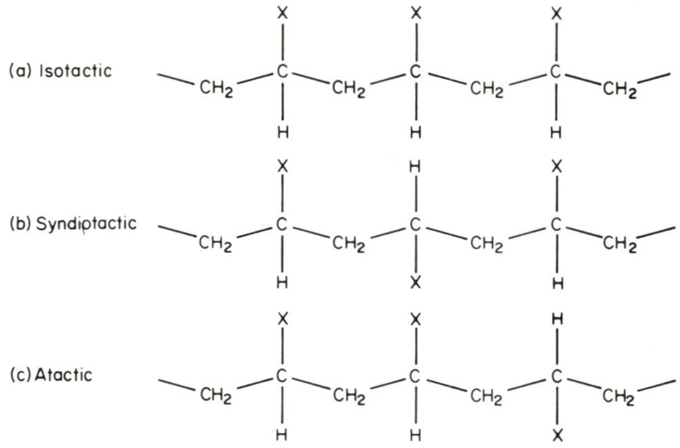

Figure 9.1 Stereo-regular isomers of an α-olefin chain

stereochemical character. The statistics of stereochemical sequences in such polymers are difficult to establish, high-resolution n.m.r. spectroscopy being the only satisfactory technique at the present time. Nevertheless, these statistics are important because chains containing long isotactic sequences can crystallise and so too can chains containing long syndiotactic sequences. But samples containing long atactic sequences have chain structures too irregular to allow crystallisation. Such materials (e.g. free-radical polymerised styrene or methyl methacrylate) are only encountered in the amorphous glass or rubber states.

In this review we shall be able to deal only with the three extreme forms of tacticity, though we must recognise that in a real polymer chain there may be a distribution of the three types of sequence and the ability to crystallise depends on the actual fraction of monomer units present in iso- or syndiotactic sequences.

With all of these reservations in mind we now consider the properties of polymer chains in isolation and then the three states of matter, composed of assemblies of polymer chains.

9.2 STATICS AND DYNAMICS OF INDIVIDUAL POLYMER CHAINS

9.2.1 Chain statistics

A polymer chain has numerous rotational isomers by virtue of internal rotation about main-chain σ-bonds. In the crystal the conformation adopted by the chain[1] is usually the one of lowest internal energy consistent with efficient packing of chains in the lattice. In solution or in the melt the conformation of the chain is continually fluctuating between all possible conformations, and a statistical analysis is required in order to specify the moments of the required physical quantities (usually end-to-end distance or radius of gyration) over the total population of conformations. In the glass the conformations are random but time independent.

In the literature there is confusion between the two terms 'configuration' and 'conformation' as applied to polymer chains. In this review 'conformation' applies specifically to the spatial distribution of the chain, i.e. rotational isomerism; 'configuration' is used to denote stereochemical isomerism.

9.2.1.1 Phantom chains

The simplest analysis of the conformations of a polymer chain treats a linear molecule made up of n links of length l. There are no restrictions on bond angles and internal rotations and a 'phantom' chain is assumed so that it can pass through itself and different units can occupy the same point in space at the same time. To the approximation that the chain obeys Gaussian statistics[2,3], if we fix one end of the chain at the origin of a Cartesian coordinate system the probability that the other end is located in a volume element dx dy dz at x, y and z in space is

$$p(x,y,z)\,\mathrm{d}x\,\mathrm{d}y\,\mathrm{d}z = (b^3/\pi^{\frac{3}{2}}) \exp[-b^2(x^2 + y^2 + z^2)]\,\mathrm{d}x\,\mathrm{d}y\,\mathrm{d}z \quad (9.1)$$

where

$$b^2 = \tfrac{3}{2}nl^2$$

This function may be resolved into three independent component probabilities of the type

$$p(x)\,\mathrm{d}x = (b/\pi^{\frac{1}{2}}) \exp(-b^2x^2)\,\mathrm{d}x \quad (9.2)$$

whose form is reproduced in Figure 9.2, implying that the probability of a given x component of chain end-to-end distance is independent of its y and z components.

Equation (9.1) gives the probability of finding the chain with an end-to-end distance $r^2 = x^2 + y^2 + z^2$, in the interval r to r + dr and *in a fixed direction in space* defined by x, y, z. The probability of a given end-to-end distance r *irrespective of direction* is obtained by considering all chain ends lying on the surface of a sphere of radius r within a shell of thickness dr, *viz.*

$$P(r)\,\mathrm{d}r = (4b^3/\pi^{\frac{1}{2}})r^2 \exp(-b^2r^2)\,\mathrm{d}r \quad (9.3)$$

From Figures 9.2(a) and 9.2(b) the component probabilities $p(x)$, etc. have their maxima at the origin. The function $P(r)$ has its maximum at

$$r_{\max} = b^{-1} \quad (9.4)$$

From equation (9.3) we can obtain various moments of the distribution of end-to-end distances available to our phantom, freely-jointed random-flight chain, e.g.

$$\langle r \rangle = \int r\,P(r)\,\mathrm{d}r = (8/3\pi)^{\frac{1}{2}}\,n^{\frac{1}{2}}l \quad (9.5)$$

$$\langle r^2 \rangle = \int r^2P(r)\,\mathrm{d}r = \tfrac{2}{3}b^2 = nl^2 \quad (9.6)$$

or

$$(\langle r^2 \rangle)^{\frac{1}{2}} = n^{\frac{1}{2}}l \quad (9.7)$$

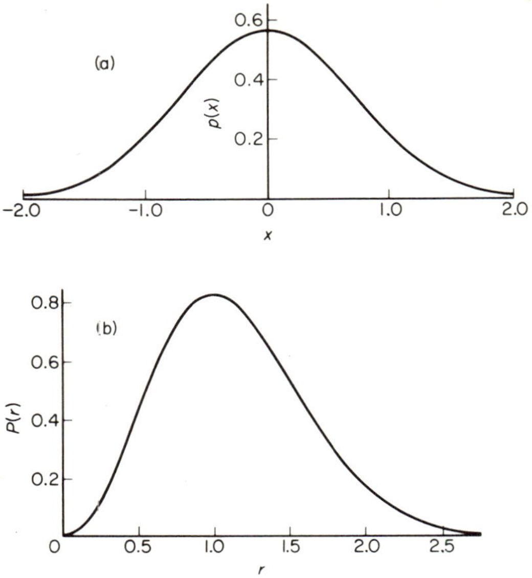

Figure 9.2 (a) Gaussian density distribution of the x-component of the chain end-to-end vector. (b) Radial distribution function of the end-to-end vector

This treatment involves an approximation which assumes that the range of end-to-end distance r is very much less than the fully extended length of the chain, nl. A defect of Gaussian statistics in this context is that the probability function extends out to $r = \infty$, whereas $P(r) = 0$ for $r > nl$ for a real chain. More accurate non-Gaussian treatments[4] avoid these difficulties. If the bond lengths in the chain are not identical, then l^2 in equation (9.6) is replaced by the mean-square bond length.

This model is made more realistic by introduction of a fixed bond angle of supplement θ. A simple chain is then characterised by n, l and θ. In the limit of large n,

$$\langle r^2 \rangle_{\text{free}} = nl^2\,(1 + \cos\theta)/(1 - \cos\theta) = nl^2 f(\theta) \tag{9.8}$$

A real chain does not have free rotation: there is a rotational function $V(\phi_i)$ influencing rotation about bond i and dependent on the azimuthal angle between bonds $i - 1$ and $i + 1$. In a phantom chain we can assume that $V(\phi_i)$ about the bond *is independent* of rotations about neighbouring bonds and hence

$$\langle r^2 \rangle = nl^2\left(\frac{1 + \cos\theta}{1 - \cos\theta}\right)\left(\frac{1 + \eta}{1 - \eta}\right) \equiv nl^2 f(\theta)\; f(\phi_i, T) \tag{9.9}$$

where

$$\eta = \frac{\displaystyle\int_0^{2\pi} \cos\phi_i \exp\left[-V(\phi_i)/kT\right] \mathrm{d}\phi_i}{\displaystyle\int_0^{2\pi} \exp\left[-V(\phi_i)/kT\right] \mathrm{d}\phi_i}$$

The introduction of $V(\phi_i)$ represents a very important step since the chain dimensions may now be temperature dependent.

9.2.1.2 *Real chains*

Conformations of real chains depart from their phantom model in a most important respect, namely that they cannot cross their own paths. Out of all the phantom conformations only a small fraction are free from self-intersections and are therefore acceptable for a real chain. Average dimensions of the real chain, e.g. $\langle r^2 \rangle$, are thus increased. The problem for a real chain can be separated into two parts. Firstly there are the short-range interactions between atoms and groups which are neighbours or near neighbours along the chain. Secondly there are long-range interactions involving pairs of units which are widely separated in the chain sequence but which come near to one another in space in certain chain conformations. By definition we take $\langle r^2 \rangle_0$ to denote the value of $\langle r^2 \rangle$ which obtains in the absence of long-range effects. A factor a was introduced by Flory[5] to describe the increase in a linear dimension of the average conformations as a result of long-range effects (i.e. the excluded volume effects); thus

$$\langle r^2 \rangle = a^2 \langle r^2 \rangle_0 \tag{9.10}$$

9.2.1.3 *Short-range interactions*

Much of Flory's second book[6] is devoted to the mathematical description of the equilibrium set of conformations corresponding to $\langle r^2 \rangle_0$ available to real chains. The rotational isomeric state model is used and interdependent rotational potentials are included. The total conformational energy is expressed as a sum of energies of first-neighbour pairs of bonds. The statistical weights (u) corresponding to the energies (E) of pairs of bonds in given orientations are obtained from

$$u = \exp\left[-E/kT\right] \tag{9.11}$$

and these can be obtained for all the states ζ of a particular bond i. The statistical weight of a conformation of the chain of n links as a whole is

$$\Omega = \prod_{i=2}^{n-1} u_{\zeta i} \tag{9.12}$$

and the conformational partition function Z is obtained from a summation over all conformations. The energy E has to be estimated from knowledge of barriers to internal rotation and non-bonded interactions and there are usually insufficient data available to perform *a priori* calculations. Furthermore, this direct evaluation of Z would be prohibitive for a chain having $n > 15$. Alternative methods requiring less computational effort are outlined in the text[2].

In the case of the polymethylene chain it is possible to use the method to make an *a priori* estimate of $\langle r^2 \rangle_0$ or $C = \langle r^2 \rangle_0 / nl^2$ and $\mathrm{d}ln\langle r^2 \rangle_0/\mathrm{d}T$.

At 140 °C, C becomes substantially independent of n for $n > 200$ at a value of 6.87, which compares with phantom chain estimates of $C = 2.0$ for freely rotating links with tetrahedral bond angles and $C = 3.2$ for independent hindered rotations. For other structures there are at present insufficient data on potential functions to allow the selection of the appropriate rotational isomeric states at or near potential minima and also the evaluation of the statistical weights of the chosen states. In these cases experimental values of $\langle r^2 \rangle_0$, dipole moments, etc. and the temperature coefficients of these values are used to calculate the statistical weights. These results can then be compared with qualitative estimates of the form of the conformational energy based on inferences from analogous molecules and approximate estimates of the principal contributions to the energy. Polymers analysed in this manner include poly(tetramethylene oxide), poly(methylene oxide), poly(ethylene oxide), poly(trimethylene oxide), poly(tetramethylene oxide), poly(dimethylsiloxane), polyamides, polyesters, poly-1,4-butadiene, poly-1,4-isoprene, polyisobutene and stereo-regular and stereo-irregular forms of polyvinyl chains. Williams and Flory[7] have shown that an analysis of the random conformations of the bisphenol-A polycarbonate leads to a value of $\langle r^2 \rangle_0 / nl^2 f(\theta) = 1.0$. The calculations imply that such a value, close to that expected for a chain without hindered internal rotation, is to be expected for other polymers containing polyphenylene units, and this has been found to be so for 2,6-disubstituted poly(phenylene oxides)[8-10] and a polysulphone[11].

9.2.1.4 Long-range interactions

Over the past ten years important progress has also been made with the excluded volume problem. Physically it is easy to see that the effect of finite volume must be to expand and broaden the distribution of end-to-end distances. Monte Carlo calculations[12] based on various lattice models, including a tetrahedral one appropriate to polymethylene, lead to the result that

$$\langle r^2 \rangle \propto n^{1+\gamma}$$

where $0.22 > \gamma > 0.18$ for long chains. However, Monte Carlo calculations are subject to uncertainty associated with the enrichment procedure used to combat the rapid attrition of successful non-intersecting walks as n increases. Domb[13] has calculated exact values of $\langle r^2 \rangle$ for n-alkanes up to C_{12} based on geometrical analyses of all possible conformations. Extrapolation to infinite chain length gives a value for γ similar to that derived from Monte Carlo methods. But again there is uncertainty since the extrapolated result is derived from calculations made on chains far too short to display excluded volume effects comparable with those experienced in a long polymer chain. More recently Edwards[14] has succeeded in obtaining an asymptotic solution in a closed form for the position of the nth link in an infinite chain. In the limit $n \rightarrow \infty$ this leads to a mean-square end-to-end distance

$$\langle r^2 \rangle \propto n^{6/5}$$

Thus the exponent gives $\gamma = 0.20$, in agreement with the Monte Carlo result, and in fact the result anticipated many years ago by Flory. A higher exponent 4/3, i.e. $\gamma = 0.333$, has been favoured by other workers but there is most support for the lower value 6/5.

Edwards has extended his analysis to the behaviour of a single polymer chain in the critical 'θ' region[15], to topological constraints experienced both by an infinitely long chain[16] and by chains crosslinked into a network[17], and to the entropy of a confined polymer chain[18].

9.2.1.5 *Experimental studies of chain conformation*

In the crystal, x-ray diffraction methods still provide by far the most extensive and detailed information on chain conformation. An extensive compilation of crystallographic data on polymers is available through the table prepared by Miller[19].

Laser Raman spectroscopy and neutron incoherent inelastic scattering and i.r. spectroscopy have been used. In general, however, vibrational spectroscopy is a secondary tool in the determination of chain conformation in crystals. It is difficult to obtain unambiguous evidence of a particular conformation from vibrational studies, and, despite advances in technique and interpretation, x-ray diffraction is still the primary technique yielding bond angles and lengths in addition to molecular symmetry.

Most of the conformational data on the random coil comes from dilute solution light scattering or viscosity measurements under θ-conditions or from extrapolations from measurements made in moderately good solvents. $\langle r^2\rangle_\theta$ or $\langle r^2\rangle_\theta/nl^2$ can be compared with the value of $\langle r^2\rangle$ calculated on the assumption of free rotation about fixed angles for the phantom chain. The conformational parameter $\sigma^2 = \langle r^2\rangle_\theta/\langle r^2\rangle_{\mathrm{free}}$ is the most widely quoted parameter. Typical results are given in Table 9.1. In most cases estimates of $\mathrm{d}ln\langle r^2\rangle_\theta/\mathrm{d}T$ have been made. In rubbers, stress–temperature coefficients can be used to determine the temperature coefficient of unperturbed dimensions[20]. It must be noted that estimates of $\mathrm{d}ln\langle r^2\rangle/\mathrm{d}T$, especially from solution

Table 9.1 Dimensions of polymer coils in bulk polymers

Polymer	*Molecular weight*	R_g/Å *from neutron scattering in bulk polymers**	R_g/Å *in θ-solvent*
Poly(methyl methacrylate)	2.5×10^5	126[23]	110
Polystyrene	2.1×10^4	37[24]	41
	5.7×10^4	63[24]	71
	9.0×10^4	83[24]	87
	9.7×10^4	90[25]	84
	1.7×10^5	115[24]	117
	3.2×10^5	145[24]	159
	5.3×10^5	195[24]	212
	8.9×10^5	280[25]	250
	11.0×10^5	295[24]	
Poly(ethylene oxide)	3.3×10^3	26[26]	23

* Experimental errors on $\langle R_g\rangle$ are in the range $\pm$ 5–10%

measurements, are subject to large experimental error. Indeed, overoptimistic estimates of the experimental errors pertinent to measurements of $\langle r^2 \rangle_\theta$ are also characteristic of this field.

A most important recent development is the use of small-angle neutron scattering to measure the radii of gyration of protonated polymer molecules dispersed at low concentrations in their deuterated matrix.

A considerable body of evidence has now accumulated to show that in amorphous polymers, particularly in glasses, polymer chains display dimensions which are experimentally indistinguishable from the values obtained in θ-solvents, i.e. unperturbed dimensions. A comparison of results currently available is given in Table 9.1.

Thus a long-standing debate seems to have been resolved regarding chain conformation, since this conclusion supports another proposition made 20 years ago by Flory[21] which has become the basis of many 'theorems' relating to polymer chains in amorphous states. The fact that the coils obey essentially Gaussian statistics carries the added implication that the organisation of such assemblies of polymer chains in the bulk is essentially random. However, just as in liquids[22], there is short-range order over distances comparable with the diameters of the repeat units.

9.2.1.6 *Entropy of a single chain*

A definite probability can be assigned to the likelihood of a good polymer chain having a given end-to-end distance and therefore we can discuss the concept of the entropy of a single chain. The entropy is derived from the Gaussian probability function by the application of the Boltzmann relation between the entropy of the chain in a given set of conformations having end-to-end distances located at xyz in the volume element $dx\,dy\,dz$, and the probability of encountering that value of r, i.e.

$$s = k \ln p(r) \tag{9.13}$$

where k is the Boltzmann constant. Application of equation (9.1) gives for the entropy of the chain

$$s = c - kb^2(x^2 + y^2 + z^2) = c - kb^2r^2 \tag{9.14}$$

where c is a constant. Note that the entropy (like the probability) is a maximum when the two ends of the chain are coincident and that the difference in entropy between any two states is proportional to the difference in the two values of r^2. This result is used in Section 9.3.1.1 to provide a molecular basis for the mechanics of rubber elasticity.

9.2.2 The dynamics of individual chains

The physical properties of bulk polymers are determined by the mobility of their molecular chains, especially by the segmental motions of the polymer chain backbone. This type of motion depends upon intramolecular phenomena, such as dipole–dipole interactions and steric hindrance between groups adjacent to the bond about which internal rotation is to occur, and also upon intermolecular interactions, such as entanglements with other chains. Consequently it is desirable to study these effects separately. Since gas-phase studies of macromolecules are impossible, the best way to minimise the interactions between different polymer chains is to study them in dilute solution. Dynamic studies of polymers in solution not only yield information on molecular mobility, but, in the case of stiff or rigid polymer chains, offer unique information about the equilibrium chain conformation.

9.2.2.1 Modes of motion of polymer chains

The dynamic properties of individual polymer chains arise from two main types of motion. The first involves the long-range translational and rotational motions of whole polymer chains. The second involves the worm-like motion of fairly short segments of the chain backbone.

In its simplest form the long-range motion requires all segments of the molecule to move in unison, and so requires a clearly defined inter-segmental geometry. This criterion is satisfied by rigid rod-like molecules, such as polypeptides coiled in a helix, but may also be satisfied by certain random coil polymers when the time required for overall movement of the coil is shorter than the time required for segmental relaxation.

The segmental type of motion gives rise to two relaxation regions in solid amorphous polymers, normally termed the α and β processes[27]. The α process, which is associated with the glass transition of the bulk polymer (*cf.* Section 9.3.2) is observed both in solution and in the rubbery state. The relationship between α and β processes is considered more fully in Section 9.3.2.5.

A useful compilation of papers dealing with the varied experimental techniques for studying the dynamics of individual polymer chains is contained in the 1970 Faraday Discussion on Polymer Solutions[28]. The techniques include dynamic light scattering and neutron scattering; viscoelastic, dielectric, ultrasonic and n.m.r. relaxation; and fluorescence depolarisation of dye-tagged polymers. Some of these techniques are sensitive to only one type of motion. Thus viscoelastic relaxation of *solutions* studies the long-range type of motion[28] and Doppler broadening studies of Rayleigh light scattering ordinarily provide information on translational diffusion coefficients only. However, depolarised light can, in some circumstances, provide additional information on rotational contributions to the diffusion of macromolecules[29].

It is not possible to give a detailed discussion here of all these techniques, and therefore attention will be focused primarily on the technique of dielectric relaxation, which gives a clear indication of both long-range and short-range modes of motion of polar macromolecules in solution[30].

9.2.2.2 Relaxation of rod-like molecules

The study of the dynamic behaviour of rod-like molecules is of considerable interest. Not only do cer' in naturally occurring polymers fall into this class, but for all polymers with finite energy barriers to internal rotation it must be possible to reduce the chain length to a size where the molecule is essentially rod-like. Conversely, any real rod must have a degree of flexibility, so that if the length can be increased sufficiently, a point must be reached when even slight curvature can build up to cause an overall coil-like conformation. Consequently, for any polymer there should be in principle a chain length at which the observed behaviour changes from 'rod-like' to 'coil-like'.

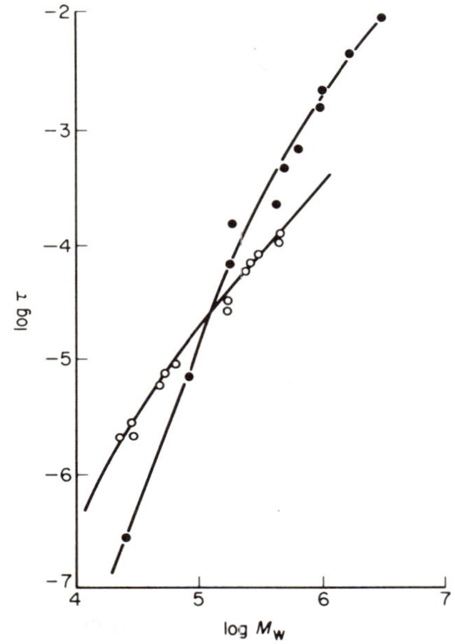

Figure 9.3 Variation of relaxation time with weight average molecular weight: ○, poly(γ-benzyl-L-glutamate) in benzene with ε-caprolactam as deaggregant; ●, poly(n-butyl isocyanate) in benzene[38]

The dynamic behaviour of dissolved rod-like polymers has been studied by a variety of techniques. In the case of visco-elastic studies, when a dilute solution of such rod-like molecules is subjected to steady shear flow, the distribution of orientations departs from spherical symmetry. The resultant orientation is opposed by Brownian motion, and the relaxation to the equilibrium position may be characterised by a diffusion coefficient D, or by a relaxation time $\tau = D/6$. The various theoretical treatments[31-33] of this behaviour show that the hydrodynamic characteristics depend very strongly on the length-to-diameter ratio of the rod. For a constant rod diameter the dependence on rod length is almost cubic. A number of theoretical treatments can be generalised[33] to give the following expression for the end-over-end relaxation time:

$$\tau = \frac{\pi \eta_s L^3}{6kT(ln(L/b) - \gamma)} \tag{9.15}$$

η_s is the solvent viscosity, L is the length of the rod, and b and γ have slightly different values in the various approaches. Thus although measurement of these relaxation times does not reveal the effect of chemical structure on the rate of conformational changes, it is possible to obtain information on the overall molecular shape, such as the rod length and diameter, and on the 'flexibility' of the rod.

The two rigid rod polymers which have been most thoroughly investigated are the polypeptide poly(γ-benzyl-L-glutamate) (PBLG)[34, 35] and poly(n-butyl isocyanate) (PNBIC)[36, 37]. Figure 9.3 shows the results for these polymers obtained by North *et al.*[35, 37, 38]. The cubic dependence of relaxation time on molecular weight is observed for PNBIC for $M_w < 10^5$; however for $M_w > 10^5$, τ varies as $(M_w)^{3/2}$. This implies the onset of coil-like behaviour. An intermediate situation seems to exist in solutions of PBLG, where τ varies approximately as $M_w{}^2$.

9.2.2.3 The normal modes of flexible chains

The 'whole chain' mode of overall motion for flexible chain polymers may be examined in terms of normal coordinate analysis. The essence of this approach is that the random coil is resolved into three one-dimensional arrays of beads each connected to its neighbour by a volumeless Hookean spring, and each experiencing a frictional drag when moving in its environment. Each bead corresponds to a chain submolecule, the dimensions of which obey Gaussian statistics. In the original treatments[39-41] the interconnecting spring was interpreted in terms of the entropy change which results from extension of the single chain (*cf.* Section 9.2.1.5). These linear springs have been replaced in other formulations by damped torsional oscillators[42] and, more recently, by a generalised stochastic model of chain motion[43].

For the simplest treatment, due to Rouse[39], the $3m$ differential diffusion equations are of the form

$$\eta x_i + (3kT/zl^2)(2x_i - x_{i-1} - x_{i+1}) = 0 \tag{9.16}$$

where m is the number of submolecules, l is the length of each link in the chain and z is the number of links in the submolecule; η is the coefficient of friction defining this viscous interaction between the beads and the solvent. Analogous equations exist for y_i and z_i. Normal coordinate transformation of these equations allows the motion of the whole assembly to be described in terms of a number of normal modes of motion (Figure 9.4). The zeroth mode ($k = 0$) corresponds to translation. The higher modes correspond to successively more complex harmonic distortions in which the mode number (k) corresponds to the number of nodes in the assembly. A relaxation time for each mode of motion is implicit in this model, and the predominant mode is $k = 1$. The calculation of these relaxation times requires an assumption to be made as to the way in which the presence of the beads and springs affects the flow of the solvent. The methods of Rouse[39] and Bueche[40] assume that there is no effect (the free draining coil approximation) whereas that of Zimm[41] assumes that the solvent flow velocity is reduced in regions of high bead-density (the non-free draining coil). The resulting expressions are:

(a) for the free draining coil[39, 40]

$$\tau_k = \frac{6(\eta - \eta_s)M}{\pi^2 k^2 cRT} \tag{9.17}$$

(b) for the non-free draining coil[41]

$$\tau_k = \frac{1.71\,(\eta - \eta_s)M}{\lambda_k\, cRT} \tag{9.18}$$

The significant feature of these expressions is that they allow calculation of the relaxation times τ_k, for the kth mode of motion, from expressions involving only solute concentration c, polymer molecular weight M, and zero shear-rate viscosity coefficients (for solution η and solvent η_s) as parameters; λ_k are a series of constants (eigenvalues) for each mode of motion.

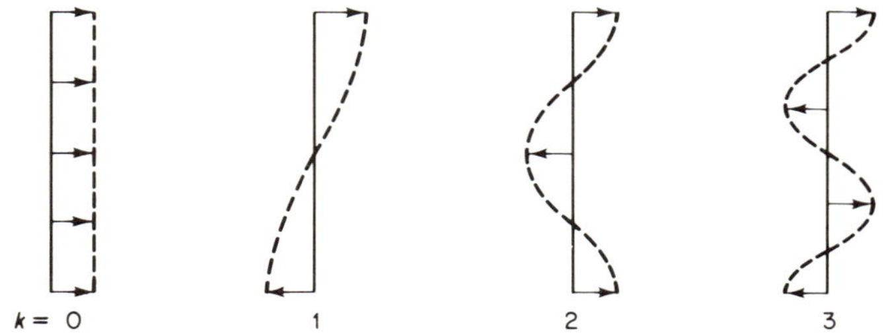

Figure 9.4 Normal modes of motion for a linear polymer. The solid vertical line represents the chain contour, which, of course, really follows a randomly kinked curve in space for any given conformation. The dashed lines indicate relative magnitudes of the displacements. Arrows are placed at positions of maximum amplitude

This type of normal-mode analysis has been extended to dielectric relaxations by various workers[43-46]. The basic problem is to ascribe an electrical charge distribution down the polymer chain which is both physically reasonable and mathematically tractable. Probably the most successful model of the charge distribution is the sawtooth function employed by Stockmayer and Baur[46]. The solution of the normal coordinate equations then yields a polarisation characterised by a single low-frequency dispersion with a relaxation time close to that for the first normal visco-elastic mode.

Normal mode analysis has successfully interpreted the low-frequency visco-elastic relaxation behaviour of polymer solutions. However, in the case of dielectric studies, relaxation effects due to 'whole molecule' modes of

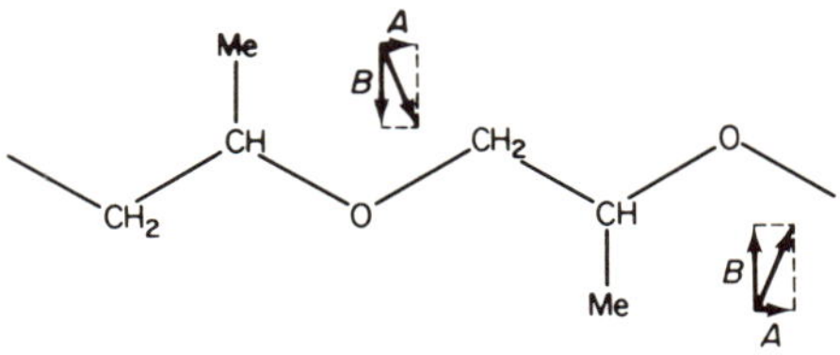

Figure 9.5 A head-to-tail poly(propylene oxide) chain indicating parallel (A) and perpendicular (B) dipole components in each repeat unit

motion are not observed for all chains of connected dipoles. The essential requirement is that each repeat unit of the chain should have a component of its dipole moment lying unidirectionally along the chain contour. This condition is satisfied, for example, in poly(propylene oxide) (PPO), which has been studied by Baur and Stockmayer[47] (Figure 9.5).

The polymer has a predominantly head-to-tail structure, and so the first normal mode ($k = 1$) was observed. If the sequence of units was reversed in the centre of the chain, the second normal mode ($k = 2$) would be observed. Figure 9.6 shows the dielectric absorption for PPO. The low-frequency peak arises from the normal mode motion, while the higher-frequency peak is due to the relaxation of the component of the dipole moment perpendicular to the main chain[48] by segmental motion.

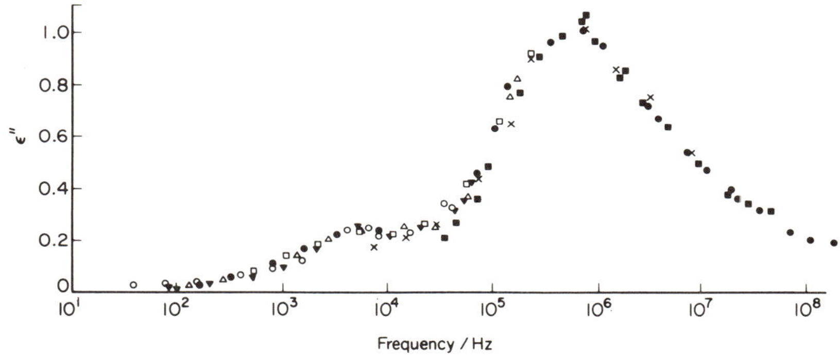

Figure 9.6 A reduced plot of the dielectric absorption for atactic poly(propylene oxide at various temperatures[47]

In these circumstances when two relaxation peaks are observed, an unambiguous assignment of the processes can be made. The higher-frequency peak is due to segmental motion and its location is essentially independent of molecular weight. The lower-frequency peak is due to 'whole molecule' motion and its location is strongly dependent upon molecular weight, as expected from the theoretical equations for the normal mode relaxation times [equations (9.17) and (9.18)]. Stockmayer *et al.*[49] have recently observed similar behaviour for poly(ε-caprolactone) with head-to-tail structure $[-(CH_2)_5CO_2-]_x$, in dioxane solution. The low-frequency loss peak shifts markedly in frequency with changing molecular weight, while the high-frequency peak, due to segmental relaxation of the perpendicular component of the dipole moment, is molecular weight independent.

9.2.2.4 Segmental motions of flexible chains

The reference to the relaxation behaviour of poly(propylene oxide) in the previous section has illustrated the fact that dielectric studies of polymers containing a fixed dipole moment component perpendicular to the main chain may yield information about the backbone segmental motion. Thus a dielectric loss peak will be observed due to segmental motion in this type of polymer unless the molecular weight is so low that 'whole molecule' motion of the coil takes place more rapidly than segmental motion. One example where this evidently occurs is with the fairly stiff chain of poly(hexene 1-sulphone)[50].

The experimental evidence for solutions of polymers such as poly(vinyl bromide)[51], poly(ethylene oxide)[52] and poly(*p*-chlorostyrene)[48] shows that the dielectric relaxation times are independent of molecular weight, as is expected for true segmental motion. Concentration dependence of relaxation times may also be quite small for highly mobile chains at temperatures well above the glass transition region[48]. Chain flexibility evidently has a pronounced effect upon segmental relaxation times, and the quantification of such effects is one of the principal reasons for such studies. Table 9.2 shows, for example, that the frequencies of maximum loss, $f_m = (2\pi\tau)^{-1}$, for poly(ethylene oxide) and poly(*p*-chlorostyrene) in the same solvent at the same temperature differ by a factor of about 500.

Table 9.2 Activation energies of loss processes

Polymer	*Frequency*/Hz *of maximum loss*	*Activation energy* /kJ mol^{-1}
Poly(ethylene oxide)[52]	1.5×10^{10}	10.3
Poly(vinyl chloride)[53]	2×10^{8}	9.2
Poly(vinyl bromide)[51]	3×10^{7}	—
Poly(*p*-chlorostyrene)[48]	3×10^{8}	20.2
Poly(*p*-fluorostyrene)[48]	4×10^{8}	20.2

In general it is possible to explain the increase in relaxation times in terms of increasing steric or electrostatic barriers to backbone rotation[38]. Probably these relaxation times should also depend upon polymer stereoregularity and configuration. However, a surprising feature of the results in Table 9.2 is the low values obtained for the activation enthalpies for relaxation. The values for poly(ethylene oxide) and poly(vinyl chloride) are comparable with the activation energy for viscous flow of solvent and to the barrier to internal rotation in ethane. Even chains with bulky substituents such as the poly(*p*-halogenostyrenes) have low activation enthalpies. The probable reason for these low values is connected with the small perturbation in average angular orientation imposed on each dipole by the electric field in the dielectric experiment. The orientation angle should be as low as one radian, and so the motion of a representative dipole may be accommodated by partial rotation of neighbouring bonds, without crossing the highest energy barriers in the energy–angle profile[38]. However, acoustic studies[54] of the high-energy to low-energy state transition in polystyrene solutions yields an activation enthalpy of 27.7 kJ mol^{-1}. This is not much higher than the dielectric activation energy observed for poly(*p*-halogenostyrenes)[48]. The rates of segmental motions may also be studied indirectly by measuring the rates of diffusion-controlled reactions involving macromolecules. Again there is a difference between the exact type of segmental motion involved here and that studied in dielectric relaxation, but in many cases both rate processes show a similar variation with the chain backbone structure[38].

9.3 POLYMERS IN BULK

In this section we shall discuss the basic principles under-pinning the properties of assemblies of polymer chains. The section is sub-divided into three major parts dealing separately with rubbers, glasses and partially crystalline materials. For each state the review of static phenomena concentrates mainly on thermodynamic properties and this is followed by a survey of dynamic properties.

9.3.1 Rubbers

We have noted in the introduction that the rubbery state is essentially the liquid state of a high polymer; thus in order to prevent flow it is necessary to cross-link assemblies of polymer chains into a three-dimensional network. In order to discuss the equilibrium properties of these materials it is necessary to concentrate on networks rather than raw polymers, though in passing one should note that properties such as the coefficients of thermal expansion, compressibility and thermal pressure are typically of the same order of magnitude as those of high molecular weight liquids, both for raw and uncross-linked materials. For example:

	α_P	β_T	γ_V
Natural Rubber at 25 °C	6.6×10^{-4} deg^{-1}	6.0×10^{-5} atm^{-1}	10.9 atm deg^{-1}
Polydimethylsiloxane at 25 °C	8.3×10^{-4} deg^{-1}	$1.1_9 \times 10^{-4}$ atm^{-1}	7.0 atm deg^{-1}

9.3.1.1 Kinetic theory and thermodynamics of rubber elasticity

Several theoretical models have been used to calculate the elastic properties of a network but all embody the same basic physical conception and lead to substantially the same conclusions[55]. The origin of long-range rubber elasticity lies of course in the ability of the segments of the chains between cross-links to uncoil and the essence of the kinetic theory of rubber elasticity is that it relates change in molecular dimensions to the macroscopic deformation to which the sample is subjected. By introducing the relation between entropy and vector length for a single chain (*cf.* Section 9.2.1.6) the entropy of the total network is calculated first in the unstrained state and then in the strained state. The difference yields the entropy of deformation.

It is assumed that the *xyz* components of the chain vector length change on deformation in the same ratio as the corresponding dimensions of the bulk rubber. In the simplest treatment a second assumption is made, *viz.* that the material deforms without change of volume. Thus for the general type of pure strain in which the three principal extension ratios of the sample are λ_1, λ_2, λ_3, we have $\lambda_1\lambda_2\lambda_3 = 1$. If we take the principal axes of strain to be parallel to the *xyz* axes used to define the conformation of individual chains, the components of length of the individual chains change on deformation from equation (9.14):

$$\Delta s = -kb^2\{(\lambda_1^2 - 1)x^2 + (\lambda_2^2 - 1)y^2 + (\lambda_3^2 - 1)z^2\} \tag{9.19}$$

This expression has to be summed over the assembly of N chains contained in unit volume of the network. Since in the unstrained state the molecules are randomly oriented we have $\langle x^2 \rangle = \langle y^2 \rangle = \langle z^2 \rangle = \frac{1}{3}\langle r_0^2 \rangle$, where $\langle r_0^2 \rangle$ is the mean-square chain vector length in the unstrained state. Putting this equal to $\langle r_0^2 \rangle$ for a corresponding set of chains in the free or uncross-linked state, $(3/2b^2)$, as given by equation (9.6), we obtain on summation of equation (9.19) the entropy of deformation ΔS for the network in the form

$$\Delta S = -\tfrac{1}{2}Nk(\lambda_1^2 + \lambda_2^2 + \lambda_3^2 - 3) \tag{9.20}$$

At this stage it is usually assumed that all states of deformation have the same internal energy, i.e. $\Delta U = 0$ for the deformation process. Then the change in Helmholtz energy for the isothermal reversible process is

$$\begin{aligned}\Delta A &= -T\Delta S \\ &= \tfrac{1}{2}NkT(\lambda_1^2 + \lambda_2^2 + \lambda_3^2 - 3)\end{aligned} \tag{9.21}$$

If we now restrict the argument to simple extension which implies equal contractions in the two transverse dimensions, the condition of constant volume requires that

$$\lambda_1 = \lambda,\ \lambda_2 = \lambda_3 = \lambda^{-\frac{1}{2}} \tag{9.22}$$

Insertion in equation (9.21) gives

$$\Delta A = \tfrac{1}{2}NkT(\lambda^2 + 2\lambda^{-1} - 3) \tag{9.23}$$

To obtain the force supported by the rubber network let us consider the specimen to be a cube of unit edge length in the unstrained state. Then if f is the force per unit unstrained cross-sectional area

$$f = \frac{\mathrm{d}\Delta A}{\mathrm{d}\lambda} = NkT\left(\lambda - \frac{1}{\lambda^2}\right) \tag{9.24}$$

The form of this theoretical stress–strain curve is shown in Figure 9.7 along with experimental results for a typical sample of natural rubber.

It will be noticed that the equation gives a reasonable account of the property of the rubber at low extension ratios but the experimental curve deviates considerably at values of $\lambda > 3$. This deviation almost certainly arises from non-Gaussian effects, i.e. the effects of real chains in the network being stretched beyond the regime of applicability of the Gaussian probability function. There is a considerable literature describing the theoretical treatment of these effects. Basically, however, the kinetic theory of rubber elasticity gives a simple description of the properties of elastomers in a variety of forms of strain and also explains their behaviour when swollen with solvents.

Of course the simple theory is based on phantom chains with freely jointed

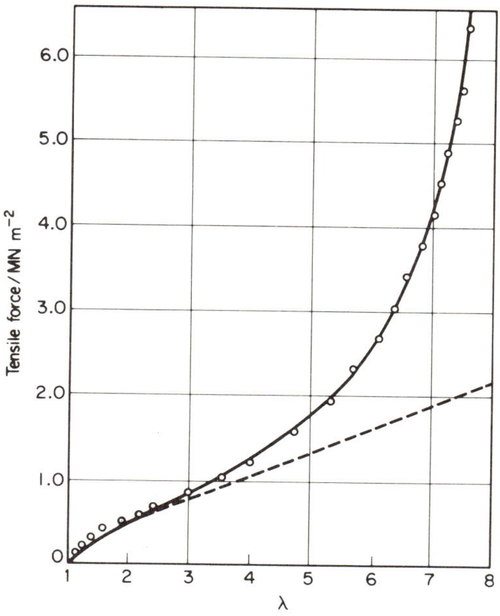

Figure 9.7 Experimental force–extension curve for a rubber network, compared with the Gaussian theoretical curve (– – – –)

links. In 1958 Flory[56] introduced the concept of hindered internal rotation and consequently the idea that $\Delta U \neq 0$ for the deformation process because of the difference in energy between rotamers involved in chain conformations. The end result of his analysis is that equation (9.24) is re-cast in the form

$$f = NkT \frac{\langle r_i^2 \rangle}{\langle r_0^2 \rangle} \left(\lambda - \frac{1}{\lambda^2} \right) \tag{9.25}$$

The additional factor $\langle r_i^2 \rangle / \langle r_0^2 \rangle$ is the ratio of the mean-square length $\langle r_i^2 \rangle$ of the network chains in the undistorted state of the network (at volume V) to the mean-square length $\langle r_0^2 \rangle$ of an identical set of free chains.

The significance of the factor $\langle r_i^2 \rangle / \langle r_0^2 \rangle$ arises because in the elementary theory it is assumed that $\langle r_i^2 \rangle = \langle r_0^2 \rangle$, i.e. chains in the unstrained network have the same mean-squared length as the corresponding set of free chains. Even if this were true at one particular temperature, for example the temperature at which the network was formed, it would not be true at any other temperatures since $\langle r_i^2 \rangle$ is determined by the volume of the network whereas $\langle r_0^2 \rangle$ is temperature-dependent in its own right. Differentiation of equation (9.25) leads directly to a relation between the internal energy change on deformation of the rubber and the temperature dependence of the unperturbed dimensions of the chain, i.e.

$$\left(\frac{\delta U}{\delta l} \right)_{V,T} = f - T \left(\frac{\delta f}{\delta T} \right)_{V,l} = fT \frac{\mathrm{d} \ln \langle r_0^2 \rangle}{\mathrm{d}T} \tag{9.26}$$

Using this result Flory showed that ΔU was not in fact zero for real rubbers. Flory's result depends, of course, on the validity of equation (9.25). Allen *et al.*[57] have made rigorous direct measurements of the temperature coefficient of stress at constant volume and proved that Flory's result is substantially correct. A summary of the results for natural rubber is given in Table 9.3.

Table 9.3 Values of relative internal energy contribution (f_e/f) to the stress for natural rubber

f_e/f	*Reference*
0.18	Flory (1961)[56]
0.25 ± 0.11	Roe and Krigbaum (1962)[123]
0.2	Allen *et al.* (1963)[57]
0.15 ± 0.3	Shen (1969)[124]
0.126 ± 0.016	Boyce and Treloar (1970)[125]
0.12 ± 0.02	Allen *et al.* (1971)[57]

At present[58] neutron scattering measurements are being made in an attempt to determine separately $\langle r_i^2 \rangle$ and $\langle r_0^2 \rangle$ for a poly(ethylene oxide) network, cross-linked with polyfunctional isocyanates. The measurements involve the use of a small amount of protonated polymer dissolved in its deuterated matrix as described in Section 9.2.1.5. When these measurements are completed then the remaining parameter in equation (9.25) which needs to be tested is the relationship between the stress supported by the rubber network and the number of chains. This has proved unexpectedly difficult and yet it is important because there are two possible forms of equation (9.21):

$$\Delta A = ANkT(\lambda_1^2 + \lambda_2^2 + \lambda_3^2 - 3) + BNkT \ln \lambda_1 \lambda_2 \lambda_3 \qquad (9.27)$$

The derivation of Flory and Wall[59] yields

$$A = 1, B = 1$$

whereas James and Guth[60] and Edwards[61] obtain results which correspond to $A \to \frac{1}{2}$ and $B = 0$. Recently Allen *et al.*[62] have investigated the moduli of polystyrene gels of carefully controlled degrees of cross-linking and have produced results consistent with the value $A \to \frac{1}{2}$. Even so the precise value of A is not established because of the influence of physical entanglements which enhance the observed modulus.

Finally there remains the question of whether the simple Gaussian equation of state embodied in equations (9.25) and (9.26) really applies to simple networks. An equation of the form

$$f = \left(2C_1 + \frac{2C_2}{\lambda}\right)\left(\lambda - \frac{1}{\lambda^2}\right) \qquad (9.28)$$

has been favoured by other workers[63], but even this does not adequately represent the behaviour of real networks over a wide range of extension ratios. Indeed there is now evidence that the deviations from the kinetic theory of rubber elasticity are likely to be specific for each particular network. One hopes that, as in the case of the behaviour of a non-ideal gas, some useful general statistical treatment will emerge to include, in the case of rubbers, chemical cross-links, physical entanglements and the formation of redundant loops.

9.3.1.2 Dynamic properties

When a bulk polymer sample is subjected to a given mechanical strain, s, the resultant stress, σ, decreases as a function of time; a time-dependent tensile stress-relaxation modulus, $E(t)$, may be defined as

$$E(t) = \sigma(t)/s \tag{9.29}$$

Curves of the type illustrated in Figure 9.8 are obtained[64] when log $E(t)$ is plotted as a function of log(time) for each temperature at which the experiment is performed.

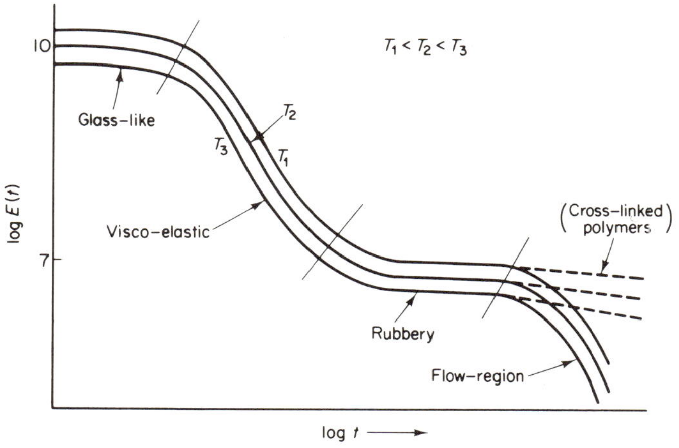

Figure 9.8 Schematic modulus–time curves for a polymer at different temperatures

At very short times and/or at low temperatures the polymer has the high modulus of $\sim 10^{10}$ dyn cm^{-2} characteristic of the glassy state. On a considerably longer time scale and/or at higher temperatures, the polymer has a modulus three or four orders of magnitude lower and exhibits rubbery behaviour. There is a clearly identifiable transition region in which the properties of the material are markedly time-dependent. This is the glass transition region in which the material shows visco-elastic behaviour. At still longer times and/or higher temperatures the molecules begin to undergo large-scale translational motion and the modulus decreases even further, unless this is prevented, by the presence of permanent chemical cross-links of the type discussed in the foregoing section. The *time* at which the major 'transition' from 'glass' to 'rubber'-like behaviour takes place depends upon the *rate* of the long-range conformational motion in the main chain of the polymer, for a given temperature (and pressure, etc.). Clearly the rates of segmental motion in amorphous polymers are of fundamental importance as they determine the bulk physical characteristics of the polymer. Accordingly these motions have been widely investigated by dynamic-mechanical[64], dielectric[27] and n.m.r.[65] techniques. More recently, correlation spectroscopy utilising incoherent quasi-elastic neutron[66] and light scattering[67] techniques have been deployed on this problem.

Mechanical and dielectric relaxation experiments have been performed both as a function of time and frequency. In principle[68], frequency-dependent properties can be obtained as the appropriate Laplace or Fourier transform of the corresponding time-dependent property, and the inverse frequency to time transforms also exist. For example, the complex tensile modulus $E^*(\omega)$ is obtained from the Fourier transform:

$$E^*(\omega) \equiv E'(\omega) + jE''(\omega) = E_r + j\omega \int_0^\infty E(t)\, e^{-j\omega t}\, dt \qquad (9.30)$$

E_r is the equilibrium, or relaxed, component of the storage modulus $E'(\omega)$, and $E''(\omega)$ is the loss modulus. The stress–relaxation modulus is given by the inverse transform:

$$E(t) = \frac{2}{\pi}\int_0^\infty \frac{[E'(\omega) - E_r]}{\omega} \sin \omega t\, d\omega = \frac{2}{\pi}\int_0^\infty \frac{E''(\omega)}{\omega} \cos \omega t\, d\omega \qquad (9.31)$$

Dielectric measurements on polymers can be made over a very wide frequency range, from 10^{-4} to 10^{12} Hz. In this case the measurable frequency-dependent property is the complex permittivity

$$\varepsilon^*(\omega) = \varepsilon' - j\varepsilon''$$

The corresponding time-dependent property is $\phi(t)$, the macroscopic decay function for a parallel-plate sample. $\phi(t)$ is given by[69]

$$\phi(t) = \frac{2}{\pi}\int_0^\infty \left(\frac{\varepsilon_0 - \varepsilon'(\omega)}{\varepsilon_0 - \varepsilon_\infty}\right) \sin \omega t \frac{d\omega}{\omega} = \frac{2}{\pi}\int_0^\infty \left(\frac{\varepsilon''(\omega)}{\varepsilon_0 - \varepsilon_\infty}\right) \cos \omega t \frac{d\omega}{\omega} \qquad (9.32)$$

where ε_0 and ε_∞ are the limiting low and high frequency values of the permittivity. Figure 9.9 illustrates[70] some typical data for the long-range segmental motion (α process) observed for poly(ethyl acrylate) in the rubbery state. The data were obtained at various applied pressures at a fixed temperature. The decay function obtained from these data via equation (9.32) conforms[69] with the empirical relationship

$$\phi(t) = \exp[-(t/\tau_0)^\beta] \qquad 0 < \beta \leqslant 1 \qquad (9.33)$$

for a value of $\beta = 0.38$. This purely empirical decay function characterises[71] the α relaxation observed for many amorphous polymer systems.

Recently a fresh source of information on the segmental motions of elastomeric chains has come from neutron scattering measurements. This technique offers a unique probe for the investigation of segmental motion

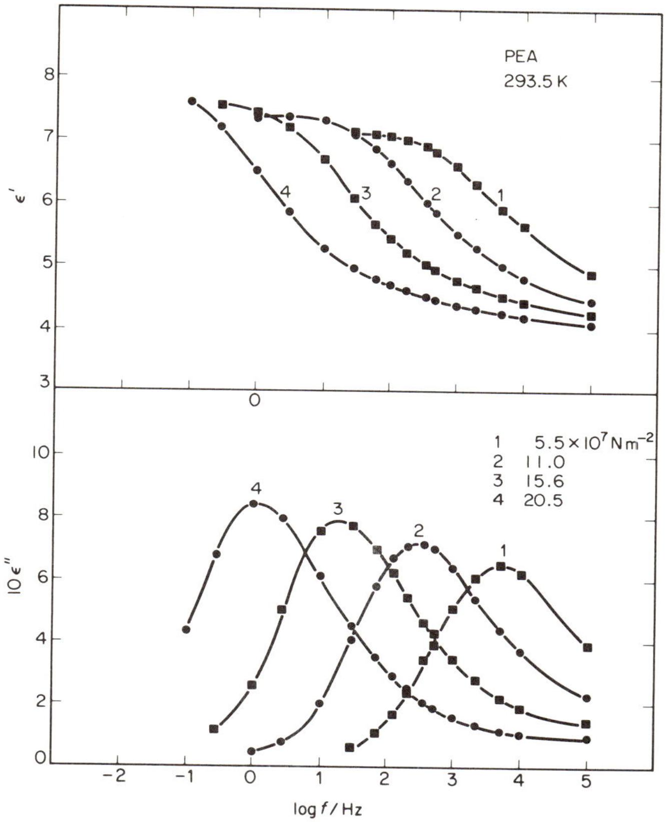

Figure 9.9 Permittivity (ε') and loss factor (ε'') as a function of log frequency for amorphous poly(ethyl acrylate) at 293.5 K and at various applied pressures[70]

since the wavelength of cold neutrons ($\lambda \approx 5$ Å) is of similar magnitude to that of the repeat unit in the polymer. Neutrons may undergo transfer of very small amounts of energy in the scattering process, which results in 'Doppler broadening' of the quasi-elastic peak. This broadening should arise predominantly from the low-energy local motions of the polymer chain. Allen *et al.*[72] have studied poly(dimethylsiloxane) and its low molecular weight precursors, and have obtained values for the self-diffusion coefficient D from the broadening of the quasi-elastic peak, by assuming the validity of a simple diffusion model. Figure 9.10 shows the variation of D with the number of silicon atoms in the chain. A significant aspect of the results is that the motion contributing to D persists in the high molecular weight limit and is similar in magnitude for the larger chains and rings. For this reason it is thought to arise from the diffusive motion of short segments of chain. The activation energy for the process is only 8 kJ mol^{-1} and is not strongly dependent upon intermolecular effects. A similar Doppler broadening has been observed in a poly(propylene oxide) elastomer[73]. Thus while the method of data analysis for this technique is in need of refinement, it has opened up the way to observe the motions of small units. This complements the results from other relaxation techniques which observe segmental motion over a longer chain length.

There are two main aspects to existing theories of the segmental motion of the chain backbone. Firstly, the temperature and pressure dependence of the average relaxation time $\tau = (2\pi f_m)^{-1}$ has been interpreted on the basis of theories of the glass–rubber transition (*cf.* Section 9.3.2.3). Secondly, attempts have been made to explain the shape of the mechanical and dielectric loss peaks.

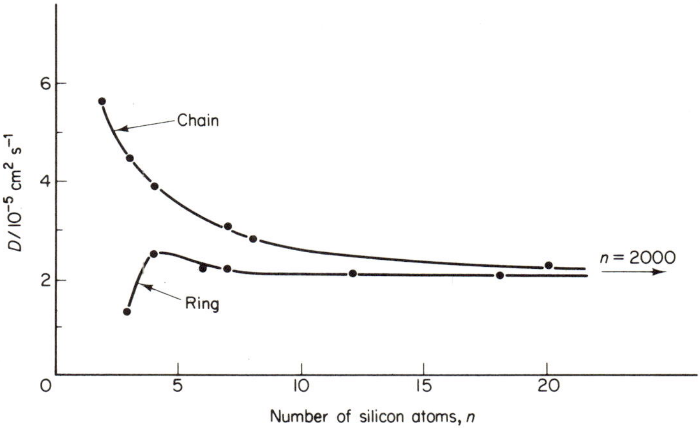

Figure 9.10 Self-diffusion constant D obtained from neutron scattering studies for ring and linear chain poly(dimethylsiloxane)

Relaxation times for segmental motion may be analysed in terms of their dependence upon temperature and pressure to yield activation energies and activation volumes[74]. Constant pressure activation energies, Q_p, for the α process are about 200–400 kJ mol^{-1}, while constant volume activation energies, Q_v, are about 20–30% lower than these values[70]. Activation volumes are of the order of 100–200 cm^3 mol^{-1}. Great caution should be exercised in interpreting these values; they do not correspond to the energy or volume requirements of a single unit in relaxation. The relaxation process is co-operative and involves the concerted motion of many backbone units. The experimental observation that Q_v is a major part of Q_p, for several polymers, favours energy-barrier, rather than free-volume, theories of the a (T_g) process in polymers (*cf.* Section 9.3.2.3). The equation for the dependence of relaxation time on free volume V_f may be written[27]

$$\tau = \tau_0 \exp(B/V_f) \tag{9.34}$$

If V_f is defined as the difference between the actual volume V and the occupied volume, V_0, then since $(\delta \log \tau/\delta T)_V$ is negative, it follows that $(\delta V_0/\delta T)_V$ is negative, an unlikely conclusion.

The shape of loss peaks measured in the frequency domain is determined by the nature of the decay function for relaxation, that is $\phi(t)$, in the dielectric case [see equation (9.32)]. Now when $\phi(t)$ has a form which is non-exponential in linear time, it is always possible[69] to express $\phi(t)$ in terms of a distribution of relaxation times $H(\tau)$

$$\phi(t) = \int_0^\infty H(\tau) \exp(-t/\tau)\, dt \tag{9.35}$$

However, such a procedure may not be necessary to account for the experimental results. Nevertheless, in terms of this approach the normal-mode type of theory, outlined in Section 9.2.2.3, has been employed to account for some of the longer relaxation times in the glass–rubber transition region. The failure to account for the shorter relaxation times results from the fact that motions of units shorter than the sub-molecule are not considered. Theories involving more detailed models of the chain dynamics have been proposed by Kirkwood and Fuoss[76], and by Yamafuji and Ishida[77], to account for the width of the dielectric loss curves. However, these theories also employ parameters such as segmental diffusion coefficients and friction coefficients, which relate to phenomenological rather than molecular models. Furthermore, the approximations involved in the evaluation of these models make it difficult to assess the physical significance of the final equations.

In the case of the dielectric decay function it is possible to formulate the problem in a more tractable manner with the aid of correlation functions. For moderately polar media, the observed decay function $\phi(t)$ is essentially equivalent[69] to $\Gamma(t)$, the molecular dipole correlation function. This is a quantity of fundamental importance in time-dependent statistical mechanics[78], and may be calculable from an appropriate model of the chain dynamics. Neglecting the possible effects due to tacticity, $\Gamma(t)$ for segmental motion may be written as[69]

$$\Gamma(t) = \frac{\gamma_{ii}(t) + \sum \gamma_{ij}(t)}{\mu^2 + \sum \gamma_{ij}(0)} \tag{9.36}$$

where $\gamma_{ii}(t) = \langle \boldsymbol{\mu}_i(0) \cdot \boldsymbol{\mu}_i(t) \rangle$ is the auto-correlation function for a reference dipole i in a polymer chain, and $\gamma_{ij}(t) = \langle \boldsymbol{\mu}_i(0) \cdot \boldsymbol{\mu}_j(t) \rangle$ is the cross-correlation function between dipole i and a further dipole j in the same polymer chain.

Since $\phi(t) \approx \Gamma(t)$, it follows that $\Gamma(t)$ has the empirical form $\exp[-(t/\tau_0)^\beta]$. Williams *et al.*[69] have suggested that the form of this correlation function is not necessarily due to an underlying distribution of relaxation times, but may arise from a non-exponential auto-correlation function $\gamma_{ii}(t)$ due to the co-operative nature of molecular motion in polymer chains above T_g. Support for this concept has come from the fact that $\Gamma(t)$ reduces to $\gamma_{ii}(t)$ when certain assumptions[79] are made concerning the dipolar relaxation, and from the fact that a decay function of the form $\exp[-(t/\tau_0)^\beta]$ has been observed for certain supercooled liquids and solutions of small molecules[79, 80]. This suggests that the segmental mobility of chains above T_g may not be primarily controlled by the intramolecular chain connectivity, but by the intermolecular environment permitting co-operative motion.

With regard to the detailed mechanism for the segmental motions responsible for the α relaxation, an obvious approach would be to attempt to generate the correlation functions by computer simulation of the motions. Although this has been achieved by Monnerie *et al.*[81] for the correlation function $\langle 3\cos^2 \theta(t) - 1 \rangle/2$, which is appropriate to n.m.r. relaxation and fluorescence depolarisation, the long-time segmental motions of polymers are outside the present time range of molecular dynamics studies[81a]. However, the defect diffusion model of Phillips *et al.*[82], based on the earlier work of Glarum[83], has recently been found[84] to be consistent with the relaxation of molecules in supercooled liquids. This may also be an appropriate model for co-operative segmental relaxation in polymers. Edwards'[85] self-consistent approach using continuum models of the polymer chain promises to bring together equilibrium and dynamic properties. This will still leave the problem of extending the model to take into account the detailed effect of molecular structure.

9.3.2 Glasses

The word glass is often associated with transparent materials produced by cooling inorganic melts. In fact many organic polymers at ordinary temperatures exist in a similar physical state. A glass is hard and brittle and is easily distinguished from a rubbery or fluid polymer. Time is important as well as temperature in assessing their properties and there is a well-defined visco-elastic interval which intervenes between the rubber and glass states and which has been described in the preceeding Section 9.3.1.2.

9.3.2.1 The glass temperature

Descriptions of the glassy state are based on consideration of the thermodynamic extensive properties, e.g. volume V, enthalpy H or entropy S as a function of temperature; for a polymer which can be obtained either crystalline or as a glass, curves of the types shown in Figure 9.11 are obtained.

In the range AB, the substance is a rubber with $a_T \approx 7 \times 10^{-4}\ \text{deg}^{-1}$. Since many polymers do not crystallise at all, in the C region a_T reduces continuously to $\sim 2 \times 10^{-4}\ \text{deg}^{-1}$ and becomes constant at that value as the material is transformed to a glass. Crystallisation, where it occurs, is not sharp (Figure 9.11) so that the melting point (T_{m}) must be defined as the break in the curve at B. However, crystallisation is usually very far from complete, and a transition is observed in the neighbourhood of C similar to that for the amorphous polymer. This is a glass transition in the amorphous part of the semi-crystalline polymer.

Wide use is made of the term glass transition temperature (T_{g}), implying that there is a well-defined temperature at which the material transforms suddenly from rubber to glass. In fact the transition is gradual. The term is given a precise meaning by extrapolation of the time-independent parts of the curve to meet at T_{g} or T_{g}' or T_{g}'', as shown in Figure 9.12. This T_{g} becomes dependent on the experimental procedure as well as on the chemical nature of the substance. The timescale of a typical dilatometer experiment is usually long enough to give a 'static' value, i.e. if the timescale is increased by an order of magnitude there is little effect on the estimated value of T_{g}. Tabulations of T_{g} values for polymers usually refer to these conditions.

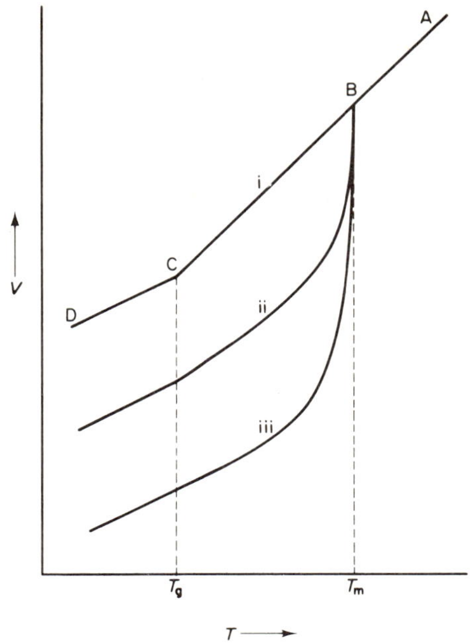

Figure 9.11 Dilatometric curves for melting and glass transitions in a polymer: (i) no crystallisation; (ii) partial crystallisation; (iii) complete crystallisation

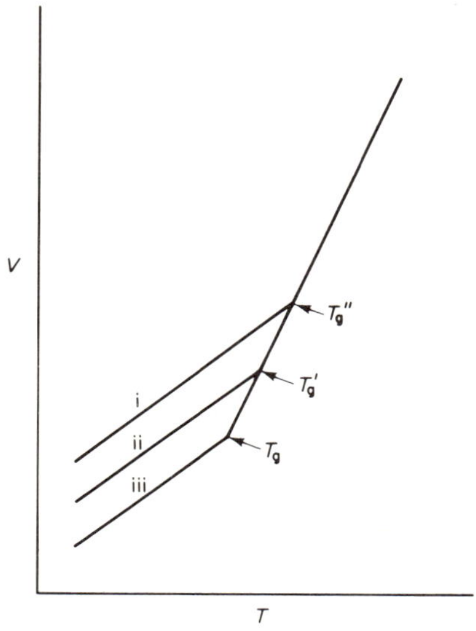

Figure 9.12 The effect of rate of cooling on a glass transition: (i) fast cooling; (ii) slower cooling; (iii) slow cooling

9.3.2.2 *Effect of stress on T_g*

The glass temperature is influenced by mechanical stresses, for example increase in hydrostatic pressure raises T_g. The influence of pressure P on T_g raises the question as to whether a glass transition can be regarded as a thermodynamic transition. This arises from the similarity between the curve ABCD of Figure 9.11 and the thermodynamic definition of a second-order transition, i.e. one in which there is a discontinuity in the second derivatives of the free energy, e.g. the coefficient of cubical expansion $a = (\delta \ln V/\delta T)_P$, compressibility $\kappa = -(\delta \ln V/\delta P)_T$, heat capacity $c_P = (\delta H/\delta T)_P$, etc.

If Δa, $\Delta\kappa$ and Δc_P are the discontinuities (liquid–glass) at the transition temperature T_g (assumed to be sharp), the pressure coefficient of a second-order transition is given by a set of equations[86], of which two are:

$$\frac{dT_g}{dP} = \frac{\Delta\kappa}{\Delta a} \tag{9.37}$$

$$= \frac{TV\Delta a}{\Delta c_P} \tag{9.38}$$

These relationships can be tested by evaluating Δa, etc. from measurements made by avoiding the region of T_g. Thus Δa is obtained by extrapolating values for the rubber from higher temperatures and for the glass from lower temperatures. This procedure is consistent with the method used to define T_g. An analysis[87] of measurements on polystyrene gave the results:

$$\frac{dT_g}{dP} = 3100, \frac{\Delta\kappa}{\Delta a} = 7100, \frac{TV\Delta a}{\Delta c_P} = 3600 \text{ kN}^{-1} \text{ m}^2$$

The first and third agree within experimental error; the second differs significantly.

The interpretation of such anomalies is simpler than is often suggested. Equations (9.37) and (9.38) provide an indirect comparison between two glasses, obtained by cooling from A (Figure 9.11) through the glass transition (i) at atmospheric pressure and (ii) at higher pressure. If the two glasses have the same density, equation (9.37) holds; if they have the same enthalpy and entropy, then equation (9.38) applies. We may use the fact that (9.38) applies to infer that glasses produced under high pressures typically have higher-than-normal densities, but only normal enthalpies. There is now direct evidence of this densification[88, 89].

9.3.2.3 *Static theories of the glass transition*

Free volume theories: The total volume of the rubber is assumed to be divided into the part occupied by the molecules and the part in which the molecules are free to move. The latter is termed the free volume and is communal. As the rubber is cooled, both the occupied and free volumes contract. The glass is distinguished in two ways: (a) the free volume remains constant, independent of temperature, and (b) redistribution of free volume no longer occurs.

This view of the glass transition can be translated into mathematical form[90] in terms of the coefficients of expansion a_l and a_g of rubber and glass. a_g represents the expansion of the occupied volume; thus $a_l - a_g = \Delta a$ must represent the expansion of the free volume. Taking the total volume V_g at T_g as the reference point, then the free volume V_f at $T > T_g$ is given by

$$V_f = V_{fg} + V_g \Delta a(T - T_g)$$

where V_{fg} is the free volume of the glass. The same argument can be developed for the effect of pressure. Now if $\Delta\kappa$ is the difference in compressibility $\kappa_l - \kappa_g$ and κ_g is assumed to be the compressibility of the occupied volume, the effect of both temperature and pressure on the free volume of the rubber is represented by

$$V_f = V_{fg} + V_g \{\Delta a(T - T_g^0) - P\Delta\kappa\} \tag{9.39}$$

where T_g^0 is the glass temperature at $P = 0$. If the glass temperature is identified as the temperature at which the free volume of the rubber tends to V_{fg}, then

$$(T_g - T_g^0)/P = \Delta\kappa/\Delta a \tag{9.40}$$

in agreement with equation (9.37).

Enthalpy and entropy theories: Goldstein[91] notes that just as the volume of a rubber can be divided arbitrarily into 'occupied' and 'free', one may also associate an excess of enthalpy H_e with the molecular freedom which is present in the rubber but which is frozen in the glassy state. If glass formation in a substance always occurs when H_e reaches a certain value, one can show, by an argument similar to that based on a constant free volume, that $dT_g/dP = TV\Delta a/\Delta c_P$. This equation appears to hold reasonably well. Goldstein's hypothesis is consistent with the evidence. Similarly an argument can be formulated in terms of an excess entropy, and yields the same equation.

Kauzmann's classic paper[92] on glasses points to another kind of entropy theory. For a polymer which can be obtained in either crystalline or glassy states the S–T curve for the rubber when extrapolated below T_g cuts the S–T curve for the crystalline material at a temperature between 0 and T_g ($= T^1$, say). Kauzmann suggested that T^1 may be considered to be a thermodynamic glass temperature, since one could not readily imagine a rubber having a lower entropy than the crystal. Gibbs and DiMarzio[93] have developed this concept for linear polymer chains. T^1 is defined by requiring that the conformational entropy shall vanish at that temperature. In physical terms the population of different rotamers is temperature dependent, the rotamer of the lowest energy is favoured increasingly as the temperature falls and the number of conformations available declines. Using a lattice model Gibbs and DiMarzio obtained a relationship between T^1 and the energy difference between rotamers. Since T^1 is not accessible experimentally this cannot be directly checked, but the qualitative conclusion is clearly that T_g increases with increase of conformational flexibility and is independent of intermolecular forces.

9.3.2.4 Transition temperatures and chemical structure

There is both theoretical and technological interest in the problem of relating the glass temperature of a polymer to its chemical structure. The temperature range in which the polymer can be used for a given purpose is limited by the location of T_g. Thus it is desirable to be able to define the factors of which this depends[94]. Two obvious factors are the molecular cohesion and the rigidity of the polymer chain. The former can be assessed by two procedures, based on the cohesive energy density or the internal pressure[95], $P_i = (\delta U/\delta V)_T$. The cohesive energy density (c.e.d.) represents the total energy to separate all the molecules in unit volume and for a liquid is obtainable directly from the molar heat of evaporation, ΔH_{vap}, and volume, V_m: c.e.d. $= (\Delta H_{vap} - RT)/V_m$. This is of course inaccessible for a polymer, but a value can be obtained indirectly from the swelling or solubility of the polymer in a range of liquids of known c.e.d. (It is assumed that the polymer has the same c.e.d. as the liquids in which it swells or dissolves most readily.) The method has to be applied with caution, but gives sensible results in most cases. The other factor is the flexibility of the individual chain and this is not easily quantified. In the free volume theory it is related to the statistical volume element in which mobility has to be generated and the c.e.d. is used as a measure of cohesion.

The entropy theory of Gibbs and co-workers focuses attention entirely on the chain stiffness. This neglect of intermolecular cohesion is not supported by the evidence as the results assembled in Table 9.4 demonstrate. On the other hand, empirical rules based on free volume arguments are successful only within a limited range of polymers.

Table 9.4 **Glass transition temperatures, cohesive energy densities (c.e.d.) and internal pressures (P_i) of some polymers**

Polymer	T_g^0 /°C	c.e.d. /cal cm^{-3}	P_i *at* 20°C
Poly(2,6-dimethylphenylene oxide)	210	—	—
Poly(methyl methacrylate)	105	88	105
Poly(styrene)	100	76	110
Poly(vinyl acetate)	20	~90	—
Poly(methyl acrylate)	0	90	—
Poly(isobutylene)	−70	63	80
Poly(natural rubber)	−72	64	88
Poly(propylene oxide)	−73	80	90
Poly(dimethylenesiloxane)	−120	58	57

9.3.2.5 Dynamic properties

Although the main chain of a polymer molecule in the glass is frozen in one set of random conformations, one or more loss peaks are frequently observed in the glassy state by the various relaxation techniques (these are designated β, γ, δ . . . loss processes). The major secondary transition, the β process, is typically characterised by a very broad loss peak and low intensity, relative to the main-chain α process. Many amorphous polymers contain side-groups capable of undergoing hindered rotation independent of the main chain[27] and the β process can often be ascribed to such rotations. However, some linear polymers, considered to be largely amorphous, such as poly(vinyl chloride) show β peaks which cannot possibly result from side group rotations. Hence some limited local motion of the chain backbone may occur in the glassy state. In fact the magnitudes of the β processes in polymers without flexible side chains are generally lower than β processes of polymers with flexible side chains. The β process is not yet fully understood, but it is significant that the α and β processes converge and frequently coalesce at high temperatures for a number of polymers[27,96] (see Figure 9.13 for polyisobutylene). This may indicate a strong interrelation in their origins even when they are well resolved on the log frequency axis.

The plot of log f_{max} *versus* $1/T$ for β processes is usually a straight line and activation energies of 40–80 kJ mol^{-1} are typical.

A simple model for the dielectric α and β processes has been proposed by Williams and Watts[70] in which the β process is attributed to *partial* relaxation

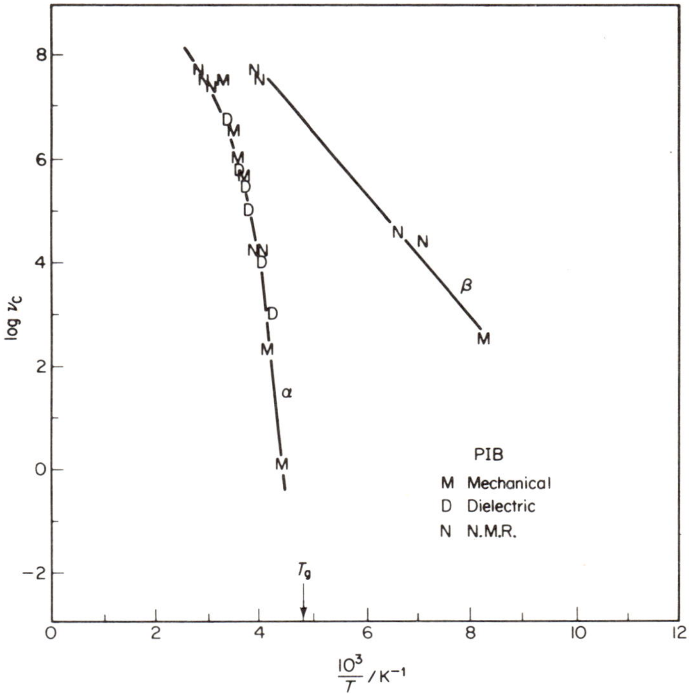

Figure 9.13 Temperature dependence of the correlation frequency for the α and β processes in polyisobutylene

of the chain dipole vector in a range of possible local environments. At longer times and/or higher temperatures, these environments may themselves be relaxed by the onset of the α process. At high temperatures the local environments relax so quickly that the local motions are not detected. This explains the convergence of the primary and secondary processes in the rubbery state, and several other features of the β process. This model does not specify the detailed mechanism for local motions. However, Johari and Goldstein[80] have observed a very broad β process in a number of small molecule glass-forming systems which appears to be quite similar to that observed in polymers. This implies that the local environment is a key factor in local relaxation.

A very clear example of the identification of a β process with a detailed molecular mechanism lies in the work of Heijboer[97], who studied the viscoelastic relaxation behaviour of several methacrylate copolymers containing the cyclohexyl group in the ester side chain [see (1)]. In Figure 9.14 it can

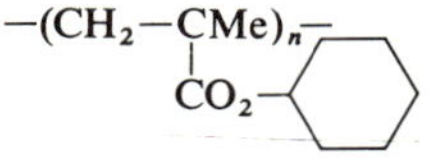

(1)

be seen that the magnitude but not the temperature of this low-temperature relaxation is affected by the concentration of cyclohexyl groups. The observed activation energy for this relaxation is 48.3 kJ mol^{-1}, identical with that observed for the mechanical relaxation in many low molecular weight

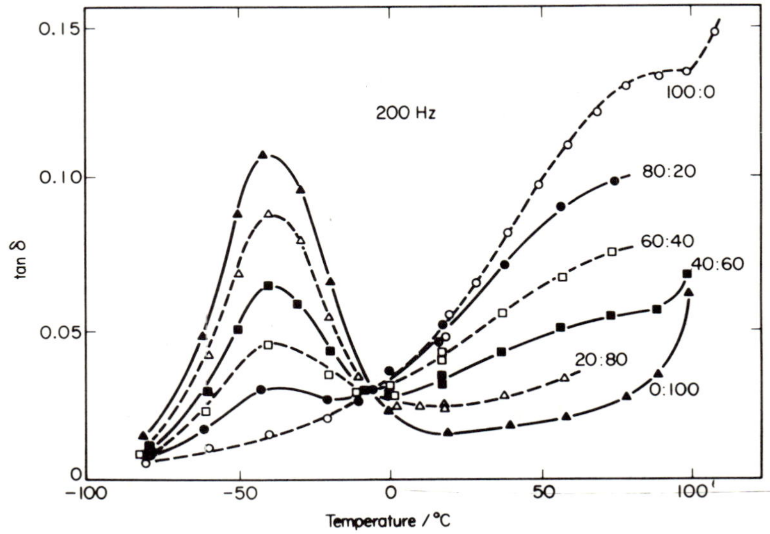

Figure 9.14 Temperature dependence of tan δ for copolymers of methyl methacrylate and cyclohexyl methacrylate, at different methyl:cyclohexyl percentage compositions

cyclohexyl derivatives. This leads to an identification of the polymeric relaxation with the flipping of the substituted cyclohexyl ring from one chain conformation to the other.

A number of detailed molecular theories[27] have been proposed to account for the secondary processes in polymers without flexible side chains. In the 'local mode' theories the origin of these relaxation processes is attributed to damped torsional oscillations about carbon–carbon bonds. The amplitudes of these vibrational modes are predicted to range from low values at the highest frequencies to large values at the lower frequencies. The breadth of the mechanical and dielectric loss curves in the β relaxation region can be readily understood in terms of the wide frequency distribution of the vibrations. The normal Arrhenius dependence of relaxation frequency and the insensitivity of some β relaxations to pressure can also be deduced from the theory. Calculation of the activation energy for the local mode relaxation in polyethylene yields a figure of about 40 kJ mol^{-1}, in reasonable agreement with experiment.

In the particular case of methyl side groups in glasses, weak mechanical losses occur at very low temperatures (δ processes)[98]. Activation energies are in the range 4–30 kJ mol^{-1}, and these correlate well with the potential barriers hindering internal rotation, V_3, which can be calculated from the torsional fundamental frequency, as observed by neutron scattering[99].

9.3.3 Crystalline polymers

In the introduction we noted that crystalline polymers are usually only partially crystalline. This arises because of the difficulties of sorting out the tangled array of long molecules present in the rubbery state as the crystal regions grow. Inevitably as nucleation occurs at different sites in an isotropic material, different parts of the same chain may become involved in crystal growth in different regions of the matrix of the polymer. Crystallinity then develops to the point at which the boundaries of the ordered regions are separated by hopelessly entangled amorphous domains from which further chains cannot be extracted. As we shall see, the balance of amorphous and crystalline material can be predicted by pre-orienting or post-orienting the sample in relation to the crystallising process. As a result one encounters crystalline polymers not only as thermoplastic materials of the type used in many domestic articles but also in highly drawn forms such as synthetic fibres of various kinds and tapes. Above the glass transition the crystalline polymer consists of crystals embedded in a rubbery matrix. Below the glass transition the matrix is a glass. Clearly the overall properties of the material are dependent on the nature of the amorphous component and on the degree of crystallinity achieved. In this part of the review we concentrate on the properties of the crystalline region. In general we assume that the crystalline domains have developed to a point at which their properties are independent of size. We also assume that the chemical structure of the polymer so far as its stereochemical isomers are concerned is pure, and that we are dealing with the limiting properties of isotactic or syndiotactic forms where these exist.

9.3.3.1 Relation between chemical structure and crystal structure

We noted in Section 9.2.1.4 that the conformations of polymer chains in the crystal have been determined by x-ray diffraction methods. Broadly speaking the chains[100] are arranged as assemblies of planar zig-zags for polyethylene and nylon type structures (see Figure 9.15). For other polymers[100] helical or folded form spredominate. Structures such as —$(CH_2O)_n$— and —$(CH_2$—$CH_2O)_n$— crystallise in complicated helices and exist in several different crystal forms, none of which contains a planar zig-zag structure. This is explained in terms of the lower energies of the gauche rotational conformations around the C—O bonds and their electrostatic repulsion. Isotactic structures of the type —CH_2CHR— invariably crystallise as helices too, but here the structure of the helix is regular and is determined largely by the size of the side group R[101]. In Table 9.5 we summarise the number of monomers per turn of the helix for a range of α-olefin polymers.

Table 9.5 Helical conformations of α-olefin polymer chains in crystals

Monomer units per turn	R
Planar zig-zag	H
3:1	—Me, —Et, —$CH{=}CH_2$
	—$CH_2CH_2CHMe_2$
	—OMe, —OCH_2CHMe_2
7:2	—CH_2CHMe_2
	—$CH_2CHMeCH_2Me$
4:1	—$CHMe_2$

It is interesting to observe that helices are a common feature of the structure of synthetic polymers which do not possess hydrogen bonds and therefore one cannot argue in the case of proteins that the presence of intramolecular hydrogen bonds is absolutely essential to the conformation of the α-helix, though clearly hydrogen bonds are a stabilising influence.

The relationship between the melting points of crystalline polymers and chemical structure has been a matter of considerable debate. One can try to correlate melting points with structure in terms of the thermodynamic relationship

$$T_m = (\Delta H_m/\Delta S_m) \qquad (9.41)$$

A priori one might consider that the enthalpy of fusion is dominated by intermolecular forces and that the entropy of fusion is determined by the randomness of the conformations which the polymer chains adopt in the melt. Unfortunately this leaves out the contribution from volume changes on melting. Several workers have investigated the contribution to the enthalpy and entropy of fusion arising (a) from the randomness of the structure at constant volume and (b) the increase in volume from that of the crystal to the melt. Unfortunately it appears[102, 103] that the two contributions are comparable in size and so there is no very simple correlation to be made.

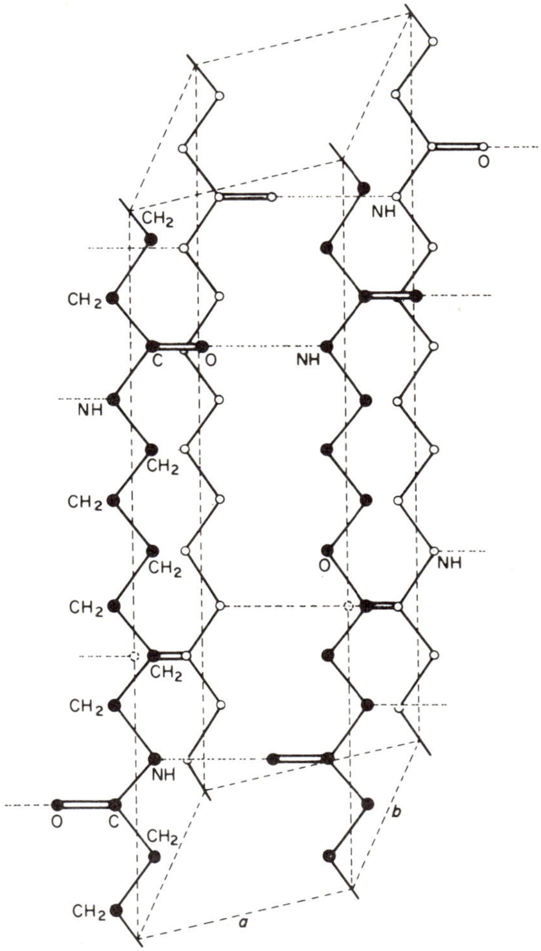

Figure 9.15 Structure of Nylon 66 ($[NH(CH_2)_6$-$NHCO(CH_2)_4CO]_n$)

However, within a given class of polymers there is generally a relationship between melting point and structure which corresponds to the degree of conformational flexibility the polymer chain can adopt in the melt. The more σ-bonds there are in the repeat unit, the larger the entropy of fusion tends to be and hence the lower the observed melting point. Such trends are born out broadly by the entries in Table 9.6.

Table 9.6   Heats, entropies and temperatures of fusion

Polymer	T_m /°C	ΔH_f /cal g^{-1}	ΔS_t /cal deg^{-1} g^{-1}
Ethylene	139	66	0.16
Propylene	176	62	0.13
But-1-ene	126	59	0.10
$-(CH_2)_2-OCO-\langle\bigcirc\rangle-OCO-$	265	30	0.56
$-(CH_2)_4-OCO-\langle\bigcirc\rangle-OCO-$	232	34	0.67
$-(CH_2)_6-OCO-\langle\bigcirc\rangle-OCO-$	160	33	0.76
$-(CH_2)_{10}-OCO-\langle\bigcirc\rangle-OCO-$	138	36	0.88

9.3.3.2   Morphology

Crystallisation in a polymer can arise in two different circumstances: (a) when the amorphous melt is held at a temperature below the melting point but above the glass transition temperature to give thermally induced crystallisation, and (b) when the rubbery material is strained to give stress-induced crystallisation.

In the case of thermo-crystallisation the crystal growth is random and the x-ray patterns consist of Debye Scherrer rings. In the stress-induced samples they usually consist of arcs. Additionally the fibre-type pattern also results when thermo-crystallised material is stretched, e.g. in the cold drawing process which is used to produce nylon fibres. In this section we wish to examine the texture of the crystalline material, i.e. the shape and size of the crystal entities and their forms and aggregation.

Since this review is primarily concerned with polymers in bulk we are concerned exclusively with melt crystal polymers and immediately a major problem presents itself in that the examination of the fine structure of the partially crystalline polymer is inaccessible to most present day techniques, e.g. electron microscopy. The use of surface replicas, whilst being one of the

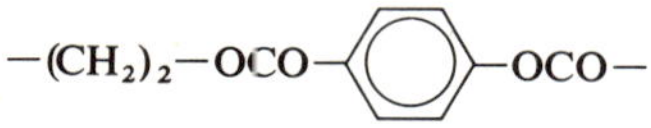

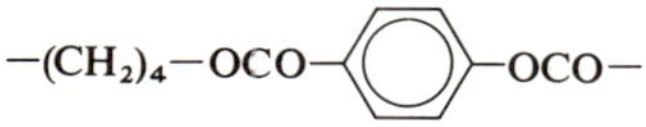

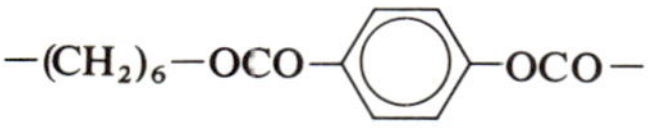

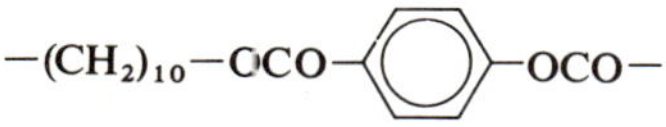

few methods of investigation available, is often open to question and only in a few instances has the investigation of fracture surfaces provided decisive information. In metallurgy the solvent etching technique has proved very useful, but in investigating the morphology of polymers it has proved of little value because of swelling effects at the surfaces. However, in the past decade two methods have been particularly useful. In 1964 Palmer and Cobbold[104] introduced the use of nitric acid to disperse bulk polyethylene into its crystalline constituent by selectively oxidising the amorphous material binding them together. This treatment is of course drastic and must be used with care, and even now the amount of structural change caused by the technique is not quite clear. The second method[105] used crystallisation through increasingly concentrated solutions to extrapolate towards the behaviour of melts in the form of thin films.

Although a vast amount of work has been done, at the present time the only comprehensive body of information refers to crystallisation of polyethylene[106] and for this reason the morphological picture presented here relies very heavily on the results obtained for this particular polymer.

In the case of polyethylene the crystalline units are lamellae. They are elongated and bounded by (110) and (100) faces with the direction of links along the *B* faces. Much of the work of the last decade suggests, but by no means proves conclusively, that the lamellae include folded chains, as is the case for lamellae deposited from dilute solution.

Dilute solutions[107] were first used to produce single crystals of polymers because it was considered necessary to separate the molecules sufficiently far apart for one molecule only to be involved in the growth of a single crystal. However, in melt-grown crystals the possibility of molecular links between lamellae is high. In fact the number of chains joining one lamellar to the next compared with those which stay in a given lamellar by repeated folding is likely to be an important morphological characteristic of the polymer. Such interlamellar chains have been identified in crystalline mixtures of polyethylene and normal paraffins and a concensus view is that generally speaking the interlamellar links are more often found in the form of fibrous crystals consisting of a few hundred chains rather than individual molecular threads.

The lamellae in crystalline polymers grow together to generate a fibular structure which is ribbon-like and often helically wound. Under the optical microscope one then sees this fibrillating crystal growth leading to spherulites. It must be emphasised at this point that the morphological features of spherulites are not confined to polymers. They occur commonly in the crystal growth behaviour of many simple materials and even for these there are still many unsolved problems, particularly those relating to the space filled by the twisting fibrils and the origin of the twist itself.

There has been much discussion[107] of the role of impurities in generating spherulite growth, but even this is equivocal. Some non-polymeric organic substances have been subjected to the most rigorous refining techniques and they still produced materials with spherulite morphology, and thus one cannot attribute all spherulite growth in polymers to the presence of impurities. The origin of non-crystallographic branching which must inevitably be present in the spherulite is still obscure, though very recently it has been observed that an intermediate structure between fibrilar and spherulite, e.g.

hedritic growth, does occur at a lower rate than spherulite growth. There is considerable evidence to suggest that the chain-folding mechanism which is operative for crystallisation from solution is valid for crystallisation from the melt.

In both solution and melt crystallised polymers it appears that the highly folded chain is not the most stable thermodynamic form of the system[108]. Extended chain crystals, or at any rate crystals with a very long folded length, are more stable. Such crystals can be obtained in two ways. Crystallisation at high pressures[109] tends to give extended-chain polyethylene crystals which are quite different in structure from the folded-chain crystals obtained from solution and from crystallisation at ordinary pressures. Under other conditions[110] 'shish kebab' type structures are generated in which the axis of the 'shish kebab' consists of fibrular crystals with extended chains and there is a lamellar over-growth which results finally in the 'shish kebab' appearance of the polyethylene.

Mention of extended chain crystals raises the question of structure–property relations in synthetic fibres. Generally the fibre is spun as a crystalline material and then usually cold drawn to enhance the mechanical properties (especially tensile strength and modulus) in a process which gives a high degree of molecular alignment. However, the ultimate modulus along the fibre axis is always much less than the theoretical modulus which can be calculated from the crystallographic structure. This is due to chain discontinuities along the fibre axis. Such calculations show that polyethylene should be the synthetic fibre with highest modulus since this structure contains the largest number of molecular chains per unit cross-section. Two very important notes have appeared which bear on this point. Mackley *et al.*[111] noted that dilute solutions of polyethylene in xylene produced material with a 'shish kebab' morphology when the crystal deposit was obtained by impinging two jets. By injecting molten polymer through two mutually opposed orifices a longitudinal velocity gradient is generated which results in fibrous crystallisation of the polyethylene which may well have different properties from materials obtained hitherto. More recently, Ward and Capaccio[112] have reported a study of the effect of draw ratio on partially crystalline polyethylene which has given fibres of unusually high modulus—within a factor of four of the theoretical value.

9.3.3.3 Dynamic properties

Since polymers crystallised from the bulk normally contain a finite proportion of amorphous material, these essentially composite materials exhibit complex relaxation behaviour. Thus it is necessary to determine which of the observed relaxation processes originate from the crystalline regions before relating these processes to specific molecular mechanisms. The relaxation behaviour of the amorphous phase can be a complicated factor because the rates of segmental motion are often altered with respect to pure amorphous samples by the presence of crystallites. For example, the glass transition temperature of polyethylene is a highly controversial subject, and estimates in the literature[113] cover a range of about 100 °C. For polymers which cannot be obtained in a pure amorphous solid form, it is desirable to compare the high-frequency relaxation behaviour of the melt with that of the semi-crystalline solid. This has been achieved recently for poly(oxymethylene)[114], poly(ethylene oxide)[115] and poly(hexamethylenesebacamide)[116].

When it is possible to study the relaxation behaviour as a function of percentage crystallinity (χ), observed processes which increase in intensity as χ increases are identified as belonging to the crystal. Supporting evidence has come in a number of instances, especially for polyethylene, from dynamic studies of solution-grown single crystals[75], and from crystals of extended chain n-alkanes consisting of from 10 to 100 —CH_2— units[75]. Solvent extraction of components of lower molecular weight and controlled variations in chain structure (e.g. branching or tacticity), have also been used. The detailed morphological information which is now available enables the effect on relaxation behaviour of varying lamellar thickness and defect type and concentration to be evaluated[75].

On the basis of current morphological information, the following types of crystalline process are possible[75, 117]: (a) co-operative motions of the molecular chains along the length of the crystallite—these relaxations will be related to the lamellar thickness; (b) motions associated with defects such as end groups incorporated in the crystal, which could also involve chain movement; (c) motions associated with morphological details, such as chain folds, or of large morphological features such as interlamellar shear or interfibrillar shear; (d) motions of restricted chain lengths such as local motions, also postulated to exist in the glassy amorphous state.

In evaluating these models it must be remembered that different relaxation techniques respond with different sensitivity to the various types of molecular motion. Thus, for example, dielectric loss peaks are observed for polyethylene and these are attributed to the presence of polar groups formed by oxidation at the chain ends. Relaxation processes due to chain ends are therefore more likely to originate from motions of type (b) above, than from motions within the crystal lattice. However[118], the copolymer of ethylene with carbon monoxide evidently incorporates the carbonyl group within the polyethylene crystal lattice, and therefore the dielectric technique is used to detect the libration of the chain within the lattice of this polymer.

The motional processes postulated to explain the observed relaxation behaviour for crystals of polyethylene and of n-alkanes were analysed by Hoffman *et al.*[75] in terms of simple site models. Their $a_c - B$ model for extended chain crystals is summarised in Figure 9.16. The figure illustrates the chain-relaxation process from site 1, involving the rotation of the chain through 180° together with its simultaneous *translation* along its axis to reach a new equilibrium position, site 2, of slightly higher energy. The reverse process takes place over the free energy barrier $\Delta Q^*(n)$ in the same way. For such a two-site model it may be shown that the relaxation time is

$$\tau = [2(\omega_{12} + \omega_{21})]^{-1} \approx A^{-1} \exp [\Delta Q^*(n)/kT] \tag{9.42}$$

where ω_{21} and ω_{12} are transition probabilities. The activation free energy for a chain of n carbon atoms may be written in terms of enthalpy and entropy of activation, so that

$$\Delta Q^*(n) = \Delta H^*(n) - T\Delta S^*(n) \tag{9.43}$$

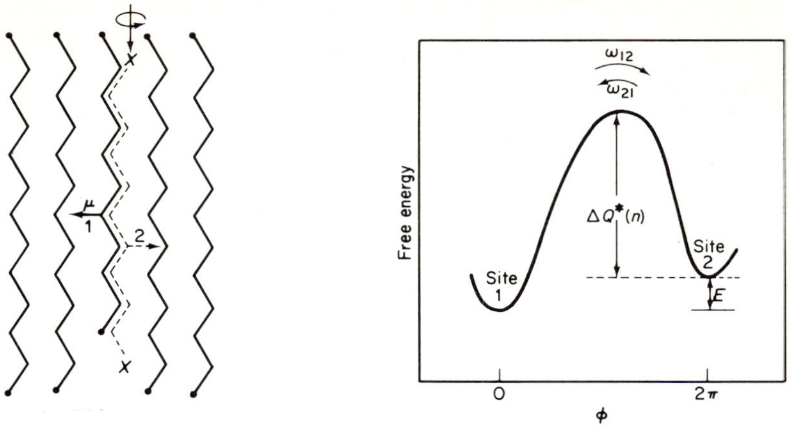

Figure 9.16 $a_c - B$ model for relaxation in extended-chain crystal

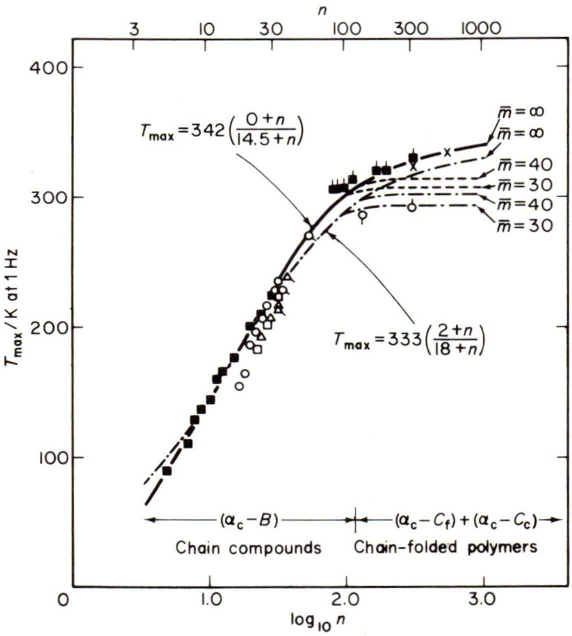

Figure 9.17 The temperature of maximum dielectric or mechanical loss at 1 Hz plotted against log (no. of carbon atoms) for extended-chain crystals and chain-folded crystals, together with theoretical curves [equation (9.45)]. The dashed lines ($\bar{m}$ = 40 and $\bar{m}$ = 30) correspond to the modification of the rigid-rod approximation to include the effect of chain-twisting

Then if the chain rotation is identical to the rotation of a *rigid* rod,

$$\Delta H^*(n) = 2\Delta H^*_{end} + (n - 2)\,\Delta H^*_{CH_2} \tag{9.44}$$

where ΔH^*_{end} and $\Delta H^*_{CH_2}$ refer to the enthalpy of each end group and that of each segment, respectively. A similar expression may be given for $\Delta S^*(n)$, and combination of these equations yields the expression

$$T_{max} = T_0\,(a + n)/(b + n) \tag{9.45}$$

T_{max} is the temperature of maximum loss, for a fixed frequency of measurement, T_0 is the value of T_{max} for very long chains ($n \rightarrow \infty$) and a and b are factors involving the activation parameters and the relaxation time. A value of T_{max} is obtained which rises rapidly for small n and levels off to the value T_0 as n becomes large (see Figure 9.17). However, the rigid-rod approximation does not hold for high values of n, and so the theory has been modified to allow for the effect of chain twisting and Figure 9.17 shows the good fit obtained with experimental data. The intensities of relaxation processes have also been calculated on the basis of these site models[75].

In addition to relaxation processes, the vibrational motions of crystals can be studied by optical[119] and, more recently[120], neutron scattering techniques. Of particular interest are phonons, or elastic waves in a crystal, which involve longitudinal and transverse vibrations of the entire lattice. The longitudinal motion of a model one-dimensional chain, containing two species of atoms of masses m_1 and m_2,

$$\begin{array}{c} —\mathrm{O}—\mathrm{O}—\mathrm{O}—\mathrm{O}—\mathrm{O}—\mathrm{O}—\mathrm{O}— \\ m_1 \quad m_2 \quad m_1 \quad m_2 \quad m_1 \quad m_2 \end{array}$$

yields a dispersion relation involving two branches: (i) the *acoustical* modes, in which adjacent particles move together with equal amplitude, as when a sound wave is transmitted along the chain, and (ii) the *optical* modes, in which adjacent particles move against each other but the centre of mass does not move.

The shapes of the optical and acoustical dispersion curves can be predicted if the molecular force field is known. Strong intramolecular coupling along the chain produces curved, strongly phase-dependent dispersion behaviour. Comparison of calculated spectra with optical spectra is limited by the optical selection rules which apply to infrared and Raman spectra, but in incoherent inelastic neutron scattering spectroscopy all modes of motion are observable. Thus the density-of-states spectrum $g(\omega)$ can be observed, and compared directly with $g(\omega)$ computed from the potential function of the molecule. In addition, coherent neutron scattering is used to map these phonon dispersion curves as a function of momentum transfer[120].

In proton-containing polymers such as polyethylene, the strong incoherent scattering usually masks the coherent component and so completely deuterated samples must be used in order to observe the coherent scattering. A comprehensive study of deuterated polyethylene has been made by White *et al.*[120] in which measurements were made both parallel and perpendicular to the oriented chain axes. The velocity of sound was calculated from the observed acoustic phonons for each crystallographic direction. An estimate of the elastic moduli of the polyethylene crystal can be made from these results, on the assumption that the molecule can be treated as a uniform elastic rod. Then

$$E = \rho V^2$$

E is Young's modulus and V the velocity of sound in a given direction; ρ is the crystallographic density. In the direction parallel to the polyethylene chain, $E_{\parallel} = 3.0 \times 10^{12}$ dyn cm^{-2}, which compares well with the value obtained from Raman spectra. In the direction perpendicular to the chain axis, $E_{\perp} = 15 \times 10^{10}$ dyn cm^{-2}. Of course, a simple mechanical measurement made on a laboratory sample would give an isotropic value of E which differs both from $E_{\parallel}$ and $E_{\perp}$ because it measures the overall response of the partially crystalline matrix. Measurement of these anisotropic values for the essentially perfect crystal aids in understanding how the macroscopic E value results from the averaged dynamic response of the amorphous and crystalline components.

An expression for the modulus E of a polyethylene chain in a perfect crystal can be obtained[121] from a calculation of the force required to extend the planar zig-zag molecule

$$E = \frac{l \cos \theta}{A}\left[\frac{\cos^2 \theta}{k_l} + \frac{\sin^2 \theta}{4k_p}\right]^{-1}$$

where l is the bond length, A is the cross-sectional area of the chain, θ is the angle between the bonds and the overall axis, k_l is the force constant for separation of the carbon atoms and $k_p = (k_\alpha/l^2)$, where k_α is the valence angle force constant. Treloar[121] calculated a value for E of 1.82×10^{12} dyn cm^{-2}; Shimanouchi *et al.*[122], using more exact values for the force constants derived from a Urey–Bradley potential, obtained $E = 3.4 \times 10^{12}$ dyn cm^{-2}. This latter value is in fairly good agreement with the neutron scattering results. Other calculations of this type suggest that it is now possible to calculate the elastic moduli in the direction parallel to the chain axis to within about 20% for most polymer crystal structures.

9.4 CONCLUSION

In covering such a wide field this review has omitted many aspects of structure/property relations in polymers. The properties of polymer solutions and of gels are scarcely mentioned, nor are electrical properties. However, we hope the reader will see that progress in the study of the states of polymerised matter has reached a similar maturity to the study of the states of matter of simple substances. Admittedly molecular weight distributions complicate the issue for polymers but the statistical methods available are adequate to deal with this complication. In fact, barriers to progress in the study of polymeric materials are now abreast of those for simple substances. Obvious examples given in the review include the theory of the glassy state and the limits to our understanding of the dynamics of both large and small molecules. Experiments of simple specification on well-characterised polymers are capable of interpretation as precise as for simple molecules.

The future of polymer science has two particularly exciting aspects. First, theoretical physicists like de Gennes and Edwards have established a more fundamental basis for the theory of polymerised materials from which considerable progress will be made. We can expect self consistent equations which incorporate both dynamic and static features of rubbers and gels. (The glassy state promises to be more difficult.) To match theoretical advances an array of scattering experiments is just beginning to produce results to test these theories. Neutron scattering (both coherent and incoherent), Rayleigh scattering and the use of synchrotron radiation are gaining in importance and they will call for samples prepared with even more care (e.g. much lower levels of dust and other impurities). Second is the convergence and cross-fertilisation of the understanding of synthetic polymers and naturally occurring polymeric systems occurring *in vivo* and *in vitro*. Synthetic analogues of natural systems and the use of synthetic polymers in biological environments offer much scope for material science and biological engineering.

We look forward to an era when the physics of systems of large molecules is an integral part of chemical physics—and not something tacked on as a special topic—and to a time when reviews on the structure/property relationship in polymers draw examples freely from the science of synthetic *and* natural macromolecules.

References

1. Bunn, C. W. (1958). *Discuss. Faraday Soc.*, **25,** 95
2. Flory, P. J. (1969). *Statistical Mechanics of Chain Molecules* (New York: Interscience)
3. Volkenstein, M. V. (1963). *Configuration Statistics of Polymeric Chains*, (New York: Interscience)
4. Treloar, L. R. G. (1958). *Physics of Rubber Elasticity* (Oxford: University Press)
5. Flory, P. J. (1953). *Principles of Polymer Chemistry* (Ithaca: Cornell University Press)
6. Ref. 2, p. 19
7. Williams, A. J. and Flory, P. J. (1968). *J. Polymer Sci. A2*, **6,** 1945
8. Akers, P. J., Allen, G. and Bethell, M. (1968). *Polymer*, **9,** 575
9. Borrales-Rienda, J. M. and Pepper, D. C. (1967). *Eur. Polymer J.*, **3,** 535
10. Opshoor, A. (1968). *Polymer*, **9,** 599
11. Allen, G. and McAinsh, J. (1969). *Eur. Polymer J.*, **5,** 319
12. Wall, F. T. and Erpenbeck, J. J. (1959). *J. Chem. Phys.*, **30,** 634
13. Domb, C. (1963). *J. Chem. Phys.*, **38,** 2957
14. Edwards, S. F. (1965). *Proc. Phys. Soc.*, **85,** 613
15. Edwards, S. F. (1965). *Critical Phenomena, NBS Misc. Publication*, **273,** 225
16. Edwards, S. F. (1967). *Proc. Phys. Soc.*, **92,** 9
17. Edwards, S. F. (1969). *J. Phys. C*, **2,** 1
18. Edwards, S. F. and Freed, J. K. (1969). *J. Phys. A*, **2,** 145
19. Miller, R. L. (1966). *Polymer Handbook Section 3*, (J. Brandrup and E. H. Immergut, editors) (New York: Interscience)
20. Mark, J. E. (1973). *Rubber Chem. Technol.*, **46,** 593
21. Ref. 5, p. 602
22. Egelstaff, P. A. (1969). *Introduction to the Liquid State* (London: Academic Press)
23. Kirste, R. G., Kruse, W. A. and Schelten, J. (1973). *Makromol. Chem.*, **162,** 299
24. Benoit, H., Cotton, J. P., Decker, D., Farnoux, B., Higgins, J. S., Jannink, G., Ober, R. and Picot, C. (1973). *Nature, Phys. Sci.*, **245,** 13
25. Ballard, D. G. H., Wignall, G. D. and Schelten, J. (1973). *Eur. Polymer. J.*, **9,** 965
26. Allen, G., Higgins, J. S., Maconnachie, A., Mobbs, R. and Wright, C. J. (1974). To be published
27. McCrum, N. G., Read, B. E. and Williams, G. (1967). *Anelastic and Dielectric Effects in Polymeric Solids*, (New York: John Wiley)
28. (1970). *Polymer Solutions, Discuss. Faraday Soc.*, **49**

29. Wada, A., Suda, N., Tsuda, T. and Soda, K. (1969). *J. Chem. Phys.*, **50,** 31
30. Block, H. and North, A. M. (1970). *Advan. Mol. Relaxation Processes*, **1,** 309
31. Burgers, J. M. (1938). *Verhandel. Koninkl. Ned. Akad. Wetenschap. Afdel, Natuurk., Sec. 1*, **16,** 113
32. Broersma, S. (1960). *J. Chem. Phys.*, **32,** 1626
33. O'Konski, C. T. and Haltner, A. J. (1956). *J. Amer. Chem. Soc.*, **78,** 3604
34. Wada, A. (1961). *Polamino acids, Polypeptides and Proteins*, 131 (M. A. Stahman, editor) (Madison, Wisconsin: University of Wisconsin Press)
35. Block, H., Hayes, E. F. and North, A. M. (1970). *Trans. Faraday Soc.*, **66,** 1095
36. Yu, H., Bur, A. J. and Fetters, L. J. (1966). *J. Chem. Phys.*, **44,** 2568
    Bur, A. J. and Roberts, D. E. (1969). *J. Chem. Phys.*, **51,** 406
37. Lockhead, R. Y., Dev, S. B. and North, A. M. (1970). Ref. 26, p. 244
    Lockhead, R. Y. and North, A. M. (1972). *J. Chem. Soc. Faraday Trans. II*, **68,** 1089
38. North, A. M. (1972). *Chem. Soc. Rev.*, **1,** 49
39. Rouse, P. E. (1953). *J. Chem. Phys.*, **21,** 1272
40. Bueche, F. (1954). *J. Chem. Phys.*, **22,** 603
41. Zimm, B. H. (1956). *J. Chem. Phys.*, **24,** 269
42. Tobolsky, A. V. (1968). *Polymer Sci. A2*, **6,** 1177
43. Stockmayer, W. H., Gobush, W., Chikahisa, Y. and Carpenter, D. K. (1970). *Discuss. Faraday Soc.*, **49,** 182
44. Van Beek, L. K. H. and Hermans, J. J. (1957). *J. Polymer. Sci.*, **23,** 211
45. Kastner, S. (1961). *Kolloid Z.*, **178,** 24, 119; *Z. Electrochem.*, **65,** 669
46. Stockmayer, W. H. and Baur, M. E. (1964). *J. Amer. Chem. Soc.*, **86,** 3485
47. Baur, M. E. and Stockmayer, W. H. (1965). *J. Chem. Phys.*, **43,** 4319
48. Stockmayer, W. H. (1967). *Pure Appl. Chem.*, **15,** 539
49. Jones, A. A., Brehm, G. A. and Stockmayer, W. H. (1973). Paper presented at the Tobolsky Memorial Symposium, Princeton, N.J., 1973. To be published in *J. Polymer Sci. C*
50. Bates, T. W., Ivin, K. J. and Williams, G. (1967). *Trans. Faraday Soc.*, **63,** 1964
51. Kryszewski, M. and Marchal, J. (1958). *J. Polymer. Sci.*, **29,** 103
52. Davies, M. M., Williams, G. and Loveluck, G. D. (1960). *Z. Electrochem.*, **64,** 575
53. de Brouchere, L. and van Nechel, R. (1952). *Bull. Soc. Chim. Belges*, **61,** 261, 452
54. Bauer, H.-J., Hässler, H. and Immendörfer, M. (1970). *Discuss. Faraday Soc.*, **49,** 238
55. Treloar, L. R. G. (1958). *Physics of Rubber Elasticity* (Oxford: Clarendon Press)
56. Flory, P. J. (1961). *Trans. Faraday Soc.*, **57,** 829
57. Allen, G., Bianchi, U. and Price, C. (1963). *Trans. Faraday Soc.*, **59,** 2493
    Allen, G., Kirkham, M. C. ,Padget, J. C. and Price, C. (1971). *Trans. Faraday Soc.*, **67,** 1278
58. Allen, G., Higgins, J. S., Maconnachie, A., Wright, C. J. (1974). To be published
59. Flory, P. J. and Rehner, J. (1943). *J. Chem. Phys.*, **11,** 512, 521
    Wall, F. T. (1942). *J. Chem. Phys.*, **10,** 485
60. James, H. M. and Guth, E. (1943). *J. Chem. Phys.*, **11,** 455
61. Edwards, S. F. and Freed, K. (1970). *J. Phys. C*, **3,** 739, 750, 760
62. Allen, G., Walsh, D. J. and Holmes, P. (1974). To be published
63. Rivlin, R. S. (1948). *Phil. Trans. Roy. Soc. A*, **241,** 379
64. Ferry, J. D. (1969). *Viscoelastic Properties of Polymers* (New York: John Wiley)
65. Slichter, W. P. (1971). *N.M.R. Basic Principles and Progress*, **4,** 209 (Berlin: Springer-Verlag)
66. Allen, G., Higggins, J. S. and Wright, C. J. (1974). *J. Chem. Soc. Faraday Trans. II*, **70,** 348
67. Pecora, R. (1972). *Ann. Rev. Biophys. Bioeng.*, **1,** 257
    Wun, K. L., Feke, G. T. and Prins, W. (1974). *Discuss. Faraday Soc.*, **57,** paper 11
    McAdam, J. D. G., King, T. A. and Knox, A. (1974). *Chem. Phys. Lett.*, in the press
68. Gross, B. (1953). *Mathematical Structure of the Theories of Viscoelasticity* (Paris: Hermann)
69. Cook, M., Watts, D. C. and Williams, G. (1970). *Trans. Faraday Soc.*, **66,** 2503
70. Williams, G. and Watts, D. C. (1971). *N.M.R. Basic Principles and Progress*, **4,** 271 (Berlin: Springer-Verlag); (1971). *Trans. Faraday Soc.*, **67,** 1971
71. Williams, G., Watts, D. C., Dev, S. B. and North, A. M. (1971). *Trans. Faraday Soc.*, **67,** 1323

72. Allen, G., Brier, P. N., Goodyear, G. and Higgins, J. S. (1972). *Faraday Symp. Chem. Soc.*, **6**, 169
73. Allen, G., Higgins, J. S. and Wright, C. J. (1974). *J. Chem. Soc. Faraday Trans. II*, **70**, 348
74. Williams, G. (1964). *Trans. Faraday Soc.*, **60**, 1548
75. Hoffman, J. D., Williams, G. and Passaglia, E. (1966). *J. Polymer Sci. C*, **14**, 173
76. Kirkwood, J. G. and Fuoss, R. M. (1941). *J. Chem. Phys.*, **9**, 329
77. Yamafuji, K. and Ishida, Y. (1962). *Kolloid Z.*, **183**, 15
78. Zwanzig, R. (1965). *Ann. Rev. Phys. Chem.*, **16**, 67
Berne, B. J. and Harp, G. D. (1970). *Advan. Chem. Phys.*, **17**, 63
79. Williams, G. and Hains, P. J. (1972). *Faraday Symp. Chem. Soc.*, **6**, 14
80. Johari, G. P. and Goldstein, M. (1970). *J. Chem. Phys.*, **53**, 2372
81. Dubois-Violette, E., Geny, F., Monnerie, L. and Parodi, O. (1969). *J. Chim. Phys.*, **66**, 1865
Geny, F. and Monnerie, L. (1969). **66**, 1872
81a. Berne, B. J. (1971). *An Advanced Treatise on Physical Chemistry*, Vol. 8*B*, *The Liquid State*, (H. Eyring, W. Jost and D. Henderson, editors) (London: Academic Press)
82. Phillips, M. C., Barlow, A. J. and Lamb, J. (1972). *Proc. Roy. Soc.*, **A329**, 193
83. Glarum, S. (1960). *J. Chem. Phys.*, **33**, 639
84. Davies, M., Hains, P. J. and Williams, G. (1973). *J. Chem. Soc. Faraday Trans II*, **69**, 1785
85. Edwards, S. F. (1970). *Discuss. Faraday Soc.*, **49**, 43; (1974). *Discuss. Faraday Soc.*, **57**, paper 9
86. O'Reilly, J. M. (1962). *J. Polymer Sci.*, **57**, 429
87. Gee, G. (1966). *Polymer*, **7**, 177
88. Allen, G., Ayerst, R. C., Cleveland, J. R., Gee, G. and Price, C. (1968). *J. Polymer Sci.*, **23**, 127
89. Price, C., Williams, R. C. and Ayerst, R. C. (1972). *Amorphous Materials*, 117 (R. W. Douglas and B. Ellis, editors) (New York: Wiley)
90. Simba, R. and Boyer, R. F. (1962). *J. Chem. Phys.*, **37**, 1003
91. Goldstein, M. (1963). *J. Chem. Phys.*, **39**, 3369
92. Kauzmann, W. (1948). *Chem. Rev.*, **43**, 219
93. Gibbs, J. H. and DiMarzio, E. A. (1958). *J. Chem. Phys.*, **28**, 373
94. Boyer, R. F. (1963). *Rubber Chem. Technol.*, **36**, 1303
95. Allen, G., Gee, G., Mangaraj, D. and Wilson, G. (1960). *Polymer*, **1**, 467
96. McCall, D. W. (1969). *Molecular Dynamics and the Structure of Solids, National Bureau of Standards, Special Publication 301* (Washington: N.B.S.)
97. Heijboer, J. (1956). *Kolloid Z.*, **148**, 36
98. Sauer, J. A. (1971). *J. Polymer Sci. C*, **32**, 69
99. Allen, G., Wright, C. J. and Higgins, J. S. (1974). *Polymer*, in the press
100. Bunn, C. W. and Holmes, D. R. (1958). *Discuss. Faraday Soc.*, **25**, 95
101. Natta, G. and Corradine, P. (1960). *Nuovo Cimento Suppl.*, **15**, 9
102. Starkweather, H. W. and Boyd, R. H. (1960). *J. Phys. Chem.*, **64**, 410
103. Allen, G. (1964). *J. Appl. Chem.*, **14**, 1
104. Palmer, R. P. and Cobbold, A. (1964). *Makromol. Chem.*, **74**, 174
105. Bassett, D. C., Keller, A. and Mitsihashi. (1963). *J. Polymer Sci. A*, **1**, 763
106. Keller, A. (1968). *Rep. Progr. Phys.*, **31**, 623
107. Keller, A. (1957). *Phil. Mag.*, **2**, 1171
108. Keller, A. and Bassett, D. C. (1960). *Proc. Roy. Microsc. Soc.*, **19**, 243
109. Anderson, F. R. (1963). *J. Polymer Sci. C*, **3**, 123
110. Pennings, A. (1967). *J. Polymer Sci. C*, **16**, 1799
111. Mackley, M. R., Keller, A. and Frank, F. C. (1973). *Polymer*, **14**, 16
112. Ward, I. M. and Capaccio, V. (1973). *Nature Phys. Sci.*, **243**, 143
113. Ref. 27, p. 372; Boyer, R. F. (1973). *Plastics and Polymers*
114. Porter, C. H., Lawler, J. H. L. and Boyd, R. H. (1970). *Macromolecules*, **3**, 308
115. Porter, C. H. and Boyd, R. H. (1971). *Macromolecules*, **4**, 589
116. Boyd, R. H. and Porter, C. H. (1972). *J. Polymer Sci. A2*, **10**, 647
117. Ward, I. M. (1971). *Mechanical Properties of Solid Polymers*, 180 (London: Wiley Interscience)
118. Phillips, P. J., Kleinheiss, G. and Stein, R. S. (1972). *J. Polymer Sci. A2*, **10**, 1593

119. Schaufele, R. F. (1970). *Macromol. Rev.*, **4,** 67
120. Holliday, L. and White, J. W. (1971). *Macromolecules Leiden, IUPAC*, 545 (London: Butterworths)
Twistleton, J. F. and White, J. W. (1972). *Neutron Inelastic Scattering*, 301 (Vienna: International Atomic Agency)
121. Treloar, L. R. G. (1960). *Polymer*, **1,** 95
122. Shimanouchi, T., Asahina, A. and Enomoto, S. (1962). *J. Polymer Sci.*, **59,** 93
123. Roe, R. J. and Krigbaum, W. R. (1962). *J. Polymer Sci.*, **61,** 167
124. Shen, M. C. (1969). *Macromolecules*, **2,** 358
125. Boyce, P. H. and Treloar, L. R. G. (1970). *Polymer*, **11,** 21

Index

Acetonitrile
  molecular beam spectroscopy, 64
Acetylene
  molecular beam spectroscopy, 56, 58
  polarisability data, 186
Actinides
  relativistic hyperfine operators, 257
Activation energy
  poly(*p*-chlorostyrene), loss processes in, 364
  poly(ethylene oxide), loss processes in, 364
  poly(*p*-fluorostyrene), loss processes in, 364
  poly(vinyl bromide), loss processes in, 364
  poly(vinyl chloride), loss processes in, 364
Alkalies (*see also* Lithium hydroxide; Rubidium hydroxide)
  dimers with inert gases, 113
Alkali halides (*see also* specific salts)
  electric dipole moments, 38
  field gradients, 41
  rotational spectroscopy, 35
Alkali metals (*see also* Lithium; Potassium; Rubidium; Sodium)
  dimers, molecular beams, 72–75
Aluminium
  L-bridged dimers, asymmetry parameters, 275
Ammonia
  inversion in, determination by high-resolution spectroscopy, 21
  molecular beam spectroscopy, 61–63
Angular overlap model
  for bonding distortion, 208–211
Argon
  complex with chlorine fluoride, molecular beam spectroscopy, 110
    with hydrochloric acid, molecular beam spectroscopy, 110
    with hydrofluoric acid, molecular beam spectroscopy, 110
  dimer, Lennard Jones parameter, 97
    Morse curve parameter, 97
    u.v. spectroscopy, 107
Argon *continued*
  polarisability data, 186
  polymers, electron diffraction, 113
    mass spectrometry, 109
  van der Waals complex with hydrogen, energy level diagram, 100
    i.r. spectroscopy, 105
  van der Waals complex with oxygen, i.r. spectroscopy, 106
Argon chloride fluoride
  structural constants, 81
Argon hydrogen chloride
  van der Waals molecule spectroscopic constant, 80
Atoms
  electric dipole polarisabilities, 149–194
Average structure
  determination by high-resolution spectroscopy, 15–17

Bacitracin
  n.m.r., 130
Barium oxide
  electric dipole moment, 38
Barium sulphide
  electric dipole moment, 38
Bending vibration
  determination by high-resolution spectroscopy, 22
Benzene
  polarisability data, 186
Bonding
  in magnetochemical models, 195–237
Boron
  L-bridged dimers, asymmetry parameter, 275
Boron trifluoride
  dimer, molecular beam spectroscopy, 112
Bromine
  structure factor calculation for liquid, 341

Cadmium
  dimer, u.v. spectroscopy, 108

Caesium
dimer, electric field gradient at, 74
Caesium bromide
diatomic potential constants, 35
Caesium chloride
diatomic potential constants, 35
electric dipole moment, 38
Caesium fluoride
diatomic potential constants, 35
electric dipole moment, 38
spin–spin interaction, 44
Caesium hydroxide
bending vibration, determination by high-resolution spectroscopy, 22
Caesium iodide
diatomic potential constants, 35
Calcium
dimer, u.v. spectroscopy, 108
Carbon dioxide
dimer, i.r. spectroscopy, 106
Lennard Jones parameter, 97
molecular beam spectroscopy, 112
molecular beam spectroscopy, 56, 58
polarisability data, 186
polymers, electron diffraction, 113
second virial coefficient, 336
Carbon disulphide
molecular beam spectroscopy, 56, 58
polarisability data, 186
Carbon monoxide
electric dipole moment, 38
$^3\Pi$, molecular beam spectroscopy, 50
polarisability data, 186
Carbon tetrachloride
polarisability data, 186
Carbon tetrafluoride
polarisability data, 186
Carbonyl sulphide
effective structure, determination by high-resolution spectroscopy, 10
equilibrium bond distances, 8
molecular beam spectroscopy, 56, 58, 83, 84
polarisability data, 186
Chlorine
structure factor calculation for liquid, 340
Chlorine fluoride
complex with argon, molecular beam spectroscopy, 110
electric dipole moment, 38
Chloroform
polarisability data, 186
Cholesteric phases
n.m.r. in, 125
Chromium, μ-oxo[bispentamine]-
magnetic data, 223–227
Clathrates
n.m.r., 125
Clausius–Mossotti, relation, 166, 181
Cobalt
octahedral complexes, dinuclear, magnetic exchange in, 232–235
quadrupole coupling data for, 265
trigonal-bipyramidal carbonyl complexes, Mössbauer spectroscopy, 284
——, *trans*-dimesitylbis(diethylphenylphosphine)-
magnetochemistry, 199
——, dimesityldiphosphine-
magnetochemistry, 200
——, *trans*-tetrapyridinebis(isothiocyanato)-
molecular susceptibility calculation, 218
Complexes
weakly bound, molecular beam spectroscopy, 70–82
Compressibility
isothermal, calculation, 337, 338
liquids, 324
Computers
simulation, equilibrium properties of molecular fluids and, 332–334
Configuration
distribution functions, equilibrium statistical mechanics and, 304–306
of polymer chains, 351–353
Configurational energy
molecular fluids, 336, 337
Conformation
polymer chains, 353
Covalent molecules
quadrupole coupling in, 267–292
Crystalline polymers
structure/property relationships, 381–390
Crystalline state
high polymers, 351
Crystallisation
model for phase transition, 301
Crystal structure
high polymers, chemical structure and, 382–384
Cuprates, tetrachloro-
magnetochemistry, 200
Cyanamide
inversion in, determination by high-resolution spectroscopy, 21
Cyanogen chloride
average structure, determination by high-resolution spectroscopy, 16
effective structure, determination by high-resolution spectroscopy, 10
Cyclopropane
polarisability data, 186

Detectors
for molecular beam electric resonance spectroscopy, 29
universal beam, for molecular beam spectroscopy, 53

Deuterium (*see also* Hydrogen)
  nuclear quadrupole moments, 40
  polarisability, anisotropy, 37
    calculation, 184
    data, 186
  quadrupole coupling constant, 69
Deuterium chloride
  /hydrochloric acid dipole derivatives, 39
  quadrupole constant, 42
Deuterium fluoride
  electric dipole moment, 38
  quadrupole constant, 42
Diamagnetic molecules
  n.m.r., 121–133
Dielectric constant
  electric dipole polarisability and, 165, 166
  molecular fluids, 310
  second virial coefficient and, 180
Dielectric polarisation
  gases, 151
Dielectric relaxation
  dynamics of high polymer chains by, 359
Dielectric saturation, 179
Dimers
  molecular beams, electric deflection of, 71
Dispersion
  electric dipole polarisability and, 163–165
Distortion
  of bonding, magnetochemical models for, 203–221
  from ideal geometry, hyperfine interactions and, 290–292
Dynamic light scattering
  dynamics of high polymer chains by, 359
Dynamic properties
  in crystalline polymers, 386–390
  high polymer glasses, 378–381
  rubber, 369–374
Dynamics
  of high polymer chains, 358–365

Effective structure
  determination by high-resolution spectroscopy, 9, 10
Elasticity
  rubber, kinetic theory and thermodynamics of, 365–369
Electron diffraction
  molecular structure determination by, 17–19
Electrical properties
  virial coefficients, 180, 181
Electric dipole moment
  diatomic molecules, interaction with electric polarisability, 36–39
Electric dipole polarisability
  atoms, 149–194
  molecules, 149–194
  perturbation theory, 152–162
Electric field induced second harmonic generation, 175–177
Electric hyperfine interactions, 245, 246
Electric multipole expansion, 243, 244
Electric polarisability
  diatomic molecules, interaction with electric dipole moment, 36–39
Electron diffraction
  van der Waals molecules, 113
Electron paramagnetic resonance
  hyperfine structure, 242
Electro-optic effects, 175–179
ENDOR
  hyperfine structure, 242
Entropy
  of high polymer chains, 358
  molecular fluids, 336, 337
Entropy of fusion
  high polymers, 384
Enzymes
  lanthanide complexes, n.m.r., 139
  paramagnetic, n.m.r., 141
*E*0 nuclear volume effects, 250–253
Equation of state
  molecular fluids, 336, 337
Equilibrium properties
  molecular fluids, 299–347
Equilibrium structure
  determination by high-resolution spectroscopy, 7, 8
Ethane
  polarisability data, 186
Ethylene
  polarisability data, 186
Europium
  chelates, n.m.r., 136

Ferrocyanides
  magnetochemistry, 201
Ferromagnetism, 221
Ferrous fluorosilicate
  magnetochemistry, 202
Ferrous Tutton salt
  magnetochemistry, 202
Fibula
  in crystalline polymers, 385
Fluids
  polyatomic, equilibrium properties, 301
Fluorescence depolarisation
  dynamics of high polymer chains by, 359
Fluorides
  spin–spin interaction, 44
Fluorine chloride
  polarisability anisotropy, 37
Formaldehyde
  molecular beam spectroscopy, 64
——, fluoro-
  molecular beam spectroscopy, 68

Formamide
inversion in, determination by high-resolution spectroscopy, 21
molecular beam spectroscopy, 68
Formic acid
molecular beam spectroscopy, 68
Furan
molecular beam spectroscopy, 67

Gadolinium
chelates, n.m.r., 136
Gallium
L-bridged dimers, asymmetry parameters, 275
Gases
low density, equilibrium properties, 311
Germanium oxide
electric dipole moment, 38
Glasses
structure/property relationships, 374–381
Glass state
high polymers, 350
Glass temperature
structure and, 378
Glass transitions
high polymers, 375
static theory of, 376, 377

Halogens (*see also* Bromine; Chlorine)
Heat of fusion
high polymers, 384
Heavy water
molecular beam spectroscopy, 66
Hedritic growth
in crystalline polymers, 386
Helium
dimer, u.v. spectroscopy, 107
polarisability, calculation, 184
data, 186
High-resolution spectroscopy
molecular structure determination by, 1–25
Hund's rules, 222
Hydrochloric acid
complex with argon, molecular beam spectroscopy, 110
/deuterium chloride dipole derivatives, 39
dimer, i.r. spectroscopy, 103
electric dipole moment, 38
polarisability anisotropy, 37
quadrupole constant, 42
Hydrocyanic acid
effective structure, determination by high-resolution spectroscopy, 10
molecular beam spectroscopy, 56, 66, 84
Hydrofluoric acid
complex with argon, molecular beam spectroscopy, 110
Hydrofluoric acid *continued*
dimer, i.r. spectroscopy, 104
molecular beam spectroscopy, 75
polarisability anisotropy, 37
polymers, electron diffraction, 113
Hydrogen (*see also* Deuterium)
dimer, i.r. spectroscopy, 105
molecular beam magnetic resonance spectroscopy, 31
$C^3\Pi_u$, molecular beam spectroscopy, 50
polarisability, anisotropy, 37
calculation, 184
data, 186
spin–spin interaction, 44
van der Waals complex with argon, energy level diagram, 100
i.r. spectroscopy, 105
Hydrogen bonds
in molecular beams, 75–79
Hydrogen selenide
molecular beam spectroscopy, 66, 67
Hydrogen sulphide
molecular beam spectroscopy, 66
Hydroxides
electric dipole moment, 38
Hyperfine interactions
molecular structure and, 239–297
Hyperpolarisability
calculations, 185
Hyper-Raman scattering
electric dipole polarisability and, 172–174
Hyper-Rayleigh scattering
electric dipole polarisability and, 172–174

Indium
L-bridged dimers, asymmetry parameters, 275
Inert gases (*see also* Argon; Helium; Krypton; Neon; Xenon)
dimers, Raman spectroscopy, 106
u.v. spectroscopy, 107
with alkalies, 113
Inertial defect, 5
Infrared spectroscopy
high polymer chain conformation by, 357
molecular fluids, 310
molecular structure determination by, 2
van der Waals molecules, 103–106
Integral equations
equilibrium properties of molecular fluids and, 311–314
Interaction coefficients
molecular fluids, 342, 343
Interactions
long-range, in high polymer chains, 356, 357
short-range, in high polymer chains, 355, 356

Intermolecular forces
electric dipole polarisabilities and, 182, 183
of molecular fluids, 302, 303
Inversion
determination by high-resolution spectroscopy, 21
Iron
low spin complexes, Mössbauer Zeeman studies, 280
pentacarbonyl, molecular beam spectroscopy, 63
trigonal-bipyramidal carbonyl complexes, Mössbauer spectroscopy, 284
Isocyanic acid
molecular beam spectroscopy, 68
Isomer shift
Mössbauer spectroscopy and, 252
Isotope shift
Mössbauer spectroscopy and, 252
Isotropic potential
effective, molecular fluids, 325, 326

Kerr constant
molecular fluids, 310
Kerr effect, 152
electro-optic, 177–179
optical, 179
second virial coefficient and, 180
Kinetic theory
rubber elasticity, 365–369
Kramers–Krönig relationship, 164
Krypton
dimer, Lennard Jones parameter, 97
morse curve parameter, 97
u.v. spectroscopy, 107
polarisability data, 186

Landé interval rule, 221
Lanthanides (*see also* Europium; Gadolinium)
n.m.r., 135–139
relativistic hyperfine operators, 257
Lattice
quadrupole coupling in covalent molecules and, 267–271
Lennard Jones potential, 95
Liquid-crystalline solvents
n.m.r. in, 125
Liquids
structure factor, 339–341
Lithium
dimer, electric field gradient at, 74
Lithium bromide
diatomic potential constants, 35
electric dipole moment, 38
field gradient, 41
nuclear quadrupole moments, 40
Lithium chloride
diatomic potential constants, 35
electric dipole moment, 38
field gradient, 41
Lithium deuteride
quadrupole constant, 42
Lithium fluoride
diatomic potential constants, 35
electric dipole moment, 38
molecular beam spectroscopy, 58
nuclear quadrupole moments, 40
spin–spin interaction, 44
Lithium hydride
electric dipole moment, 38
nuclear quadrupole moments, 40
Lithium hydroxide
molecular beam spectroscopy, 56, 57, 85
Lithium iodide
diatomic potential constants, 35
electric dipole moment, 38
Lithium oxide
electric dipole moment, 38
2Π, high resolution spectroscopy, 49
Lithium sodium
polarisability anisotropy, 37
London dispersion forces
in molecular beams, 79
Lorentz–Lorenz relation, 162
Low temperature
magnetic property measurement at, 219–221
Lysozyme
n.m.r., 130

Magnesium
dimer, u.v. spectroscopy, 108
Magnetic dipole coupling, 254, 255
Magnetic exchange
model for, 221–235
Magnetic hyperfine interactions, 253–256
Magnetic octupole effects, 254
Magnetic properties
diatomic molecules, 46
Magnetic susceptibility
model for, 215–219
Magnetochemical models
bonding in, 195–237
Many-electron systems
calculations with angular overlap model, 211–215
Mass spectrometry
van der Waals molecules, 109, 110
Mean spherical model
for equilibrium properties of molecular fluids, 314
Mean squared force
molecular fluids, 310
Mean squared torque
in equilibrium statistical mechanics of molecular fluids, 309, 310

Mean squared torque *continued*
liquids, calculated, 339
Melting
high polymers, 375
Mercury
dimer, u.v. spectroscopy, 108
Methane
molecular beam spectroscopy, 59
n.m.r., 128
polarisability data, 186
second virial coefficient, 336
structure, determination by electron diffraction and spectroscopy, 18
substituted, molecular beam spectroscopy, 63, 64
——, chloro-
molecular beam spectroscopy, 64
polarisability data, 186
——, deutero-
molecular beam spectroscopy, 59
——, dichloro-
polarisability data, 186
——, difluoro-
molecular beam spectroscopy, 66, 67
polarisability data, 186
——, fluoro-
molecular beam spectroscopy, 63
n.m.r., 128
polarisability data, 186
——, tetrachloro-—*see* Carbon tetrachloride
——, tetrafluoro-—*see* Carbon tetrafluoride
——, trichloro-—*see* Chloroform
——, trifluoro-
molecular beam spectroscopy, 64
polarisability data, 186
Methanol
molecular beam spectroscopy, 64
Methyl isocyanide
molecular beam spectroscopy, 64
Microwave spectroscopy
hyperfine structure, 242
molecular structure determination by, 3
Models
for intermolecular forces in molecular fluids, 302, 303
Molecular beam absorption spectroscopy, 33
Molecular beam deflection, 33
Molecular beam electric resonance spectroscopy, 29–31
Molecular beam magnetic resonance spectroscopy, 31, 32
Molecular beam maser spectroscopy, 32
Molecular beam spectroscopy
hyperfine structure, 242
molecules from, 27–92
van der Waals molecules, 110–113
Molecular dynamics
in computer simulation of equilibrium properties of molecular fluids, 332–334
Molecular fluids
equilibrium properties, 299–347
Molecular geometry
determination by high-resolution spectroscopy, 4
Molecular structure
determination, by high-resolution spectroscopy, 1–25
by n.m.r., 119–147
high polymers, 349–394
hyperfine interactions and, 239–297
Molecular weight
distribution in high polymers, 351
structure/property relationships and, 390
Molecules
asymmetric top, molecular beam spectroscopy, 64–68
diatomic, molecular beam spectroscopy, 34–53
electric dipole polarisabilities, 149–194
from molecular beam spectroscopy, 27–92
polyatomic, linear, molecular beam spectroscopy, 56–58
polyatomic, molecular beam spectroscopy, 53–70
pyramidal, molecular beam spectroscopy, 61–63
rod-like, relaxation of, 359–361
tetrahedral, molecular beam spectroscopy, 59–61
van der Waals—*see* van der Waals molecules
Monte Carlo calculations
in computer simulation of equilibrium properties of molecular fluids, 332–334
in perturbation theory for equilibrium properties of molecular fluids, 321
Morphology
in crystalline polymers, 384–386
Mössbauer spectroscopy
hyperfine structure and, 242
quadrupole coupling and, 262
Multiphoton absorption, 179

Nematic phases
n.m.r. in, 125
Neon
dimer, Lennard Jones parameter, 97
Morse curve parameter, 97
u.v. spectroscopy, 107
polarisability data, 186
Neutron diffraction
molecular fluids, 308
Neutron inelastic scattering
high polymer chain conformation by, 357
Neutron scattering
dynamics of high polymer chains by, 359

Neutron scattering *continued*
in high polymers, 391
Nickel
tetracarbonyl, molecular beam spectroscopy, 63
Nitric oxide
dimer, i.r. spectroscopy, 104
molecular beam spectroscopy, 112
polarisability data, 186
Nitrogen
$a^3\Sigma_u^+$, molecular beam spectroscopy, 50
polarisability data, 186
structure factor calculation for liquid, 340
thermodynamic functions, calculated, 337
Nitrogen oxide
electric dipole-moment, 38
$^2\Pi$, high resolution spectroscopy, 49
Nitrous oxide
effective structure, determination by high-resolution spectroscopy, 10
polarisability data, 186
structure determination by high-resolution spectroscopy, 19
Nozzle beam sources
for molecular beam spectroscopy, 53
Nuclear decay rates, 260
Nuclear hyperfine interactions
diatomic molecules, 39, 40
Nuclear magnetic multiple moments, 253
Nuclear magnetic resonance
dynamics of high polymer chains by, 359
hyperfine structure, 242
molecular structure determination by, 119–147
Nuclear Overhauser effect, 132, 133
Nuclear quadrupole coupling, 246–250
Nuclear quadrupole resonance
hyperfine structure, 242
Nylon 66
structure, 383

Optical polarisability, 163
Optical processes
non-linear, 157
Optical properties
non-linear, electric dipole polarisabilities and, 172–180
virial coefficients, 180, 181
Optical pumping
hyperfine structure and, 242
Optical rectification, 179
Orbital angular momentum
exchange in, 227–231
Orbital reduction factor
in magnetochemical models, 198
Oxygen
dimer, Lennard Jones parameter, 97
$X^3\Sigma_g^-$, molecular beam spectroscopy, 53
polarisability data, 186
Oxygen *continued*
van der Waals complex with argon, i.r. spectroscopy, 106
Ozone
structure, determination by high-resolution spectroscopy, 9

Paramagnetic molecules
n.m.r., 133–141
Percus–Yevick theory
equilibrium properties of molecular fluids and, 312, 313
Perturbation theory
electric dipole polarisabilities, 152–162
for equilibrium properties of molecular fluids, 318–332
Phantom chains
in high polymers, 353–355
Phase transitions
nematic–isotropic, model for, 317
Phosphine
molecular beam spectroscopy, 61–63
Phosphorus nitride
electric dipole moment, 38
nuclear quadrupole moments, 40
Pockels effect, 175
Polarisability (*see also* Hyperpolarisability)
dynamic, calculation, 185
static, calculation, 183, 184
Poly(γ-benzyl-L-glutamate)
relaxation time in, 360
Poly(but-1-ene)
heat, entropy and temperature of fusion, 384
Poly(n-butyl isocyanate)
relaxation time in, 360
Poly(*p*-chlorostyrene)
activation energy of loss processes in, 364
Poly(2,6-dimethylphenylene oxide)
glass transition temperature, 378
Poly(dimethylenesiloxane)
glass transition temperature, 378
Polyethylene
crystalline, lamellae in, 385
heat, entropy and temperature of fusion, 384
Poly(ethylene oxide)
activation energy of loss processes in, 364
polymer coil dimensions, 357
Poly(*p*-fluorostyrene)
activation energy of loss processes in, 364
Polyglycine
random-coil, n.m.r., 130
Polyisobutylene
dielectric α and β processes in, 379
glass transition temperature, 378
Polymer chains
statics and dynamics, 352–365

Polymers
bulk, structure/property relationships, 365–390
structure/property relationships, 349–394
Poly(methyl acrylate)
glass transition temperature, 378
Poly(methyl methacrylate)
glass transition temperature, 378
polymer coil dimensions, 357
Polymyxin
n.m.r., 130
Poly(natural rubber)
glass transition temperature, 378
Poly(propylene)
heat, entropy and temperature of fusion, 384
Poly(propylene oxide)
glass transition temperature, 378
Polystyrene
glass transition temperature, 378
polymer coil dimensions, 357
Polyterephthalates
heat, entropy and temperature of fusion, 384
Poly(vinyl acetate)
glass transition temperature, 378
Poly(vinyl bromide)
activation energy of loss processes in, 364
Poly(vinyl chloride)
activation energy of loss processes in, 364
Potassium
dimer, electric field gradient at, 74
molecular beams, laser and white-light induced fluorescence, 73
Potassium argon
molecular beam spectroscopy, 82
Potassium bromide
diatomic potential constants, 35
Potassium cloride
diatomic potential constants, 35
electric dipole moment, 38
field gradient, 41
Potassium fluoride
diatomic potential constants, 35
electric dipole moment, 38
field gradient, 41
spin–spin interaction, 44
Potassium iodide
diatomic potential constants, 35
Propane
substitution structure, determination by high-resolution spectroscopy, 12
——, 2-chloro-
structure determination by high-resolution spectroscopy, 14
Proteins
structure, n.m.r., 130
Proton resonance enhancement
in n.m.r. of paramagnetic molecules, 140
Pseudoquadrupole interaction, 245
Pyrrole
molecular beam spectroscopy, 67

Quadrupole coupling
in covalent molecules, 267–292
molecular structure and, 261–267
Quadrupole coupling constants
diatomic molecules, 42
Quadrupole coupling interaction
diatomic molecules, 40–42

Raman effect
stimulated, 179
Raman scattering (*see also* Hyper-Raman scattering)
electric dipole polarisability and, 166–170
Raman spectroscopy
high polymer chain conformation by, 357
molecular fluids, 310
molecular structure determination by, 3
van der Waals molecules, 106, 107
Rayleigh depolarisation ratios
second virial coefficient and, 180
Rayleigh scattering (*see also* Hyper-Rayleigh scattering) 151
electric dipole polarisability and, 166–170
in high polymers, 391
Refraction, 151
Refractive index
electric dipole polarisability and, 162, 163
intensity dependent, 179
second virial coefficient and, 180
Relativistic hyperfine operators, 256, 257
Relaxation
of rod-like molecules, 359–361
Relaxation methods
n.m.r., 129–132
in n.m.r. of paramagnetic molecules, 139–141
Resonant optical harmonic conversion, 179
Ribonuclease A
n.m.r., 130
Rings
puckering, determination by high-resolution spectroscopy, 22
Rotation
first order polarisability and, 181, 182
internal, determination by high-resolution spectroscopy, 21
Rotational spectroscopy
diatomic molecules, 34–36
Rotation
–vibration interaction, 5
Rubber
structure/property relationships, 365–374
Rubber state
high polymers, 350

Rubidium
dimer, electric field gradient at, 74
Rubidium bromide
diatomic potential constants, 35
Rubidium chloride
diatomic potential constants, 35
electric dipole moment, 38
Rubidium fluoride
diatomic potential constants, 35
electric dipole moment, 38
field gradient, 41
spin–spin interaction, 44
Rubidium hydroxide
bending vibration, determination by high-resolution spectroscopy, 22
Rubidium iodide
diatomic potential constants, 35

Scaled particle theory
for equilibrium properties of molecular fluids, 314–318
Second harmonic generation
non-linear optical phenomena, 172–174
Silane, tetramethyl-
n.m.r., 128
Silicon oxide
electric dipole moment, 38
Smectic phases
n.m.r. in, 125
Sodium
dimer, electric field gradient at, 74
laser and white-light induced fluorescence, 73
Sodium bromide
diatomic potential constants, 35
electric dipole moment, 38
field gradient, 41
Sodium chloride
diatomic potential constants, 35
electric dipole moment, 38
field gradient, 41
Sodium fluoride
diatomic potential constants, 35
electric dipole moment, 38
field gradient, 41
spin–spin interaction, 44
Sodium iodide
diatomic potential constants, 35
electric dipole moment, 38
Solids
diamagnetic molecules, n.m.r., 123–125
Spectroscopy (*see also* Infrared spectroscopy; Microwave spectroscopy; Molecular beam absorption spectroscopy; Molecular beam electric resonance spectroscopy; Molecular beam magnetic resonance spectroscopy; Molecular beam maser spectroscopy; Molecular beam spectroscopy; Mössbauer spectroscopy; Nuclear magnetic resonance; Nuclear quadrupole resonance; Raman spectroscopy; Rotational spectroscopy; Ultraviolet spectroscopy)
Spherical harmonic expansion
intermolecular forces in molecular fluids, 303
Spherulites
in crystalline polymers, 385
Spin–rotation interaction
diatomic molecules, 42, 43
Spin–spin interaction
diatomic molecules, 43, 44
Stark effect, 166
quadratic, 151
electric dipole polarisability and, 170, 171
Statics
of high polymer chains, 352–358
Statistical mechanics
equilibrium, 304–311
Sternheimer effect, 248
Stress
glass temperature and, 375
Strontium oxide
electric dipole moment, 38
Substitution structure
determination by high-resolution spectroscopy, 10–13
Sulphur dioxide
polarisability data, 186
Sulphur hexafluoride
polarisability data, 186
Supersonic nozzles
molecular beam source, 29
van der Waals molecule formation in, 98, 99
Symmetry
molecular, electric dipole polarisability and, 160–162
quadrupole coupling and, 263
Symmetry axis
determination by high-resolution spectroscopy, 4
Synchrotron radiation
of high polymers, structure/property relationship and, 391

Temperature (*see also* Low temperature)
high polymer glasses and, 374–376
Temperature of fusion
high polymers, 384
Thallium fluoride
electric dipole moment, 38
spin–spin interaction, 44
Thermodynamics
for bulk phase molecular fluids, 307, 308
for interfacial molecular fluids, 308
rubber elasticity, 365–369

Thiols
electric dipole moment, 38
Third harmonic generation
non-linear optical phenomena, 174, 175
Tin
five-coordinate, regression method, 281–284
Triose phosphate isomerase
n.m.r., 130
Tungsten
octacyano anions, Mössbauer spectroscopy, 287–290

Ultrasonic relaxation
dynamics of high polymer chains by, 359
Ultraviolet spectroscopy
molecular structure determination by, 3
van der Waals molecules, 107–109

Valence
quadrupole coupling in covalent molecules and, 267–271
van der Waals molecules, 93–118
energy levels, 99–102
equilibrium properties of molecular fluids and, 330–332
formation, 95–99
molecular beam spectroscopy, 79–82
Vibrating rotor
theory, 5–7
Vibration (*see also* Bending vibration)
in crystalline polymers, 389
first order polarisability and, 181, 182
–rotation interaction, 5
Virial coefficients
of electrical and optical properties, 180, 181
molecular fluids, 335, 336
Viscoelastic relaxation
dynamics of high polymer chains by, 359

Water (*see also* Heavy water)
dimer, molecular beam spectroscopy, 75
liquid, molecular beam spectroscopy, 70
molecular dynamic simulation, 334
molecular beam spectroscopy, 65

Xenon
dimer, Lennard Jones parameter, 97
Morse curve parameter, 97
u.v. spectroscopy, 108
polarisability data, 186
polymers, electron diffraction, 113
X-ray diffraction
high polymer chain conformation by, 357
molecular fluids, 308

Zeeman effect
ammonia, 61
Zeeman studies
diatomic molecules, 44–48
Zinc
dimer, u.v. spectroscopy, 108